engineering economy

Sixth Edition

Engineering Economy

E. PAUL DeGARMO

Registered Professional Engineer
Professor of Industrial Engineering and Mechanical Engineering Emeritus
University of California, Berkeley

JOHN R. CANADA

Registered Professional Engineer
Professor of Industrial Engineering
North Carolina State University at Raleigh

WILLIAM G. SULLIVAN

Registered Professional Engineer
Associate Professor of Industrial Engineering
University of Tennessee, Knoxville

Macmillan Publishing Co., Inc.
NEW YORK

Collier Macmillan Publishers
LONDON

Earlier editions entitled *Introduction to Engineering
Economy*, by E. P. DeGarmo and B. M. Woods, copyright
1942 and 1953 by Macmillan Publishing Co., Inc. Earlier
editions entitled *Engineering Economy* © 1960 and copyright © 1967 and 1973 by Macmillan Publishing Co., Inc.

Macmillan Publishing Co., Inc.
866 Third Avenue, New York, New York 10022

Collier Macmillan Canada, Ltd.

Library of Congress Cataloging in Publication Data

DeGarmo, Ernest Paul,
    Engineering economy.

    Bibliography: p.
    Includes index.
    1. Engineering economy.   I. Canada, John R., joint
author. II. Sullivan, William G.,   joint author.
III. Title.
TA177.4.D43   1979   658.1′596   78-2265
ISBN 0-02-328160-X

Printing: 4 5 6 7 8   Year: 1 2 3 4 5

preface

This textbook and basic reference is intended primarily for engineering students who are studying their first formal course in engineering economy. It is also useful for students in business-related curricula concerned with economic analysis of alternatives by widely used nonmathematically sophisticated methods. It assumes that the students have had little formal exposure to economics, accounting, and statistics and have a very limited knowledge of the common financial practices of business.

Inasmuch as all aspects of the field of engineering economy cannot be covered in depth in a one-quarter or one-semester first course, the book is not intended to impress the professionals in the field. Rather, it is written from the viewpoint and needs of the students who will use it. Its purpose is to give them a sound understanding of the basic aspects of the subject and some insight into approaches that can be used for making sound economic decisions concerning the types of problems they are likely to encounter in other engineering courses and in their careers. At the same time they should acquire a solid base for further study after graduation, which will permit them to understand and use more advanced, and constantly developing, procedures needed or helpful in analyzing the more complex economic problems that they may encounter in their later careers.

Although the book undoubtedly will be used by those other than college students, we ask their pardon for placing them second to students in organizing the contents. As in previous editions, no single economy study method or pattern is advocated. Instead, several methods are presented and used in the belief that different procedures can be used advantageously in various situations, provided the analyst understands the basic principles involved.

That costs and the economic use of capital remain a major concern to engineers was reemphasized by a survey of chief engineers and vice presidents in engineering of 85 companies in the United States. When queried as to whether their young engineers, during their first five years after graduation, were required to make economy studies of proposed projects, over 97% replied "yes" and 80% added "frequently." Seventy-five percent believed that engineering economy should be a required subject in all undergraduate engineering curricula. Companies engaged in space and defense work expressed the same views in only slightly lower percentages but often more emphatically.

To some extent the subject matter included in this edition has been influenced by the results of the above-mentioned survey. There was rather general agreement that capital budgeting, interest and money timing, decisions among alternatives, depreciation, financing, breakeven analyses, replacement analysis, and effects of risk and uncertainty should be included in a basic course in engineering economy. The table of contents makes it evident that other subjects also have been included. Over half of the numerous problem exercises at the end of each chapter are completely new, and virtually all of the remaining exercises are different from those in the previous edition.

The sequence of chapters in this edition differs somewhat from the fifth edition. This edition is divided into three sections as follows: Part I, "Background and Tools for Engineering Economic Analysis," which contains four chapters, ending with "Interest and Money–Time Relationships"; Part II, "Applications of Engineering Economic Analysis," contains ten chapters in approximately the same order of progression as found in most texts in this field, including separate chapters on "Economy Studies for Public Projects" and "Economy Studies for Public Utilities"; Part III, "Other Useful Methods for Minimizing Resource Requirements" contains four succinct chapters to acquaint the student with such methods as "Value Engineering," "Linear Programming," and "Critical Path Economy."

As a guide to instructors or students engaging in self-study, the sequencing of chapters is intended to follow a logical pattern. With the following exceptions, the chapters should be used sequentially. Chapters 2 and 3 can be made optional according to the emphasis desired. Chapters 15 through 18 are mutually independent and can be made optional. Finally, Chapters 12 and 13 may be omitted by persons not interested in economy studies for public agencies and utilities. Because students will recognize a number of situations where computers can be used advantageously in solving problems, a simple computer program for solving internal rate of return problems is included in Appendix C.

The presentation assumes some facility with linear algebra, but differential calculus is needed in several limited sections. Some understanding of elementary probability is needed for a small section of Chapter 6, and a speaking acquaintance with linear programming concepts is useful for Chapter 17. Adequate references are given for these topics.

For the many suggestions from colleagues, students, and practicing engineers who have used the previous editions, we express our deep appreciation. We hope they will find this volume to be equally helpful. Special acknowledgment and thanks are due Mrs. Martha Jackson, *secretary extraordinaire*, for again rescuing the manuscript from various forms of unintelligibility with great ability.

E. Paul DeGarmo
John R. Canada
William G. Sullivan

contents

part I
Background and Tools for Engineering Economic Analysis

chapter 3 selections in present economy   41

chapter 4 interest and money—time relationships   63

part II
Applications of Engineering Economic Analysis

chapter 5 basic methods for making economy studies 111

chapter 6 risk, uncertainty, sensitivity, and multiple attributes 146

chapter 7 selections among alternatives   183

chapter 8 estimating, inflation, and costs   229

chapter 9 depreciation and valuation 270

chapter 10 the effects of income taxes in economy studies 296

chapter 11 replacement studies   327

chapter 12 economy studies for public projects   366

part III
Other Useful Methods for Minimizing Resource Requirements

appendixes

part I

Background and Tools for Engineering Economic Analysis

chapter 1

introduction

This book is concerned with the evaluation of alternative uses of capital in engineering and business projects. An outstanding phenomenon of present-day industrialized civilizations is the extent to which engineers and business managers, through the use of capital, are able to multiply the effectiveness of efforts to harness resources to satisfy the needs and wants of people. Consequently, capital, in the form of money for people, machines, and materials, is an economic necessity in virtually all engineering and business projects.

In the United States an average of over $20,000 of capital is required for each worker in industry. In the more highly mechanized or automated industries, the investment per worker often exceeds $80,000. Naturally, those who supply the capital, and who ultimately control the spending of it, are concerned that it be used to best advantage. Likewise, those who design projects and make the managerial decisions as to whether they should be undertaken and how they should be operated are equally concerned that the available capital be used effectively.

In our capitalistic economy the success of engineering and business projects is more commonly measured in terms of financial efficiency than in any other way. It is unlikely that a project will achieve maximum financial success unless it is also properly planned and operated with respect to its technical, social, and financial requirements. Because engineers are most likely to understand the technical requirements of a project, they are very frequently called upon to make a study combining the technical and financial details, as well as social and aesthetic values, of a project and thus to provide the analysis upon which they or others can base sound managerial decisions.

Engineers play a unique and important role in the conception of new ideas and projects that will require the expenditure of capital to reach the hardware or operational stage. In fact, it is the necessity for considering properly such a combination of factors that distinguishes in large part the work of engineers from that of theoretical scientists.

3

engineering economy and decisions among alternatives

Economic analyses that primarily involve engineering and technical projects commonly are called *engineering economy* studies. In recent years the techniques that largely were originally developed in the field of engineering economy have been expanded and adapted for use in a much broader spectrum of business situations. Consequently, broader titles frequently are attached to these studies, as, for example, *economic analyses for decision making*. There is no clear line of demarcation between the various types of studies, and the matter of name is of little consequence.

Virtually all engineering problems can be solved in more than one way. Most projects can be carried out in more than one way. Almost all business decisions involve doing one thing or another, even though one alternative is merely to do nothing or to maintain the status quo. Thus *economy studies deal with the differences in economic results from alternatives*. This concept of economic differences between alternatives is basic and most important in making economy studies; if there is no alternative, there is no need to make an economy study.

The scope and importance of modern engineering projects make it essential that all the significant factors involved in the economy of an undertaking not only be considered but handled in an accurate, correct manner, so that the results will be satisfactory from all the viewpoints touched by the project. The alternative chosen should never be a matter of guess. Hunches and intuition are not reliable enough. Some will raise their voices to protest that much of the information must be based upon estimates, and therefore the final answer can be found by guesswork. Estimates made after careful study and based on all the information that is available should be considerably better than haphazard guesses. While a carefully prepared estimate may also be a guess, it is an enlightened guess, because the available information has been considered and used. The accuracy of an estimate, and indeed entire economy studies, is usually limited only by the amount of time and effort that one is willing to devote to making the analysis.

engineering and management

Business continues to become more technical. Consequently, engineers play an increasingly important role in management. More and more decision-making managers of businesses are engineers. Some of these decisions are made primarily on the basis of the economic factors involved. More often there are many other factors that must be weighed, and sometimes these may prevail over purely economic considerations. When managers are not engineers, they increasingly call upon engineers to make technical-economic analyses and to provide the data and recommendations upon which managerial decisions can be based. In such situations an engineer is essentially in the position of a consultant to management, and must combine technical and economic knowledge to provide sound conclusions and recommendations. With the recent developments in mathematical, statistical, and computer techniques, which permit the quantitative handling of more complex economic problems than before, the engineer has an opportunity to play an even more important role in the decision-making process. Not only does he have the mathematical and scientific background for understanding and using such techniques, but he has the engineering background that enables him to recognize the practical limitations of these techniques and the effect of the lack of information that usually exists in real situations. He is thus in a position to make the necessary compromises and adjustments that will enable a realistic, although possibly not perfect, solution to be achieved.

measures of financial effectiveness

In most technical courses studied by the engineer or engineering student, he is concerned with obtaining the most effective utilization of materials and/or energy. The degree of effectiveness of the utilization is measured by the familiar equation

$$\text{efficiency} = \frac{\text{output}}{\text{input}} \tag{1-1}$$

When dealing with energy relationships and ordinary materials, the engineer knows that the efficiency can never exceed 100%. However, when dollars are considered as the material, a different situation exists. This may be expressed as

$$\text{financial efficiency} = \frac{\text{dollars income}}{\text{dollars spent}} \tag{1-2}$$

It may readily be seen that, unless the financial efficiency can and does exceed 100%, a project is not desirable from a purely financial viewpoint.

Engineers will immediately recognize that Equation 1-1 can be used in at least two ways. The first, and less common, use involves the *total* output and input over the life of a project. The second, and more frequent, use involves *instantaneous* values of output and input. The use of electrical meters to determine the input and output of an electric motor is an example of this. This latter type of efficiency is more useful, inasmuch as we do not wish to wait until a piece of equipment has worn out at the end of its life before being able to determine what its efficiency has been. In the same manner, a shorter period of time is used in obtaining a more usable measure of financial efficiency.

The most commonly used measure of financial efficiency is

$$\text{annual rate of return} = \frac{\text{annual net profit}}{\text{invested capital}} \tag{1-3}$$

which is expressed in percent. Obviously, there will be a difference in the computed rate of return, depending upon whether the annual net profit is calculated before or after income taxes have been paid. Thus it is wise to specify whether a computed rate of return is before or after taxes.

Although the annual rate of return undoubtedly is the most universally used measure of financial efficiency, we should be aware that other measures are very commonly used in certain situations. In certain instances *least cost*, either first cost or annual cost, may be a measure of financial effectiveness. However, in most cases the annual cost must include profit (return) on the capital used in order to be meaningful. Another measure that can be used in some cases is the *profit per sales dollar*. However, this measure must be used very carefully in order not to be misleading.

nonmonetary values

Few decisions, either personal or business, are made solely on the basis of financial considerations. Furthermore, the financial efficiency of a project may be affected to a considerable extent by *nonmonetary* values. If a person goes to a store to buy a new suit of clothes and finds that one in plain black can be obtained for $10 less than one of comparable quality in attractive colors, he probably will not make the decision solely on the basis of price. The financial efficiency of the manufacturer of the garments will, without question, be affected by the desires of the customers for color. Thus satisfactory decisions and recommendations regarding the feasibility of engineering projects must take into account all factors, monetary and nonmonetary, that will affect the undertaking.

Some of the most common nonmonetary (also called *intangible* and *irreducible*) factors that must be considered are economic laws, general business conditions, social and human values, consumer likes and dislikes, and governmental regulations. All business operates within an economic system that functions in accordance with certain general rules. To attempt to operate an engineering project in violation of basic economic laws is to court disaster. Whether a product or service is socially desirable, acceptable, or needed can easily determine the success or failure of a venture. Similarly, whether a product or service meets the existing likes and dislikes of the public, which may change with time, may be all-important in the success or failure of an enterprise.

equity and debt capital

The capital that is used for financing engineering and business ventures may be classified in two fundamental categories. *Equity capital* is owned by those who will use it—those who own the venture. The owners of equity capital venture it in the hope of receiving a profit. *Debt capital*, often called *borrowed capital*, is obtained by those who will use it in a venture by borrowing it from its owners. In return, the owners receive interest from the borrowers. They do not receive any other benefits that may accrue from the use of the capital in the venture it finances. On the other hand, neither do they participate as fully in the risks of the venture. Thus the return to the owners of equity capital is profit; the return to the lenders of borrowed capital is interest.

the relationship of economy studies and accounting

Economy studies are made for the purpose of determining whether capital should be invested in a project or whether it should be utilized in a different manner than it presently is being used. Economy studies always deal, at least for one of the alternatives being considered, with something that currently is not being done. Economy studies thus provide information upon which investment and managerial decisions about future operations can be based. Thus the engineering economic analyst might be termed an *alternatives fortune-teller*.

After a decision to invest capital in a project has been made and the capital has been invested, those who supply and manage the capital want to know the financial results. Therefore, procedures are established so that financial events relating to the investment can be recorded and summarized and financial efficiency determined. At the same time, through the use of proper financial information, controls can be established and utilized to aid in guiding the venture toward the desired financial goals. General accounting and cost accounting are the procedures that provide these necessary services in a business organization. Accounting studies thus are concerned with *past* and *current* financial events. Thus, the accountant might be termed a *financial historian*.

The accountant is somewhat like a data recorder in a scientific experiment. Such a recorder reads the pertinent gauges and meters and records all the essential data during the course of an experiment. From these it is possible to determine the results of the experiment and to prepare a report. Similarly, the accountant records all significant financial events connected with an investment, and from these data he can determine what the results have been and can prepare financial reports. Just as an engineer can, by taking cognizance of what is happening during the course of an experiment and making suitable corrections, gain more information and better results from the experiment, managers must also rely on accounting reports to make corrective decisions in order to improve the current and future financial performance of the business.

Accounting is generally a source of much of the past financial data that are needed in making estimates of future financial conditions. Accounting is also a prime source of data for *postmortem*, or after-the-fact, analyses that might be made regarding how well an investment project has turned out compared to the results that were predicted in the economy study. Because the accountant has the benefit of historical facts and 20–20 hindsight, he has a distinct advantage over the economy-study analyst, who must deal with estimates of the future.

A proper understanding of the origins and meaning of accounting data is needed in order to properly use or not use that data in making projections into the future and in comparing actual versus predicted results.

accounting fundamentals

This section contains an extremely brief and simplified exposition of the elements of accounting in recording and summarizing transactions affecting the finances of the enterprise. These fundamentals apply to any entity (such as an individual, corporation, governmental unit, etc.), called here just a "firm."

All accounting is based on the *fundamental accounting equation*, which is

$$\text{Assets} = \text{Liabilities} + \text{Ownership} \qquad (1\text{-}4)$$

where "Assets" are those things of monetary value that the firm *possesses*, "Liabilities" are those things of monetary value that the firm owes, and "Ownership" is the worth of what the firm owns (also referred to as "equities," "net worth," etc.).

The fundamental accounting equation defines the format of the *balance sheet*, which is one of the two most common accounting statements, and which shows the financial position of the firm at any given point in time.

Another important, and rather obvious, accounting relationship is

$$\text{revenue} - \text{expenses} = \text{profit (or loss)} \qquad (1\text{-}5)$$

This relationship defines the format of the *income statement* (also commonly known as a "profit-and-loss statement"), which summarizes the revenue and expense results of operations *over a period of time*.

A useful analogy is that a balance sheet is like a "snapshot" of the firm at an instant in time, while an income statement is a summarized "moving picture" of the firm over an interval of time. It is also useful to note that a revenue serves to increase the ownership amount for a firm, while an expense serves to decrease the ownership amount for a firm.

To illustrate the workings of accounts in reflecting the decisions and actions of a firm, suppose that XYZ firm decides to undertake an investment opportunity and that the following sequence of events occurs over a period of 1 year:

1. Organize a firm and invest $3,000 cash as capital.
2. Purchase equipment for a total cost of $2,000 by paying cash.
3. Borrow $1,500 through note to bank.
4. Manufacture year's supply of inventory through the following:
   (a) Pay $1,200 cash for labor.
   (b) Incur $400 account payable for material.
   (c) Recognize the partial loss in value (depreciation) of the equipment amounting to $500.

5. Sell on credit all goods produced for year, 1,000 units at $3 each. Recognize that the accounting value of these goods is $2,100, resulting in an increase in equity (through profits) of $900.
6. Collect $2,200 of account receivable.
7. Pay $400 account payable and $1,000 of bank note.

A simplified version of the accounting entries recording the same information in a format that reflects the effects on the fundamental accounting equation (with a "+" denoting an increase and a "−" denoting a decrease) is shown in Table 1-1. A summary of results is shown in Figure 1-1.

It should be noted that the profit for a period serves to increase the value of the ownership in the firm by that amount. Also, it is worth noting that the net cash flow from operation of $1,400 (= $3,000 − $1,200 − $400) is not all profit. This was recognized in transaction 4c, in which capital consumption (depreciation) for equipment of $500 was declared. Thus the profit was $900, or $500 less than the net cash flow.

FIGURE 1-1. Balance sheet and income statement as a result of transactions shown in Table 1-1.

XYZ FIRM
BALANCE SHEET
AS OF DEC. 31, 19xx

Assets		Liabilities and Ownership	
Cash	$2,100	Bank note	$ 500
Accounts receivable	800	Equity	3,900
Equipment	1,500	TOTAL	$4,400
TOTAL	$4,400		

XYZ FIRM
INCOME STATEMENT
FOR YEAR ENDING DEC. 31, 19xx

Operating revenues (Sales)		$3,000
Operating costs (Inventory depleted)		
Labor	$1,200	
Material	400	
Depreciation	500	
		$2,100
Net income (Profits)		$ 900

table 1-1

accounting effects of transactions —XYZ firm

	Account	1	2	3	4	5	6	7	Balances at End of Year
					Transaction				
Assets	Cash	+$3,000	-$2,000	+$1,500	-$1,200		+$2,200	-$1,400	+$2,100
	Account receivable					+$3,000	- 2,200		+ 800
	Inventory				+ 2,100	- 2,100			0
	Equipment		+ 2,000		- 500				+ 1,500
equals									**$4,400**
Liabilities	Account payable				+ 400			- 400	
	Bank note			+ 1,500				- 1,000	+ 500
plus									
Ownership	Equity	+ 3,000				+ 900			+ 3,900
									$4,400

One very important indicator of after-the-fact financial performance that can be obtained from Figure 1-1 is "annual rate of return," defined in Equation 1-3. If the invested capital is taken to be the owners' (equity) investment, the annual rate of return can be found to be $900/$3,900 = 23%.

Figures 1-2 and 1-3 provide rather detailed, realistic examples of balance sheets and income statements, respectively. They include explanations of the major categories of accounts shown, but they are unusual because figures are shown for both a large industrial firm and a large utility firm and the figures are adjusted so that the total assets for each firm is $100 million.

Financial statements are usually most meaningful if figures are shown for 2 or more years (or other reporting period such as quarters or months), or for two or more individuals or firms. In so doing the figures can be used to reflect trends or comparative financial indications which are useful in enabling investors and management to determine the effectiveness of investments *after* they have been made.

One should not assume that the figures contained in accounting reports are absolutely correct and indicative, even though they have been prepared with the utmost care by highly professional accountants. This is because accounting procedures often must include certain assumptions that are based on subjective judgment. For example, the years of life on which depreciation expense for a particular asset is based has to be estimated or assumed and the estimate may turn out to have caused unrealistic depreciation expenses and book values in accounting reports. Also, there are many accepted practices in accounting that may provide unrealistic information for management control purposes. For example, the net book value of an asset is generally declared in the balance sheet at the original first cost price minus any accumulated depreciation, even though it may be recognized that the true value of the asset at a particular time is far above or below this reported book value.

capital flow

Figure 1-4 is a schematic diagram of the flow of capital within an enterprise. It shows how investment capital of several main types is converted into goods and services, thus incurring costs and revenues and (hopefully) interest and profits as payment for the use of capital. Note that the investment and the costs are each divided into several categories which are common for management accounting purposes.

cost accounting

Cost accounting is a phase of accounting that is of particular importance to the economy-study analyst because it is the source of much of the cost data that are needed in making economy studies. Modern cost accounting may satisfy

BALANCE SHEETS
AS OF DEC. 31, 19xx
(All Figures in Millions of Dollars)

ASSETS

	Industrial Firm	Utility Firm	Explanation The Firm has:
Current Assets			
Cash	$ 2.8	$ 1.9	Checking accounts and cash
Marketable securities	5.4	4.9	Treasury bonds, notes, bills
Notes and accounts receivable	15.2	4.7	Amounts owed to firm by customers and others
Inventories	8.2	0.5	Materials, work in progress and finished goods
Prepaid expenses	1.4	0.9	Amounts paid for goods and services not yet used
Fixed Assets			
Property, plant, and equipment	101.5	112.0	Land, buildings, machinery at cost price
Less accumulated depreciation	35.8	25.3	Sum of all depreciation charges for fixed assets over years of use
Net book value	65.7	86.7	
Other Assets	1.3	0.4	Copyrights, goodwill, investments in subsidiaries, etc.
Total Assets	$100.0	$100.0	Total value of property possessed by firm

Balance Sheet (Cont'd)

LIABILITIES + EQUITIES

	Industrial Firm	Utility Firm	Explanation
			The Firm owes:
Current Liabilities			
⎫ Notes and loans payable	$ 20.4	7.0	Obligations which have been incurred
⎬ Accounts payable and accrued liabilities			but not yet paid
⎬ Income taxes payable			
⎬ Advance billing and customer's deposits			
⎭ Dividends payable			
Fixed Liabilities			
Long-term debt	12.4	36.1	Borrowed funds
Other Liabilities			
⎫ Annuity, insurance, and other reserves	28.9	1.4	Funds set aside for later payments
⎬ Deferred income tax credits			or benefits received which have not
⎭ Other deferred credits			yet been paid
Total Liabilities	41.3	44.5	
Equities			
Common and preferred stock	13.3	34.7	What has been paid in by owners
Reinvested earnings	45.4	20.7	Accumulation of prior earnings minus dividends
Total Equities	58.7	55.4	
Total Liabilities + Equities	$100.0	$100.0	

FIGURE 1-2. **Typical Balance Sheets or Statements of Financial Condition with data for an industrial firm and a utility firm.**

INCOME STATEMENTS
FOR YEAR ENDING DEC. 31, 19xx
(All Figures in Millions of Dollars)

	Industrial Firm	Utility Firm	Explanation
Income			
Sales revenue	$74.2	$35.7	Cash or obligations to pay received from customers
Expenses			
Personnel, materials, etc. for operations	46.4	16.7	Cash paid or obligations incurred for wages, material, fuel, property supplies, etc.
Depreciation and depletion	5.1	5.3	Non-cash allocation of portion of cost of plant and equipment
Total expenses from operations	51.5	22.0	
Net Income From Operations	22.7	13.7	Profits from normal business operations
Add: Dividends, interest, and misc. received	+2.4	+0.9	Payments received for securities held in other firms
Deduct: Interest paid	−1.1	−1.6	Payments made for borrowed money
Net Taxable Income	24.0	13.0	Profits before income taxes deducted
Income Taxes	13.0	7.0	Cash paid for income taxes
Net Income After Taxes	11.0	6.0	Profits after taxes

FIGURE 1-3. Typical Income Statements or Profit and Loss Statements, with data for an industrial firm and for a utility firm.

14

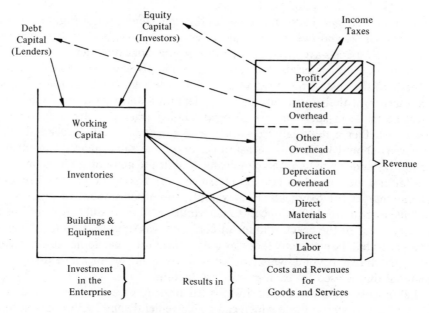

FIGURE 1-4. Enterprise capital flow.

any or all of the following objectives:

1. Determination of the actual cost of products or services.
2. Provision of a rational basis for pricing goods or services.
3. Provision of a means for controlling expenditures.
4. Provision of information on which operating decisions may be based and by means of which operating decisions may be evaluated.

Although the basic objectives of cost accounting are simple, the exact determination of costs usually is not. As a result, some of the procedures used are arbitrary devices that make it possible to obtain reasonably accurate answers for most cases but which may contain a considerable percentage of error in other cases.

the elements of cost

One of the first problems in cost accounting is that of determining the elements of cost that arise in the production of an article or the rendering of a service. A study of how these costs occur gives an indication of the accounting procedure that must be established to give satisfactory cost information. Also, an understanding of the procedure that is used to account for these costs makes it possible to use them more intelligently and correctly.

From an engineering and managerial viewpoint in manufacturing enterprises, it is common to consider the general elements of cost to be *direct materials, direct labor*, and *overhead*. These are shown in the right-hand part of Figure 1-4. Such terms as *burden* and *indirect costs* are often used synonymously with overhead, and overhead costs are often divided into several subcategories.

Ordinarily, the materials that can be conveniently and economically charged directly to the cost of the product are called direct materials. However, it is at once apparent that this is not a precise definition and that arbitrary decisions must be made in such matters. Several guiding principles are used when we decide whether a material is classified as a direct material. In general, direct materials should be readily measurable, be of the same quantity in identical products, and be used in economically significant amounts. Those materials that do not meet these criteria are classified as *indirect materials* and are a part of the charges for overhead. For example, the exact amount of glue used in making a chair would be difficult to determine. Still more difficult would be the measurement of the exact amount of coal that was used to produce the steam that generated the electricity that was used to heat the glue. Some reasonable line must be drawn beyond which no attempt is made to measure directly the material that is used for each unit of production.

Labor costs also are ordinarily divided into *direct* and *indirect* categories. Direct labor costs are those which can be conveniently and easily charged to the product or service on which worked. Other labor costs, such as for supervisors and material handlers, are charged as indirect labor and are thus included as part of overhead costs. It is often imperative to know what is included in direct labor and direct material cost data before attempting to use them in economy studies.

In addition to indirect materials and indirect labor there are numerous other cost items that must be incurred in the production of products or the rendering of services. Property taxes must be paid; accounting and personnel departments must be maintained; buildings and equipment must be purchased and maintained; supervision must be provided. It is essential that these necessary *overhead* costs be allocated to each unit produced in proper proportion to the benefits received. Proper allocation of these overhead costs is not easy, and some factual, yet reasonably simple, method of allocation must be used.

As might be expected where solutions attempt to meet conflicting requirements, such as exist in overhead-cost allocation, the resulting procedures are empirical approximations that are accurate in some cases and less accurate in others.

There are many methods of allocating overhead costs among the products or services produced. The most commonly used methods involve allocation in proportion to direct labor cost, or direct labor hours, or direct materials cost, or sum of direct labor and direct materials cost, or machine hours. In each of these methods it is necessary to know what the total of the overhead costs have been if postmortem costs are being determined, or to estimate what the total overhead costs will be if predicted or standard costs are being determined. In either case, *the total overhead costs will be associated with a certain level of production.* This is an important condition that should always be remembered when we are dealing with unit-cost data. They can only be correct for the conditions for which they were determined.

To illustrate one method of allocation of overhead costs, consider the method which assumes that overhead is incurred in direct proportion to the cost of direct labor used. With this method the overhead rate (overhead per dollar of direct labor) and the overhead cost per unit would be

$$\text{overhead rate} = \frac{\text{total overhead in dollars for period}}{\text{direct labor in dollars for period}}$$

$$\text{overhead cost per unit} = \text{overhead rate} \times \text{direct labor cost per unit}$$

Suppose that for a coming period (say quarter) the total overhead cost is expected to be $100,000 and the total direct labor cost is expected to be $50,000. From this, the overhead rate = $100,000/$50,000 = $2 per dollar of direct labor cost. Suppose further that for a given unit of production (or job) the direct labor cost is expected to be $60. From this, the overhead cost for the unit of production would be $60 × $2 = $120.

This method obviously is simple and easy to apply. In many cases it gives quite satisfactory results. However, in many other instances, it gives only very approximate results because some items of overhead, such as depreciation and taxes, have very little relationship to labor costs. Quite different total costs may be obtained for the same product when different procedures are used for the allocation of overhead costs. The magnitude of the difference will depend on the extent to which each method produces or fails to produce results that are in accord with the facts. The choice of an overhead allocation method should be based on what seems most reasonable under the circumstances.

the use of accounting costs in economy studies

When we recognize that accounting costs are linked to a definite set of conditions, and that they are the result of certain arbitrary decisions concerning overhead-cost allocation, it is apparent that they should not be used without modification in cases where the conditions are different from those for which they were determined. Economy studies invariably deal with situations that now are *not* being done. Thus ordinary accounting costs normally cannot be used, without modification, in economy studies. However, if we understand how the accounting costs were determined, we should be able to break them down into their component elements and then, often, we find that these cost elements will supply much of the cost information that is needed for an economy study. Thus an understanding of the basic objectives and procedures of cost accounting will enable the economy-study analyst to make optimum use of available cost information and to avoid needless work and serious mistakes.

multiple objectives and subjectivity in decision analysis

Although the primary concern of this book is techniques for considering economic or monetary desirability, it should be emphasized that the usual decision between alternatives involves objectives other than those which can be reasonably reduced to monetary terms. For example, a limited listing of objectives other than profit maximization or cost minimization that may be important to a firm are

1. Improvement of safety record to a certain level.
2. Minimization of risk of loss.
3. Maximization of service quality.
4. Minimization of cyclic fluctuations in production.
5. Maintenance of a desired public image.
6. Reduction of pollutants to a certain level.
7. Maximization of employee satisfaction.

Economic analyses formally provide only for consideration of those objectives or factors that can be reduced to monetary terms. The results of these analyses should be weighed together with other nonmonetary (intangible) objectives or factors before a final decision is made. There exist analytical methods to accomplish this if the appropriate estimates can be made, but these are beyond the scope of this book. In general, one should pursue formal analysis as far as is technically and economically feasible, and then depend upon the judgment and experience of the decision maker and his staff to consider factors not included in the formal analysis.

the decision-making process

Rational decision making is a complex process. Following is a list of six interrelated phases which make up that process, together with brief comments on each:

1. Recognition of the problem—A problem is any condition or circumstance that may impede one from achieving his goals or objectives. It may merely be an opportunity for improvement.
2. Definition of the goal or objective—This may be specific, such as to produce X products; or general, such as to reduce costs or to increase profits.
3. Identification of feasible alternatives—This phase often is not given sufficient attention. Unless the best alternative is considered, the result of comparisons will always be suboptimal.
4. Selection of a criterion or criteria for judging which alternative is best—For studies in which only monetary factors need to be considered, a sole economy study criterion such as present worth or rate of return is sufficient. Otherwise, multiple criteria, to correspond with objectives, should be used.

5. Prediction of the outcomes for each alternative—This estimating phase is difficult and crucial to the soundness of the analysis.
6. Choice of the best alternative to achieve the objective—This wraps it up, normally.

Although this list implies a logical ordering of the phases, one may well cyclically use information from one phase to further develop work on one or more other phases several times before reaching a final choice. Indeed, the choice reached may be unsatisfactory in some way(s), thus prompting reconsideration of all previous phases.

role of the engineer and importance of economic decision studies

Economic analyses and decisions between alternatives can be made by the engineer considering alternatives in the design of equipment, facilities, or people–machine systems. However, the decisions are more commonly made by a manager acting upon a number of investment opportunities and alternatives within each opportunity. Whenever the alternatives involve technical considerations, the engineer can serve to provide estimates and judgments for the analyses which provide a basis for final managerial decisions.

Engineering economic analyses are most appropriate for problems that are sufficiently important to warrant serious thought and have economic aspects that are of significant importance in reaching a final choice between alternatives. A vast number of problems that one will encounter in the business world (and in one's personal life) meet these criteria. Hence economic decision studies can be an extremely valuable tool in a great many situations.

problems

1-1. How would you define "engineering economy"?
1-2. Discuss the concept of financial efficiency and explain why it is different from thermal efficiency, for example.
1-3. Explain why differences between alternatives are of major interest in engineering economy studies.
1-4. Why do engineers get involved with economy studies? Why would it not be better for an accountant to perform economy studies?
1-5. Describe three situations in which monetary comparisons among alternatives could be *less important* than nonmonetary differences among them.
1-6. What is the difference between equity capital and debt capital? What constitutes "net worth" in the fundamental accounting equation? Give examples.
1-7. In your own words, explain the difference between the basic purposes of *engineering economy* and *accounting*.

1-8. What would occur if, through legislation, all profit were eliminated from a certain industry? Would this necessarily assure a decrease in the selling price of the product of this industry?

1-9. Explain why ordinary accounting data may not be useful in an economy study.

1-10. Explain the differences in the purposes of the two primary financial statements (balance sheets and income statements).

1-11. You have started your own business producing precision-wound induction coils. At the end of the first year the following sequence of events has occurred.
   (a) Organized firm and invested $2,000 as capital.
   (b) Purchased equipment worth $1,500 financed by a note with the bank.
   (c) Manufactured year's inventory by:
      (1) $1,200 labor expense.
      (2) $500 materials expense on account.
      (3) $350 depreciation.
   (d) All of the 5,000 units produced were sold on account for $0.50 apiece.
   (e) Collected $2,000 of accounts receivable.
   (g) Paid $500 on accounts payable and $750 on the bank note.
   Prepare (a) a balance sheet for the end of the year and (b) an income statement for the year.

1-12. Does a company that makes a single, multipart product need a good cost accounting system? Why?

1-13. What are the three common elements of cost in cost accounting? Why must some cost accountants also be able to do *cost estimating* work?

1-14. How would you attempt to deal with multiple objectives in an engineering economy problem?

1-15. What is rational decision making, and why is flipping a coin not necessarily a rational way to make a decision?

1-16. List the six phases of the decision making process and explain why cycling back through the process may be required in some situations.

1-17. One product of a multiproduct company was produced by a group of four skilled and experienced workers, who were paid $4 per hour. They worked 8 hours per day, 250 days per year, in an area of 400 square feet in a single-story building of light construction. They required very little supervision, and the labor turnover was almost zero. The cost of the material in the product was $8 per unit.

The equipment used by these workers cost $3,200 and had an expected life of 4 years. It consumed 1 kilowatt of electricity per hour. The company followed a policy of charging overhead to all its products on the basis of 125% of direct labor cost.

The company was considering dropping this item from its product line because the selling price was only slightly greater than the computed cost. Comment on this situation.

1-18. Refer to Figures 1-2 and 1-3. For the industrial firm and the utility firm, compute and compare the following financial ratios.
   (a) Current assets ÷ current liabilities (a relatively high ratio, say 2:1, indicates satisfactory ability to meet current financial obligations)
   (b) Total expenses from operations ÷ sales revenue (a low ratio is desirable)
   (c) Sales revenue ÷ inventory (a large ratio reflects fast turnover of goods through a company's operations)

1-19. The DEF Corporation wishes to purchase all the stock of the much smaller LOM Company, which has 200,000 shares of stock outstanding. The stock currently is selling on the market at $25 per share. The corporation believes it can acquire all of the stock if it makes an offer midway between the present market price and the book value of the equities at the close of the previous year's operations. The balance sheet of the LOM Company at that time showed the following:

Current assets	$6,200,000
Fixed assets	4,100,000
Other assets	300,000
Current liabilities	3,150,000
Fixed liabilities	1,150,000
Earned surplus	6,300,000

What price should be offered for the stock?

1-20. Under normal conditions, the ABC Company makes products, A, B, and C in such quantities that the direct labor costs per day are $400, $500, and $600 for the three products, respectively, and the total overhead is just absorbed (covered) when it is charged at the rate of 150% of direct labor. Because of a change in market conditions, the sales of product C have been reduced by 50% and those of product A have doubled. The total overhead has not changed. To what percentage of direct labor cost must the overhead charge be reduced in order to avoid overcharging for overhead? (Assume that direct labor and direct material costs are directly proportional to output.)

chapter 2

the economic environment and cost concepts

There are certain economic principles that frequently must be taken into account in economy studies. In many instances, economic concepts aid materially in managerial decision making.

In general terms, economics deals with the interactions between people and wealth. Because people, as individuals, are not all alike as to their reactions, the subject of economics necessarily must deal with these interactions in generalized terms. Also, in most instances we cannot quantify the results of these interactions in absolute, monetary terms. Thus they must be dealt with as "nonmonetary" or "intangible" considerations. Nevertheless, their effects on the financial aspects of a venture can be just as real and certain as can the law of gravity on the physical aspects. The purpose of this chapter is to examine, briefly, some of these basic economic concepts and to indicate how they may be factors for consideration in making economy studies and managerial decisions.

consumer and producer goods and services

The goods and services that are produced and utilized may be divided conveniently into two classes. *Consumer goods and services* are those products or services that are directly used by people to satisfy their wants. Food, clothing, homes, cars, television sets, haircuts, opera, and medical services are examples. The producers and vendors of consumer goods and services must be aware of, and are subject to, the whims and caprices of the people to whom their products are sold. At the same time, the demand for such goods and services is directly related to people and in many cases, as will be discussed later, may be determined with considerable certainty.

Producer goods and services are used to produce consumer goods and services or other producer goods. Machine tools, factory buildings, buses, and farm machinery are examples. Although, in the long run, producer goods serve to satisfy human wants, they are the means to that end. Thus the amount of producer goods needed is determined indirectly by the amount of consumer goods or services that are demanded by people. However, because the relationship is much less direct than for consumer goods and services, the demand for, and production of, producer goods may greatly precede or lag behind the demand for the consumer goods that they will produce. For example, a company may build and equip a factory to manufacture a product for which there presently is no demand, in the hope that it will be able to create such a demand through advertising and sales effort. On the other hand, a company may wait until the demand for an existing product is well established before deciding to build facilities to produce the product. Thus the decision to buy or build producer goods is not a direct result of the demand for the consumer goods or services they can produce, but is related directly to management's estimate of the profit potential of the facilities. This is quite a different situation from that for many consumer goods—food, for example—where the consumption cannot be deferred indefinitely.

It can be seen that if an economy study involves investment in production facilities, the problem of estimating the demand for the product or service will be quite different, depending on whether producer goods or services or consumer goods or services are involved. Furthermore, the effects of variations in economic conditions could be quite different for the two cases.

measures of economic worth

Goods and services are produced, and desired, because directly or indirectly they have *utility*—the power to satisfy human wants and needs. Thus they may be used or consumed directly, or they may be used to produce other goods or services that may, in turn, be used directly. Utility most commonly is measured in terms of *value*, expressed in some medium of exchange as the *price* that must be paid to obtain the particular item.

Much of our business activity revolves around increasing the utility (value) of materials and products by changing their form or location. Thus iron ore, worth only a few dollars per ton, may be increased in value to several dollars a pound by processing it and combining it with suitable alloying elements and converting it into razor blades. Similarly snow, worth almost nothing when high in distant mountains, can be made quite valuable when it is delivered in melted form several hundred miles away to dry Southern California.

necessities, luxuries, and price-demand

Goods and services may be divided into two types, *necessities* and *luxuries*. Obviously, these terms are relative, because for most goods and services what one person may consider to be a necessity may be considered by another to be a luxury. Economic status is an important factor in one's views regarding luxuries and necessities. Other factors also may be determining. For example, a man living in one community may find that an automobile is an absolute necessity for him to be able to get to and from his place of employment. If the same man lived and worked in a different city, he might have adequate public transportation available, and an automobile would be strictly a luxury. Also, the classification of goods and services into luxuries and necessities is less easy in the case of producer goods than for consumer goods. However, it is clear that for all goods and services there is a relationship between the price that must be paid and the quantity that will be demanded, or purchased. This general relationship is depicted in Figure 2-1; as the selling price is increased, there will be less demand for the product; and as the selling price is decreased, the demand will increase. The relationship between price and demand can be expressed as a linear function:

$$P = a - bD \quad \text{for} \quad 0 \le D \le \frac{a}{b} \tag{2-1}$$

where a and b are constants. It follows, of course, that

$$D = \frac{a - P}{b} \tag{2-2}$$

Although Figure 2-1 illustrates the general relationship between price and demand, it would be expected that this relationship would be different for

FIGURE 2-1. General price-demand relationship. Note that "price" is considered to be the independent variable but is shown as the vertical axis. This convention is commonly used by economists; because it may be somewhat confusing to engineers, who usually show the independent variable as the abscissa, an asterisk has been added before the independent variable.

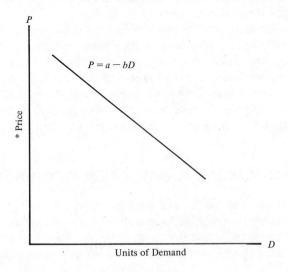

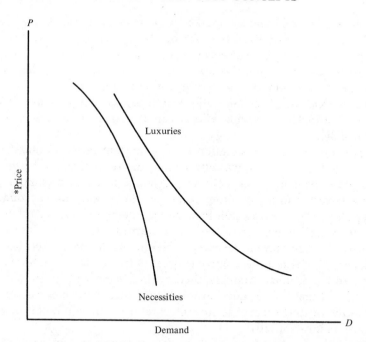

FIGURE 2-2. Generalized price–demand relationship for luxuries and necessities. Note that "price" is considered to be the independent variable but is shown as the vertical axis.

necessities and luxuries. Consumers can readily forgo the consumption of luxuries if the price is greatly increased, but they find it more difficult to reduce their consumption of true necessities. Also, they will use the money saved by not buying luxuries to pay the increased cost of the necessities. Figure 2-2 shows how the demand curves for luxuries and necessities might differ.

The extent to which price changes influence demand varies according to the *elasticity* of the demand. The demand for products is said to be *elastic* when a decrease in the selling price results in a considerable increase in sales. On the other hand, if a change in selling price produces little or no effect on the demand, the demand is said to be *inelastic*. It is clear, from Figure 2-2, that luxury items have greater elasticity of demand than do necessities.

competition

Because economic laws are general statements regarding the interaction of people and wealth, they will be affected by the economic environment in which the people and the wealth exist. Most general economic principles are stated for situations in which *perfect competition* exists.

Perfect competition occurs in a situation in which any given product is supplied by a very large number of vendors and there is no restriction against additional vendors entering the market. Under such conditions, there is assurance of complete freedom on the part of both buyer and seller. Actually, of course, perfect competition may never exist because of a multitude of factors that impose some degree of limitation upon the actions of buyers or sellers, or both. However, with conditions of perfect competition assumed, it is easier to formulate general economic laws. When deviations from perfect competition are known to exist, their probable economic effects can be taken into account, at least approximately.

The existing competitive situation is an important factor in most economy studies. It will have a very real effect upon decisions that are made. Unless information is available to the contrary, it should be assumed that competition does or will exist, and the resulting effects should be taken into account.

Monopoly is at the opposite pole from perfect competition. A perfect monopoly exists when a unique product or service is available from only a single vendor and that vendor can prevent the entry of all others into the market. Under such conditions the buyer is at the complete mercy of the vendor as to the availability and price of the product. Actually, there seldom is a perfect monopoly. This is due to the fact that few products are so unique that substitutes cannot be used satisfactorily, or to the fact that governmental regulations prohibit monopolies if they are unduly restrictive.

A monopoly may be of great benefit to a producer in that he may be able to control the supply and the price to provide the maximum profit. Although this might result in high prices for the product, higher *total* profits may be obtained from lower prices and wider distribution. Under some conditions, a monopoly may avoid costly duplication of facilities and thus make possible lower prices for products and services. This situation is recognized by governing bodies in granting public utilities exclusive rights to render service in a given territory. Such a practice, for example, avoids having two electric power companies duplicate power-distribution lines and equipment in the same city. When vendors are granted such a monopolistic position, the governmental body also regulates the rates that the utility can charge for its services to assure that the customers are not overcharged. Thus governmental regulation takes the place of competition in determining prices.

A *semimonopoly* may be permitted for a limited time through the issuance of patents or copyrights that help to maintain the uniqueness of a product by preventing identical products from being marketed.

An *oligopoly* exists when there are so few suppliers of a product or service that action by one will almost inevitably result in similar action by the others. Thus if one of the only three gasoline stations in an isolated town raises the price of its product by 1 cent per gallon, the other two would probably do the same because they could do so and still retain their previous competitive positions.

It is apparent that many conditions between those of perfect competition, perfectly monopoly, and oligopoly can and do exist. Thus it is to be expected that *actual* relationships among supply, price, and demand will be somewhat different from those derived for the ideal conditions. Also, it would be extremely difficult to state or depict such relationships for all possible conditions. Therefore, it is most practicable to derive these relationships for the case of perfect competition; the changes that would result from nonperfect conditions are usually apparent.

the total-revenue function

The total revenue, TR, that will result from a business venture during a given period is the product of the selling price per unit and the number of units sold. Thus

$$\text{TR} = \text{price} \times \text{demand} \tag{2-3}$$

If the relationship between price and demand as given in Equation 2-1 is used,

$$\text{TR} = (a - bD)D = aD - bD^2 \quad \text{for } 0 \le D \le \frac{a}{b} \tag{2-4}$$

If we neglect any cost functions, the relationship between total revenue and demand, *for the condition expressed in Equation 2-4* may be represented by the curve shown in Figure 2-3. Under these conditions, the maximum total revenue also would produce maximum total profit. From the calculus, the demand, D^*, that will produce maximum total revenue can be obtained by solving:

$$\frac{d\text{TR}}{dD} = a - 2bD = 0 \tag{2-5}$$

Thus†

$$D^* = \frac{a}{2b} \tag{2-6}$$

FIGURE 2-3. Total-revenue function as a function of demand.

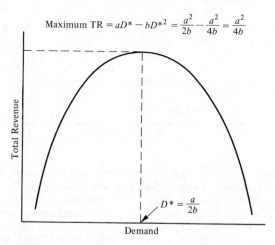

$$\text{Maximum TR} = aD^* - bD^{*2} = \frac{a^2}{2b} - \frac{a^2}{4b} = \frac{a^2}{4b}$$

$$D^* = \frac{a}{2b}$$

Total Revenue

Demand

† D^* means the optimal value of D.

It must be emphasized that, because of cost–volume relationships, most businesses would not obtain maximum profits by maximizing revenue. Thus the cost–volume relationship must be considered and related to revenue.

At this point, attention is called to the derivative of the total revenue with respect to volume (demand), $d\text{TR}/dD$, which is called the *incremental* or *marginal revenue*; this will be discussed later in more detail.

cost-volume relationships

In virtually all businesses, there are certain costs that remain constant over a wide range of activity as long as the business does not permanently discontinue operations. Such costs as property taxes, interest on borrowed capital, and many of the overhead costs are of this type. Such costs commonly are referred to as *fixed costs*. There are other costs, however, that vary more or less directly with the volume of output, and these are called *variable costs*. These include the costs of such items as materials, labor, and power.† Using this concept of fixed and variable costs, their relationship to the total-cost function is portrayed in Figure 2-4.‡ Thus at any demand D,

$$C_{\text{total}} = C_F + C_v \qquad (2\text{-}7)$$

FIGURE 2-4. Typical fixed, variable, and total costs as functions of volume.

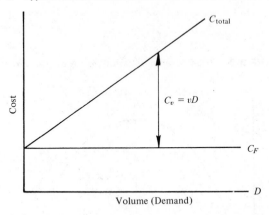

† In a later chapter it will be pointed out how some costs that may be fixed in one range of conditions may be variable in another and that costs that are variable for one range of conditions may be considered to be fixed for other conditions.

‡ In actual cases the variable costs frequently are not exactly proportional to volume, and thus the total-cost line is not exactly linear. However, in most cases the actual total-cost line does not deviate much from a straight line, and a straight line relationship is used here for simplification.

where C_F and C_v denote fixed and variable costs, respectively. For the linear relationship, assumed here,

$$C_v = (v)(D) \tag{2-8}$$

where v is the variable cost per unit.

When the total revenue–demand relationship, as depicted in Figure 2-3, and the total cost–demand relationship, as depicted in Figure 2-4, are combined, Figure 2-5 results for the case where $(a - v) > 0$, and the volumes for which profit and loss result are clearly evident. We obviously are interested in the condition for which maximum profit will be obtained. Since

$$\text{profit} = (aD - bD^2) - (C_F + vD)$$

$$= -C_F + (a - v)D - bD^2 \quad \text{for} \quad 0 \le D \le \frac{a}{b}$$

$$\frac{d(\text{profit})}{dD} = a - v - 2bD = 0$$

For maximum profit,

$$D = \frac{a - v}{2b} \tag{2-9}$$

If $(a - v) \le 0$, the profit will be maximized when $D = 0$.†

The condition for maximum profit as expressed in Equation 2-9 becomes more meaningful and useful when it is related to the incremental revenue,

FIGURE 2-5. Combined cost and revenue functions as functions of volume, and their effect on profit.

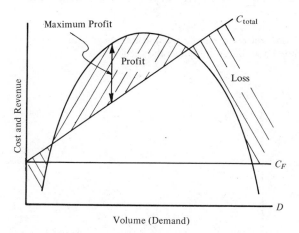

† This relationship simply says that if the variable cost per unit is equal to or greater than the price that will produce *no* demand, nothing should be produced; thus losses are minimized.

Equation 2-5. If we substitute Equation 2-9, the condition for maximum profit, into Equation 2-5, we obtain

$$\frac{dTR}{dD} = a - \frac{2b(a - v)}{2b} = v \tag{2-10}$$

Because v is the variable cost per unit, Equation 2-10 means that in order to obtain maximum profit, we should increase the output as long as the incremental revenue exceeds the incremental production cost, stopping when they are equal.

the law of supply and demand

As was discussed previously and illustrated in Figure 2-1, under competitive conditions there is a relationship between the price customers must pay for a product and the amount that they will buy. There is a similar relationship between the price at which a product can be sold and the amount that will be made available. If the price they can get for their products is high, more producers will be willing to work harder, or perhaps risk more capital, in order to reap the greater reward. If the price they can obtain for their products declines, they will not produce as much because of the smaller reward that they can obtain for their labor and risk. Some will stop producing and turn their efforts to other endeavors. This relationship between price and the volume of product produced can be portrayed by the curve shown in Figure 2-6.

If Figures 2-1 and 2-6 are combined, Figure 2-7 results. Figure 2-7 illustrates the basic economic *law of supply and demand*, which states that under conditions of perfect competition, the price at which a given product will be supplied and purchased is the price that will result in the supply and the demand being equal.

Because many economy studies deal with investments that will increase the

FIGURE 2-6. General price–supply relationship. Note that "price" is considered to be the independent variable but is shown as the vertical axis.

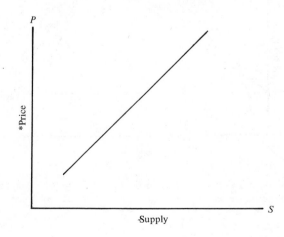

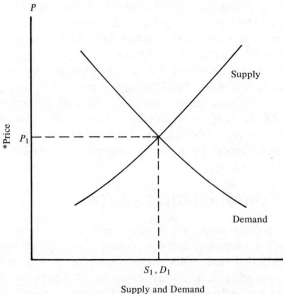

FIGURE 2-7. Price–supply–demand relationship, showing equal supply and demand at a given price. Note that "price" is considered to be the independent variable but is shown as the vertical axis.

FIGURE 2-8. Price–supply–demand relationship, showing how the addition of supply, at a given price, will cause a new and lower price to be established. Note that "price" is considered to be the independent variable but is shown as the vertical axis.

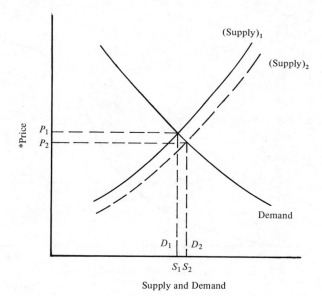

amount of a given product that will be available on the market, we are interested in what will happen to the selling price of the commodity under the proposed conditions. If a producer is willing to supply additional volume of a product to the market at existing prices, this means that a new price–supply condition has been created. As shown in Figure 2-8, at the original price P_1, an additional amount of a product is made available and a new price–supply curve exists. Because there has been no change in the price–demand relationship meanwhile, the intersection of the new supply curve results in a new and lower price, P_2, corresponding to the new demand, D_2. Obviously, a reverse situation results from a decrease in the amount of a product offered to the market at a given price.

the law of diminishing returns

The law of diminishing returns is another basic economic principle that frequently is a factor in economic decision making and helps to provide an understanding of conditions that often are encountered in economy studies. All engineers and engineering students have encountered the results of this principle, but often they do not realize that what they have encountered is related to this law. Figure 2-9, for example, is such a case. Basically, the law refers to the amount of *extra output* that results from successive *additions* of equal units of *variable* input to some *fixed* (or limited) amount of some *other input factor*. The law may be stated as follows: When one or more of the input factors of production

FIGURE 2-9. Input–output curve for a $3\frac{1}{2}$-horsepower constant-speed motor.

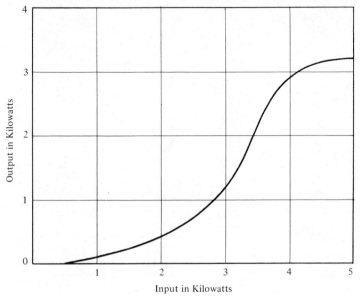

Input in Kilowatts

is limited, either by absolute quantity or by increasing cost, adding more of the variable input factors will result in an output's being reached beyond which such additions will result in a less-than-proportionate increase in output.

Although the earlier concepts concerning the principle of diminishing returns apparently were derived from experiences relating to the use of agricultural land, where the amount of land was the obvious fixed factor, a little reflection makes it apparent that it is equally applicable to many engineering and business situations. Table 2-1 shows the numerical data from which Figure 2-9 was obtained. It will be noted that for the early increases in input, up to an input of 4.0 kilowatts (kW), the actual increase in output is greater than proportional†; beyond this point, the output is less than proportional. In this case the fixed input factor is the electric motor. As more and more of the variable factor—electrical energy—is added, each kilowatt has a smaller portion of the motor on which it can act, thus reaching an ultimate point beyond which the efficiency decreases.

As mentioned previously, the results of the principle of diminishing returns are found with great frequency. Some examples are all types of prime movers, office and apartment buildings, and production facilities where equipment cannot be expanded in capacity. In each case there is an input factor that is limited by increasing cost or absolute quantity. It should also be recognized that capital and markets may also constitute limited input factors because of the increasing cost of obtaining them. Thus diminishing returns is an important consideration in many investment decisions.

Some ventures and types of businesses are more subject to the diminishing-returns principle than others because of the fixed factors involved. Such industries

table 2-1

load test data for $3\frac{1}{2}$-horsepower constant-speed motor showing proportional and actual output per unit of input

(1)	(2)	(3)	(4)	(5)
	Input (kW)	Actual Output (kW)	Proportional Output (kW)	Output per Unit of Input
A	0.5	0.00	—	0.000
B	1.0	0.10	—	0.100
C	1.5	0.25	0.15	0.167
D	2.0	0.41	0.33	0.205
E	2.5	0.70	0.51	0.280
F	3.0	1.20	0.84	0.400
G	3.5	2.20	1.40	0.628
H	4.0	2.92	2.52	0.730
I	4.5	3.15	3.28	0.700
J	5.0	3.20	3.50	0.640

† In this range each successive increase of 0.5 kilowatt in input results in a greater increment of output than did the previous 0.5 kilowatt.

sometimes are called *increasing-cost industries*; this means that they tend to encounter increased unit costs as they attempt to increase their volume of activity. Others, which can readily expand without this difficulty, are called *decreasing-* or *normal-cost industries*.

breakeven charts

The total-revenue function expressed in Equation 2-4 and illustrated in Figure 2-3 was a generalized relationship involving a straight-line relationship between price and demand, as illustrated in Figure 2-1. For a specific business the revenue function is the product of price and volume sold and thus is essentially a straight line over a considerable range of volume change. For this condition the relationship between volume and fixed costs, variable costs, and revenue is shown in Figure 2-10. These commonly are called *breakeven charts*. Breakeven charts are very useful in portraying and understanding the effects of variations in fixed and variable costs on the profitability of a business. Thus they may be used to portray the effects of proposed changes in operational policy.

In breakeven charts the fixed costs, variable costs, and revenue are plotted against output, either in units, dollar volume, or percent of capacity. Thus in Figure 2-10 the line $F'F$ represents the fixed costs of production. The line $F'V$ shows the variation in total variable cost with production; because its starting

FIGURE 2-10. Typical breakeven chart for a business enterprise.

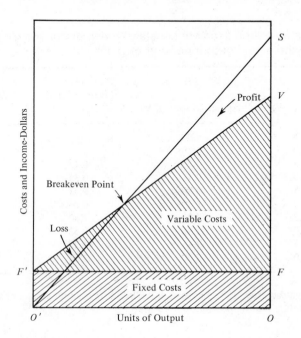

point is at F', it actually represents the sum of all production costs. The gross revenue from sales is represented by the line $O'S$. Inasmuch as $F'V$ represents the total costs of production and $O'S$ the total revenue from sales, the intersection of these two lines is the point at which revenue is exactly equal to costs and is often called the *breakeven point*. With the rate of production at that point, the business will make no profit and have no loss. If the production rate is greater than that at the breakeven point, a profit will result. When the rate is less than that at the breakeven point, a loss will be sustained. By representing the revenue and costs of a business in this manner, we can easily determine the possibility of profit for any rate of production. Because these breakeven charts show the relationship between revenue and costs for all possible volumes of activity, they are, in effect, continuous income (profit and loss) statements.

One of the most satisfactory uses of breakeven charts is to show the relative effects of changes in fixed and variable costs of a business. This is illustrated in Figure 2-11. Figure 2-11a shows the breakeven chart for a certain business that has a sales capacity of $300,000 per year. For the values of revenue and cost shown, the breakeven point occurs at 50% of sales capacity. For any volume of sales over $150,000, the business will make a profit.

The chart in Figure 2-11b is drawn to determine what the effect will be if the variable costs are decreased 10%, with all other factors remaining constant. The profits are increased by $20,000 when the business is operated at 100% of sales capacity. However, a more significant effect is the change in the breakeven point. The 10% decrease in variable costs lowers the breakeven point to approximately 41% of sales capacity. This means that not only will the company make greater profits if the plant is operated at capacity, but, more important, some profit will be earned if operations are greater than 41% rather than 50% of capacity. In times of recession it is usually more important that a business be able to earn some profit, or not incur a loss, when operating at partial capacity

FIGURE 2-11. Effect of changes in fixed and variable costs on the location of the breakeven point. Maximum sales capacity (100%) corresponds to $300,000 per year. (a) Before change, (b) with a 10% decrease in variable costs, and (c) with a $20,000 decrease in fixed costs.

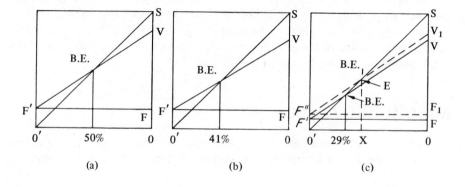

than it is for it to be able to earn a very large profit if operated at a capacity that is entirely beyond the realm of possibility.

The third breakeven chart, Figure 2-11c, was drawn to determine to what extent the breakeven point will be shifted if the same saving of $20,000 is effected out of the fixed costs. The solid lines in Figure 2-11c give the solution to this question. It is shown that this change in fixed costs will lower the breakeven point from 50% to approximately 29% of sales capacity. This makes the importance of controlling fixed costs very apparent. The saving of $20,000 in fixed costs lowers the breakeven point nearly 12% more than would be the case if the same saving were made in variable costs at maximum sales capacity.

Unfortunately, it is usually easier to effect savings in variable costs than in fixed costs. We might wish to know what saving in fixed costs would give the same result as a greater economy in variable costs, and the dashed lines in Figure 2-11c provide this information. A vertical line XE is drawn upward from a point corresponding to 41% of sales capacity until it intersects the revenue line $O'S$ at point E. Through point E a line $F''V_1$ is drawn parallel to $F'V$. $F''F_1$ is the fixed-cost line that will give the required breakeven point of 41% of sales capacity. The fixed costs are determined by the ordinate $O'F''$, corresponding to $42,000. This means that in this case a saving of only $8,000 in fixed costs will lower the breakeven point as much as a saving of $20,000 in variable costs. Thus the effect of fixed costs upon the breakeven point is evident.

It should be noted that in each case illustrated in Figure 2-11b and c the profit at 100% of capacity is the same. However, the percentage of capacity at which *some* profit can be earned differs. It should also be remembered that the changes in percentages apply only to the conditions of the example given, although other cases will show a similar change.

cost concepts for economic analysis

The word *cost* has many meanings in many different settings. The kind of cost concept that should be used depends upon the business decision to be made. Financial records resulting from the firm's accounting function aim at describing what has happened in the past, whereas useful decision-making concepts of cost aim at projecting what is expected to happen in the future as a result of alternative courses of action. Indeed, different combinations of cost ingredients are appropriate for various kinds of management problems. The following sections contain descriptions of some additional cost concepts that are most important in the making of economic analyses.

opportunity costs

An *opportunity cost* is a cost which, although hidden or implied, is incurred because of the use of limited resources in such a manner that the chance or opportunity to use those resources to monetary advantage in an alternative use is forgone. As an example suppose that a project involves the use of firm-owned warehouse space that is presently vacant. The cost for that space which should be charged to the project in question should be the income or savings that possible alternative uses of the space may bring to the firm. In other words, the cost for the space for purposes of an economy study should be the *opportunity cost* of the space. This may be more than or less than the average cost of that space which might be obtained from accounting records.

As another example, consider a student who could earn $8,000 for working during a year and who chooses instead to go to school and spend $3,000 to do so. The total cost of going to school for that year is $11,000: $3,000 cash outlay and $8,000 for income forgone. (*Note*: This neglects the influence of income taxes and assumes that the student has no earning capability while in school.)

Opportunity Cost in Determination of Interest Rates for Economic Analysis

A very important use of the opportunity cost principle is in the determination of the interest (minimum attractive return) to use in analyses of proposed capital investment projects. The proper interest rate is not just the amount that would be paid for the use of borrowed money, but is rather the opportunity cost, that is, the return forgone or expense incurred because the money is invested in that project rather than in possible alternative projects. Suppose that a firm always has available certain investment opportunities, such as expansion or bond purchases, which will earn a minimum of, say, $X\%$. This being the case, the firm would be unwise to invest in other alternative projects earning less than $X\%$. The $X\%$ may be thought of as the opportunity cost of not investing in the readily available alternatives.

In economy studies it is necessary to recognize the time value of money irrespective of how the money is obtained, whether it be through debt financing, through owners' capital supplied, or through reinvestment of earnings generated by the firm. Interest on project investments is a cost in the sense of an opportunity forgone, an economic sacrifice of a possible income that might have been obtained by use of that money elsewhere.

Opportunity Cost in Replacement Analysis

As another illustration of the opportunity cost principle, suppose that a firm is considering replacing an existing piece of equipment which originally cost $50,000, presently has an accounting book value of $20,000, and can be salvaged now for $5,000. For purposes of an economic analysis of whether or not to replace the existing piece of equipment, the investment in that equipment should be considered as $5,000; for by keeping the equipment, the firm is giving up the *opportunity* to obtain $5,000 from its disposal. This principle is elaborated upon in Chapter 11.

sunk costs

Sunk costs are costs resulting from past decisions and which are therefore irrelevant to the consideration of alternative courses of action. Thus, sunk costs should not be considered directly in economic analyses, but should be recognized because of the human tendency to be unduly influenced by them.

As an example, suppose that you bought a stock several years ago for $10 a share. The stock performed badly, and 1 year ago it was selling for $1. You have kept the stock and now it is selling for $3. Your question is: Should you sell or keep the stock? You might reason that selling would result in a ($10 − $3)/$10 = 70% loss, or in a ($3 − $1)/$1 = 200% gain over the low selling price. Actually, both the $10 and the $1 are in the past and are thus sunk costs. The valid viewpoint for answering the question is to consider the likely future performance of $3 in the stock (an opportunity cost if it is kept) compared to the same $3 in some other investment.

A classical example of a sunk cost occurs in the replacement of assets. Suppose that the piece of equipment discussed in a previous section, which originally cost $50,000, presently has an accounting book value of $20,000 and can be salvaged now for $5,000. For purposes of an economic analysis, the $50,000 is a sunk cost. However, the viewpoint can be taken that the sunk cost should be considered to be the difference between the accounting book value and the present realizable salvage value, which is called *book loss* or *capital loss*. According to this viewpoint, the sunk cost is $20,000 minus $5,000, or $15,000. Neither the $50,000 nor the $15,000 should be considered in an economic analysis, except for the manner in which the $15,000 affects income taxes, as will be discussed in Chapter 10.

fixed costs versus incremental costs

In most any change that is subject to an economic analysis (like buying a new machine, changing volume of production, etc.), some costs are affected and other costs are not affected. Those costs that are not affected by the change (i.e., remain constant) are often referred to as *fixed costs*, while those costs that are affected by the change are referred to as *incremental costs*. When describing costs for different volumes of production, the term "variable" is often used in place of incremental.† When it is desired to describe changes in costs for a small change in volume of production, the terms *differential* and *marginal* are often used in place of variable.

If one is making an economic analysis of a proposed change, it can be remembered that only the incremental costs need to be considered, since only prospective differences between alternatives need to be taken into account.

---

† The term *variable* was used earlier in this chapter in the section "Cost–Volume Relationships."

cash costs versus book costs

Costs that involve payments of cash or increases in liability are called *cash costs* to distinguish them from noncash (book) costs. Other common terms for cash costs are "out-of-pocket costs" or costs that are "cash flows." Book costs are costs that do not involve cash payments, but rather represent the amortization of past expenditures for items of lengthy durability. The most common examples of book costs are depreciation and depletion charges for the use of assets such as plant and equipment. In economic analyses, only those costs that are cash flows or potential cash flows need to be considered. Depreciation, for example, is not a cash flow and is important only in the way it affects income taxes, which are cash flows.

consideration of inflation in economic analyses

One increasingly recognized factor in economic analyses is *inflation*, which is the increase in the number of units of currency (dollars, for example) necessary to buy a given unit of goods or service. The economic analysis methods and examples explained in the next several chapters may seem to ignore inflation, but this is not necessarily true. All the methods can be used so as to correctly consider inflation provided that the types of estimates and interest rates (rates of return) are consistent with respect to either including or not including inflation. Inflation and two methods of explicitly accounting for it in economy studies will be described in depth in Chapter 8.

problems

2-1. Describe the difference between consumer goods and producer goods. Why is it more difficult to estimate the demand for producer goods?

2-2. Discuss the concept of economic utility. How does the utility of necessities differ from the utility of luxuries for low-income groups in this country?

2-3. Explain why the elasticity of demand for a luxury product (e.g., tennis rackets) may be relevant in an economy study of an investment in proposed equipment for manufacturing this product.

2-4. How would the elasticity of demand for necessities differ from the elasticity of demand for luxuries? Draw a simple graph to illustrate your answer.

2-5. Explain why perfect competition is in reality an ideal that is difficult to attain in the United States. List several business situations in which perfect competition is approached.

2-6. What is the difference between a monopoly and an oligopoly? Give at least one example of each.

2-7. In the context of industrial production, define and illustrate a fixed cost and a variable cost. List as many cost components of each as you can.

2-8. Explain how the law of supply and demand would operate if a bumper crop of corn were produced this year (assume that the Federal government does not buy any surpluses).

2-9. Make up a situation that illustrates the law of diminishing returns. How could this phenomenon affect an engineering economy study of two material handling systems?

2-10. Try to list as much information as possible that can be graphically presented to a decision maker in terms of a breakeven chart.

2-11. What is an opportunity cost and how will it be used in various engineering economy studies? What are a few of the opportunity costs of taking a job in Buffalo, New York, instead of a job in Manhattan, Kansas, that would have paid $2,000 per year less?

2-12. What is an incremental, or differential, cost?

2-13. Why should sunk costs be ignored in economy studies when income taxes are not considered? Make up an example of a sunk cost that you encounter every day.

2-14. A company produces and sells a product and thus far has been able to control the volume of the product just about as the company desires by varying the selling price. The company is seeking to maximize its net profit. It has concluded that the relationship between price and demand, per month, is approximately $D = 500 - 5P$, where P is the price in dollars. The company's fixed costs are $1,000 per month and the variable costs are $20 per unit. What number of units, D, should be produced and sold to maximize the net profit? (Obtain the answer both mathematically and graphically.)

2-15. A company estimates that as it increases its sales by decreasing the selling price of its product, the revenue changes by the relationship

$$revenue = aD - bD^2$$

where D represents the units of demand per month, with $0 \leq D \leq a/b$. The company has fixed costs of $1,000 per month, and the variable costs are $4 per unit. If $a = $ $6 and $b = $ $0.001, determine the sales volume for maximum profit and the maximum profit per month.

2-16. The ABC Corporation has a sales capacity of $1,000,000 per month. Its fixed costs are $350,000 per month, and the variable costs over a considerable range of volume, are $0.50 per dollar of sales. (a) What is the breakeven point volume? (b) What would be the effect of decreasing the variable costs by 25% if the fixed costs thereby increased by 10%? (c) What would be the effect if the fixed costs were decreased by 10% and the variable costs increased by the same percentage?

chapter 3

selections in present economy

This chapter is devoted to economy studies for cases in which interest and money–time relationships do not need to be considered. Chapter 4 and subsequent chapters concentrate on economy studies in the more general (and somewhat more difficult) cases in which alternatives have differing amounts of money at different points in time, so that the time value of money must be taken into account.

Studies that do not require consideration of time–money relationships are sometimes called *present-economy studies* and generally involve selection between alternative designs, materials, or methods. The effects of interest do not enter into the solution of these problems because of one or more of the following conditions:

1. There is no investment of capital; only out-of-pocket costs are involved. As an example, should a business man make a trip by air, requiring 3 hours of his time, or by train, requiring 12 hours? Here there is only a balancing of the difference in fares against the value of his time.

2. After any first cost is paid, the long-term costs will be the same for all alternatives, or will be proportional to the first cost. Thus the alternative that has the lowest first cost also will be the most economical in the long run. An example of this type may be found in the construction of a highway overpass. Whether conventional or prestressed-concrete construction is used, the maintenance costs would be the same and the ownership costs would be proportional to the first costs.

3. The alternatives will have identical results upon any capital investment involved. For example, in the making of a certain screw-machine product, it is found that either free-cutting aluminum or screw-machine steel can be used and can be machined in the same amount of time. The decision will be determined by the costs of the two materials.

In none of these cases is time a significant factor. The first case involves selection between alternative methods; the second involves selection between

41

alternative designs; the third concerns selection between materials. Because most parts, products, and structures can be designed in several ways to achieve a given function, and because so many different materials and processes are available for use, it is inevitable that many present-economy studies must be made. Such studies tend to have certain characteristics. First, they usually are relatively simple in form. Second, the required decision usually is quite easily made, once the cost data have been organized; often the totals of the alternative-cost columns make the correct decision virtually obvious. Third, the major effort in such studies is in the collection or computation of the various cost elements. In this regard, when there is an absence of desired, factual cost information— and this too often is the case—a considerable amount of experience may be necessary in the field involved so that accurate estimates can be made, or we must rely on someone who has such experience. Otherwise, important, but seemingly insignificant, items may be overlooked.

Another problem that often faces one in present-economy studies is the fact that a large and sometimes almost unlimited number of alternatives may be available, or may be made available, for study. This is particularly true in connection with designs and methods. In some cases, if it were desired, the analyst could originate new designs almost without limit. However, at some point he or she must recognize that the basic objective is to arrive at an economic decision. This means that a cutoff point must be set regarding the number of alternatives to be considered. Clearly, we do not wish to miss considering those alternatives that offer the prospect of real economic advantage. On the other hand, we do not wish to devote much time to making detailed analyses of alternatives that quickly can be determined as not offering much prospect for economic gain, thus possibly delaying the adoption of an economic alternative.

Handling such situations requires the exercise of good judgment. One should quickly gain a general idea of the probable economic advantages that appear to be possible in those alternatives that seem to have definite differences. This usually will require rather brief, preliminary economy studies to be made of a number of alternatives. Such studies will not be highly accurate, but they will give the student an idea of the general magnitude of the probable economic differences, and thus they provide a rather good guide as to which alternatives should be dropped from consideration and which should be studied in detail. This commonsense approach is merely the application of two very basic principles relating to engineering practice: (1) there is no virtue in spending $10 to save $1; and (2) sooner or later a decision must be made so that the work can proceed.

the necessity for equivalent results

One important factor that must be kept in mind in present-economy studies is that the results of the various alternatives must be substantially *equivalent*.†

---

† Here "equivalent" should be taken to mean that the economic results have roughly the same effect or meaning to the firm.

For example, if we were considering the alternatives of digging a ditch by pick and shovel or by a ditchdigging machine and found that the narrowest ditch that could be dug by the machine was 1 foot wider than was wanted, the results could hardly be said to be equivalent. Thus we must be certain that each alternative being considered will produce results that are satisfactory. This concept of equivalence does not mean that the results must always be identical. If it did not matter that the ditch were somewhat wider than a certain minimum, the extra width produced by the machine would be neither an advantage nor a disadvantage. Frequently, materials or methods must be compared that are not in all ways identical but that may be used so as to produce equivalent results. For example, a mild-steel bolt with a cross-sectional area of 1.0 square inch has an ultimate strength in tension of approximately 60,000 pounds. A nickel–steel alloy bolt of the same size would have a strength of 90,000 pounds. If strength were a factor in the design, it would not be correct to compare the cost of mild-steel and nickel–steel bolts of the same size. Yet a fair selection could be made between a mild-steel bolt with a cross-sectional area of 1.0 square inch and a nickel–steel bolt having a cross section of only 0.667 square inch, inasmuch as both have the same strength, provided that tensile strength is the sole design criterion.

Although there must be at least *minimal* equivalence among the alternatives being considered, this does not mean that one alternative cannot exceed another in a requisite property. For example, if it were found, in the example just cited, that the same size bolts of mild-steel and nickel–steel could be purchased for the same price, in most instances the extra strength of the nickel–steel bolt would be no detriment, and this fact would be no bar to its adoption. But at the same time we should not assign added economic advantage to a superior quality that cannot in fact be used or put to actual economic gain.

selection among materials

One of the most frequent types of present-economy studies is that in which selection must be made among materials. As more new and specialized materials become available, such problems become more numerous. These studies are frequently complicated by the fact that materials usually differ from each other in more than one property. For example, consider the choice between using ordinary low-carbon steel and a low alloy/high yield strength steel in an application where yield strength is the design criterion upon which the selection must be based. The properties of the two materials are as follows:

	Low-Carbon Steel	Low Alloy/High Yield Strength Steel
Yield strength (psi)	32,000	52,000
Ultimate strength (psi)	64,000	75,000
Cost per pound	$0.24	$0.30

These materials may be compared on the basis of the cost to obtain 1 pound of yield strength in the following manner:

Low-carbon steel:

$$\frac{\$0.24}{32,000} = \$0.0000075$$

Low alloy/high yield strength steel:

$$\frac{\$0.30}{52,000} = \$0.00000577$$

Thus for this application the low alloy/high yield strength steel would be cheaper.

In making selections between two or more materials, we must make certain that the proper criterion is used. For example, if we were selecting between steel and aluminum for a given product, both rigidity and strength might be of importance. It would therefore be necessary to consider both the strength and the modulus of elasticity of each material in order to arrive at a proper decision. Because materials usually differ with respect to more than one property, we must be certain that true equivalence is achieved.

total cost in material selection

In a large proportion of cases, economic selection between materials cannot be based solely on the costs of the materials. Very frequently, a change in materials will affect the processing costs, and often shipping costs may be altered. A good example of this is the part illustrated in Figure 3-1. The part was produced in considerable quantities on a high-speed turret lathe, using 1112 screw machine steel costing $0.10 per pound. A study was made to determine whether it might be cheaper to use brass screw stock, costing $0.70 per pound. Since the weight of steel required per piece was 0.0353 pound and of brass was 0.0384 pound, the material cost per piece was $0.0035 for steel and $0.0269 for brass. However, when the manufacturing and standards departments were con-

FIGURE 3-1. Small screw-machine product.

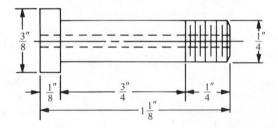

sulted, it was found that, although 57.1 parts per hour were being produced using steel, if brass were used, the output would be 102.9 parts per hour. Inasmuch as the machine operator was paid $3.25 per hour and the overhead cost for the turret lathe was $2.50 per hour, the total-cost comparison for the two materials was as follows:

	1112 Steel		Brass	
Material	$0.10 × 0.0353 =	$0.0035	$0.70 × 0.0384 =	$0.0269
Labor	$3.25/57.1 =	0.0569	$3.25/102.9 =	0.0316
Overhead[a]	$2.50/57.1 =	0.0436	$2.50/102.9 =	0.0243
TOTAL COST PER PIECE		$0.1040		$0.0828

Saving per piece by use of brass = $0.1040 − $0.0828 = $0.0212

[a] A given overhead rate applied to different alternatives without modification may be invalid for economic analyses even though useful in after-the-fact accounting allocations for whatever alternative is used. Unless explicitly stated otherwise, any given overhead rate is assumed to be applicable to all alternatives considered in the examples of this chapter and the remainder of the book.

Because a large number was made each year, the saving of $21.20 per thousand was a substantial amount. It also is clear that costs other than the cost of material were of basic importance in the economy study.

Later history of this same product illustrates the fact that shipping costs also must often be considered in the selection between materials. After the part had been made from brass stock for several years, it was found that it was desirable to supply domestic and foreign assembly plants of the company by using air freight as a shipping medium. This led to a study of the possible use of a heat-treated aluminum alloy. This material cost $0.85 pound and the cost of heat-treating each part, at an outside plant, was $0.018. Production studies indicated that the aluminum alloy could be machined at the same speeds as the brass stock.

The specific gravities of the brass and aluminum alloy are 8.7 and 2.75, respectively, and the raw and finished weights of the parts were as follows:

	Brass (lb)	Aluminum Alloy (lb)
Raw material	0.0384	(0.0384)(2.75/8.7) = 0.01213
Finished part	0.0150	(0.0150)(2.75/8.7) = 0.00474

Consequently, the comparative costs, including shipping, were as follows:

	Brass	Aluminum Alloy
Material	$0.0269	$0.0103
Labor	0.0316	0.0316
Heat treatment	—	0.0180
Overhead [a]	0.0436	0.0436
Shipping	0.0119	0.0038
TOTAL COST PER PIECE	$0.1140	$0.1073

[a] See footnote on preceding page.

A decision was made to use the aluminum alloy for those parts that were to be shipped by air to the subsidiary plants and brass for the parts that were consumed at the local plant, inasmuch as there was a small saving in favor of brass when shipping costs could be omitted.

Care should be taken in making economic selections between materials to assure that any differences in yields or resulting scrap are taken into account. Very commonly, alternative materials do not come in the same stock sizes, such as sheet sizes and bar lengths. This may considerably affect the yield obtained from a given weight of material; similarly, the resulting scrap may differ for different materials. This factor can have serious economic implications when one of the materials is considerably more costly than another. Determination of these effects is an instance where experience may be most helpful.

economy of location

Many problems of location, particularly those of alternative locations for manufacturing plants, warehouses, retail stores, and so on, involve time because of differences in construction costs, taxes, insurance rates, and operating costs. Such examples will be considered in later chapters. However, some location problems are not affected by time. The following is a typical example.

In connection with surfacing a new highway, the contractor has a choice of two sites on which to set up his asphalt-treating equipment. He will pay a subcontractor $0.20 per cubic yard per mile for hauling the mixed material from the mixing plant to the job site. Factors relating to the two sites are as follows:

	Site A	Site B
Average hauling distance	3 miles	2 miles
Monthly rental	$100	$500
Cost to set up and remove equipment	$1,500	$2,500

If site B is selected, there will be an added charge of $36 per day for a flagman.

The job involves 50,000 cubic yards of mixed material. It is estimated that 4 months (17 weeks of 5 working days per week) will be required for the job.

The two sites may be compared in the following manner:

	Site A	Site B
Rental	4 × $100 = $ 400	4 × $500 = $ 2,000
Setup cost	= 1,500	= 2,500
Flagman		5 × 17 × $36 = 3,060
Hauling	50,000 × $0.20 × 3 = 30,000	50,000 × $0.20 × 2 = 20,000
TOTAL	$31,900	$27,560

Thus site B is clearly less costly. There are some location problems that can be handled as present-economy studies but that are of such complexity that they can be analyzed more readily by the application of linear programming procedures. These are discussed in Chapter 17.

problems involving makeready and put-away times

There are a substantial number of operations that involve a combination of makeready and put-away times that precede and follow the production portion of the cycle. Such problems frequently do not have a least-cost or maximum-profit alternative within practicable limits. However, proper analysis can provide information on which reasonable decisions can be based.

This type of problem can be illustrated by the situation where typed material is duplicated repeatedly in offices. When the typed stencils are to be rerun, they must be obtained from the files and placed on the duplicating machine, and the machine must be inked preparatory for operation. After the run is completed, the stencil must be removed from the duplicator, cleaned, and replaced in the file. For a particular case, this makeready and put-away time is 11 minutes. Once the duplicator has been started, it can turn out the work at the rate of 0.6 second per copy. Consequently, the total time required to produce any number of copies is as shown in Figure 3-2.

It is apparent that the total time required is the sum of a fixed amount (11 minutes) plus a variable amount equal to $0.01N$, where N is the number of copies. It is also evident that the operator cost per copy produced will be affected by the number of copies that are run each time the duplicator is set up. If the operator of the machine receives $3.60 per hour, the operator cost per copy, as a function of the number of copies in a single run, would vary as follows:

Copies per Run	Time Required (min)	Operator Cost per Copy
10	11.1	$0.06660
50	11.5	0.01380
100	12.0	0.00720
200	13.0	0.00390
300	14.0	0.00280
500	16.0	0.00192
1,000	21.0	0.00126

These results are shown in graphical form in Figure 3-3.

It becomes apparent from an examination of Figure 3-3 that neither this curve nor the data given in the table provide an answer as to how many copies should be run in a single setup of the duplicator. However, they do supply information upon which a decision may be based. It is apparent that we should endeavor not to run less than about 200 copies at each setup in order to avoid the steep portion of the cost curve. However, before we can arrive at a reasonable decision, a quick determination should be made of the approximate amount of money that may be saved by running various quantities. For example, if 1,000 copies were made in five runs of 200 each, the total cost would be $3.90. If the

FIGURE 3-2. Time required to set up and mimeograph copies of material.

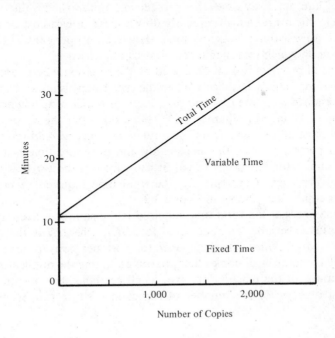

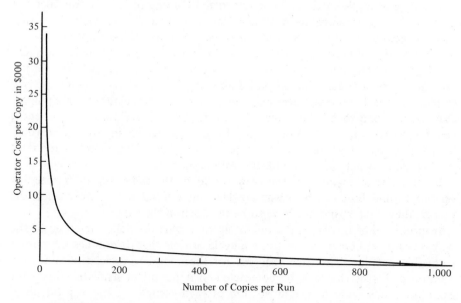

FIGURE 3-3. Cost for mimeographing runs of various numbers of copies.

entire 1,000 copies were run at one time, the cost would be $1.26—a possible saving of $2.64. Should this be considered to be a worthwhile amount (and storage and inventory costs are neglected), we might decide to run the 1,000 copies in one lot. On the other hand, if it were held that changes in method should not be made unless a saving of at least, say, $4.00 would result, a change from 200 to 1,000 copies per run would not be justified. Thus in problems of this type, we must decide what magnitude of saving (or cost or profit) is significant. When this is coupled to cost versus run size data, such as portrayed in Figure 3-3, a rational decision can be reached.

It must be emphasized that the foregoing analysis assumes that the amount of money tied up in unused product between runs is negligible and that the productive value of the machine during the time required for setup may be neglected. In many cases these factors cannot be neglected. Under such conditions, time must be taken into account, and such problems will be considered in Chapter 15.

alternative machine speeds

Machines frequently can be operated at different speeds, resulting in different rates of product output. However, this usually results in different frequencies of machine downtime to permit servicing or maintaining the machine, such as resharpening or adjusting tooling. Such situations lead to economy studies to determine the optimum or preferred operating speed.

A simple example of this type involved the planing of lumber. Lumber put through the planer increased in value by $0.02 per board foot. When the planer was operated at a cutting speed of 5,000 feet per minute, the blades had to be sharpened after 2 hours of operation, and lumber could be planed at the rate of 1,000 board feet per hour. When the machine was operated at 6,000 feet per minute, the blades had to be sharpened after $1\frac{1}{2}$ hours of operation, and the rate of planing was 1,200 board feet per hour. Each time the blades were changed, the machine had to be shut down for 15 minutes. The blades, unsharpened, cost $5 per set and could be sharpened 10 times before having to be discarded. Sharpening cost $1 per set. The crew that operated the planer changed and reset the blades. At what speed should the planer be operated?

Because the labor cost for the crew would be the same for either speed of operation, and because there was no discernible difference in wear upon the planer, these factors did not have to be included in the study.

In problems of this type, the operating time plus the delay time due to the necessity for tool changes constitute a cycle of time that determines the output from the machine. The time required for the complete cycle determines the number of cycles that can be completed in a period of available time—for example, 1 day—and a certain portion of each complete cycle is productive. The actual productive time will be the product of the productive time per cycle and the number of cycles per day.

	Value per Day
At 5,000 feet per minute	
Cycle time = 2 hours + 0.25 hour = 2.25 hours	
Cycles per day = 8 ÷ 2.25 = 3.555	
Value added by planing = 1,000 × 3.555 × 2 × $0.02 = $142.20	
Cost of resharpening blades = 3.555 × $1 = $3.56	
Cost of blades = 3.555 × $5/10 = 1.78	
TOTAL COST	5.34
Net increase in value per day	$136.86
At 6,000 feet per minute	
Cycle time = 1.5 hours + 0.25 hour = 1.75 hours	
Cycles per day = 8 ÷ 1.75 = 4.57	
Value added by planing = 4.57 × 1.5 × 1,200 × $0.02 = $164.50	
Cost of resharpening blades = 4.57 × $1 = $4.57	
Cost of blades = 4.57 × $5/10 = 2.29	
TOTAL COST	6.86
Net increase in value per day	$157.64

Thus it was more economical to operate at the higher speed, in spite of the more frequent sharpening of blades that was required.

It should be noted that this analysis assumes that the added production can be used. If, for example, the maximum production needed is equal to or less than that obtained by the slower machine (1,000 × 3.555 cycles × 2 hours = 7,110 board-feet per day), then the value added would be the same for each speed, and the decision then should be based on which speed minimizes total cost.

This type of study is of great importance in connection with metal-cutting machine tool operations. Changes of cutting speeds can have a great effect upon tool life. In addition, because the cost of machine tools and wage rates has increased, it is important that productivity be maintained at as high a level as possible. Under these conditions, it has frequently been found that increased cutting speeds give greater overall economy, even though the cutting-tool life is considerably less than was accepted practice in former years. This is particularly true if rapid means can be devised for changing tools when required.

economy of crew size

Another quite common present-economy problem involving selection between methods is that of proper crew size. Very often a particular job can be done by a work crew having different numbers of men. What number will result in the least cost?

As an example, a certain overhaul job was done with considerable frequency. It was customary to use a three-man crew. However, when such a crew was used it was found that, because of limitations in work space, two of the men were idle 33% of the time; one of these men was also idle an additional 50% of the time. Thus all three men worked simultaneously only 17% of the time. Each workman costs $3.17 per hour. Each job required the use of a kit of tools and equipment that was worth $3.00 per hour and would be required regardless of crew size. If the three-man crew required 4 hours to do the job, what crew size would be most economical?

Figure 3-4 shows the working and idle time of each crew member. It is apparent that the total work content of the job is

$$4(1.00 + 0.67 + 0.17) = 7.36 \text{ hours}$$

Inasmuch as any portion of the work can be done by a single man, the crew could consist of any desired size. If there were only one man in the crew, he

FIGURE 3-4. Working and idle times of the members of a three-man repair crew.

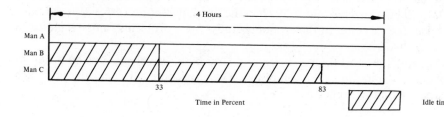

would have 7.36 hours of work and no delay. If two men were used, only one of them could work during 0.33 × 4 hours, or 1.32 hours. This would leave 7.36 − 1.32 = 6.04 man-hours of work content to be performed by both men, which would require 3.02 hours of elapsed time. Thus the total time required for a two-man crew would be 3.02 + 1.32 = 4.34 hours. For one-, two-, and three-man crews, the cost of doing the job would then be as follows:

Number of Men in Crew	Hours Required	Labor Cost (= hours required × no. men × $3.17)	Tool Cost (= hours required × $3.00)	Total
1	7.36	$23.33	$22.08	$45.41
2	4.34	27.52	13.02	40.54
3	4.0	38.04	12.00	50.04

It is thus apparent that a two-man crew is the most economical. It should be noted that if there were any cost associated with nonproductive time while the machine is overhauled, this cost should be added to determine the most economic crew size.

processing materials having limited and unlimited supply

Some present-economy studies involve alternative methods of processing raw materials that sometimes are limited in quantity. When this restriction is present, certain precautions must be observed that need not be considered when the supply is unlimited.

As an example of this type, consider two methods that are available for processing a certain ore to recover the contained metal. The ore, as it comes from a certain mine, contains 22% of the metal. Two methods of treatment for the ore are available. Each uses the same equipment, but they involve the use of different chemicals and grinding to different degrees of fineness. To process a ton of ore by method A costs $14.35 and recovers 84% of the contained metal. Processing by method B costs $16.70 per ton of ore, and 91% of the metal is recovered. If the recovered metal can be sold for $0.06 per pound, which method of processing should be used?

There is only a certain amount of ore in the mine; thus the supply of raw material obviously is limited. However, the fact that a different total amount of metal will be recovered and sold when the two different processes are used must be taken into account in the economy study. Such problems can be handled in either of two ways.

Probably the most foolproof method is to base the calculations on a unit of raw material—in this case a ton of ore. This procedure automatically puts each alternative on the proper basis.

Method A:

profit per ton of ore = 2,000 × 0.22 × 0.84 × $0.06 − $14.35

= $7.83

Method B:

profit per ton of ore = 2,000 × 0.22 × 0.91 × $0.06 − $16.70

= $7.32

Thus, although method B would result in more metal being recovered and sold, the profit obtained would not be as great as if method A were used.

A second method of working this type of problem is either to determine the cost of producing a unit of output—1 pound of metal in this case—by each method of processing, or to determine the profit obtained from the sale of 1 pound. However, we must then take into account the fact that, because of the different recoveries from the same amount of ore, the two methods will not produce the same total number of pounds of the product.

When the supply of the raw material is not limited, the procedure for determining which method of processing will give the lowest cost for producing a unit of output is a satisfactory, and usually somewhat simpler, procedure. Since there is no limitation on the amount that can be produced, we need be concerned only with obtaining each unit of product at the least cost. For the above example, the cost per pound of metal produced would be

Method A: $14.35/2,000 × 0.22 × 0.84 = $0.0388
Method B: $16.70/2,000 × 0.22 × 0.91 = $0.0417

Thus method A is the least costly, which in this case happens to correspond with the choice based on profit per ton of ore.

the proficiency of labor

Workers of different proficiencies frequently produce different amounts of output in a given time. Through the use of various types of incentives, either monetary or nonmonetary, workers of all types may be induced to develop and exhibit different degrees of proficiency, resulting in various levels of output. Present-economy studies are useful in establishing and maintaining the proper incentives.

As an example of this type, consider the case of a product that was made by hand in a small factory. The workers were paid $0.40 per acceptable piece produced. It was found that if a worker produced 80 pieces per day, 5% would be rejected. If 90 pieces were produced per day, 10% would be rejected, and at a rate of 100 pieces per day, 20% would be rejected. The cost for materials was $0.50 per piece, and the materials in any rejected products had to be thrown away. There was a fixed overhead expense of $10.00 per day per worker, regardless of considerable change in output. Three questions arose concerning the situation: (1) at which of the three outputs did the workers make the highest wages; (2) at which output did the factory achieve the lowest unit cost; and (3) was some adjustment in the wage-payment situation desirable?

The wages obtained by the workers for the three output rates were as follows:

	Pieces Produced per Day		
	80	90	100
Rejects	4	9	20
Acceptable pieces per day (N)	76	81	80
Earnings ($N \times \$0.40$)	$30.40	$32.40	$32.00

It was thus evident that it was to the workers' advantage to produce at the rate of 90 pieces per day.

From the viewpoint of unit cost, the situation was as follows:

	Rate per Day		
	80	90	100
Material cost	$40.00	$45.00	$50.00
Labor cost	30.40	32.40	32.00
Overhead	10.00	10.00	10.00
TOTAL	$80.40	$87.40	$92.00
Cost per acceptable piece	$ 1.058	$ 1.079	$ 1.150

The calculations showed that the lowest unit cost would be obtained at an output rate of 80 pieces per day. It was thus evident that the degree of proficiency that was likely to result from the incentive wage system would not be best from the viewpoint of the company. A change in the wage structure was therefore desirable.

economic selection of a beam

A typical design selection may be found in the choice of wooden beams used to support a floor that will sustain a uniformly distributed load of 2,200 pounds over a span length of 12 feet. Using structural-grade Douglas fir and limiting the deflection to $\frac{1}{300}$ of the span, the analyst found that there are three common sizes that will meet the required conditions—4 by 8 inches, 3 by 10 inches, and 2 by 12 inches.

Obviously, beams of these three sizes do not contain the same number of board feet of material. Also, when the cost is determined, it is learned that these three sizes do not cost quite the same per board-foot. The 4- by 8-inch and 3- by 10-inch sizes list at $175 per thousand board-feet, and the 2- by 12-inch size can be obtained for only $150 per thousand board-feet. A final cost comparison of the beams is as follows:

Nominal Size (in.)	Cost per 1000 Board-Feet	Board-Feet per Beam	Cost per Beam
4 by 8	$175	(4 × 8 × 144)/144 = 32	$5.60
3 by 10	175	(3 × 10 × 144)/144 = 30	5.25
2 by 12	150	(2 × 12 × 144)/144 = 24	3.60

It is thus clear that for this particular situation a design that utilizes 2- by 12-inch beams would be advantageous.

economic span length for bridges

The design of a multiple-span bridge is an example that can be used to illustrate (1) the effect of design on costs, (2) the use of approximate cost curves, and (3) some of the limitations of such curves and data.

In many cases of design selection, some of the cost factors increase as certain design parameters are changed, and others decrease. The problem is one of selecting the design that affords the lowest total cost. Figure 3-5 shows the cost curves relating to such a condition. These cost curves show the cost per foot of structure for low-level combined bridges on sand foundations 200 feet deep. As might be expected, the cost of the substructure decreases somewhat as the span length increases. On the other hand, the cost of the steelwork increases quite rapidly as the span length increases. The greatest economy would be achieved by using span lengths of about 325 feet.

Now consider the application of these cost curves to the design of a bridge having a total length of 1,600 feet and the added requirement that at or adjacent to the center of the bridge there must be one span of at least 400 feet. Two designs are being considered: (1) four equal spans of 400-foot length, and (2) one center span of 400 feet and two 300-foot spans on each side of the center span.

Directly applying the cost data contained in Figure 3-5, we obtain the following figures:

Alternative	Length of Spans (ft)	Number of Spans	Cost per Foot	Cost
1	400	4	$2,880	$4,608,000
2	400	1	2,880	1,152,000
	300	4	2,760	3,312,000
TOTAL COST				$4,464,000

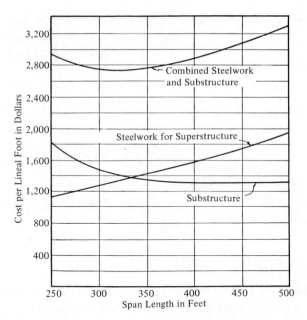

FIGURE 3-5. Costs per lineal foot of structure for low-level combined bridges on sand foundations 200 ft deep. (*Economics of Bridgework* by J. A. L. Waddell, published by John Wiley & Sons, Inc., New York.)

It thus appears that alternative 2 will be more economical. However, before coming to a final decision, we should consider two factors related to the use of the cost curves of Figure 3-5. First, these cost curves assume that *all* spans are of equal length. Thus, when these curves are used in estimating the cost for alternative 2, some error would be introduced. Second, it should be asked whether the construction methods and materials used in connection with determining the cost relationships portrayed in Figure 3-5 were the same as would be involved in building the bridge being considered. The effect of changed materials and methods could be considerable. Therefore, when using historical-cost data, either in tabular or curve form, considerable care and judgment must be exercised. For preliminary economy studies such data may be quite adequate; for final studies upon which important decisions are to be based, such data may not be sufficiently accurate.

the relation of design tolerances and quality to production cost

Design alternatives very commonly must be considered from the viewpoint of economy of manufacture. Some designs contain features that inherently are more costly to produce than others. Unnecessarily tight dimensional tolerances are prime culprits in this respect. Not only does increased accuracy cost money, but the relaxation of accuracy requirements may make it possible to use different and less expensive processes. Figure 3-6 shows a set of curves relating tolerance and increased cost for several casting methods.

An example of how relaxation of tolerances can affect costs is found in the case of a sand-cast part that required considerable machining so that the specified tolerances could be obtained. The cost for the finished part was as follows:

Cost of casting	$ 9.50
Tooling cost, per part	0.50
Machining	8.25
TOTAL	$18.25

FIGURE 3-6. Relative costs for maintaining specified tolerances with various casting processes.

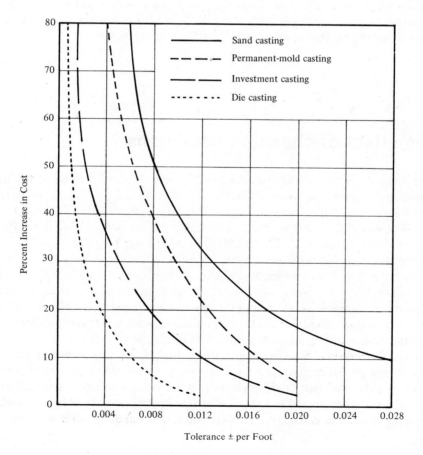

When the tolerance requirements were reconsidered, it was found that by relaxing the tolerances the part could be used as cast if it were made by die casting, except for tapping some holes. An economy study of the die cast part showed the cost to be as follows:

Cost of casting	$2.25
Tooling cost, per part	1.75
Machining	3.10
TOTAL	$7.10

Thus, in making design studies and economy studies of alternative designs, we should not overlook the possibility of making more economical production methods possible by changes in design.

Closely related to the matter of tolerances is the cost of quality. Inspection and control procedures, which are necessary to ensure quality as well as the cost of better quality itself, add to the cost of a product. In many instances it is found to be more economical to permit a small, assured number of defects to pass the inspection barriers and pay the necessary costs for replacements and repairs on these defective products than to pay the added cost that would be required to eliminate all defects. Obviously, customer satisfaction is an important intangible that must be considered in such decisions.

the effect of changing conditions

One word of caution should be added concerning the results obtained through selections in present economy. The correct solution today may not be correct tomorrow if the conditions upon which the data were based have changed. An excellent example of this is found in the effect on the gold-mining industry of the increase in price of gold from $20.67 to $35.00 per ounce in 1934. Prior to that date it was not profitable to use certain processes that were necessary to recover the gold from low-grade ores. As a result, large piles of tailings, which were considered worthless, were accumulated at many mines. After the price of gold was increased, it was found that a considerable portion of the low-grade ores and tailings could be put through more expensive recovery processes so that the gold could be recovered and a substantial profit earned.

Thus, because we have made a correct analysis of a problem today and selected the proper method, material, or design, we should not be unmindful of new methods, materials, and processes that are being developed and that may have a vital effect upon the economy of our operation. It is indeed rare that anything is done so efficiently that time will not reveal even better methods.

problems

3-1. An automatic machine can be operated at three speeds with the following results:

Speed	Output (pieces per hour)	Time Between Tool Grinds (hr)
A	400	15
B	480	12
C	540	10

A set of unground tools costs $150 and can be ground 30 times. The cost of each grinding is $12. The time required to change and reset the tools is $1\frac{1}{2}$ hours, and such changes are made by a tool-setter who is paid $5.75 per hour. Overhead on the machine is charged at the rate of $3.75 per hour, including tool-change time. All output produced can be used. At which speed should the machine be operated?

3-2. A company finds that, on the average, two of its engineers spend 60 hours each month in flying time, using commercial airlines, in making service calls to customers' plants, always traveling together. It finds that the cost for airline tickets, airport buses, car rental, and so on, costs approximately $2,000 per person per month. An air charter service offers to supply a small business jet and pilot on 24-hour notice at a cost of $1,200 per month plus $125 per hour of flying time and $25 per hour for waiting time on the ground at the destination. It states that experience for similar situations has shown that total travel time will be reduced by 50% by using the charter service.

The company estimates that the cost of car rental at destinations probably would amount to about $250 per month if the charter service is used, and the average "waiting time" will be about 40 hours per month. It also estimates that each engineer's time is worth $40 per hour to the company. Should the charter service be used?

3-3. A duplicating service will make electrostatic copies of documents up to $8\frac{1}{2}$ by 11 inches in size for 4 cents per copy for the first 10 and 3 cents per copy for all additional. It also will reproduce such documents by offset printing at a cost of 30 cents for the master and 1 cent per copy, with a minimum quantity of 30 being required. (a) Assuming that each process gives satisfactory copies, which will be more economical for 25 copies? (b) At what quantity are the two processes equal in cost?

3-4. For a construction job, a contractor can haul rock by means of an ordinary 10-cubic-yard dump truck, or he can use a combination rig where a 10-cubic-yard truck also pulls an 8-cubic yard trailer. When this combination is used, the trailer slides into the empty dump truck body for the return trip. The total equivalent operating cost, including driver, amortization of capital, interest on the investment, fuel, and maintenance, is $12.50 per hour for the dump truck and $17.50 per hour for the tandem rig. For the required 10-mile haul, the loading, unloading, hauling, and return times are as follows:

	Time (min)	
	Conventional	Double Rig
Loading	3	5
Unloading	4	10
Hauling	24	28
Return	20	24

(a) Which type of equipment will be more economical? (b) If hauling and return times are directly proportional to the haul distance, for what length of haul would the two types of equipment provide equal costs?

3-5. In the design of an automobile engine part, an engineer has a choice of either a steel casting or an aluminum alloy casting. Either material provides the same service. However, the steel casting weighs 8 ounces as compared with 5 ounces for the aluminum casting. Every pound of extra weight in the automobile has been assigned a penalty of $6 to account for increased fuel consumption during the expected life of the car. The steel casting costs $3.20 per pound while the aluminum alloy can be cast for $7.40 per pound. Machining costs per casting are $5.00 for steel and $4.20 for aluminum. Which material should the engineer select and what is the difference in unit costs?

3-6. For a certain application, a stainless steel, having a yield strength of 117,000 psi, a modulus of elasticity of 29,000,000 psi, and a specific gravity of 7.6 can be used. It costs $2.80 per pound. As an alternative, an aluminum alloy, having a yield strength of 67,000 psi, a modulus of elasticity of 10,000,000 psi, and a specific gravity of 2.8 could be used. This alloy costs $1.90 per pound. If the selection can be based solely on the yield strength, which material would be more economical?

3-7. Seawater contains 2.1 pounds of magnesium per ton. By using the processing method A, 85% of the metal can be recovered at a cost of $3.25 per ton of water pumped and processed. If process B is used, 70% of the available metal is recovered, at a cost of only $2.60 per ton of water pumped and processed. The two processes are substantially equal as to investment costs and time requirements. (a) If the extracted metal can be sold for $2.40 per pound, which processing method should be used? (b) At what selling price for the metal would the two processes be equally economical?

3-8. One method for developing a mine will result in the recovery of 62% of the available ore deposit and will cost $23 per ton of material removed. A second method of development will recover only 50% of the ore deposit, but it will cost only $15 per ton of material removed. Subsequent processing of the removed ore recovers 300 pounds of metal from each ton of processed ore and costs $40 per ton of ore processed. The ore contains 300 pounds of metal per ton, which can be sold for $0.80 per pound. Which method for developing the mine should be used?

3-9. A repetitive repair operation normally is done by a crew of 3 men, who utilize a pickup truck and equipment that have an hourly cost of $4.00. Each man is paid $3.75 per hour. With this combination of men and equipment, the task requires 3 hours, but one man is idle 25% of the time because of space limitations and, during an additional 10% of the time two men are idle, also because of space limitations. The job can be done by one or more men, but in all cases the pickup truck and equipment would be required. What crew size would be most economical?

3-10. The assembly of a certain component for television sets has been done by two skilled workers, each of whom is paid $4.00 per hour and can produce 10 units per hour (as a group). Of the 10 units, 1 will be defective and will have to be discarded at a material loss of $2.50. It has been proposed that this operation could be done by three persons of lesser skill, who would be paid only $3.50 per hour, and who, as a group, could complete 13 units per hour. It is estimated that the number of defects per hour would not be changed if this alternative procedure is used. Which procedure would you recommend?

3-11. An operator of a fleet of diesel trucks has been using oil-filter cartridge A on each truck. This type of filter cartridge costs $5.00 per unit, and it has been replaced each 5,000 miles, with 1 quart of oil being added each 1,000 miles thereafter until the next oil change. A salesperson for filter cartridge B, which costs only $2.00, claims that if this unit is used and replaced each 2,000 miles, with the oil also being changed, no oil will have to be added between changes, and oil costing only $0.90 per quart can be used in place of the presently used oil, which costs $1.25 per quart. If the motors require 6 quarts of oil when the oil is changed, which type of filter would be more economical?

3-12. The bottler of a popular soft drink can purchase glass bottles for $0.04 each or all-aluminum cans for $0.05 each. If the aluminum cans are used, she believes she can achieve a saving of $0.005 per can in shipping and handling costs, owing to the reduced weight. In addition, by setting up a reclamation drive, she believes that she can recover about 50% of the cans by offering to pay $0.10 per pound for returned cans at four return centers which she will establish in the area her company serves. The used cans weigh 20 to the pound, and she can sell the used cans to the can manufacturer for $0.15 per pound. She estimates that the overhead cost of operating the return centers will not exceed 20% of the cost of repurchasing the used cans. In addition, she believes that she will obtain favorable response from ecology-minded customers. Which type of beverage container would you recommend that she use?

3-13. In cooperation with a rehabilitation program of a veterans' hospital, a company has offered to employ a certain number of partially disabled veterans on the basis that they will be paid an hourly, or piece, wage which will permit the company to obtain the same unit cost as it achieves with nondisabled workers. On a certain job, it has been paying regular workers $3.60 per hour and production has been 40 units per hour, of which 4% are rejected as defective. The direct-material cost is $2.00 per unit and is not recoverable from defective units. A test has shown that a certain type of trained but partially disabled worker can produce 32 units per hour, with only 1% defective. The overhead cost of the space and equipment used is $2.50 per hour per worker. (a) What hourly pay rate should be established for the disabled workers? (b) Comment on the resulting situation.

3-14. A manufacturer of wooden shipping containers has lost the business of a major customer who, for economic reasons, has converted to the use of foamed plastic containers for his product, which is shipped by air freight. This customer ships three sizes of product, weighing 25, 50, and 100 pounds, at a rate of $1 per pound. The company pays $30, $42, and $50 for the three sizes of plastic containers, which weigh 7, 12, and 25 pounds, respectively.

The manufacturer of the wooden containers, which weigh four times as much as the plastic containers, is attempting to regain the lost business by pricing her product so as to be competitive. The prices for the three sizes have been $8, $12, and $22, respectively.

(a) At what prices would she have to sell her products in order to compete?

(b) She estimates that decreasing the weight of the wooden containers by one-third could be achieved without causing the selling price to be increased more than 25% over the present price. Would this procedure produce favorable results?

3-15. A recent engineering graduate was given the job of determining the best production rate for a new type of casting in a foundry. After experimenting with many combinations of hourly production rates and total production cost per hour, he summarized his findings:

Total Cost/Hour	$1,000	$2,600	$3,200	$3,900	$4,700
Castings Produced/Hour	100	200	300	400	500

The engineer then talked to the firm's marketing specialist, who provided these estimates of selling price per casting as a function of production output:

Selling Price/Casting	$20.00	$17.00	$16.00	$15.00	$14.50
Castings Produced/Hour	100	200	300	400	500

(a) What production rate would you recommend to maximize total profits?

(b) How sensitive is the rate in part (a) to changes in total production cost per hour?

chapter 4

interest and money–time relationships

The majority of economy studies involve commitment of capital (expressed in money) for a period of time of such duration that the effect of time on the capital must be considered. To account for this effect, we must understand that capital is a dynamic and productive factor in business and thus is entitled to a "wage," consequently causing a cost when it is used. We then can take this cost factor into account in our economy studies. It is the purpose of this chapter to provide an understanding of the cost of capital and to show how the calculations relating to its proper consideration may be made.

the return to capital

In the privately financed sectors of a capitalistic economy, the suppliers of capital expect to receive some recompense for supplying it and permitting it to be used. This recompense for the use of capital often is referred to by economists as the *return to capital*, being analogous to wages, which are the recompense of labor, and to the cost paid for materials. It thus represents payment for the use of capital. Even when the capital is supplied by governmental agencies, there is a requirement, for many types of projects, that a return for the use of the capital be demanded; this is because the agency either borrows the capital from the public or obtains it from the public through taxation and recognizes that the individual taxpayer could have obtained a return from the capital had the agency permitted him to keep it.

There are several reasons why a return to capital is essential and thus must be considered in economy studies and economic decision making. First, the return to capital pays the supplier for forgoing the use of his money (or property) during the time the user has it. Second, the return is payment for the risk the supplier takes in permitting another person, or an organization, to use his or her capital. Third, the fact that the supplier can get a return by making capital available acts as an incentive for him to accumulate capital and to make it

63

available. Thus *whenever capital is required in engineering and business projects, it is essential that proper consideration be given to the required return to capital in economy studies.*

when must interest and profit be considered?

If capital for financing an enterprise must be borrowed, that which is paid for its use is called *interest*. On the other hand, if a person or corporation owns sufficient capital to finance a proposed project, the return that accrues is called *profit*. Since there is no borrowed money in the true meaning of the term, there is no interest expense. However, in this case if the owner of the capital decides to invest it in the proposed venture, he or she must forgo using it for some other profitable purpose—even if the other purpose is merely leaving it in a bank where it could draw interest. Thus he must decide whether the expected profit, usually measured in terms of an annual rate of return, is sufficient to justify investment in the proposed venture. Although no interest cost is involved, if the capital is invested in the project, he would expect, as a minimum, to receive profit at least equal to the amount he has sacrificed by not using it in some other equally attractive and available opportunity. This profit, which is lost or forgone, is called the *opportunity cost* of using capital in the proposed venture. Thus whether equity or borrowed capital is used, there is a cost for the capital used in the sense that the venture must provide a return on capital sufficient to make it profitable to suppliers of the capital.

In order to determine whether the expected capital return (profitability) is sufficient, it usually is necessary to compare the expected rate of profit with the rate that could be obtained from using the same capital in some other manner. Thus profit is a factor that must be considered in a large proportion of economy studies, although it is not a cost in the normal sense when equity capital is used.

There are some cases in which interest or profit may be omitted reasonably from economy studies. If the time involved in the study is not more than 1 year, interest is usually neglected. In cases where the interest rate is quite low and the time involved is, say, only 2 or 3 years, the omission of interest will introduce relatively little error. Chapter 3 was devoted to problems in which interest is not a factor. Good judgment must be exercised in deciding whether interest can be neglected in a particular situation. When in doubt, it should be included.

origins of interest

Like taxes, interest has existed from the time of man's earliest recorded history. Records reveal its existence in Babylon in 2000 B.C. In the earliest instances interest was paid for the use of grain or other commodities that were borrowed; it was also paid in the form of grain or other goods. Many of the existing interest practices stem from the early customs in the borrowing and repayment of grain and other crops.

History also reveals that the idea of interest became so well established that a firm of international bankers existed in 575 B.C., with home offices in Babylon. Its income was derived from the high interest rates it charged for the use of its money for financing international trade. Interest is not only one of our oldest institutions, but its uses have suffered little change through the years.

simple interest

When the total interest earned or charged is directly proportional to the principal involved, the interest rate, and the number of interest periods for which the principal is committed, the interest and interest rate are said to be *simple*.

When simple interest is applicable, the total interest, I, earned or paid may be computed by the formula

$$I = (P)(N)(i) \tag{4-1}$$

where P = principal amount lent or borrowed
N = number of interest periods (e.g., years)
i = interest rate per interest period

Thus, if \$100 is loaned for 3 years at a simple interest rate of 5% per annum, the interest earned will be

$$I = \$100 \times 0.05 \times 3 = \$15$$

The total amount owed at the end of 3 years would be \$100 + \$15 = \$115.

Simple interest is not used frequently in commercial practice in modern times, but it is of interest to contrast with compound interest, which is explained below.

compound interest

Whenever the interest charge for any interest *period* (a *year*, for example) is based on the remaining principal amount plus any accumulated interest charges up to the beginning of that period, the interest is said to be *compound*. The effect of compounding of interest can be shown by the following table for \$100 loaned for 3 periods at an interest rate of 5% compounded per period.

Period	(1) Amount Owed at Beginning of Period	(2) = (1) × 5% Interest Charge for Period	(3) = (1) + (2) Amount Owed at End of Period
1	\$100.00	\$5.00	\$105.00
2	105.00	5.25	110.25
3	110.25	5.51	115.76

Thus \$115.76 would be due for repayment at the end of the third period. If the length of a period is 1 year, the \$115.76 at the end of three periods (years) can be compared directly with the $115.00 given earlier for the same problem with simple interest. The difference is due to the effect of *compounding*. This difference would be much greater for larger amounts of money, higher interest rates, or greater numbers of years. Compound interest is much more common in practice than is simple interest and is used throughout the remainder of this book.

equivalence

To better understand the mechanics of interest and to introduce the concept of equivalence, let us consider a situation in which we borrow $8,000 and agree to repay it in 4 years together with 10% annual interest. There are a great many ways in which the debt might be repaid, but for simplicity we have selected four specific ways for our example. Table 4-1 tabulates the four plans.

In plan 1, $2,000 will be paid at the end of each year plus the interest due at the end of the year for the use of money to that point. Thus, for the first year we shall have had the use of $8,000. The interest owed is 10% × $8,000 = $800. The end-of-year payment is therefore $2,000 principal plus $800 interest, for a total payment of $2,800. At the end of the second year, another $2,000 principal will be repaid plus interest on the money used during the year. This time the amount owed has declined from $8,000 to $6,000 because of the $2,000 principal payment at the end of the first year. The interest payment is 10% × $6,000 = $600, making the end-of-year payment a total of $2,600. As indicated in Table 4-1, the series of payments continues each year until the loan is fully repaid at the end of the fourth year.

Plan 2 is another way to repay $8,000 in 4 years with interest at 10%. This time the end-of-year payment is limited to the interest due, with no principal payment. Instead, the $8,000 owed is repaid in a lump sum at the end of the fourth year. In this plan the end-of-year payment in each of the first 3 years is 10% × $8,000 = $800. The fourth year the payment is $800 interest plus the $8,000 principal, for a total of $8,800.

Plan 3 calls for equal end-of-year payments of $2,524 each. At this point we have not shown how the amount of $2,524 was computed. By following the computations in Table 4-1, we see that this series of four payments of $2,524 repays a $8,000 debt in 4 years with interest at 10%.

In plan 4, no payment is made until the end of the fourth year, when the loan is completely repaid. Note what happens at the end of the first year. The interest due for the first year (10% × $8,000 = $800) is not paid. Instead, it is added to the debt. So the second year the debt has increased to $8,800, and the end-of-the-second-year interest is 10% × $8,800 = $880. It is again unpaid and so it, too, is added to the debt, increasing it to $9,680. At the end of the fourth year the total sum due has grown to $11,713 and is paid at that time.

When we are indifferent whether we have a quantity of money now or the assurance of some other sum of money, or series of future sums of money, in the future, we say that the present sum of money is *equivalent* to the future sum or series of future sums. For example, if a firm believed that 10% was an appropriate interest rate, it would have no particular preference as to whether it received $8,000 now or the payments shown in column 6 of Table 4-1 for either plan 1, 2, 3, or 4. Thus all four repayment plans can be described as equivalent to each other and to $8,000 now.

Equivalence is an essential factor in engineering economic analysis. For example, consider the following schedule of payments for two alternatives:

Year	Alternative A	Alternative B
1	$ 2,800	$ 800
2	2,600	800
3	2,400	800
4	2,200	8,800
TOTAL	$10,000	$11,200

If you were given your choice between the two alternatives, which would you choose? Alternative A requires that there be larger payments in the first 3 years, but the total payments are smaller than the sum of the alternative B payments. To make a decision, the cash flows must be manipulated so that they can be compared. The technique of equivalence is the key. We can determine by mathematical manipulation an equivalent value at some point or points in time for A and a comparable equivalent value for B. Then we can judge the relative attractiveness of the two alternatives, not from their cash flows, but from comparable equivalent values. In this example we know from Table 4-1 that A and B are actually plan 1 and plan 2, respectively, and that they each repay a present sum of $8,000 with interest at 10%. Therefore, the two alternatives are equivalent when the interest rate is 10%. Since they both are equivalent to $8,000 now, the alternatives are equally attractive. This could not have been deduced by mere examination of the given cash flows. It was necessary to learn this by determining the equivalent values (at the present, or at some other point in time) for each alternative.

One way to rationalize that the plans in Table 4-1 are, indeed, equivalent would be to plot the amount owed at the beginning of each year (column 2) versus the year. The area under such a curve would be the dollar-years money is owed. When this is compared against the total interest paid (bottom of column 3), one finds that the relationship is constant:

Plan	Area Under Curve (dollar-years)	Total Interest Paid ($)	Ratio: $\dfrac{\text{Total Interest Paid}}{\text{Area Under Curve}}$
1	20,000	2,000	0.10
2	32,000	3,200	0.10
3	20,960	2,096	0.10
4	37,130	3,713	0.10

table 4-1
four plans for repayment of $8,000 in 4 years with interest at 10%

(1) Year	(2) Amount Owed at Beginning of Year	(3) = 10% × (2) Interest Owed for Year	(4) = (2) + (3) Total Money Owed at End of Year	(5) Principal Payment	(6) = (3) + (5) Total End-of-Year Payment
Plan 1: At end of each year pay $2,000 principal plus interest due					
1	$8,000	$ 800	$8,800	$2,000	$ 2,800
2	6,000	600	6,600	2,000	2,600
3	4,000	400	4,400	2,000	2,400
4	2,000	200	2,200	2,000	2,200
		$2,000		$8,000	$10,000
Plan 2: Pay interest due at end of year and principal at end of 4 years					
1	$8,000	$ 800	$8,800	$ 0	$ 800
2	8,000	800	8,800	0	800
3	8,000	800	8,800	0	800
4	8,000	800	8,800	8,000	8,800
		$3,200		$8,000	$11,200

Plan 3: Pay in 4 equal end-of-year payments

1	$8,000	$ 800	$8,800	$1,724	$ 2,524
2	6,276	628	6,904	1,896	2,524
3	4,380	438	4,818	2,086	2,524
4	2,294	230	2,524	2,294	2,524
		$2,096		$8,000	$10,096

Plan 4: Pay principal and interest in one payment at end of 4 years

1	$ 8,000	$ 800	$ 8,800	$ 0	$ 0
2	8,800	880	9,680	0	0
3	9,680	968	10,648	0	0
4	10,648	1,065	11,713	8,000	11,713
		$3,713		$8,000	$11,713

Thus the ratio of total interest paid to area under curve is 10% for all plans. From the calculation we now understand why the repayment plans can be equivalent and yet require repayment of different total sums of money. The answer is, of course, that the four repayment plans provide the borrower with different quantities of dollar-years.

Finally, it should be recognized that equivalence is dependent on interest rate. The four plans for repayment illustrated in Table 4-1 are equivalent only at an interest rate of 10%. Calculations that we shall learn how to do later will show us that the total present worths of the end-of-year payments (in column 5) will vary among the various plans for any interest rate other than 10%.

notation and cash flow diagrams

The following notation is used throughout this book for compound interest calculations:

i = interest rate per interest period
N = number of compounding periods
P = present sum of money: the *equivalent* worth of one or more cash flows at a relative point in time called the present
F = future sum of money: the *equivalent* worth of one or more cash flows at a relative point in time called the future
A = end-of-period cash flows (or equivalent end-of-period values) in a uniform series continuing for a specified number of periods
G = uniform period-by-period increase or decrease in cash flows or amounts (the arithmetic gradient)

The use of cash flow (time) diagrams is strongly recommended for situations in which the analyst needs to clarify or visualize what is involved when flows of money occur at various times. Indeed, the usefulness of the cash flow diagram for economic analysis problems can be analogized to that of the free-body diagram for mechanics problems. Figure 4-1 shows a cash flow diagram for the problem in the preceding section that involved finding the worth at the end of 3 years of $100 loaned at an interest rate of 5% compounded annually.

FIGURE 4-1. Cash flow diagram for example problem with compound interest (viewpoint of lender).

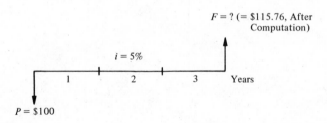

The cash flow diagram employs several conventions:

1. The horizontal line is a *time scale* with progression of time moving from left to right. The period (or year) labels are applied to intervals of time rather than points on the time scale. Note that the end of period 2 is coincident with the beginning of period 3. Only if specific dates are employed should the points in time rather than intervals be labeled.
2. The arrows signify cash flows; normally downward arrows represent disbursements (negative cash flows or cash outflows) and upward arrows represent receipts (positive cash flows or cash inflows).
3. The cash flow diagram is dependent upon point of view. For example, the problem shown in Figure 4-1 was based on cash flow as seen by the lender. If the direction of both arrows had been reversed, the problem would have been as seen by the borrower.

interest formulas for discrete compounding and discrete payments

Table 4-2 provides a summary of the six most common discrete compound interest factors and symbols. These factors are derived and explained by example problems in the following sections. These formulas are for *discrete compounding*, which means that the interest is compounded at the end of each finite-length period, such as a month or a year. Interest formulas for the counterpart of discrete compounding, called *continuous compounding*, will be discussed at the end of this chapter.

interest formulas relating present and future worths of single amounts

Figure 4-2 shows a cash flow diagram involving a present single sum, P, and a future single sum, F, separated by N periods with interest at $i\%$ per period. Two formulas relating these amounts are presented next.

FIGURE 4-2. General cash flow diagram relating present worth and future worth of single payments.

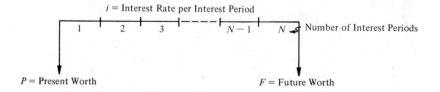

i = Interest Rate per Interest Period

N = Number of Interest Periods

P = Present Worth

F = Future Worth

table 4-2
discrete compounding interest factors and symbols[a]

To Find:	Given:	Factor by Which to Multiply "Given"	Factor Name	Factor Functional Symbol[b]
For single payments:				
F	P	$(1 + i)^N$	Single payment compound amount	$(F/P, i\%, N)$
P	F	$\dfrac{1}{(1 + i)^N}$	Single payment present worth	$(P/F, i\%, N)$
For uniform series (annuities):				
F	A	$\dfrac{(1 + i)^N - 1}{i}$	Uniform series compound amount	$(F/A, i\%, N)$
P	A	$\dfrac{(1 + i)^N - 1}{i(1 + i)^N}$	Uniform series present worth	$(P/A, i\%, N)$
A	F	$\dfrac{i}{(1 + i)^N - 1}$	Sinking fund	$(A/F, i\%, N)$
A	P	$\dfrac{i(1 + i)^N}{(1 + i)^N - 1}$	Capital recovery	$(A/P, i\%, N)$

[a] Key: i = interest rate per interest period
  N = number of interest periods
  A = uniform series amount (occurs at the end of each interest period),
  F = future worth
  P = present worth.
[b] The functional symbol system is used throughout this book.

Finding F When Given P

If an amount of P dollars exists at a point in time and $i\%$ is the interest (profit or growth) rate per period, the amount will grow to a future amount $F = P + Pi = P(1 + i)$ by the end of one period; by the end of two periods, the amount will grow to $P(1 + i)(1 + i) = P(1 + i)^2$; by the end of three periods, the amount will grow to $P(1 + i)^2(1 + i) = P(1 + i)^3$; and by the end of N periods the amount will grow to

$$F = P(1 + i)^N \qquad (4\text{-}2)$$

The quantity $(1 + i)^N$ is commonly called the *single payment compound amount factor*. Numerical values for this factor are given in the second column of the tables in Appendix E, for a wide range of values of i and N. In this book we shall use the functional symbol $(F/P, i\%, N)$ for $(1 + i)^N$. Hence Equation 4-2 can be expressed as

$$F = P(F/P, i\%, N) \qquad (4\text{-}3)$$

where the factor in parentheses denotes the unknown and known, the interest rate per period, and the number of interest periods, respectively. Note that the sequence of F and P in F/P is the same as in the initial part of Equation 4-3. This sequencing of letters is true of all functional symbols used in this book, which makes them easy to remember.

Examples of this type of problem, together with a cash flow diagram and solution, are given in the first part of the body of Table 4-3. Note in Table 4-3 that for each of the six common discrete compound interest circumstances covered, two problem statements are given: (a) in borrowing–lending terminology, and (b) in equivalence terminology, but that they both represent the same cash flow situation. Indeed, there are generally many ways in which a given cash flow situation can be expressed.

In general, a good way to interpret a relationship such as Equation 4-3 is that the calculated amount, F, at the point in time at which it occurs, *is equivalent to* (has the same weight or value as) the known value, P, at the point in time at which it occurs, for the given interest or profit rate, i.

Finding P When Given F

From Equation 4-2, $F = P(1 + i)^N$. Solving this for P gives the relationship

$$P = F\left(\frac{1}{1 + i}\right)^N = F(1 + i)^{-N} \qquad (4\text{-}4)$$

The quantity $(1 + i)^{-N}$ is called the *single payment present worth factor*. Numerical values for this factor are given in the third column of the tables in Appendix E for a wide range of values of i and N. We shall use the functional symbol $(P/F, i\%, N)$ for this factor. Hence

$$P = F(P/F, i\%, N) \qquad (4\text{-}5)$$

table 4-3
discrete compounding example problems and solutions

Example Problems: (all using interest or profit rate of $i = 10\%$ compounded annually)

To Find:	Given:	(a) In Borrowing-Lending Terminology:	(b) In Equivalence Terminology:	Diagram	Solution
For single payments					
F	P	A firm borrows $1,000 for 8 years. How much must it repay in a lump sum at the end of the eighth year?	What is the equivalent worth at the end of 8 years of $1,000 at the beginning of those 8 years?	$P = \$1,000$ $i = 10\%$ 1 ⋯ 8 $F = ?$	$F = P(F/P, 10\%, 8)$ $= \$1,000(2.1436)$ $= \$2,143.60$
P	F	A firm desires to have $1,000 6 years from now. What amount should be deposited now to provide for it?	What is the equivalent present worth of $1,000 6 years from now?	$F = \$1,000$ $i = 10\%$ 1 ⋯ 6 $P = ?$	$P = F(P/F, 10\%, 6)$ $= \$1,000(0.5645)$ $= \$564.50$

For uniform series

F	A	If three annual deposits of $2,000 each are placed in an account, how much money has accumulated immediately after the last deposit?

$$F = A(F/A, 10\%, 3)$$
$$= \$2,000(3.3100)$$
$$= \$6,620$$

P	A	How much should be deposited in a fund now to provide for nine end-of-year withdrawals of $100 each.

$$P = A(P/A, 10\%, 9)$$
$$= \$100(5.7590)$$
$$= \$575.90$$

A	F	What uniform annual amount should be deposited each year in order to accumulate $10,000 at the time of the fifth annual deposit?

$$A = F(A/F, 10\%, 5)$$
$$= \$10,000(0.1638)$$
$$= \$1,638$$

A	P	What is the size of 10 equal annual payments to repay a loan of $1,000? First payment due 1 year after receiving loan.

$$A = P(A/P, 10\%, 10)$$
$$= \$1,000(0.1627)$$
$$= \$162.70$$

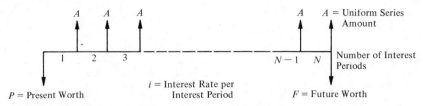

FIGURE 4-3. General cash flow diagram relating uniform series (ordinary annuity) to its present worth and future worth.

Examples of this type of problem, together with a cash flow diagram and solution, are given in the second part of the body of Table 4-3.

interest formulas relating a uniform series (annuity) to its present and future worths

Figure 4-3 shows a general cash flow diagram involving a series of uniform payments, each of amount A, occurring at the end of each period for N periods with interest at $i\%$ per period. Such a uniform series is often called an *annuity*. It should be noted that the formulas and tables below are derived such that:

1. P (present worth) occurs one interest period before the first A (uniform payment).
2. F (future worth) occurs at the same time as the last A, and N periods after P.

The timing relationship for P, A, and F can be observed in Figure 4-3. Four formulas relating A to F and P are developed below.

Finding F When Given A

If A dollars exist at the end of each period for N periods and $i\%$ is the interest (profit or growth) rate per period, the future worth, F, at the end of the Nth period is obtained by summing the future worths of each of the payments of amount A. Thus, if A_1 is the payment at the end of the first period, A_2 is the payment at the end of the second period, $\ldots$, A_{N-1} is the payment at the end of the $(N-1)$th period, and A_N is the payment at the end of the Nth period, then

$$F = A_1(1 + i)^{N-1} + A_2(1 + i)^{N-2} + A_3(1 + i)^{N-3} + \cdots + A_{N-1}(1 + i)^1$$
$$+ A_N(1 + i)^0$$
$$= A[(1 + i)^{N-1} + (1 + i)^{N-2} + (1 + i)^{N-3} + \cdots + (1 + i)^1 + (1 + i)^0]$$

The term in brackets is a geometric series having a common ratio of $(1 + i)^{-1}$, which reduces to

$$F = A\left[\frac{(1 + i)^{N-1} - (1 + i)^{-1}}{1 - (1 + i)^{-1}}\right]$$

By multiplying numerator and denominator by $(1 + i)$, this reduces to

$$F = A\left[\frac{(1 + i)^N - 1}{i}\right] \qquad (4\text{-}6)$$

The quantity $\{[(1 + i)^N - 1]/i\}$ is called the *uniform series compound amount factor*. Numerical values for this factor are given in the fourth column of the tables in Appendix E for a wide range of values of i and N. We shall use the functional symbol $(F/A, i\%, N)$ for this factor. Hence Equation 4-6 can be expressed as

$$F = A(F/A, i\%, N) \qquad (4\text{-}7)$$

Examples of this type of problem, together with a cash flow diagram and solution, are given in the third row of the body of Table 4-3.

Finding P When Given A

From Equation 4-2, $F = P(1 + i)^N$. Substituting for F in Equation 4-6, one determines that

$$P(1 + i)^N = A\left[\frac{(1 + i)^N - 1}{i}\right]$$

Dividing both sides by $(1 + i)^N$,

$$P = A\left[\frac{(1 + i)^N - 1}{i(1 + i)^N}\right] \qquad (4\text{-}8)$$

Thus Equation 4-8 is the relation for finding the equivalent present worth (as of the beginning of the first period) of a uniform series of end-of-period payments of amount A. The quantity in brackets is called the *uniform series present worth factor*. Numerical values for this factor are given in the fifth column of the tables in Appendix E for a wide range of values of i and N. We shall use the functional symbol $(P/A, i\%, N)$ for this factor. Hence

$$P = A(P/A, i\%, N) \qquad (4\text{-}9)$$

Examples of this type of problem, together with a cash flow diagram and solution, are given in the fourth row of the body of Table 4-3.

Finding A When Given F

Taking Equation 4-6 and solving for A, one finds that

$$A = F\left[\frac{i}{(1 + i)^N - 1}\right] \qquad (4\text{-}10)$$

Thus Equation 4-10 is the relation for finding the amount, A, of a uniform series of payments occurring at the end of each of N interest periods which would be equivalent to (have the same value as) its future worth, F, occurring at the end of the last period. The quantity in brackets is called the *sinking fund factor*. Numerical values for this factor are given in the sixth column of the tables in Appendix E for a wide range of values of i and N. We shall use the functional symbol $(A/F, i\%, N)$ for this factor. Hence

$$A = F(A/F, i\%, N) \tag{4-11}$$

Examples of this type of problem, together with a cash flow diagram and solution, are given in the fifth row of the body of Table 4-3.

Finding A when Given P

Taking Equation 4-8 and solving for A, one finds that

$$A = P\left[\frac{i(1 + i)^N}{(1 + i)^N - 1}\right] \tag{4-12}$$

Thus Equation 4-12 is the relation for finding the amount, A, of a uniform series of payments occurring at the end of each of N interest periods which would be equivalent to (have the same value as) the present worth, P, occurring at the beginning of the first period. The quantity in brackets is called the *capital recovery factor*. Numerical values for this factor are given in the seventh column of the tables in Appendix E for a wide range of values of i and N. We shall use the functional symbol $(A/P, i\%, N)$ for this factor. Hence

$$A = P(A/P, i\%, N) \tag{4-13}$$

Examples of this type of problem, together with a cash flow diagram and solution, are given in the last row of the body of Table 4-3.

If, at this point, the reader feels the need for example applications of the six common formulas, review Equations 4-8 through 4-13. Table 4-3 should be studied carefully.

deferred annuities (uniform series)

All annuities (uniform series) discussed to this point involved the first payment being made at the end of the first period, and are called *ordinary annuities*. If the payment does not begin until some later date, the annuity is known as a *deferred annuity*. If the annuity is deferred J periods, the situation is as portrayed in Figure 4-4. It should be noted in Figure 4-4 that the entire framed ordinary annuity has been removed from "time present" or "time 0" by J periods. It must be remembered that in an annuity deferred for J periods the first payment is made at the end of the $(J + 1)$ period, assuming that all periods involved are equal in length.

The present worth (one period before the first uniform payment) of an annuity with payments of amount A is, from Equation 4-9, $A(P/A, i\%, N)$. The worth of the single amount $A(P/A, i\%, N)$ as of J periods previous to time 0 will be $A(P/A, i\%, N)(P/F, i\%, J)$.

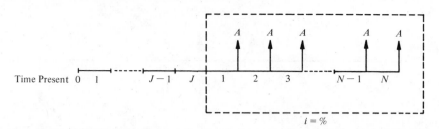

FIGURE 4-4. General cash flow representation of a deferred annuity (uniform series).

EXAMPLE 4-1

To illustrate the discussion above, suppose that a father, on the day his son is born, wishes to determine what lump amount would have to be paid into an account bearing interest at 5% compounded annually to provide payments of $2,000 on each of the son's 18th, 19th, 20th, and 21st birthdays.

Solution

The problem is represented in Figure 4-5. One should first recognize that an ordinary annuity of four payments of $2,000 each is involved, and that the present worth of this annuity occurs at the 17th birthday. It often is helpful to use a subscript with P or F to denote the point in time. Hence

$$P_{17} = A(P/A, 5\%, 4) = \$2,000(3.5460) = \$7,092$$

Note the dashed arrow in Figure 4-5, denoting P_{17}. Now that P_{17} is known, the next step is to calculate P_0. With respect to P_0, P_{17} is a future worth, and hence it could also be denoted F_{17}. Money at a given point in time, such as 17, is the same regardless of whether it is called a present worth or a future worth. Hence

$$P_0 = F_{17}(P/F, 5\%, 17) = \$7,092(0.4363) = \$3,094$$

which is the amount that the father would have to deposit. ∎

EXAMPLE 4-2

As an addition to the problem in Example 4-1, suppose that it is desired to determine the equivalent worth of the four $2,000 payments as of the son's 24th birthday. Physically, this could mean that the four payments never were withdrawn

FIGURE 4-5. Cash flow diagram of the deferred annuity problem in Example 4-1.

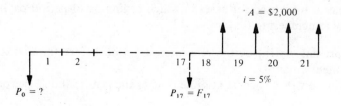

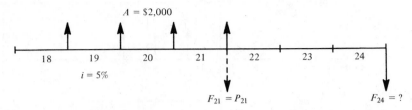

FIGURE 4-6. Cash flow diagram for the deferred annuity problem in Example 4-2.

or that possibly the son took them and immediately redeposited them in an account also earning interest at 5% compounded annually. Using our subscript system, we desire to calculate F_{24} as shown in Figure 4-6.

Solution

One way to work this is to calculate

$$F_{21} = A(F/A, 5\%, 4) = \$2,000(4.3101) = \$8,620$$

To determine F_{24}, F_{21} becomes P_{21}, and

$$F_{24} = P_{21}(F/P, 5\%, 3) = \$8,620(1.1576) = \$9,978$$

Another quicker way to work the problem is to recognize that the $P_{17} = \$7,092$ and $P_0 = \$3,094$ are each equivalent to the four $2,000 payments. Hence one can find F_{24} directly given P_{17} or P_0. Using P_0, we obtain

$$F_{24} = P_0(F/P, 5\%, 24) = \$3,094(3.2251) = \$9,978$$

which checks with the previous answer. ∎

uniform series with beginning-of-period payments

It should be noted again that all the interest formulas and corresponding tabled values for uniform series have assumed end-of-period payments. These same tables can be used for cases in which beginning-of-period payments exist merely by remembering that:

1. P (present worth) occurs one interest period before the first A (uniform payment amount).
2. F (future worth) occurs at same time as the last A, and N periods after P.

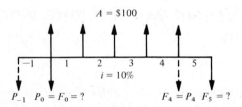

FIGURE 4-7. Cash flow diagram of uniform series with beginning-of-period payments in Example 4-3.

EXAMPLE 4-3

Figure 4-7 is a cash flow diagram depicting a uniform series of five beginning-of-period payments of $100 each. Thus the first payment is at the beginning of the first period (time 0), and the fifth payment is at the beginning of the fifth period, which coincides with the end of the fourth period (time 4). If the interest or profit rate is 10%, it is desired to find (a) the worth of the uniform series at the beginning of the first period, P, and (b) the worth at the end of the fifth period, F_5.

Solution

(a) There are several ways to work these types of problems. To find P_0, one way is first to calculate

$$P_{-1} = A(P/A, 10\%, 5) = \$100(3.7908) = \$379.08$$

Note that the P is at time -1 because that is one period before the first A payment. Also note that the interest factor is for five periods because there were five payments. Next, the problem is to find P_0 given P_{-1}. In this case P_0 becomes a future worth and could be denoted F_0. Hence

$$P_0 = F_0 = P_{-1}(F/P, 10\%, 1) = \$379.08(1.100) = \$416.99$$

(b) To find F_5 without use of the fact that $P_0 = \$416.99$, the first logical step is to calculate

$$F_4 = A(F/A, 10\%, 5) = \$100(6.1051) = \$610.51$$

Note that the F is at time 4 because that is at the same time as the last A payment. Also note that the interest factor is again for five periods, corresponding to the number of payments. Next, the problem is to find F_5 given F_4. In this case F_4 becomes a present worth and could be denoted P_4. Hence

$$F_5 = P_4(F/P, 10\%, 1) = \$610.51(1.10) = \$671.56 \; \blacksquare$$

Another, easier way to determine F_5 would be to start with $379.08 as of time -1, or $416.99 as of time 0, and to calculate the future worth at time 5. Indeed, once the equivalent worth of one or more payments is found as of a certain point in time, the equivalent worth of those same payments can be found as of any other point in time as long as the interest or profit rate, i, is known.

equivalent present worth, future worth, and annual worth

EXAMPLE 4-4

Figure 4-8 depicts an example problem with a series of year-end payments extending over 8 years. The payments are $100 for the first year, $200 for the second year, $500 for the third year, and $400 for each year from the fourth through the eighth. These could represent something like the expected maintenance expenditures for a certain piece of equipment or payments into a fund. Note that the payments are shown at the end of each year, which is a standard assumption for this book and for economic analyses in general unless one has information to the contrary. It is desired to find the equivalent (a) present worth, (b) future worth, and (c) annual worth of these payments if the annual interest or profit rate is 20%.

Solution

To find the equivalent present worth, P_0, one needs to sum the worth of all payments as of the beginning of the first year (time 0).

$$
\begin{aligned}
P_0 &= F_1(P/F, 20\%, 1) & &= \$100(0.8333) & &= \$83.33 \\
&+ F_2(P/F, 20\%, 2) & &+ \$200(0.6944) & &+ \$138.88 \\
&+ F_3(P/F, 20\%, 3) & &+ \$500(0.5787) & &+ \$289.35 \\
&+ A(P/A, 20\%, 5)(P/F, 20\%, 3) & &+ \$400(2.9906)(0.5787) & &+ \$692.26 \\
& & & & \Sigma\, P_0 &= \$1{,}203.82
\end{aligned}
$$

To find the equivalent future worth, F_8, one can sum the worth of all payments as of the end of the eighth year (time 8). However, since the equivalent present worth is already known to be $1,203.82, one can calculate directly

$$F_8 = P_0(F/P, 20\%, 8) = \$1{,}203.82(4.2998) = \$5{,}176.19$$

The equivalent annual worth of the irregular series can be calculated directly from either P_0 or F_8, as follows:

$$A = P_0(A/P, 20\%, 8) = \$1{,}203.82(0.2606) = \$313.72$$

FIGURE 4-8. Example 4-4 for calculating the equivalent present worth, future worth, and annual worth.

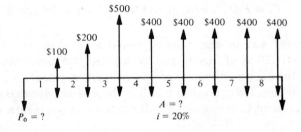

or

$$A = F_8(A/F, 20\%, 8) = \$5176.18(0.0606) = \$313.68$$

Thus one finds that the irregular series of payments shown in Figure 4–8 is equivalent to \$1,203.82 at time 0, \$5,176.19 at time 8, or a uniform series of \$313.68 for each of 8 years. ∎

interest formulas relating uniform (arithmetic) gradient series to its present and annual worths

Some economic analysis problems involve receipts or disbursements that are projected to increase or decrease by a uniform amount each period, thus constituting an arithmetic series. For example, maintenance and repair expenses on specific equipment may increase by a relatively constant amount each period.

Figure 4-9 is a cash flow diagram of a series of end-of-period payments increasing by a constant amount, G, each period. The G is known as the *gradient amount*. Note that the timing of payments on which the derived formulas and tabled values are based is as follows:

End of Year	Payment
1	0
2	G
3	$2G$
⋮	⋮
$N - 1$	$(N - 2)G$
N	$(N - 1)G$

FIGURE 4-9.   General cash flow diagram for uniform gradient series of G dollars per period.

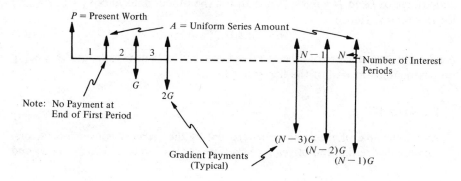

Finding P When Given G

The present worth, P, of this series is

$$P = G\left[\frac{1}{(1 + i)^2}\right] + 2G\left[\frac{1}{(1 + i)^3}\right] + \cdots + (N - 2)G\left[\frac{1}{(1 + i)^{N-1}}\right]$$

$$+ (N - 1)G\left[\frac{1}{(1 + i)^N}\right]$$

$$P = G \times \frac{1}{i}\left[\frac{(1 + i)^N - 1}{i(1 + i)^N} - \frac{N}{(1 + i)^N}\right] \tag{4-14}$$

The term

$$\frac{1}{i}\left[\frac{(1 + i)^N - 1}{i(1 + i)^N} - \frac{N}{(1 + i)^N}\right]$$

is called the *gradient to present worth conversion factor*. It also can be expressed as $(1/i)[(P/A, i\%, N) - N(P/F, i\%, N)]$. Numerical values for this factor are given in Table 21 of Appendix E for a wide range of i and N values. We shall use the functional symbol $(P/G, i\%, N)$ for this factor. Hence

$$P = G(P/G, i\%, N) \tag{4-15}$$

Finding A When Given G

To obtain a uniform series of amount A that is equivalent to the same gradient series, one need only multiply the present worth in Equation 4-15 by $(A/P, i\%, N)$. Hence

$$A = P(A/P, i\%, N) = G(P/G, i\%, N)(A/P, i\%, N)$$

$$= G \times \frac{1}{i}\left[\frac{(1 + i)^N - 1}{i(1 + i)^N} - \frac{N}{(1 + i)^N}\right]\left[\frac{i(1 + i)^N}{(1 + i)^N - 1}\right]$$

$$= G \times \left[\frac{1}{i} - \frac{N}{(1 + i)^N - 1}\right] \tag{4-16}$$

The term in brackets is called the *gradient to uniform series conversion factor*. Numerical values for this factor are given in Table 22 of Appendix E for a wide range of i and N values. We shall use the functional symbol $(A/G, i\%, N)$ for this factor. Hence

$$A = G(A/G, i\%, N) \tag{4-17}$$

Again, note that the use of these gradient conversion factors calls for there being no payment at the end of the first period.

EXAMPLE 4-5

As an example of the straightforward use of the gradient conversion factors, suppose that certain end-of-year expenses are expected to be $1,000 for the second year, $2,000 for the third year, and $3,000 for the fourth year, and that if interest is 15% per year, it is desired to find the equivalent (a) present worth at the beginning of the first year, and (b) uniform annual worth at the end of each of the 4 years.

Solution

It can be observed that this schedule of expenses fits the model of the gradient formulas with $G = \$1,000$ and $N = 4$. Note there was no payment the first period. The present worth can be calculated as:

$$P_0 = G(P/G, 15\%, 4) = \$1,000(3.79) = \$3,790$$

The uniform annual worth can be calculated from Equation 4-17 as:

$$A = G(A/G, 15\%, 4) = \$1,000(1.3263) = \$1,326.30$$

Of course, once the present worth was known, the uniform annual worth could have been calculated as

$$A = P_0(A/P, 15\%, 4) = \$3,790(0.3503) = \$1,326.30 \blacksquare$$

EXAMPLE 4-6

As a further example of the use of gradient formulas, suppose that one has payments as follows:

End of Year	Payment
1	$5,000
2	6,000
3	7,000
4	8,000

and that one wishes to calculate their equivalent present worth using gradient interest formulas.

Solution

The schedule of payments is depicted in the top diagram of Figure 4-10. The bottom two diagrams of Figure 4-10 show how the original schedule can be broken into two subschedules of payments, a uniform series of $5,000 payments plus a uniform gradient of $1,000 that fits the general gradient model for which factors are tabled. The summed present worths of these two subschedules equal the present worth of the original problem. Thus, using the symbols shown in Figure 4-10,

$$P_{OT} = P_{OA} + P_{OG}$$

$$= A(P/A, 15\%, 4) + G(P/G, 15\%, 4)$$

$$= \$5,000(2.8550) + \$1,000(3.79) = \$14,275 + 3,790 = \$18,065$$

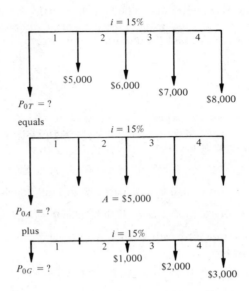

FIGURE 4-10. Example 4-6, involving a uniform gradient.

Of course, the uniform annual worth of the original payments could be calculated by the same rationale, or by merely finding A, given that the equivalent present worth is $18,065. ∎

EXAMPLE 4-7

As a further example of the use of gradient formulas, suppose that one has payments which are timed in exact reverse of the payments depicted in Example 4-6. The top diagram of Figure 4-11 shows these payments as follows:

End of Year	Payment
1	$8,000
2	7,000
3	6,000
4	5,000

Solution

The bottom two diagrams of Figure 4-11 show how these payments can be broken into two subschedules of payments. It must be remembered that the uniform gradient formulas and tables provided are for increasing gradients only. Hence, one must subtract an *increasing* gradient of payments that *did not* occur. Thus

$$P_{OT} = P_{OA} - P_{OG}$$
$$= A(P/A, 15\%, 4) - G(P/G, 15\%, 4)$$
$$= \$8,000(2.8550) - \$1,000(3.79) = \$22,840 - 3,790 = \$19,050$$

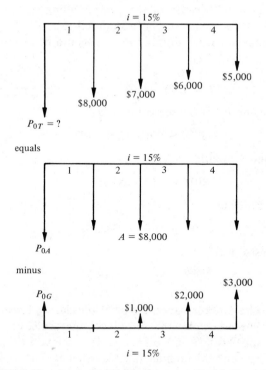

FIGURE 4-11. Example 4-7, involving a uniformly decreasing gradient.

Again, the uniform annual worth of the original decreasing series of payments could be calculated by the same rationale, or by finding A, given that the equivalent present worth is $19,050. ∎

Note from Examples 4-6 and 4-7 that the present worth of $18,065 for an increasing gradient series of payments is different from the present worth of $19,050 for a gradient of payments of like amounts but reversed timing. This difference would be even greater for higher interest rates and gradient payments, but it does exemplify the marked effect of timing of payments on equivalent worths.

nominal and effective interest rates

Very often, the interest period, or time between successive compounding, is something less than 1 year. It has become customary to quote interest rates on an annual basis, followed by the compounding period if different from 1 year in length. For example, if the interest rate is 3% per interest period and the interest period is 6 months, it is customary to speak of this rate as "6% compounded semiannually." The basic annual rate of interest is known as the *nominal rate*, 6% in this case. The actual annual rate upon the principal is not

6% but something greater, because of the compounding within 1 year. For instance, consider $100 to be invested at a nominal rate of 6% compounded semiannually. The interest earned during the year would be as follows:

First 6 months:

$$I = \$100 \times 0.03 = \$3$$

Total principal at beginning of the second period:

$$P + Pi = \$100 + \$3 = \$103$$

Interest earned second 6 months:

$$\$103 \times 0.03 = \$3.09$$

Total interest earned during year:

$$\$3.00 + \$3.09 = \$6.09$$

Actual annual interest rate:

$$\frac{\$6.09}{\$100} \times 100 = 6.09\%$$

This actual or exact rate of return upon the principal during 1 year is known as the *effective rate*. It should be noted that effective interest rates always are on an annual basis, unless specifically stated otherwise. In this text the interest rate per period is designated by i and the nominal interest rate per year by r. In the usual economy study cases in which compounding is annual, $i = r$, of course.

In general terms,

$$\text{effective rate} = (F/P, r/M, M) - 1 \tag{4-18}$$

where M is the number of compounding periods per year.

The effective rate of interest is useful for describing the compounding effect of interest earned on interest within 1 year. Table 4-4 shows effective rates for various nominal rates and compounding periods.

table 4-4
effective interest rates for various nominal rates and compounding periods

Compounding Period	Number of Periods per Year, M	Effective Rate (%) for Nominal Rate of:		
		6%	12%	24%
Annually	1	6.00	12.00	24.00
Semiannually	2	6.09	12.36	25.44
Quarterly	4	6.14	12.55	26.25
Bimonthly	6	6.15	12.62	26.53
Monthly	12	6.17	12.68	26.82
Continuously	∞	6.18	12.75	27.12

interest problems with compounding more often than once per year

Single Amounts

If a nominal interest rate is quoted and the number of compounding periods per year and number of years are known, any problem involving calculating future worths or present worths can be calculated by straightforward use of Equations 4-3 and 4-5, respectively.

EXAMPLE 4-8

For example, if $100.00 is invested for 10 years at 6% compounded quarterly. How much is it worth at the end of the 10th year?

Solution

There are four compounding periods per year, or a total of 4 × 10 = 40 periods. The interest rate per interest period is 6%/4 = 1.5%. When the values are used in Equation 4-3, one finds that

$$F = P(F/P, 1.5\%, 40) = \$100.00(1.814) = \$181.40 \quad \blacksquare$$

Uniform Series and Gradient Series

When there is more than one compound interest period per year, the formulas and tables for uniform series and gradient series can be used as long as there is a payment at the end of each interest period, as shown in Figures 4-3 and 4-9 for uniform series and gradient series, respectively.

EXAMPLE 4-9

Suppose that one has a beginning indebtedness of $10,000 which is to be repaid by equal end-of-month installments for 5 years with interest at 12% compounded monthly. What is the amount of each payment?

Solution

The number of installment payments is 5 × 12 = 60, and the interest rate per month is 12%/12 = 1%. When these values are used in Equation 4-13, one finds that

$$A = P(A/P, 1\%, 60) = \$10,000(0.0222) = \$222 \quad \blacksquare$$

EXAMPLE 4-10

Suppose that certain operating expenditures were expected to be 0 at the end of the first 6 months, $1,000 at the end of the second 6 months, and to increase by

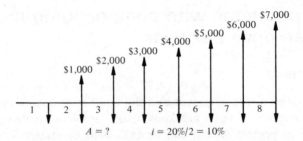

FIGURE 4-12. Example gradient with compounding more often than once per year in Example 4-10.

$1,000 at the end of each 6-month period thereafter for a total of 4 years. It is desired to find the equivalent uniform payment at the end of each of the eight 6-month periods if interest is 20% compounded semiannually.

Solution

A cash flow diagram is shown in Figure 4-12, and the solution is

$$A = G(A/G, 10\%, 8) = \$1,000(3.0045) = \$3,004.50 \ \blacksquare$$

interest problems with uniform payments less often than compounding periods

EXAMPLE 4-11

Suppose that there exists a series of 10 end-of-year payments of $1,000 each and it is desired to compute the equivalent worth of those payments as of the end of the 10th year if interest is 12% compounded quarterly. The problem is depicted in Figure 4-13.

FIGURE 4-13. Uniform series with payments less often than compounding periods in Example 4-11.

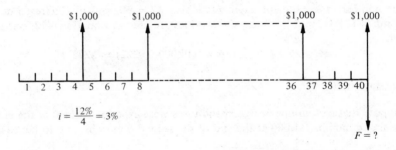

Solution

Interest is 12%/4 = 3% per quarter, but the uniform series payments are not at the end of each quarter. Hence one must make special adaptations to fit the interest formulas and tables provided.

One useful adaptation approach is to take one typical cycle of periods in which a payment occurs and convert the payment into its equivalent uniform end-of-period series of payments. The upper cash flow diagram in Figure 4-14 shows the approach applied to the first year (four interest periods) in the example of Figure 4-13. The uniform end-of-quarter payment which is equivalent to the $1,000 at the end of the year with interest at 3% per quarter can be calculated as

$$A = F(A/F, 3\%, 4) = \$1,000(0.2390) = \$239$$

Thus $239 at the end of each quarter is equivalent to $1,000 at the end of each fourth quarter (year). This is true not only for the first year but also for each of the 10 years under consideration. Hence the original series of 10 end-of-year payments of $1,000 each can be converted to a problem involving 40 end-of-quarter payments of $239 each, as shown in the lower cash flow diagram of Figure 4-14. The worth at the end of the 10th year (40th quarter) may then be computed as

$$F = A(F/A, 3\%, 40) = \$239(75.4012) = \$18,021 \quad \blacksquare$$

In general, if i is the interest rate per interest period and there is a uniform payment, X, at the *end* of each kth interest period, then the equivalent payment, A, at the end of each interest period is

$$A = X(A/F, i\%, k) \tag{4-19}$$

By similar reasoning, if i is the interest rate per interest period and there is a uniform payment, X, at the *beginning* of each kth interest period, then the equivalent payment, A, at the end of each interest period is

$$A = X(A/P, i\%, k) \tag{4-20}$$

FIGURE 4-14. Adaptation to solve Example 4-11.

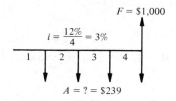

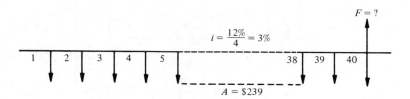

Another adaptation approach to handling uniform series with payments less often than compounding periods is to find the exact interest rate for each period in which a payment actually occurs, and then to straightforwardly apply the interest formulas and tables for that exact interest rate. For Example 4-11, interest is 3% per quarter and payments occur each year. Hence the interest rate to be found is the exact rate each year, or *effective rate*. The effective rate of 3% per quarter (12% nominal) can be found from Equation 4-18 to be

$$(F/P, r/M, M) - 1 = (F/P, 3\%, 4) - 1 = 1.1255 - 1 = 12.55\%$$

Hence the original problem in Figure 4-13 can now be expressed as shown in Figure 4-15. The future worth of this series can then be found as

$$F = A(F/A, 12.55\%, 10) = \$1,000(F/A, 12.55\%, 10)$$

Because interest factors are not commonly tabled for $i = 12.55\%$, one must either compute the $(F/A, 12.55\%, 10)$ factor by substituting $i = 12.55\%$ and $N = 10$ into its algebraic equivalent, $[(1 + i)^N - 1]/i$, or by linear interpolation in tables available.

Substitution into the algebraic equivalent will give the exact answer (same as the $18,021 by the adaptation method in Figure 4-14 except for any round-off error). Linear interpolation does not result in exact answers for nonlinear functions but normally is close enough for economic analyses.

Yet another adaptation for problems involving uniform series payments less often than compounding periods is to treat each payment as a single sum. This is usually unsatisfactory because of the number of computations involved, but it is well to recognize the possibility. Thus the solution to Example 4-11 can be computed by recognizing that the first payment is compounded 36 interest periods, the second payment is compounded 32 interest periods, and so on. Hence the problem can be solved as

$$F = \$1,000[(F/P, 3\%, 36) + (F/P, 3\%, 32) + \cdots + (F/P, 3\%, 4) + (F/P, 3\%, 0)]$$

$$= \$1,000(2.90 + 2.58 + \cdots + 1.1255 + 1.00)$$

The exact answer of $18,021 would be obtained from this calculation if there were no interpolation or round-off error.

FIGURE 4-15. Second adaptation to solve Example 4-11.

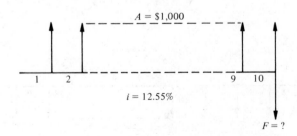

perpetuities and capitalized cost

Another type of annuity of interest to the engineer, known as a *perpetuity*, is a uniform series in which the payments continue indefinitely. If there exists a present or principal amount, $\$P$, and this earns interest at the rate of $i\%$ per period, then the end-of-period perpetual payment, $\$A$ (e.g., interest earned), that can be made from this principal is

$$A = P \times i \qquad (4\text{-}21)$$

One can then convert Equation 4-21 to show that the present worth of a perpetuity of payments of A can be found as

$$P = \frac{A}{i} \qquad (4\text{-}22)$$

This can also be seen from the relation

$$P = A(P/A, i\%, \infty) = A\left[\lim_{N \to \infty} \frac{(1 + i)^N - 1}{i(1 + i)^N}\right] = A\left(\frac{1}{i}\right)$$

The P is often spoken of as the *capitalized value* of A.

In providing for the perpetual care of some structure or the maintenance of endowed foundations, we often encounter a special type of perpetuity. A certain amount X may be needed every k periods to provide for replacement or maintenance. The owner or founder wishes to provide a fund of sufficient size so that the earnings from it will provide for this periodic demand.

To be available perpetually, X must be accumulated in k periods from the interest I that is earned by some amount of principal P, invested at rate i. Thus the period deposit toward this accumulation will be Pi. P can be computed through the relationship

$$X = I(F/A, i\%, k) = Pi(F/A, i\%, k)$$

and thus

$$P = \frac{X}{i} \times \frac{1}{(F/A, i\%, k)} = \frac{X}{i} \times (A/F, i\%, k) \qquad (4\text{-}23)$$

Another rationale for computing P in this circumstance is to reason as follows. What principal amount P, when compounded at $i\%$ per period for k periods, will at the end of the kth period equal the X needed plus P to be available for accumulating interest to provide the next X payment? Algebraically, this can be stated as

$$P(F/P, i\%, k) = X + P \qquad (4\text{-}24)$$

Thus

$$P = X/[(F/P, i\%, k) - 1]$$

EXAMPLE 4-12

What principal or capitalized amount is sufficient to provide for a replacement costing $100,000 at the end of each fifth year from now, continuing forever, if interest is 10% per year?

Solution

Using Equation 4-23,

$$P = \frac{\$100,000}{0.10} \, (A/F, 10\%, 5) = \$163,800$$

Or, using Equation 4-24, we can get the same answer:

$$P(F/P, 10\%, 5) = \$100,000 + P$$

$$P = \frac{\$100,000}{1.6105 - 1} = \$163,800 \; \blacksquare$$

If the first cost of the structure or project is added to P, the sum is known as the *capitalized cost*. Thus the capitalized cost of an article is the amount of sufficient size to purchase the article and also to provide for its perpetual maintenance and replacement as needed.

installment financing

When a series of deferred equal periodic payments is substituted for a single cash payment, as when merchandise such as an automobile is purchased, a modification of the ordinary annuity frequently is used. A finance charge is made upon the total amount owed at the beginning of the series of payments instead of only upon the unpaid balance. Such a charge is, of course, not in accord with the true nature and definition of interest. To see what the situation actually is in such cases, consider the following example.

EXAMPLE 4-13

A national finance company advertises a "6% plan" for financing the purchase of automobiles. To the amount remaining to be paid through installment payments 6% is added for each year in which money is owed. This total is divided by the number of months over which the payments are to be made, and the result is the amount of the monthly payments. For example, a woman purchases a $2,000 automobile under this plan and makes an initial cash payment of $500. She wishes to pay the balance in 24 monthly payments. What will be the amount of each payment, and what rate of interest does she actually pay?

Solution

Purchase price	= $2,000
− Initial payment	= 500
= Balance due, (P)	= $1,500
+ 6% finance charge = 0.06 × 2 years × $1,500.00 =	180
= Total to be paid	= $1,680
∴ Monthly payments = $1,680.00/24, (A)	= $ 70

Because there are to be 24 payments of $70 each, made at the end of each month, these constitute an annuity at some unknown rate of interest that should be computed only upon the unpaid balance. Therefore,

$$P = A(P/A, i\%, N)$$

$$\$1,500 = \$70.00(P/A, i\%, N)$$

$$(P/A, i\%, N) = \frac{\$1,500.00}{\$70.00} = 21.43$$

By examination of the interest tables for P/A factors for $N = 24$ which comes closest to 21.43, one finds that $(P/A, \frac{3}{4}\%, 24) = 21.8891$ and $(P/A, 1\%, 24) = 21.2434$.

Linear interpolation gives

$$i\% = 1\% - \left(\frac{21.43 - 21.2434}{21.8891 - 21.2434}\right)(1\% - \tfrac{3}{4}\%) = 0.93\%$$

Since payments are monthly, 0.93% is the interest rate per month. The nominal rate paid on the borrowed money is 0.93%(12) = 11.16% compounded monthly. This corresponds to an effective annual interest rate of $(1 + 0.0093)^{12} - 1 \cong$ 12%. ∎

discount

Two types of transactions are sometimes encountered in which *discount* is involved. In the first, the holder of a negotiable paper, such as a note or a sales contract, that is not due and payable until some future date desires to exchange the paper for immediate cash. In order to do this, he will accept a sum of cash smaller in amount than the face value of the paper. The difference between the present worth (the amount received for the paper in cash) and the worth of the paper at some time in the future (the face value of the paper or principal) is known as the *discount* for the period involved.

The second type of transaction occurs in many bank loans. As an example, a man may wish to borrow $100 from a bank for 1 year at an interest rate of 6%. The bank computes the interest, 6%, and *deducts* this amount from the $100, giving the borrower only $94. The borrower signs a note promising to repay $100 at the end of a year. The $6 that was deducted represents interest paid in advance. It also represents the difference between the present worth of the note—the \$94 received by the borrower—and the worth of the note at the end of 1 year—\$100. It is therefore the discount for the period involved. The *rate of discount* is defined as the discount of one unit of principal for one unit of time. If the rate of discount is designated by d, it is equal to the difference between 1 and its present worth (P/F):

$$d = 1 - (P/F) = 1 - \frac{1}{1 + i} = i(P/F) \qquad (4\text{-}25)$$

and

$$i = \frac{d}{(P/F)} = \frac{d}{1 - d} \qquad (4\text{-}26)$$

For the example cited, the discount was \$6. The rate of discount was

$$d = \frac{\$6}{\$100} = 0.06 = 6\%$$

The interest rate, based upon the principal actually received by the borrower, was

$$i = \frac{\$6}{\$94} = 0.0638 = 6.38\%$$

interest factor relationships

The following relationships among the six basic discrete compounding interest factors should be recognized:

$$(P/F, i\%, N) = \frac{1}{(F/P, i\%, N)} \qquad (4\text{-}27)$$

$$(A/P, i\%, N) = \frac{1}{(P/A, i\%, N)} \qquad (4\text{-}28)$$

$$(A/F, i\%, N) = \frac{1}{(F/A, i\%, N)} \qquad (4\text{-}29)$$

$$(F/A, i\%, N) = (P/A, i\%, N)(F/P, i\%, N) \qquad (4\text{-}30)$$

$$(P/A, i\%, N) = \sum_{j=1}^{N} (P/F, i\%, j) \qquad (4\text{-}31)$$

$$(F/A, i\%, N) = \sum_{j=0}^{N-1} (F/P, i\%, j) \qquad (4\text{-}32)$$

Indeed, the same relationships exist among the corresponding continuous-compounding interest factors discussed in the following sections.

interest formulas for continuous compounding and discrete payments

In most business transactions and economy studies interest is compounded at the end of discrete periods of time and, as has been discussed previously, cash flows are assumed to occur at the beginning or end of such periods. This practice will be used throughout the remaining chapters of this book. However, it is evident that in most enterprises cash is flowing in and out in an almost continuous stream. Because cash, whenever available, can usually be used profitably, this situation, in effect, produces very frequent compounding of the interest earned. So that this condition can be accounted for, the concepts of continuous compounding and continuous cash flow are used in some economy studies. Actually, the effects of these procedures compared to discrete compounding are rather small in most cases.

In the concept of continuous compounding it is assumed that cash payments occur once per year, but that compounding is continuous throughout the year. Thus, with a nominal rate of interest per year of r, if the interest is compounded M times per year, at the end of 1 year one unit of principal will amount to $[1 + (r/M)]^M$. If $M/r = k$, the foregoing expression becomes

$$\left[1 + \frac{1}{k}\right]^{rk} = \left[\left(1 + \frac{1}{k}\right)^k\right]^r$$

The limit of $(1 + 1/k)^k$ as k approaches infinity is e, the base of natural logarithms. Thus the previous expression can be written as e^r. Consequently, the *continuous compounding compound amount factor* (*single payment*) at $r\%$ nominal interest for N years is e^{rN}. Using our functional notation, we express this as

$$(F/P, r\%, N) = e^{rN} \tag{4-33}$$

Note that the symbol is directly comparable to that used for discrete compounding and discrete payments except that $\underline{r\%}$ is used to denote the nominal rate *and* the use of continuous compounding.

Since e^{rN} for continuous compounding corresponds to $(1 + i)^N$ for discrete compounding, e^r corresponds to $(1 + i)$. Equating the latter expression gives

$$i = e^r - 1 \tag{4-34}$$

By use of this relationship, the corresponding values of (P/F), (F/A), and (P/A) for continuous compounding may be obtained from Equations 4-4, 4-6, and 4-8, respectively, by substitution of $e^r - 1$ for i in these equations. Thus, for continuous compounding and discrete payments,

$$(P/F, r\%, N) = \frac{1}{e^{rN}} = e^{-rN}$$

$$(F/A, r\%, N) = \frac{e^{rN} - 1}{e^r - 1}$$

$$(P/A, r\%, N) = \frac{1 - e^{-rN}}{e^r - 1} = \frac{e^{rN} - 1}{e^{rN}(e^r - 1)}$$

Values for $(A/P, r\%, N)$ and $(A/F, r\%, N)$ may be derived through their inverse relationships to $(P/A, r\%, N)$ and $(F/A, r\%, N)$, respectively. All the continuous compounding, discrete payment interest factors and their uses are summarized in Table 4-5.

Because continuous compounding is used rather infrequently and is not used in subsequent problems in this text (except for limited exercises at the end of this chapter), detailed values for $(A/F, r\%, N)$ and $(A/P, r\%, N)$ are not given in the Appendix. However, the tables in Appendix F give values of $(F/P, r\%, N)$, $(P/F, r\%, N)$, $(F/A, r\%, N)$, and $(P/A, r\%, N)$ for a limited number of interest rates.

It is important to note that tables of interest and annuity factors for continuous compounding commonly are tabulated in two different ways. One uses effective rates of interest; the other (as done in Appendix F) uses the more common nominal annual rates of interest. In using such tables, one should make certain which method of tabulation is employed.

Example 4-14

Suppose that one has a present amount of $1,000 and it is desired to determine what uniform end-of-year payments could be obtained from it for 10 years if interest is 20% compounded continuously.

Solution

$$A = P(A/P, r\%, N)$$

Since the (A/P) factor is not tabled for continuous compounding we substitute its inverse (P/A), which is tabled. Thus

$$A = P \times \frac{1}{(P/A, \underline{20\%}, 10)} = \$1,000 \times \frac{1}{3.9054} = \$256$$

It is interesting to note that the answer to the same problem except with discrete annual compounding is

$$A = P(A/P, i\%, N) = P(A/P, 20\%, 10)$$

$$= \$1,000(0.2385) = \$239 \quad \blacksquare$$

interest formulas for continuous compounding and continuous cash flows

Continuous flow of funds means a series of payments occurring at infinitely short intervals of time; this corresponds to an annuity having infinitely short periods. In such cases the interest normally is compounded continuously. If the nominal interest rate per year is r and there are k payments per year,

table 4-5
continuous compounding and discrete cash flows interest factors and symbols[a]

To Find:	Given:	Factor by Which to Multiply "Given"	Factor Name	Factor Functional Symbol
For single payments:				
F	P	e^{rN}	Continuous compounding compound amount (single payment)	$(F/P, r\%, N)$
P	F	e^{-rN}	Continuous compounding present worth (single payment)	$(P/F, r\%, N)$
For uniform series (annuities):				
F	A	$\dfrac{e^{rN} - 1}{e^r - 1}$	Continuous compounding compounding amount (uniform series)	$(F/A, r\%, N)$
P	A	$\dfrac{e^{rN} - 1}{e^{rN}(e^r - 1)}$	Continuous compounding present worth (uniform series)	$(P/A, r\%, N)$
A	F	$\dfrac{e^r - 1}{e^{rN} - 1}$	Continuous compounding sinking fund	$(A/F, r\%, N)$
A	P	$\dfrac{e^{rN}(e^r - 1)}{e^{rN} - 1}$	Continuous compounding capital recovery	$(A/P, r\%, N)$

[a] Key: r = nominal annual interest rate, compounded continuously
N = number of periods (years)
A = uniform series amount (occurs at end of each year)
F = future worth
P = present worth.

which amount to a total of one unit per year, then, by use of Equation 4-8, for 1 year the present worth at the beginning of the year is

$$P = \frac{1}{k}\left\{\frac{[1 + (r/k)]^k - 1}{r/k[1 + (r/k)]^k}\right\} = \frac{[1 + (r/k)]^k - 1}{r[1 + (r/k)]^k}$$

The limit of $[1 + (r/k)]^k$ as k approaches infinity is e^r. By letting the present worth of one unit per year, flowing continuously and with continuous compounding of interest, be called the *continuous compounding present worth factor* (*continuous, uniform payment over one period*),

$$(P/\bar{A}, r\%, 1) = \frac{e^r - 1}{re^r} \tag{4-35}$$

where $\bar{A}$ is the amount flowing continuously over 1 year.

For $\bar{A}$ flowing each year over N years, as depicted in Figure 4-16,

$$(P/\bar{A}, r\%, N) = \frac{e^{rN} - 1}{re^{rN}} \tag{4-36}$$

which is the *continuous compounding present worth factor* (*continuous, uniform payments*).

Equation 4-35 can also be written

$$(P/\bar{A}, r\%, 1) = e^{-r}\left[\frac{e^r - 1}{r}\right] = (P/F, r\%, 1)\left[\frac{e^r - 1}{r}\right]$$

Because the present worth of 1 per year, flowing continuously with continuous compounding of interest, is $(P/F, r\%, 1)(e^r - 1)/r$, it follows that $(e^r - 1)/r$ must also be the compound amount of 1 per year, flowing continuously with continuous compounding of interest. Consequently, the *continuous compounding compound amount factor* (*continuous, uniform payment over 1 year*) is

$$(F/\bar{A}, r\%, 1) = \frac{e^r - 1}{r} \tag{4-37}$$

For N years,

$$(F/\bar{A}, r\%, N) = \frac{e^{rN} - 1}{r} \tag{4-38}$$

FIGURE 4-16.   General cash flow diagram for continuous compounding, continuous cash flows.

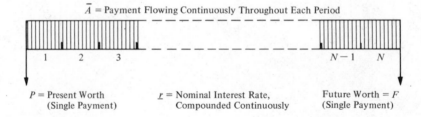

$\bar{A}$ = Payment Flowing Continuously Throughout Each Period

P = Present Worth (Single Payment) r = Nominal Interest Rate, Compounded Continuously   Future Worth = F (Single Payment)

table 4-6

continuous compounding continuous uniform cash flows interest factors and symbols[a]

To Find:	Given:	Factor by Which to Multiply "Given"	Factor Name	Factor Functional Symbol
F	$\overline{A}$	$\dfrac{e^{rN} - 1}{r}$	Continuous compounding compounding amount (continuous, uniform payments)	$(F/\overline{A}, r\%, N)$
P	$\overline{A}$	$\dfrac{e^{rN} - 1}{re^{rN}}$	Continuous compounding present worth (continuous, uniform payments)	$(P/\overline{A}, r\%, N)$
$\overline{A}$	F	$\dfrac{r}{e^{rN} - 1}$	Continuous compounding sinking-fund (continuous, uniform payments)	$(\overline{A}/F, r\%, N)$
$\overline{A}$	P	$\dfrac{re^{rN}}{e^{rN} - 1}$	Continuous compounding capital recovery (continuous, uniform payments)	$(\overline{A}/P, r\%, N)$

[a] Key: r = nominal annual interest rate, compounded continuously
  N = number of periods (years)
  A = amount of money flowing continuously and uniformly during each period
  F = future worth
  P = present worth.

This is the *continuous compounding compound amount factor* (*continuous uniform payments for N years*).

Values of $(P/\overline{A}, r\%, N)$ and $(F/\overline{A}, r\%, N)$ are given in the tables in Appendix F for various interest rates. Values for $(\overline{A}/P, r\%, N)$ and $(\overline{A}/F, r\%, N)$ can readily be obtained through their inverse relationship to $(P/\overline{A}, r\%, N)$ and $(F/\overline{A}, r\%, N)$, respectively. A summary of these factors and their use is given in Table 4-6.

EXAMPLE 4-15

What will be the worth at the end of 5 years of a uniform, continuous cash flow, at the rate of $500 per year for 5 years, with interest compounded continuously at the nominal annual rate of 8%?

Solution

$$F = \overline{A}(F/\overline{A}, \underline{8}\%, 5) = \$500 \times 6.1478 = \$3074$$

It is interesting to note that if this cash flow had been in year-end amounts of $500 with discrete compounding at 8%, the amount would have been

$$F = A(F/A, 8\%, 5) = \$500 \times 5.8666 = \$2,933$$

If the year-end payments had occurred with 8% interest compounded continuously, the amount would have been

$$F = A(F/A, \underline{8}\%, 5) = \$500 \times 5.9052 = \$2,953 \quad \blacksquare$$

problems

4-1. Draw a cash flow diagram for $10,500 being loaned out at a simple interest rate of 10% per annum over a period of 6 years. How much interest would be repaid as a lump-sum amount at the end of the sixth year?

4-2. What lump-sum amount of interest will be paid on a $1,000 loan that was made on August 1, 1976, and repaid on November 1, 1982, with ordinary simple interest at 8% per year?

4-3. How much interest would be *payable each year* on a loan of $2,000 at an annual interest rate of 8%? How much interest would have been paid over an 8-year period?

4-4. In Problem 4-3, if the interest had not been paid each year but had been allowed to compound, how much interest would be due to the lender as a lump sum at the end of the eighth year? How much extra interest is being paid here (as opposed to Problem 4-3) and what is the reason for the difference?

4-5. A future amount, F, is equivalent to $1,500 now when 8 years separates the amounts and the interest rate, compounded annually, is 12%. What is the value of F?

4-6. A present obligation of $10,500 is to be repaid in equal uniform annual amounts, each of which includes repayment of the debt (principal) and interest on the debt, over a period of 6 years. If the effective interest rate per year is 10%, what is the amount of the annual payment?

4-7. A person desires to accumulate $2,500 over a period of 7 years so that a cash payment can be made for a new roof on a summer cottage. To have this amount when it is needed, annual payments will be made to a savings account that earns 6% effective interest per year. How much must each annual payment be? Draw a cash flow diagram.

4-8. It is estimated that a certain piece of equipment can save $6,000 per year in labor and materials costs. The equipment has an expected life of 5 years and no salvage value. If the company must earn a 20% rate of return on such investments, how much could be justified for the purchase of this piece of equipment? Draw a cash flow diagram.

4-9. Deposits of $1,000 per year for 20 years will accumulate to what lump-sum amount at the end of the 20th year if the effective interest rate is 9% per annum?

4-10. Compute the effective annual interest rate in each of these situations:
(a) 10% nominal interest, compounded semiannually.
(b) 10% compounded quarterly.
(c) 10% compounded continuously.

4-11. John Q. wants his estate to be worth $65,000 at the end of 10 years. His net worth is now zero. He can accumulate the desired $65,000 by depositing $5,168 at the end of each year for the next 10 years. At what effective interest rate must his deposits be invested? Give answer to the nearest tenth of a percent.

4-12. A certain savings and loan association advertises that they pay 8% interest, compounded quarterly. What is the *effective* interest rate per annum? If you deposit $5,000 now and plan to withdraw it in 3 years, how much would your account be worth at that time?

4-13. If in Problem 4-12 you decide to deposit $800 every year for 3 years, how much could be withdrawn at the end of the third year? Suppose that, instead, you deposit $400 every 6 months for 3 years, now what would the accumulated amount be?

4-14. How long does it take a given amount of money to triple itself if the money is invested at a nominal rate of 15%, compounded monthly?

4-15. Determine the present equivalent value of $400 paid every 3 months over a period of 7 years in each of these situations:
(a) The interest rate is 12%, compounded annually.
(b) The interest rate is 12%, compounded quarterly.
(c) The interest rate is 12%, compounded continuously.

4-16. How many deposits of $100 must a person make at the end of each month if she desires to accumulate $3,350 for a new automobile? The savings account pays 9% interest compounded monthly.

4-17. Find the equivalent lump-sum amount at the end of year 9 when $P_0 = \$1,000$ and a nominal interest rate of 8% is compounded continuously.

4-18. Determine the present equivalent value of the following cash flow pattern when $i = 8\%$ per year.

End of Year	0	1	2	3	4	5	6	7
Amount ($)	−1,500	+500	+500	+500	+400	+300	+200	+100

4-19. If the *effective* annual interest rate for an automobile loan is stated as 12% and if you know continuous compounding is being used, what is the loan company's nominal interest rate?

4-20. Exactly 140 years ago, my great-grandmother deposited $190 in a large New York bank as part of a savings plan. She forgot all about the deposit during her lifetime. Two years ago the bank notified my family that the account was worth $200,000. What was the effective annual interest rate in this situation?

4-21. In a certain foreign country, a man who wanted to borrow $10,000 for a 1-year period was informed he would only have to pay $2,000 in interest (i.e., a 20% interest rate). The lender stated that the total owed to him, $12,000, would be repaid at the rate of $1,000 per month for 12 months. What was the real effective interest being charged in this transaction? Why is it greater than the apparent rate of 20%?

4-22. Suppose that you borrow $500 from the Easy Credit Company with the agreement to repay it over a 3-year period. Their stated interest rate is 5% per year. They show you the following calculation in determining the monthly payment.

Principal	$500
Total interest: 0.05(3 years)($500)	75
Loan application fee	15

They ask you to pay the interest immediately, so you leave with $425 in your pocket. Your monthly payment is calculated as follows:

$$\frac{\$500 + \$15}{36} = \$14.30/\text{month}$$

What is the effective interest rate per year?

4-23. A friend of yours just purchased a new car for $4,200. He made a $1,000 down payment on the car and then was told that to pay the remaining $3,200, a monthly payment of $160 would be required for the next 24 months. The first of these payments would be due 1 month from now. What effective annual interest rate is your friend paying on this auto loan?

4-24. Find the equivalent uniform annual amount that is equivalent to a gradient series in which the first annual payment is $500, the second annual payment is $600, the third payment is $700, and so on, and there is a total of 10 annual payments. The effective interest rate is 8%.

4-25. Determine the equivalent uniform semiannual cost of the following cash flow diagram. The interest rate is 10% compounded semiannually.

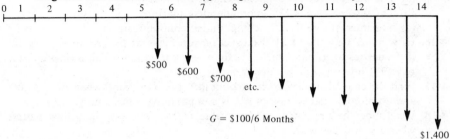

$G = \$100/6$ Months

4-26. Calculate the future worth at the end of 1983 at 8% compounded annually of this savings account:

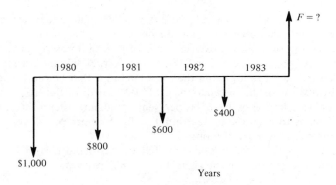

4-27. If the nominal interest rate is 10% and compounding is semiannual, what is the present worth of the following receipts?

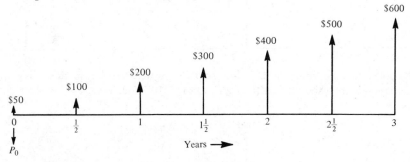

4-28. Equal end-of-year payments of $263.80 each are being made on a $1,000 loan at 10% effective interest per year.
   (a) How many payments are required to repay the entire loan?
   (b) Immediately after the second payment, what lump-sum payment would completely pay off the loan?

4-29. A firm plans to market a new minicomputer that will sell for $10,000. This financing scheme is being considered by the company: Payments of $872 each year for 20 years would be made by the purchaser after an initial down payment is made. If their interest charge is 6% compounded monthly, what down deposit should the company request?

4-30. A man deposited $2,000 in a savings account when his son was born. The nominal interest rate was 8% per year, compounded continuously. On the son's 18th birthday, the accumulated sum is withdrawn from the account. How much would this accumulated amount be?

4-31. Maintenance costs for a new bridge with an expected 50-year life are estimated to be $1,000 each year for the first 5 years, followed by a $10,000 expenditure in the 15th year and a $10,000 expenditure in year 30. If $i = 10\%$ per year, what is the equivalent uniform annual cost over the entire 50-year period?

4-32. Uncle Wilbur's trout ranch is now for sale for $40,000. Annual property taxes, maintenance, supplies, and so on, are estimated to continue to be $3,000 per year. Revenues from the ranch are expected to be $10,000 next year and then decline by $500 per year thereafter through the tenth year. If you bought the ranch, you would plan to keep it only 5 years and at that time sell it for the value of the land, which is $15,000. If your desired annual rate of return is 12%, should you become a trout rancher?

4-33. The heat loss through the exterior walls of a certain poultry processing plant is estimated to cost the owner $3,000 next year. A salesman from Superfiber Insulation, Inc., has told you, the plant engineer, that he can reduce the heat loss by 80% with the installation of $15,000 worth of Superfiber now. If the cost of heat loss rises by $200 per year (gradient) after next year and the owner plans to keep the present building for 10 more years, what would you recommend if the cost of money is 12% per year?

4-34. Solve for the value of F below so that the left-hand cash flow diagram is equivalent to the one on the right. Let $i = 8\%$ per year.

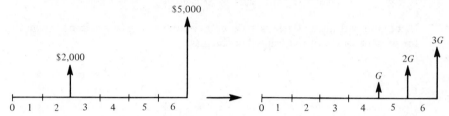

4-35. Solve for the value of G below so that the left-hand cash flow diagram is equivalent to the one on the right. Let $i = 10\%$ per year.

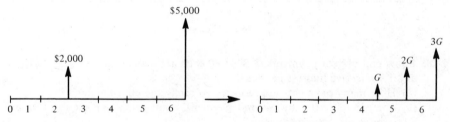

4-36. Find the value of the unknown quantity in the diagram below when $i = 10\%$ per year.

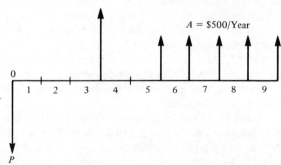

4-37. A woman arranges to repay a $1,000 bank loan in 10 equal payments at a 10% effective annual interest rate. Immediately *after* her third payment she borrows another $500, also at 10%. When she borrows the $500, she talks the banker into letting her repay the remaining debt of the first loan and the entire amount of the second loan in 12 equal annual payments. The first of these 12 payments would be made 1 year after she receives the $500. Compute the amount of each of the 12 payments.

4-38. Find the present worth at $i = 6\%$ of a $5,000 expense every 3 years, starting 10 years from the present, and continuing forever.

4-39. Mr. Gottrocks has instructed his bank to establish a trust fund that will provide $1,000 per year for the next 200 years to help pay for the upkeep on a memorial garden. The first of these payments would begin on July 1, 1986, and suppose that it is now July 1, 1982. What is the present worth in 1982 of this trust fund if the interest rate (effective) is 10%?

4-40. What is the total capitalized cost of a structure that will require construction costs of $100,000 immediately and $10,000 each year for the next 4 years, and annual year-end maintenance costs of $5,000, plus the expenditure of $25,000 at the end of each 10-year period for replacement purposes? Assume an 8% interest rate.

4-41. An expenditure of $20,000 is made to modify a materials-handling system in a small job shop. This modification will result in first-year savings of $2,000, second-year savings of $4,000 and savings of $5,000 per year thereafter. How many years must the system last if a 30% return on investment is required? Assume that the system is tailor-made for this job shop and has no salvage value at any time.

4-42. You purchase special equipment that reduces defects by $10,000 per year on item A. This item is sold on contract for the next 5 years. After the contract expires, the special equipment will save approximately $2,000 per year for 5 more years. You assume that the machine has no salvage value at the end of 10 years. How much can you afford to pay for this equipment now, if you require a 25% return on your investment?

4-43. On January 1, 1970, a government bond is purchased for $9,400. The face value of the bond is $10,000 and 8% nominal interest is paid on the face value four times each year. Thus every 3 months the bondholder receives $200 as a cash payment. When the bond matures on January 1, 1980, $10,000 is paid to the bondholder. What is the effective rate of interest being earned in this situation?

4-44. A certain heat exchanger costs $30,000 installed and has an estimated life of 6 years. By the addition of certain auxillary equipment when the heat exchanger is initially purchased, an annual saving of $1,000 in operating cost can be obtained and the estimated life of the heat exchanger can be doubled. Neglecting any salvage value for either plan and with effective annual interest at 8%, what *present* expenditure can be justified for the auxiliary equipment?

4-45. Convert the following cash flow pattern to a uniform series of end-of-year costs over a 7-year period. Let $i = 12\%$.

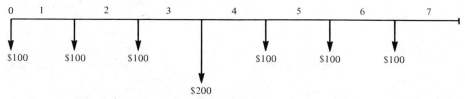

4-46. A company purchased a fleet of trucks for $156,000. Payment was to be made by an immediate cash payment of $10,000 and 12 month-end payments of $12,972 each. Another dealer offered to finance the $156,000 purchase at an interest rate of $\frac{3}{4}\%$ per month on the unpaid balance. Which offer should the company have accepted?

4-47. The membership dues of a professional society are $47 per year. However, this society encourages its members to prepay their dues by the following incentive. If the member prepays his dues for N years in a lump-sum amount, the total amount due to the society is $47 + $40($N$ − 1). Otherwise, the member would pay $47 per year at the beginning of the year for membership privileges. If an individual estimates that his personal before-tax rate of return ought to be 10%, how many years in advance should he pay his dues?

4-48. If the before-tax rate of return in Problem 4-47 is 20%, what should the decision be?

4-49. If a nominal interest rate of 8% is compounded continuously, determine the unknown quantity in each of the following situations.
    (a) What uniform end-of-year amount for 10 years is equivalent to $8,000 at the end of year 10?
    (b) What is the present equivalent value of $1,000 per year for 12 years?
    (c) What is the future worth at the end of the sixth year of $243 payments every 6 months during the 6 years? The first payment occurs 6 months from the present and the last occurs at the end of the sixth year.

4-50. (a) What is the present worth of a uniform series of annual payments of $3,500 each for 5 years if the interest rate, compounded continuously, is 10%?
    (b) What is the future worth in 6.5 years of $12,000 deposited now in a savings account paying a nominal interest rate of 7% per annum? Assume continuous compounding.
    (c) An amount of $7,000 is invested in a certificate of deposit (C.D.) and will be worth $16,000 in 9 years. What is the continuously compounded nominal interest rate for this C.D.?

4-51. (a) Determine the equivalent uniform series of annual *beginning-of-year* payments for the following cash flow pattern. The nominal interest rate is 8% compounded continuously.

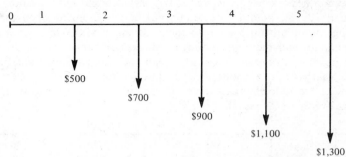

    (b) For how many years must an investment of $63,000 provide a continuous flow of funds at the rate of $16,000 per year so that a nominal interest rate of 10%, continuously compounded, will be earned?

4-52. What is the present value of the following continuous funds flow situations?
    (a) $1,000,000 per year for 4 years at 10% compounded continuously.
    (b) $6,000 per year for 10 years at 8% compounded annually.
    (c) $500 per quarter for 6.75 years at 12% compounded continuously.

part II

Applications of Engineering Economic Analysis

chapter 5

basic methods
for making
economy studies

All economy studies of capital projects should be made so as to include consideration of the return that a given project will or should produce. Because the patterns of capital investment, revenue or savings flows, and cost flows can be quite different in various projects, there is no single method for making economy studies that is ideal for all cases. Consequently, several methods, or patterns, commonly are used in practice, and all will produce equally satisfactory results and will lead to the same decision in cases where the inherent assumptions of each are applicable.

The main objective of this chapter is to demonstrate the mechanics of calculations for six basic methods for making economy studies and to briefly describe underlying assumptions and interrelationships of those methods.

Income taxes may or may not be considered in economy studies. Income taxes are an ever-present and somewhat unpleasant fact of life; most corporations pay out about half of their gross profits in the form of income taxes, both to State and Federal governments. However, in many economy studies, one obtains the same decisions or relative rankings of alternatives in studies *before* the consideration of income taxes as *after* the consideration of income taxes. All discussions and illustrations of economy study methods in this chapter and through Chapter 9 are for *before-tax* studies. Chapter 10 and parts of subsequent chapters show how to make *after-tax* studies by explicitly considering the effect of income taxes.

Economy studies can be made either from a project viewpoint or from the viewpoint of the owners of the organization. A difference occurs primarily when some borrowed capital is to be used. If we make an economy study in which we are concerned only with the potential profitability of a project without any consideration of the source of the investment funds that would be required, we do not need to consider whether owned capital or borrowed capital is to be used. For most purposes such a viewpoint is quite satisfactory. A procedure of this sort assumes that after the inherent feasibility of the project has been determined,

111

we can then consider the additional problems related to the acquisition and ownership of the required capital.

basic methods

Following are six basic methods or patterns for making economy studies, together with their abbreviations:

1. Annual worth (A.W.).
2. Present worth (P.W.).
3. Future worth (F.W.).
4. Internal rate of return (I.R.R.).
5. External rate of return (E.R.R.).
6. Explicit reinvestment rate of return (E.R.R.R.).

The first three methods convert all financial happenings into *equivalent* worths at some point or points in time using an interest rate equal to the *cost of capital*, or the *minimum attractive rate of return* (M.A.R.R.). The last three methods are different ways to calculate a rate of profit or savings (annual return as a per cent of investment) so that this can, in turn, be compared against the M.A.R.R. This chapter will describe the use of each of the methods for analyzing single projects, and later chapters will show how to use each method for studies of multiple alternatives.

the annual worth method

The term *annual worth* (A.W.) means a uniform annual series of net cash flows for a certain period of time that is equivalent in amount to a particular schedule of cash inflows (receipts or savings) and/or cash outflows (disbursements or opportunity costs) under consideration.

The criterion for this method is that as long as the *net annual worth* (i.e., annual equivalent of inflows minus outflows) is ≥ 0 the project is economically justified; otherwise, it is not justified.

If disbursements only are considered, the criterion is usually expressed as annual worth–cost (A.W.–C.), or *annual cost* (A.C.).

calculation of capital recovery cost

The *capital recovery cost* (C.R.) for a project is the equivalent uniform annual cost of the capital invested. It is an annual amount which covers the following two items:

1. Depreciation (loss in value of the asset).
2. Interest (minimum required profit) on invested capital.

As an example, consider a machine or other asset that will cost $10,000, last 5 years, and have a salvage value of $2,000. Further, the interest on invested capital, i, is 10%.

It can be shown that no matter which method of calculating depreciation is used, the equivalent annual cost of the capital recovery is the same. For example, if straight line depreciation† is used, the equivalent annual cost of interest is calculated to be $710, as shown in Table 5-1. The annual depreciation cost by the straight line method is ($10,000 − $2,000)/5 = $1,600. The $710 added to $1,600 results in a calculated capital recovery cost of $2,310.

There are several convenient formulas by which capital recovery cost may be calculated to obtain the same answer as above. Probably the easiest formula to understand is to find the annual equivalent of the investment and then subtract the annual equivalent of the salvage value. Thus

$$\text{C.R.} = P(A/P, i\%, N) - F(A/F, i\%, N) \tag{5-1}$$

where P = investment at beginning of life

F = salvage value at end of life

N = life of project

Applied to the example in Table 5-1,

$$\text{C.R.} = \$10,000(A/P, 10\%, 5) - \$2,000(A/F, 10\%, 5)$$
$$= \$10,000(0.2638) - 2,000(0.1638) = \$2,310$$

Another way to calculate the C.R. cost is to add the annual *sinking fund depreciation charge* (or deposit) to the interest on original investment (sometimes called *minimum required profit*). Thus

$$\text{C.R.} = (P - F)(A/F, i\%, N) + P(i\%) \tag{5-2}$$

table 5-1
calculation of equivalent annual cost of interest and capital recovery cost assuming straight line depreciation

Year	Value of Investment at Beginning of Year	Interest on Beginning-of-Year Investment at 10%	Present Worth of Interest at 10%	
1	$10,000	$1,000	$1,000(P/F, 10%, 1) =	$ 909
2	8,400	840	840(P/F, 10%, 2) =	694
3	6,800	680	680(P/F, 10%, 3) =	511
4	5,200	520	520(P/F, 10%, 4) =	355
5	3,600	360	360(P/F, 10%, 5) =	224
			TOTAL	$2,693

Annual equivalent interest = $2,693(A/P, 10%, 5) = $710
Total C.R. cost = annual depreciation + annual equivalent interest
= ($10,000 − $2,000)/5 + $710 = $2,310

† Explanation of various methods of depreciation will be given in Chapter 9.

Applied to the same example as in Table 5-1,

$$\text{C.R.} = (\$10,000 - \$2,000)(A/F, 10\%, 5) + \$10,000(10\%)$$
$$= \$8,000(0.1638) + \$10,000(0.10) = \$2,310$$

Yet another way to calculate C.R. cost, which seems to be the most popular in books in this field, is to add the equivalent annual cost of the depreciable portion of the investment and the interest on the nondepreciable portion (salvage value). Thus

$$\text{C.R.} = (P - F)(A/P, i\%, N) + F(i\%) \tag{5-3}$$

Applied to the same example as above,

$$\text{C.R.} = (\$10,000 - \$2,000)(A/P, 10\%, 5) + \$2,000(10\%)$$
$$= \$8,000(0.2638) + \$2,000(0.10) = \$2,310$$

Appendix 5-C describes an *approximation* formula, called the *straight line depreciation plus average profit* method, for calculating C.R. cost. Applying Equation 5-C-4 to the example in Table 5-1,

$$\text{C.R.} \cong \frac{P - F}{N} + (P - F)\left(\frac{N + 1}{2N}\right)(i) + F(i)$$

$$\cong \frac{\$10,000 - \$2,000}{5} + (\$10,000 - \$2,000)\left(\frac{6}{10}\right)(10\%) + \$2,000(10\%)$$

$$\cong \$2,480$$

EXAMPLE 5-1

An investment of $10,000 can be made in a project that will produce a uniform annual revenue of $5,310 for 5 years and then have a salvage value of $2,000. Annual disbursements will be $3,000 each year for operation and maintenance costs. The company is willing to accept any project that will earn 10% or more, before income taxes, on all invested capital. Show whether this is a desirable investment using the annual worth method.

Solution		Annual Worth
Annual revenue		$5,310
Annual disbursements	− $3,000	
C.R. cost† = ($10,000 − $2,000)(A/P, 10%, 5)		
+ $2,000(10%)	− 2,310	
TOTAL		− $5,310
NET A.W.		$ 0

Since net A.W. = $0, the project earns exactly 10% and is thus barely justified. Of course, nonmonetary considerations would probably sway the decision either way. ∎

† Uses Equation 5-3.

the present worth method

The *present worth* (P.W.) *method* for economy studies is based on the concept of equivalent worth of all cash flows relative to some base or beginning point in time called the *present*. That is, all cash inflows and outflows are discounted back at an interest rate that is generally the M.A.R.R.

The criterion for this method is that as long as the *net present worth* (i.e., present equivalent of cash inflows minus cash outflows) is ≥0, the project is economically justified; otherwise, it is not justified.

If only outflows (disbursements) are considered, the method is characterized by negative-valued present worth amounts and can be descriptively expressed as *present worth–cost* (*P.W.-C.*).

EXAMPLE 5-2

Considering the same project as in Example 5-1, show whether it is justified using the P.W. method.

Solution		Present Worth
Annual revenue: $5,310(P/A, 10%, 5)		$20,125
Salvage value: $2,000(P/F, 10%, 5)		1,245
Investment	− $10,000	
Annual disbursements: $3,000(P/A, 10%, 5)	− 11,370	
TOTAL		− $21,370
NET P.W.		$ 0

Since net P.W. = $0, the project is once again shown to be barely justified. ▌

A special case of the present worth method, when the life or study period is infinite, is the *capitalized worth method*, which is demonstrated in Chapter 7.

the future worth method

The *future worth* (F.W.) *method* for economy studies is exactly comparable to the present worth method except that all cash inflows and outflows are compounded forward to a reference point in time called the *future*.

EXAMPLE 5-3

Considering the same project as in Example 5-1, show whether it is justified using the F.W. method.

Solution

	Future Worth
Annual revenue: $5,310(F/A, 10\%, 5)$	$32,420
Salvage value	2,000
Investment: $10,000(F/P, 10\%, 5)$	$-$16,105
Annual disbursements: $3,000(F/A, 10\%, 5)$	$\underline{-\ 18,315}$
TOTAL	$-$34,420
NET F.W.	$\ \ \ \ 0

Since the net F.W. = $0, the project is again, not surprisingly, shown to be barely justified. ∎

the internal rate of return method

The *internal rate of return* (I.R.R.) *method* is the most general and widely used rate of return method for making economy studies. It commonly is called by several other names, such as *investor's method, discounted cash flow method, receipts versus disbursements method,* and *profitability index.*

The calculation of the I.R.R. for a single project involves finding the interest rate at which the present worth of the cash inflow equals the present worth of the cash outflow. (*Note:* The customary use of present worth amounts to determine the I.R.R. could just as well be annual worths or future worths.) Expressed in general, the I.R.R. is the $i'\%$† at which

$$\sum_{k=0}^{N} R_k(P/F, i'\%, k) = \sum_{k=0}^{N} D_k(P/F, i'\%, k) \qquad (5\text{-}4)$$

where R_k = net receipts for kth year
D_k = net disbursements for kth year
N = project life or maximum number of years for study

Other ways to express the same idea are: the I.R.R. is the $i'\%$ at which the P.W. of cash inflow minus the P.W. of cash outflow equals 0; or the $i'\%$ at which the P.W. of the *net cash flow* equals 0. That is, the I.R.R. is the $i'\%$ at which

$$\sum_{k=0}^{N} R_k(P/F, i'\%, k) - \sum_{k=0}^{N} D_k(P/F, i'\%, k) = 0 \qquad (5\text{-}5)$$

The method of solving this equation normally involves trial-and-error calculation until the $i'\%$ is found or can be interpolated. Example 5-4 shows a typical solution using the common convention of "+" signs for cash inflows and "−" signs for cash outflows.

---

† i' is often used in place of i to mean the interest rate that is to be determined or calculated.

EXAMPLE 5-4

Considering the same project as in Example 5-1, determine whether it is justified using the I.R.R. method.

Solution

Expressing P.W. of net cash flow and setting it equal to zero yields

$$- \$10,000 + (\$5,310 - \$3,000)(P/A, i'\%, 5) + \$2,000(P/F, i'\%, 5) i' = ?$$

If we did not already know the answer, we would probably try a relatively low i', such as 5%, and a relatively high i', such as 25%.

at $i' = 5\%$: $- \$10,000 + \$2,310(4.3295) + \$2,000(0.7835) = + \$1,568$

at $i' = 25\%$: $- \$10,000 + \$2,310(2.6893) + \$2,000(0.3277) = - \$3,132$

Since we have both a positive and a negative P.W. of net cash flow, the answer is bracketed. Linear interpolation can be used to find an approximation of the unknown $i'\%$, as shown in Figure 5-1. The answer, $i'\%$, can be obtained graphically to be approximately 11.7%, the $i\%$ at which the net P.W. = $0.

Linear interpolation for the answer, $i'\%$, can be accomplished by using the similar triangles dashed in Figure 5-1.

$$\frac{25\% - 5\%}{\$1,568 - (- \$3,132)} = \frac{i'\% - 5\%}{\$1,568 - \$0}$$

or

$$i'\% = 5\% + \left(\frac{\$1,568}{\$1,568 - (- \$3,132)} \right)(25\% - 5\%)$$

$$= 5\% + 6.7\% = 11.7\% \blacksquare$$

FIGURE 5-1.   Use of linear interpolation to find the approximation of I.R.R. for Example 5-4.

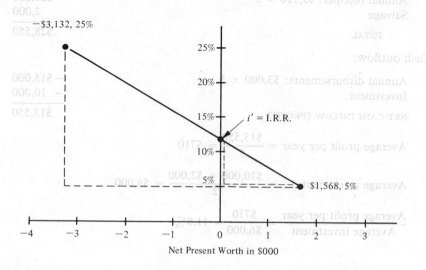

Net Present Worth in $000

The approximate solution above was merely to illustrate the trial-and-error process, together with linear interpolation. The error in this answer is due to nonlinearity of the net P.W. function and would be less if the range between interest rates used in the interpolation were smaller.

From the results of Examples 5-1 through 5-3, we happen to already know that the project is barely acceptable and that thus $i =$ M.A.R.R. $= 10\%$. We can confirm this by trying $i = 10\%$ in the net P.W. equation, as follows:

$$\text{at } i = 10\%: \quad -\$10,000 + (\$5,310 - \$3,000)(P/A, 10\%, 5)$$
$$+ \$2,000(P/F, 10\%, 5) = 0$$

An important assumption underlying, and hidden within, the I.R.R. method is that all recovered funds can be reinvested at an interest rate equal to the I.R.R. This is in contrast to the next two rate-of-return methods, which assume a specified rate for reinvestment of all recovered funds or all depreciation funds.

selecting trial rates of return when using the I.R.R. method

Users of the I.R.R. method are often perplexed as to how to select an initial and perhaps subsequent trial rate of return so as to reduce the number of trials required to obtain an acceptably accurate answer. The following is an intuitive approach.

A first trial rate of return can be obtained by finding the average annual profit without regard to timing as a percent of the average investment. For the project in Example 5-4, this can be done as follows:

Cash inflow:

Annual receipts: $5,310 × 5	$26,550
Salvage	2,000
TOTAL	$28,550

Cash outflow:

Annual disbursements: $3,000 × 5	− $15,000
Investment	− 10,000
NET CASH INFLOW (PROFIT)	$13,550

$$\text{Average profit per year} = \frac{\$13,550}{5} = \$710$$

$$\text{Average investment} = \frac{\$10,000 + \$2,000}{2} = \$6,000$$

$$\frac{\text{Average profit per year}}{\text{Average investment}} = \frac{\$710}{\$6,000} = 11.8\%$$

Thus one might start with a first trial rate of return of 12%. Substituting this into the net present worth equation for finding the I.R.R. gives a negative result (i.e., <0). It should be kept in mind that a lower interest rate will result in higher present worth factors for both single sums and uniform series; also, if the positive terms need to become larger so that the net present worth will be ≥ 0, the next trial rate should be lower.

It is recommended that the second trial rate be enough different from the first trial rate so that the answer will be bracketed between a positive and a negative net P.W. One can then interpolate or have a good basis for trying a third rate that will be close enough to the I.R.R. answer so that interpolation error then would be reasonably small.

the external rate of return method

The *external rate of return* (E.R.R.) *method* is not yet commonly used for economy studies, but it has two advantages over the I.R.R. method which will probably result in its being used increasingly in the future. These advantages are:

1. It usually can be solved for directly rather than by trial and error.
2. It is not subject to the possibility of multiple rates of return. (*Note*: The multiple rate of return problem with the I.R.R. method will be discussed in Appendix 5-B.)

The E.R.R. method involves the assumption that all recovered funds (net cash flows) can be reinvested at some specified rate of return (usually the M.A.R.R.) until the end of the life or study period for the project.

The calculation of E.R.R. for a single project involves merely finding the interest rate at which the future worth of the outflows [investment(s)] equals the future worth of the inflows. Expressed in general, the E.R.R. is the $i'\%$ at which

$$\sum_{k=0}^{N} D_k(P/F, e\%, k)(F/P, i'\%, N) = \sum_{k=0}^{N} R_k(F/P, e\%, N-k) \qquad (5\text{-}6)$$

where D_k = net outflow (excess of disbursements over receipts for kth year
R_k = net inflow (excess of receipts over disbursements) for kth year
N = project life or maximum number of years for study
e = external reinvestment rate

EXAMPLE 5-5

Considering the same project as in Example 5-1, determine whether it is justified using the E.R.R. method, assuming that funds can be reinvested at the M.A.R.R. = 10%.

Solution

$$\$10,000(F/P, i'\%, 5) = (\$5,310 - \$3,000)(F/A, 10\%, 5) + \$2,000$$
$$= \$2,310(6.105) + \$2,000$$
$$(F/P, i'\%, 5) = 1.61$$

From interest-table searching, one can find that the F/P factor for $N = 5$ is 1.61 for $i = 10\%$. Thus 10% is the E.R.R., and once again the project is shown to be marginally justified.

Another way of finding the E.R.R. $= i'\%$ is to solve directly in the algebraic equivalent of the last equation above. Thus

$$(F/P, i'\%, 5) = 1.61$$
$$(1 + i')^5 = 1.61$$
$$5 \ln (1 + i') = \ln 1.61$$
$$\ln (1 + i') = 0.0952$$
$$\ln^{-1} (1 + i') = 1.10$$
$$i = 0.10 = 10\% \quad \blacksquare$$

explicit reinvestment rate of return method

The *explicit reinvestment rate of return* (E.R.R.R.) *method* is a computationally easy means of calculating a rate of return when there is a single lump sum investment and uniform cash savings or returns at the end of each period throughout the life, N, of the investment project. The basic pattern for this rate of return method is as follows.

The calculation of the E.R.R.R. involves dividing a "net profit" amount by the initial investment, where the net profit is calculated using a depreciation charge based on the sinking fund method of depreciation. As will be explained in Chapter 9, the sinking fund depreciation charge or deposit is obtained by multiplying the depreciable investment by the sinking fund factor, $(A/F, i\%, N)$. The interest rate used for the sinking fund factor is the rate at which recovered depreciation funds are assumed to be reinvested, and this is usually the same as the M.A.R.R.

In general,

$$\text{E.R.R.R.} = \frac{R - D - (P - F)(A/F, e, N)}{P} \tag{5-7}$$

where R = uniform annual receipts or savings
$\quad D$ = uniform annual costs or disbursements
$\quad P$ = investment
$\quad F$ = salvage value
$\quad e$ = reinvestment rate

EXAMPLE 5-6

Considering the same project as in Example 5-1, determine whether it is justified using the E.R.R.R. method assuming that depreciation funds can be reinvested at the M.A.R.R. $= 10\%$.

Solution

Annual revenue		$5,310
Annual disbursements	$-\$3,000$	
Depreciation: $(\$10,000 - \$2,000)(A/F, 10\%, 5)$	$-1,310$	
TOTAL		$-4,310$
Net annual profit		$\$1,000$

$$\text{E.R.R.R.} = \frac{\text{net annual profit}}{\text{investment}} = \frac{\$1,000}{\$10,000} = 10\%$$

Thus once again the project is barely justified, because the E.R.R.R. just meets the M.A.R.R. ∎

summary comparison of economy study methods

The reader may well have wondered why six different methods for economy studies have been presented in this chapter when any one of these methods will give a valid answer for the reinvestment assumption inherent in that method. The best answer is that preferences differ between analysts and decision makers and also that often one method is easier to use than another method because of the particular cash flow patterns.

In general, equivalence methods such as A.W., P.W., and F.W. are easier to use computationally than rate of return methods. However, many decision makers prefer to think and analyze study results in terms of rates of return. Business enterprises generally adopt one, or at most two, analysis techniques for broad categories of alternatives to be analyzed.

The three equivalence methods are directly related so that one amount can be obtained from the other by the following relations:

$$\text{A.W.} = \text{P.W.}(A/P, i\%, N) = \text{F.W.}(A/F, i\%, N) \qquad (5\text{-}8)$$
$$\text{P.W.} = \text{A.W.}(P/A, i\%, N) = \text{F.W.}(P/F, i\%, N) \qquad (5\text{-}9)$$
$$\text{F.W.} = \text{A.W.}(F/A, i\%, N) = \text{P.W.}(F/P, i\%, N) \qquad (5\text{-}10)$$

These methods all involve the built-in assumption that funds can be reinvested at the $i\%$ which is normally the M.A.R.R. This same assumption is inherent in the E.R.R. method. The E.R.R.R. method uses the same assumption, except that the reinvestment applies only to recovered depreciation funds. The I.R.R. method, on the other hand, has the built-in assumption that all funds are reinvested at the particular I.R.R. rate computed for the project generating those funds.

In general, the numerical answer for a given project will be close for each of the rate of return methods. If the reinvestment rate is the same as the rate of return calculated, then I.R.R. = E.R.R. = E.R.R.R. This was true in Examples 5-4 through 5-6. This and other conditions relating answers by the three methods are summarized in the following table:

If Reinvestment Rate Is	Then E.R.R. and E.R.R.R. Are
= I.R.R.	= I.R.R.
> I.R.R.	> I.R.R.
< I.R.R.	< I.R.R.

an undesirable economy study method

There is an economy study method that is used to some extent, but it can produce misleading results, and it is definitely not recommended except as supplemental information in conjunction with a correct economy study method. It is desirable to understand what it involves so that we will not be misled when it is encountered. The *payout period method* (sometimes called *payback period method*) determines the number of years that will have to elapse in order for the invested capital to be recovered out of the net incoming cash flow. To be most meaningful, such a determination must be made on an after-tax basis. If revenue and cost estimates are uniform each year, the determination takes the following form:

$$\text{number of years for payout} = \frac{P - F}{R - D} \qquad (5\text{-}11)$$

Sometimes the salvage value is ignored, in which case the formula becomes

$$\text{number of years for payout} = \frac{P}{R - D} \qquad (5\text{-}12)$$

$R - D$ represents the net annual cash flow that can be applied toward the recovery of the invested capital.

If this method were applied to the investment project in Example 5-1, and income taxes are assumed negligible, the number of years for payout can be calculated to be $10,000/($5,310 − $3,000) = 4.33 years. This measure, by itself, does not indicate anything about the project desirability except the speed with which the investment will be recovered.

Obviously, the payout period method does not consider possible earnings from reinvested capital that is recovered during the payout period. It thus is not an exact expression. However, a more serious defect is that it does not take into consideration the economic life of the physical assets. Thus it is quite possible for one alternative that has a longer payout period than another to produce a higher rate of return on the invested capital. The use of the payout period for making investment decisions should be avoided except as a measure of how quickly invested capital will be recovered, which is an indicator of project risk.

economy studies of a new venture using various methods

The following is an example of an economy study of a new venture involving a single, initial investment and uniform revenue and cost data. It thus represents a very simple type of basic, long-term investment decision problem.

EXAMPLE 5-7

Mr. Brown has an opportunity to purchase an apartment house that has just been completed. It is located in the suburban area of a medium-size city that contains a fair amount of industrial plants. It also is within walking distance of a quite large university. The apartment house is in the process of being rented and now is more than 80% occupied, with prospects of being fully rented within a few weeks. The purchase price would be $85,000, of which $10,000 represents the value of the land. The building consists of 10 four-room apartments, a small apartment for a caretaker, and garage space for 11 automobiles. From a study of similar apartment buildings, Mr. Brown estimates that each apartment can be rented for $110 per month, with at least 95% average occupancy at all times. Heat and water are included in the rental. The operating costs are estimated to be as follows:

Caretaker	$175 per month, plus his apartment
Fuel	$400 per year
Water	$150 per year
Maintenance and repair	Equal to 1 month's rental on each rental unit per year
Taxes	$4 per $100 of assessed value; assessed value will be approximately 30% of cost of building and land
Insurance	0.5% of first cost of building, per year
Agent's commission	$2\frac{1}{2}$% of gross rental revenue

At present Mr. Brown's capital is invested in bonds that yield approximately 5% before income taxes and he feels that a project of this risk should earn at least 7% before income taxes. He estimates that the economic life of the apartment house will be at least 40 years and that the $10,000 for land will be the only salvage value at the end of that time.

The revenue each year can be computed as $10 \times \$110 \times 12 \times 0.95 = \$12,540$.

The out-of-pocket costs (disbursements) each year can be computed as:

Caretaker: $175 × 12	$2,100
Fuel	400
Water	150
Maintenance and repair: 10 × $110	1,100
Taxes: ($85,000/$100) × 0.3 × $4	1,020
Insurance: $75,000 × 0.005	375
Agent's commission: $12,550 × 0.025	314
TOTAL	$5,459

Hence the net cash inflow (revenue minus disbursements) each year is $12,540 − $5,459 = $7,081.

Show whether Mr. Brown should purchase the apartment house using the I.R.R. method.

Solution

The present worth can be written

$$- \$85,000 + \$7,081(P/A, i', 40) + \$10,000(P/F, i', 40) = 0$$

at $i' = 5\%$: $- \$85,000 + \$7,081(17.1591) + \$10,000(0.1420) \cong + \$38,000$

at $i' = 10\%$: $- \$85,000 + \$7,081(9.7791) + \$10,000(0.0221) \cong - \$15,500$

Since we are seeking the i' at which the present worth equation is zero, by linear interpolation we find that

$$i' = \text{I.R.R.} \cong 5\% + \left(\frac{\$38,000}{\$38,000 + \$15,500} \right)(10\% - 5\%) \cong 8.6\%$$

Since 8.6% is greater than 7%, the project is apparently worthy of investment. ∎

EXAMPLE 5-8

Analyze the same investment project as in Example 5-7 except use the E.R.R.R. method. Assume that he expects to reinvest accumulated depreciation funds in bonds earning 5%.

Solution

Annual revenue		$12,550
Annual costs:		
Out-of-pocket	$5,459	
Depreciation:		
$75,000(A/F, 5\%, 40) = \$75,000(0.0083)$	622	
TOTAL		$ 6,081
Net annual profit		$ 6,469

$$\text{E.R.R.R.} = \frac{\$6,469}{\$85,000} = 7.6\%$$

Since 7.6% is greater than 7%, he would probably invest in the apartment house. ∎

EXAMPLE 5-9

Analyze the same investment project as in Example 5-7 but use the annual worth method.

Solution

Annual revenue	$12,550
Annual costs:	
Out-of-pocket	$5,459
Depreciation†:	
$75,000(A/F, 7%, 40) = $75,000(0.0050)	375
Minimum required profit†: $85,000(0.07)	5,950
TOTAL	$11,784

Since the annual revenue of $12,550 exceeds the total annual costs of $11,784, the investment would be justified. ∎

EXAMPLE 5-10

Analyze the same investment project as in Example 5-7 except use the present worth method.

Solution

P.W. of inflow:	
Revenue: $12,550(P/A, 7%, 40) = $12,550(13.3317)	$167,500
Salvage of land: $10,000(P/F, 7%, 40) = $10,000(0.0668)	668
TOTAL	$168,168
P.W. of outflow:	
Investment	$ 85,000
Out-of-pocket: $5,459(P/A, 7%, 40) = $5,459(13.3317)	72,700
TOTAL	$157,700

Since $168,168 > $157,700, the project is once again shown to be worthy of investment. ∎

discussion of decision criteria to supplement economy study or studies

In order to arrive at a decision on the project for which economy studies were made in Examples 5-7 through 5-10, Mr. Brown would have to consider a number of factors, somewhat as follows. The first item undoubtedly should be the revenue estimate. Some questions to be answered would be as follows:

† Uses Equation 5-2 to determine the C.R. cost.

1. Is the assumed rental rate reasonable, particularly in relationship to similar apartments?
2. Is the assumed occupancy rate a reasonable one for this type of housing?
3. How will possible changes in economic conditions affect the rentals, both as to price and occupancy rate?

It is evident that it might be quite easy to obtain reliable answers to the first two of these questions. Answers to the third would be less precise. However, it would be possible to predict the general effect of changing economic conditions. We would also have to take into account the other types and classes of rental housing available in the area. In addition, there would be the fact that if changing economic conditions forced a reduction in the monthly rental charge, some of the expenses also probably would be reduced. By such an analysis the validity of the estimated revenue data can be determined in any economy study. In this particular case it should be possible to obtain revenue data that are quite accurate. In this study the amount of investment required is known exactly, since the finished and already rented building is to be purchased at a stated price.

Among the cost items, all can be determined without difficulty, and with some accuracy, except those for depreciation and maintenance. Obviously, the important factor in connection with depreciation is whether the assumed write-off period is reasonable. From the viewpoint of physical depreciation the assumed life of 40 years is very conservative if the building is well built. With respect to functional depreciation there would appear to be little doubt that people will still be living in apartments for 40 years. Inasmuch as the building is located near a fairly large city, and also near a large university, there would be little doubt as to the need for apartments continuing indefinitely. Thus, although 40 years is a rather long period of time, with respect to the economic life of the apartment building in question it appears to be a quite conservative write-off period. Obviously, considerable uncertainties exist with respect to any such period, but it must also be remembered that many or most of these long-range uncertainties would also apply to any use to which capital might be put. In evaluating this type of risk we must, in general, be concerned only with those risks involved in the project being considered that do not exist in any alternative investment opportunity.

Maintenance and repair costs are difficult to estimate unless historical data are available from similar projects. Such data are readily available for apartment buildings. In the absence of such data, it is well to remember that estimates of these costs are frequently too low and that they usually increase as time passes. This may be of considerable importance when long write-off periods are involved.

It thus appears that in this case all the revenue and cost data can be determined with accuracy. The primary element of risk is due to the rather long, but not excessive, write-off period. There is also the added factor that investment in the apartment would provide a hedge against inflation that does not exist in the bonds that Mr. Brown now owns. If he feels that the data have been determined carefully, he probably would make the investment.

an example with important intangibles

In many economy studies the final decision is determined primarily by the intangibles that exist in the situation. The following case illustrates this type of problem.

EXAMPLE 5-11

The management of a hotel in a city in an inland valley of California was considering the installation of an air-cooling system for all the rooms. This hotel had 150 guest rooms and was considered to be one of three first-class hotels in the city. One of the other hotels had installed such a system the previous year. A bid of $18,000 had been received for installing the system. It was estimated that the cooling system would have to be operated at full capacity for 14 weeks of each year, and at reduced capacity for 6 weeks. Operation costs at full capacity would be $17 per day and, at reduced capacity, $12 per day. The annual maintenance expense was estimated to be $125, and taxes and insurance to be $200. The life of the installation was estimated to be not less than 15 years.

If the cooling system were installed, it was estimated that 90% of the rooms would be rented during the 20 weeks of the hot weather, whereas only 80% of the rooms could be rented if no air cooling was available. These estimates were based on results of similar hotels in other cities. The existing average profit on each room that was rented was $2 per day. The owners had capital invested in stocks, paying about 6% before taxes, which could be used to finance the project. Determine if the investment should be made using the E.R.R.R. method.

Solution

Annual income: $150 \times (0.9 - 0.8) \times \$2.00 \times 7 \times 20$		$4,200
Annual expenses:		
Depreciation: $\$18,000(A/F, 6\%, 15)$	$ 774	
Out-of-pocket costs:		
Operation: $\$17 \times 7 \times 14$	1,666	
$\$12 \times 7 \times 6$	504	
Maintenance	125	
Taxes and insurance	200	
TOTAL		$3,269
Profit		$ 931

$$\text{E.R.R.R.} = \frac{\$931}{\$18,000} = 5.5\%$$

This predicted rate of return was less than that currently being received on the capital. It thus might appear that the investment should not be made. However, there was an important intangible in the situation. It was feared that unless the cooling system was installed, customers of the hotel who came to the city at various times during the year might not patronize it during the hot season, thus establishing a habit that they would follow during cold weather. The hotel might be thought of as a second-class hotel and lose a considerable amount of the patronage it now enjoyed. It was believed that the income and expense estimates were accurate and the write-off period conservative. The owners decided to invest in the cooling system, even though the predicted return as shown by the economy study was less than they were then receiving from their capital. The intangibles pointed toward greater loss if the investment were not made. ∎

an example of a proposed investment to reduce costs

EXAMPLE 5-12

A manufacturer of jewelry is contemplating the installation of a system that will recover a larger portion of the fine particles of gold and platinum that result from the various manufacturing operations. At the present time a little over $5,000 worth of these metals is being lost per year, and it is anticipated that, because of the growth of the company, this amount will increase by $500 each year for the next 10 years. The proposed system, involving a network of exhaust ducts and separators, will recover at least two-thirds of the gold and platinum that otherwise would be lost. The complete installation would cost $14,000. The best estimates for the operating costs of the system, obtained from operations of similar systems, are $1,000 per year for operating expense, $180 per year for maintenance and repairs, and 2% of the first cost annually for taxes and insurance. The company would require the investment to be written off within 10 years. The average earnings of the company, before taxes, have been about 15%. Should the recovery system be installed?

Solution

Such an investment is made to reduce some of the operating expenses, in this case the cost of the material used. Thus the saving (income) to be obtained by making an investment is almost entirely within the control of the investors. The company knows exactly what expenses have been. If the efficiency of the proposed equipment is known, the only factors that should affect the saving are the variation of production, operation, and maintenance expenses of the proposed equipment and depreciation expense. In most cases of this type these items are known or may be predicted quite accurately. The company would have a good idea of how its volume might vary. Operation and maintenance expenses can usually be estimated accurately, especially if historical data are available on the proposed equipment. Depreciation costs can be placed on the safe side by using a write-off period shorter than actual physical life.

Using the data given, we find that the cash flow situation and the internal rate of return calculation would be as shown in Table 5-2. In deciding whether or not the calculated internal rate of return of 16.8% is sufficient to justify investment, each factor that might contribute to the risk must be examined so that a measure of the total risk may be obtained. In this case it appears that the factors are quite well controlled or known. There is little reason to believe that much more risk would be involved ... would be normal operations of the company. Thus the company ... as a basis of comparison. If the company is sound and its business quite stable, a rate of 16.8% should be satisfactory, inasmuch as not average capital investor in the company is earning only 15%. ∎

It may be seen that, when capital is invested in a going concern in order to bring about reduction in costs, the risk is usually easier to determine and is often much less than when ... are involved. As a result, the rate of return required for such investments is often low.

investment where income is unknown

Decisions to invest capital to lower capital requirements are made when it is impossible to know ... where the rate ... income cases are not particularly necessary to know what the income will be. This often occurs when public or governmental improvements are made. The returns are often in nonmonetary form, yielding convenience and satisfaction to the public. When companies spend large sums of money to create customers or employee goodwill, a precise measurement of the return ... although investments must often be made.

Probably the only rule that can be established for such cases is that generally no more should be invested than is required to give a satisfactory result. Obviously, this rule cannot be followed willy-nilly. When public roads are built, for example, it may be possible to ... are certain minimum to assure permanence and low maintenance cost. In such case the rule of least annual cost becomes important. True income (or at least a large part of it) resulting from many projects that at first thought do not appear subject to the type of analysis used ... when the undertakings are analyzed fully. Whenever this can be done, it should not be neglected. Greater efficiency in the use of capital is bound to result. Some of these cases will be discussed in later chapters.

an example of an investment to reduce risk

There are cases where we consider making an investment in order to reduce, or eliminate, future costs that may result from various types of risks, usually associated with such events as fire or flood. The basic economic aspects can be illustrated by a simple example. Assume that a man agrees that once each year, for the next 20 years, he will cut a deck of 52 playing cards and that if he cuts an ace, he will pay $500 to a second individual. How much could he afford to pay, in one immediate sum, to avoid this obligation, and, further, assuming that capital is worth 9% to him?

table 5-2
tabular determination of internal rate of return for proposed gold-recovery system in example 5-12[a]

Year End N	Investment	Recovery	Costs	Net Cash Flow	$(P/F, 15\%, N)$	Net Present Worth at $i = 15\%$	$(P/F, 20\%, N)$	Net Present Worth at $i = 20\%$
0	− $14,000	—		− $14,000	1.000	− $14,000	1.000	− $14,000
1		+ $3,330	− $1,460	+ 1,870	0.870	+ 1,630	0.833	+ 1,560
2		+ 3,670	− 1,460	+ 2,210	0.756	+ 1,670	0.694	+ 1,530
3		+ 4,000	− 1,460	+ 2,540	0.658	+ 1,670	0.579	+ 1,470
4		+ 4,330	− 1,460	+ 2,870	0.572	+ 1,640	0.482	+ 1,380
5		+ 4,670	− 1,460	+ 3,210	0.497	+ 1,600	0.402	+ 1,290
6		+ 5,000	− 1,460	+ 3,540	0.432	+ 1,530	0.335	+ 1,190
7		+ 5,330	− 1,460	+ 3,870	0.376	+ 1,450	0.279	+ 1,080
8		+ 5,670	− 1,460	+ 4,210	0.327	+ 1,380	0.233	+ 980
9		+ 6,000	− 1,460	+ 4,540	0.284	+ 1,290	0.194	+ 880
10		+ 6,330	− 1,460	+ 4,870	0.247	+ 1,200	0.162	+ 790
			TOTAL	+ $19,730		+ $ 1,060		− $ 1,850

[a] I.R.R. = 15% + [$1,060/(1,060 + 1,850)] × (20% − 15%) = 16.8%

129

would be involved than is present in all the normal operations of the company. Thus the company has its own experience to use as a basis of comparison. If the company is sound and its business quite stable, a return of 16.8% should be satisfactory, inasmuch as the average capital invested in the company is earning only 15%. ▌

It may be seen that when capital is invested in a going concern in order to bring about reduction in costs, the risk is usually easier to determine and is often much less than when entirely new enterprises are involved. As a result, the rate of return required for such investments is often lower.

investment where income is unknown

Decisions to invest or not to invest capital often must be made when it is impossible to know or evaluate the return. In some cases it is not particularly necessary to know what the income will be. This often occurs when public or governmental improvements are made. The returns are often in nonmonetary form, yielding convenience and satisfaction to the public. When companies spend large sums of money to create customer or employee goodwill, a precise measurement of the return is impossible. Yet such investments must often be made.

Probably the only rule that can be established for such cases is that generally no more should be invested than is required to give a satisfactory result. Obviously, this rule cannot be followed rigidly. When public roads are built, for example, it may be advisable to spend more than a bare minimum to assure permanence and low maintenance cost. In such cases the rule of least annual cost becomes important. The income (or at least a large part of it) resulting from many projects that at first thought do not appear subject to the type of analysis under discussion can be measured if the undertakings are analyzed fully. Whenever this can be done, it should not be neglected. Greater efficiency in the use of capital is bound to result. Some of these cases will be discussed in later chapters.

an example of an investment to reduce risk

There are cases where we consider making an investment in order to reduce, or eliminate, future costs that may result from various types of risks, usually associated with such events as fires or floods. The basic economic aspects can be illustrated by a simple example. Assume that a man agrees that once each year, for the next 20 years, he will cut a deck of 52 playing cards and that if he cuts an ace he will pay $500 to a second individual. How much could he afford to pay, in one immediate sum, to avoid this obligation and risk, assuming that capital is worth 6% to him?

With four aces in a deck of 52 cards it is apparent that, on the average, an ace will be cut four times out of 52 attempts. In other words, it is probable that an ace will be cut in $\frac{1}{13}$ of the attempts. Therefore, the expected cost† to him in any one year is

$$\frac{1}{13} \times \$500.00 = \$38.46\ddagger$$

With money worth 6%, the present worth of 20 annual payments of $38.46 would be

$$\$38.46 \times (P/A, 6\%, 20) = \$38.46 \times 11.4699 = \$441.13$$

Thus, if the probability of occurrence of an event as well as the cost of that event when it occurs is known, a single or annual amount that we can afford to pay to avoid the liability resulting from its occurrence can be calculated.

The foregoing type of calculation forms the basis for all insurance. The annual premium that is paid for insurance represents the average annual loss that might occur, except that it also must include an extra charge to pay the overhead and profit costs of the insurance company.

An application of the economy study aspects of such a situation is illustrated in the following problem.

EXAMPLE 5-13

A company has a warehouse that has no automatic sprinkler system. Annual fire insurance premiums have been $1,200. If a sprinkler system costing $8,000 is installed, the insurance premium will be reduced by 50%. The system is estimated to have a life of 30 years with the annual costs for taxes and maintenance amounting to $75. Capital is worth 8%.

The company engineer estimates that the actual loss resulting from any fire, including such factors as lost orders and personnel required to arrange for repairs, would be at least 50% greater than the physical loss covered by insurance. It is assumed that the reduction in insurance premium is a good measure of the difference in the probable annual physical loss without and with a sprinkler system.

Solution

With the premium reduction being equal to $600 per year, the probable saving in actual loss due to fire is estimated to be $600 × 1.5 = $900. The costs would be

Capital recovery: $8,000($A/P$, 8%, 30) = $8,000(0.0888)	$710
Taxes and maintenance	75
TOTAL ANNUAL COST	$785

Thus the probable annual savings of $900 exceeds the annual cost of $785, so from a long-range viewpoint, the sprinkler system would be a good investment. ∎

---

† A more complete explanation of expected cost or value is contained in Chapter 6.

‡ These calculations assume that average frequency would apply to an individual. Obviously, the actual frequency might vary considerably from this for a single individual with such a limited number of attempts. If a large number of persons made the same agreement, the amount of $38.46 would be the true probable annual cost. This calculation also does not consider the degree of probability of the player's death during the 20-year period.

problems

5-1. What is the capital recovery cost of a depreciable item, stated in your own words? What is the C.R. cost of an asset that does not lose value over its lifetime (i.e., $P = F$)?

5-2. A small plant has four space heaters that were purchased 3 years ago for $1,300 each. They will last 10 more years, and will have no salvage value at that time. An expansion to the plant can be heated by two additional space heaters that will last 10 years, have no salvage value, and cost $1,800 each.

An alternative method of heating is to install a central heating system that will cost $6,400, last 10 years, and have a salvage value of $4,000. If central heating is installed, the present four space heaters can be sold for $200 each.

Compare the equivalent uniform annual cost of heating with space heaters and with central heating. Assume that fuel costs will be the same for either system and that the minimum attractive rate of return for the company is 10%. What would you recommend?

5-3. Machines A and B are to be compared on the basis of annual cost when the minimum attractive rate of return is 12%. Pertinent cost data are as follows:

	Machine A	Machine B
First cost	$10,000	$13,000
Useful life	10 years	15 years
Salvage value	$1,000	$3,000
Annual operating costs	$500	$100
Overhaul—end of 5th year	$300	$200
Overhaul—end of 10th year	None	$550

5-4. In Problem 5-3, determine the C.R. cost of machine A by all three formulas presented in the text.

5-5. A certain service can be performed satisfactorily either by process R or process S. Process R has a first cost of $8,000, an estimated service life of 10 years, no salvage value, and annual net receipts (revenues − disbursements) of $2,400. The corresponding figures for process S are $18,000, 22 years, salvage value equal to 20% of first cost, and $4,000. Assuming a minimum attractive rate of return of 18% before income taxes, find the annual worth of each process and specify which you would recommend.

5-6. A manufacturing company is considering purchasing a 10-horsepower (hp) electric motor which it estimates will run an average of 6 hours per day for 250 days per year. Past experience indicates that (1) its annual cost for taxes and insurance averages 2.5% of first cost, (2) it must make 10% on invested capital before income tax considerations, and (3) it must recover capital invested in machinery within 5 years. Two motors are offered to the company. Motor A costs $340 and has a guaranteed efficiency of 85% at the indicated operating load. Motor B costs $290 and has a guaranteed efficiency of 80% at the same operating load. Electric energy costs the company 2.3 cents per kilowatt-hour (kWh), and 1 hp = 0.746 kW.

Calculate the annual cost of each motor and indicate which motor should be purchased.

5-7. Compare the equivalent uniform annual costs of the following two electric motors if the minimum acceptable rate of return (M.A.R.R.) is 10%. The *service requirement* for the motor will be exactly 5 years, because the government contract for which the motor is needed will continue only for the next 5 years. What assumptions did you have to make in your analysis?

	Circle D Motor	Qwik Rotor Motor
First cost	$4,000	$7,000
Life of motor	3 years	7 years
Annual expenses	$800	$900
Salvage value	$0	$1,000

5-8. In the design of certain industrial facilities, the following alternatives are under consideration:

	Alternative 1	Alternative 2	Alternative 3
First cost	− $28,000	− $18,000	− $24,000
Net cash flow/year	+ $5,500	+ $2,400	+ $4,800
Salvage value	+ $1,500	0	+ $500
Service life	10 years	10 years	10 years

Assume that the interst rate (M.A.R.R.) is 15% and use the present worth method to choose the best of these three alternatives.

5-9. A house is for sale at $35,000, to be financed as follows: 10% down payment ($3,500), $1,000 closing costs (paid by the buyer upon occupancy), and a loan on the balance of $31,500 at 8% to be repaid over the next 25 years. Assume that repayments of the loan are payable once each year. Each equal end-of-year payment includes interest on the unpaid balance plus some equity.

The same house can be purchased under an alternative plan which calls for a $14,000 down payment, $200 closing costs, and a loan on the balance of $21,000 at 6% to be repaid in 25 years.

The resale value of the house and annual property taxes *do not* vary with the financing plan. If you had $15,000 cash, which method of financing would you select? Consider your minimum attractive rate of return to be 7% (a certificate of deposit in a savings and loan association) and ignore the effects of income taxes.

5-10. Compare the following two piping systems using the present worth method if M.A.R.R. = 12%.

	Black Iron	Copper
First cost	$2,000	$5,000
Life	2 years	5 years
Salvage value	0	$1,000
Annual expenses	$600	$300

What assumptions did you have to make in comparing present worths over a 10-year period?

5-11. Given: The Highridge Water District needs an additional supply of water from Steep Creek. The engineer has selected two plans for comparison.

   (a) *Gravity plan:* Divert water at a point 10 miles up Steep Creek and carry it through a pipeline by gravity to the district.

   (b) *Pumping plan:* Divert water at a point near the district and pump it through 2 miles of pipeline to the district. The pumping plant can be built in two stages, with one-half capacity installed initially and the other one-half 10 years later.

   All costs are to be repaid within 40 years, with interest at 5%. Salvage values can be ignored. During the first 10 years, the average use of water will be less than the average during the remaining 30 years. Costs are as follows:

	Gravity	Pumping
Initial investment	$2,800,000	$1,400,000
Additional investment in 10th year	None	$200,000
Operation, maintenance, and replacements	$10,000/year	$25,000/year
Power cost:		
Average first 10 years	None	$50,000/year
Average next 30 years	None	$100,000/year

   Required; Select the more economical plan for a 40-year period on the basis of present worth.

5-12. (a) Compare the two alternatives below by using the present worth method and a minimum attractive rate of return of 15%. Make the assumption that assets can be replaced with identical assets in terms of their performance and cost.

	Alternative A	Alternative B
First cost	$10,000	$15,000
Expected life	5 years	10 years
Salvage value	− $500	$4,000
Annual receipts	$8,000	$9,000
Annual disbursements	$4,000	$5,000

   (b) Determine the I.R.R. for the extra investment required to implement alternative B. (*Hint:* Find cash flow differences between B and A over a 10-year period, then find the interest rate at which the present worth of this difference is zero.)

5-13. The following estimates of *cost* apply to equipment alternatives A and B. The minimum attractive rate of return required is 20%.

	A	B
First cost	$52,000	$25,000
Operating costs	$5,000 at end of year 1 and increasing by $600 per year thereafter	$10,000 at end of year 1 and increasing by $900 per year thereafter
Overhaul costs	$5,000 every 5 years	None required
Life	20 years	10 years
Salvage value at end of life	$15,000 *if just* overhauled	Negligible

(a) Compare these alternatives by use of the present worth method.

(b) Compare the alternatives by use of the annual worth method.

5-14. Two types of small electric motors are being considered for a certain industrial application.

(a) Compute and compare their present worths when the M.A.R.R. is 20%.

	Motor R	Motor S
First cost	$700	$1,600
Expected life	5 years	15 years
Salvage value	$100	$300
Annual maintenance	$100	$50
Annual power cost	$95	$154

(b) Now compute and compare the future worths of the two motors at the end of a 15-year service period.

5-15. The PDQ Company, which produces 1,000,000 units per year of product X, finds that 8% of them are returned for correction of defects. These repairs, made by retailer dealers, cost the company an average of $0.40 each. The head of the quality control department believes that the major causes of the difficulty are poor quality of certain materials, poor workmanship, and insufficient attention to quality control. In addition, he believes that some of the old equipment should be junked and replaced by more modern equipment, at a cost of $75,000. In addition to replacing this equipment, which he proposes to replace every 5 years, he proposes to upgrade some of the materials at an annual cost of $20,000, increase some of the wage rates at an annual cost of $15,000, and institute an improved quality control program at an annual cost of $1,400. He feels confident that this total program will reduce the defects to not over 3%.

If the company wants to receive at least 15%, before taxes, on its invested capital, and property taxes and insurance will cost 2% of first cost per year on fixed assets such as equipment, calculate the future worth of alternatives identified above and recommend which one should be selected.

5-16. A major portion of the products of a manufacturer, located in Cleveland, Ohio, is sold to aircraft companies in the Los Angeles area. It has shipped the product by air freight at a total annual cost of $40,000. It finds that if it would build a small warehouse in the Los Angeles area, on land that it can lease for 10 years at an annual rental of $2,000, it would be able to supply its customers satisfactorily by shipping the product to the warehouse by truck, at an annual cost of $20,000. A small, prefabricated warehouse can be built at a cost of $20,000. The use of the warehouse would require maintaining additional inventory valued at $50,000.

One man would have to be employed at the warehouse at an annual cost of $8,000. His functions at the Cleveland plant have required only the half-time services of a similar worker. This man's released time could be utilized in other, nonrelated work. Taxes and insurance on the warehouse and inventory would be 3% of their value per year, based on the first cost of the warehouse.

It is believed that the warehouse will have a salvage value of 25% of its initial cost if the land lease cannot be renewed at the end of 10 years. The company believes that the volume of its business will continue to increase slowly, and that the technological obsolescence in the field also will increase rather rapidly. If capital used by the company typically earns 15% before taxes, calculate the future equivalent worth of alternatives identified above and recommend which one to select.

5-17. Find the internal rate of return (I.R.R.) in each of these situations:

(a)

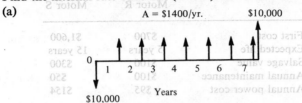

(b)

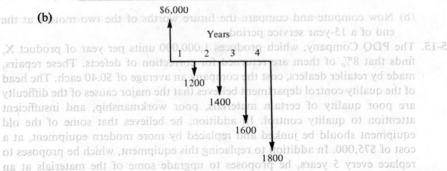

(c) You purchase a car for $4,200. After you make a $1,000 down payment on the car, the salesperson looks in her *Interest Calculations Made Simple* handbook and announces: "The monthly payments will be $160 for the next 24 months and the first payment is due 1 month from now." (Draw a cash flow diagram.)

(d)

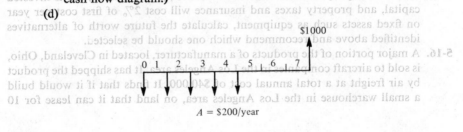

(e)

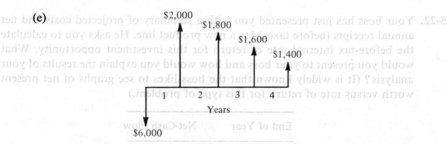

$2,000
$1,800
$1,600
$1,400

Years

$6,000

5-18. Rework part (c) and (e) of Problem 5-17 by using the E.R.R. method when e = M.A.R.R. = 8%.

5-19. A small company purchased now for $23,000 will lose $1,200 each year for the first 4 years. An additional $8,000 invested in the company during the fourth year will result in a profit of $4,000 each year from the fifth year through the fifteenth year. At the end of 15 years the company can be sold for $33,000.
(a) Determine the internal rate of return (I.R.R.).
(b) Determine the E.R.R. when e = 8%.
(c) Calculate the present worth if M.A.R.R. = 8%.

5-20. Which of these machines would you recommend? The minimum attractive rate of return is 8%.

	Machine A	Machine B
Initial cost	−$11,500	−$8,000
Annual receipts–disbursements	+$2,500	+$1,800
Life	6 years	6 years

(a) Solve by use of the I.R.R. method.
(b) Solve with the E.R.R. method when recovered depreciation capital can earn 6%.

5-21. Two types of electrical transformers are being considered for a remote location. This service requirement is expected to last indefinitely.

The transformer produced by Stepdown, Inc., will cost $3,500, has a life of 6 years, salvage value of $0, and annual operating and maintenance costs totaling $2,000 during the first year. The annual operating and repair costs are expected to be $300 higher in each succeeding year, reaching $3,500 for the sixth year.

The second transformer, manufactured by the Housingwest Company, will have an initial cost of $8,000, with an additional outlay of $4,000 at the end of the fifth year. This system is expected to have a life of 18 years with a $1,000 salvage value at the end of its life. Annual operating and repair costs are estimated to total $1,000 for each of the first 5 years, and then to increase by $100 each year through the end of year 18.
(a) Determine the prospective internal rate of return (I.R.R.) on the extra investment in the Housingwest transformer. What decision should be made?
(b) Calculate the annual worth of each transformer when M.A.R.R. = 6%. What decision should be made?

5-22. Your boss has just presented you with a summary of projected costs and net annual receipts (before taxes) for a new product line. He asks you to calculate the before-tax internal rate of return for this investment opportunity. What would you present to your boss and how would you explain the results of your analysis? (It is widely known that the boss likes to see graphs of net present worth versus rate of return for this type of problem.)

End of Year	Net Cash Flow
0	− $450,000
1	− 42,500
2	+ 92,800
3	+ 386,000
4	+ 614,600
5	− 202,200

5-23. Rework Problem 5-22 with the E.R.R. method and an external reinvestment rate of 12%. Why are answers to the two problems different?

5-24. List the advantages and disadvantages of each of the six basic methods for performing engineering economy studies.

5-25. A piece of equipment requires an investment of $10,000 and has a $1,000 salvage value at the end of its expected 9-year life. Savings attributable to the equipment will be $1,500 in the first year and will increase by $500 each year thereafter.

(a) Calculate the payout period for the equipment.

(b) Calculate the I.R.R. for the equipment.

5-26. List the major disadvantages of the payout method for making economic comparisons. In what situations might the payout method be appropriate?

5-27. Because it must have continuous electric power, a large hospital, which has its own power plant, has stand-by hookup with the local power company so it can obtain power in case of a failure in its own plant. It must pay a flat charge of $1,000 per month for this arrangement and, in addition, it purchases about $900 worth of energy each year when its plant is shut down for overhaul.

It has been proposed that the hospital should install a gas turbine and generator unit for emergency use, at a cost of $100,000. Because there would be very little use of this equipment, it is estimated that the annual maintenance cost would not exceed $700. It is estimated that the annual insurance cost would be $1,000, and that there would be an annual saving of $300 per year in power generation costs during the overhaul period of the main power unit. Because the hospital is a nonprofit organization, it pays no property taxes. The management is advised that because of the low use factor for this equipment, a 20-year estimated life would be reasonable. If borrowed money can be obtained for 7% interest, what recommendation would you make? Use the E.R.R.R. method.

5-28. The Anirup Food Processing Company is presently using an outdated method for filling 25-pound sacks of dry dog food. To compensate for weighing inaccuracies inherent to this packaging method, the process engineer at the plant has estimated that each sack is overfilled by $\frac{1}{4}$ pound on the average. A modern method of packaging is now available that would eliminate overfilling (and underfilling). The production quota for the plant is 300,000 sacks per year for the next 6 years, and a pound of dog food costs this plant $0.10 to produce. The present system has no salvage value and will last another 6 years, and the modern method has an estimated life of 4 years with a salvage value equal to 10% of its purchase price. It is also estimated that the present packaging operation will cost $2,100 per year more to maintain than the modern method. If the minimum attractive rate of return is 12% for this company, what amount could be justified for the purchase of the modern packaging method?

5-29. A firm is considering the installation of automatic data processing equipment to handle some of its accounting, payroll, and billing operations. Machines suitable for the firm's requirements may be purchased for $80,000 or leased at an annual rental of $12,000. The purchased equipment is expected to have a salvage value of $30,000 at the end of a 10-year service life. Maintenance costs are included in the rental of the leased equipment, but are estimated to run $800 annually if the equipment is purchased. Labor and operating costs for these functions will be cut from the present $40,000 per year to $20,000 with the new equipment. If the firm requires a return of 8% on investments, what would you suggest?

5-30. In medieval Europe, usury was an unpardonable sin. However, loans could be made if they were to benefit the public. For these purposes a nominal interest rate of 12% was sanctioned by public officials.

It is recorded that a Silas Smythe borrowed $90 from a local money merchant and agreed to repay the debt under the following conditions: $25 was repaid to the lender at the end of each of 4 years thereafter; and $20 was to be repaid at the end of the fifth year. This business venture was in the interest of the public.

Assuming that interest was compounded semiannually in medieval Europe, was the money merchant guilty of usury?

5-31. In order to enter the market to produce a new toy for children, a manufacturer will have to make an immediate investment of $60,000 and additional investments of $5,000 at the end of 1 year and $3,000 more at the end of 2 years. Competing toys now are being produced by two large manufacturers. From a fairly extensive study of the market, it is believed that sufficient sales can be achieved to produce year-end, before-tax, net cash flows of:

Year	Cash Flow	Year	Cash Flow
1	− $10,000	6	+ $21,000
2	+ 5,000	7	+ 21,000
3	+ 5,000	8	+ 16,000
4	+ 20,000	9	+ 15,000
5	+ 21,000	10	+ 7,000

In addition, while it is believed that after 10 years the demand for the toy will no longer be sufficient to justify production, it is estimated that the physical assets would have a scrap value of about $8,000. If capital is worth not less than 10% before taxes, would you recommend undertaking the project?

5-32. A certain project has net receipts equaling $1,000 now, costs $5,000 at the end of the first year, and then earns $6,000 at the end of the second year.

    (a) Show that multiple rates of return exist for this problem when using the I.R.R. method ($i' = 100\%, 200\%$).

    (b) If an external reinvestment rate of 10% is available, what is the E.R.R. for this project?

5-33. The prospective exploration for oil in the outer continental shelf by a small independent drilling company has produced a rather curious pattern of cash flows, as follows:

End of Year	Net Cash Flow
0	− $520,000
1–10	+ 200,000
10	−1,500,000

The $1,500,000 expense at the end of year 10 will be incurred by the company in dismantling the drilling rig.

    (a) Over the 10-year period, plot present worth versus the interest rate (i) in an attempt to discover whether multiple rates of return exist.

    (b) Based on the projected net cash flows and results in part (a), what would you recommend regarding the pursuit of this project? Customarily, the company expects to earn at least 20% on invested capital before taxes.

appendix 5-A: aids for calculation of internal rate of return

The computations to determine the internal rate of return, I.R.R., can be rather laborious, particularly if the periodic cash flows do not follow some pattern to which tabled interest factors can be readily applied.

Figure 5-2 is a form that can be used to simplify the calculations. Net cash flows for each year are entered in the column headed "CASH FLOW (0% Int. Rate)." These are the present worths at 0%. Each of these cash flows is then multiplied by the factor in the adjacent subcolumn labeled "Factor" [which are $(P/F, i\%, N)$ factors] and the result entered in the next subcolumn headed "Present Worth." These are the present worths at a 10% interest rate. The calculations are repeated for the 20%, 40%, and 60% interest rates as needed and the columns added to obtain the total present worth of disbursements (A), and the total present worth of receipts (B), for each trial interest rate. At each interest rate the total present worth of expenditures is divided by the total present worth of receipts and the result entered in the space designated "Ratio A/B."

The rate of return answer sought is the interest rate at which A equals B, or at which "ratio A/B" equals unity. If one of the interest rates used does not result in the unity ratio, we hope that the unity ratio will be bracketed so

PROJECT _____   Analyst _____   Date _____

BEFORE

Year	Period	Disbursement (0% Int. Rate)	10% Factor	10% Present Worth	20% Factor	20% Present Worth	40% Factor	40% Present Worth	60% Factor	60% Present Worth
	5th Yr		1.464		2.073		3.842		6.560	
	4th		1.333		1.728		2.744		4.100	
1982	3rd	$60,000	1.210	$72,000	1.440	$86,400	1.960		2.560	
83	2nd	5,000	1.100	5,500	1.200	6,000	1.400		1.600	
84	1st	13,000	1.000	13,000	1.000	13,000	1.000		1.000	
	TOTALS (A)	78,000		90,500		105,400				

AFTER ZERO TIME

Cal. Year	Period	Receipt (0% Int. Rate)	10% Factor	10% Present Worth	20% Factor	20% Present Worth	40% Factor	40% Present Worth	60% Factor	60% Present Worth
85	1st Yr	$18,000	0.909	$16,400	0.833	$15,100	0.714		0.624	
86	2nd	22,000	.826	18,200	.694	15,300	.510		.300	
87	3rd	23,000	.751	17,400	.579	13,400	.364		.244	
88	4th	22,000	.683	15,100	.482	10,600	.260		.152	
89	5th	18,000	.621	11,200	.402	7,200	.186		.095	
90	6th	11,000	.565	6,200	.335	3,700	.133		.059	
91	7th	5,000	.513	2,560	.279	1,400	.095		.037	
92	8th	15,000	.466	7,000	.233	3,300	.067		.023	
	9th		.424		.194		.048		.015	
	10th		.385		.162		.035		.009	
	11th		.351		.135		.024		.006	
	12th		.318		.112		.018		.004	
	13th		.290		.094		.013		.002	
	14th		.263		.078		.009		.001	
	15th		.239		.065		.006			
	16th		.217		.054		.005			
	17th		.197		.045		.003			
	18th		.180		.038		.002			
	19th		.164		.031		.002			
	20th		.149		.026		.001			
	21st		.135		.022					
	22nd		.123		.018					
	23rd		.112		.015					
	24th		.102		.013					
	25th		.092		.011					
	TOTALS (B)	$134,000		$94,060		$70,000				
	RATIO A/B	0.58		0.96		1.51				

FIGURE 5-2. Tabular determination of the internal rate of return for a proposed new venture.

141

that the answer can be interpolated. To provide for ease of interpolation, a chart is provided in Figure 5-3.

To illustrate the use of these aids, a sample problem and the associated computations are shown in Table 5-3. For the sample problem, an initial investment of $60,000 is required, and an additional investment of $20,000 will be required at the end of the second year. The project would be terminated at the end of 10 years, at which time it is expected that a recovery of $20,000 will be obtained from the salvage value of the assets. The revenues and disbursements are estimated to be as shown for the various years.

The net cash flow amounts in the right-hand column of Table 5-3 can then be transferred to the "CASH FLOW (0% Int. Rate)" column of Figure 5-2. For computational simplicity, the "zero time" is taken to be the year of the last negative cash flow, which is the end of year 1984. Computations using Figure 5-2 result in ratios of present worth of disbursements to present worth of receipts (called "Ratio A/B") of 0.58 for 0% interest, 0.96 for 10% interest,

FIGURE 5-3. Rate of return interpolation chart with entries for the example project of Figure 5-2.

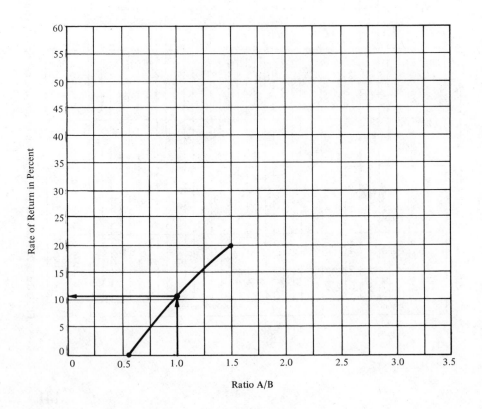

table 5-3

investments, revenues, and costs for a proposed new venture

End of Year	(A) Investment (−) or Recovery (+) of Capital	(B) Gross Revenue (+)	(C) Out-of-Pocket Costs (−)	(D) = (A) + (B) + (C) Net Cash Flow (+) or (−)
1982	− $60,000	—	—	− $60,000
1983	—	$20,000	$25,000	− 5,000
1984	− 20,000	35,000	25,000	− 13,000
1985	—	48,000	30,000	+ 18,000
1986	—	52,000	30,000	+ 22,000
1987	—	55,000	32,000	+ 23,000
1988	—	55,000	33,000	+ 22,000
1989	—	50,000	32,000	+ 18,000
1990	—	35,000	24,000	+ 11,000
1991	—	25,000	20,000	+ 5,000
1992	+ 20,000	10,000	15,000	+ 15,000

and 1.51 for 20% interest. Hence the answer interest rate (I.R.R.), at which the ratio is 1.00, is bracketed between 10% and 20%. Linear interpolation gives the answer:

$$\text{I.R.R.} = i' = 10\% + \left(\frac{1.00 - 0.96}{1.51 - 0.96}\right)(20\% - 10\%) = 10.7\%$$

Figure 5-3 is a chart for graphical interpolation. To obtain the answer for the example in Table 5-3 and Figure 5-2, note that the three plotted points form a curvilinear relationship. Such graphical interpolation, if performed on large, accurately scaled paper, can result in greater answer accuracy than mathematical linear interpolation.

It should be noted also that internal rates of return can be calculated very easily using computer programs. Such a program is given in Appendix 5-C.

appendix 5-B: the multiple rate of return problem with the I.R.R. method

Whenever the I.R.R. method is used and the cash flow reverses sign (from net outflow to net inflow or the opposite) more than once over the period of study, one should be alert to the possibility that either no answer (interest rate) or multiple answers may exist. Actually, the maximum number of possible rates of return for any given project is equal to the number of cash flow reversals over time. As an example, consider the following project for which the I.R.R. is desired.

EXAMPLE 5-B-1

Year	Net Cash Flow
0	+$ 500
1	− 1,000
2	0
3	+ 250
4	+ 250
5	+ 250

Solution

		P.W. at 35%		P.W. at 63%	
Year	Net Cash Flow	Factor	Amount	Factor	Amount
0	+$ 500	1.35	+$ 676	1.63	+$ 813
1	− 1,000	1.00	− 1,000	1.00	− 1,000
2	0				
3	+ 250	0.55	+ 137	0.38	+ 95
4	+ 250	0.41	+ 102	0.23	+ 57
5	+ 250	0.30	+ 75	0.14	+ 35
		NET P.W.	Σ = 0		Σ = 0

Thus the present worth of the net cash flows equals 0 for interest rates of 35% and 63%. Whenever multiple answers such as this exist, it is likely that neither is correct.

An effective way to overcome this difficulty and obtain a plausible answer is to manipulate cash flows as little as necessary so that there is only one reversal of the net cash flow over time. This can be done by using the minimum attractive rate of return to manipulate the funds, and then solving by the discounted cash flow method. For the example, if the minimum attractive rate of return is 10%, the +$500 at year 0 can be compounded to year 1 to be $500(F/P, 10%, 1) = +$550. This, added to the −$1,000 at year 1, equals −$450. The −$450, together with the remaining cash flows, which are all positive, now fit the condition that there be only one reversal in the net cash flow over time. The interest rate at which the present worth of the net cash flows equals 0 can now be shown to be 19%, per the following table:

		P.W. at 19%	
Timing	Net Cash Flow	Factor	Amount
1	−$450	1.00	−$450
3	+ 250	0.70	+ 177
4	+ 250	0.59	+ 150
5	+ 250	0.48	+ 123
		NET P.W.	Σ = 0

It should be noted that whenever a manipulation of net cash flows such as the above is done, the calculated rate of return will vary according to what cash flows are manipulated and at what interest rate. The less the manipulation and the closer the minimum rate of return to the calculated rate of return, the less the variation in the final calculated rate of return. ∎

Probably the most straightforward way to overcome this multiple rate of return problem in general is to use the E.R.R. method.

appendix 5-C: straight line depreciation plus average profit method

The straight line depreciation plus average profit method provides a convenient approximation of the capital recovery cost, that is, annual cost of depreciation plus interest. The only alleged advantage it provides is that it does not require use of compound interest tables. It provides a fairly close approximation of the exact capital recovery cost for low lives and interest rates, but the approximation error (understatement of exact costs) increases very rapidly with increasing lives and/or increasing interest rates.

This method involves merely adding the annual cost of depreciation using the straight line method (to be discussed in Chapter 9) to the profit or interest on the average capital to be invested. The average capital to be invested is taken as the average of the capital invested during the first and last years.

The annual cost of depreciation by the straight line method is

$$\frac{P - F}{N} \tag{5-C-1}$$

The capital invested at the beginning of the first year is P and at the beginning of the last year is the depreciation charge for 1 year plus the salvage value. The average of this is

$$\frac{P + [(P - F)/N] + F}{2} = (P - F)\left(\frac{N + 1}{2N}\right) + F \tag{5-C-2}$$

Multiplying $i\%$ times the average capital invested, as stated in Equation 5-C-2, results in the average profit formula:

$$(P - F)\left(\frac{N + 1}{2N}\right)(i) + F(i) \tag{5-C-3}$$

Adding Equations 5-C-1 and 5-C-3 gives the formula for straight line depreciation plus the average profit (approximate capital recovery cost):

$$\text{C.R.} \cong \frac{P - F}{N} + (P - F)\left(\frac{N + 1}{2N}\right)(i) + F(i) \tag{5-C-4}$$

chapter 6

risk, uncertainty, sensitivity, and multiple attributes

In the problems discussed in Chapter 5 specific assumptions were stated concerning applicable revenues and costs. Although the matter of risk and uncertainty was mentioned, procedures for studying their possible effects were purposely omitted so that attention could be concentrated on the basic methods of economic analysis and decision making. In effect, in Chapter 5 it was assumed that a high degree of confidence could be placed in all estimated values, which is sometimes called *assumed certainty*. Decisions made solely on the basis of this kind of analysis sometimes are called *decisions under certainty*. This is a rather misleading term, in that there rarely is a case in which estimated quantities can be assumed to be certain. In virtually all situations there is doubt as to the ultimate results that will be obtained from an investment. It is the purpose of this chapter to present and discuss several methods that are helpful in analyzing investment situations where such doubts exist.

the meaning of risk, uncertainty, and sensitivity

Risk frequently is defined as the variations of actual values from estimated or expected values† that are due to chance (random) causes. In economic analyses, the actual values or outcomes are known only after a project is undertaken or completed. *Uncertainty* frequently refers to variations in actual values that are due to errors in estimating, the inability to make accurate estimates because of insufficient information about the factor or the future, or the failure to consider all the factors. Although we may make a technical distinction between risk and uncertainty, both can cause results to vary from predictions, and from the economy study viewpoint there seldom is anything significant to be gained by attempting to treat them separately. Therefore, in the remainder of this book the terms will be used interchangeably.

† The statistical interpretation of *expected values* will be discussed later in this chapter.

146

In dealing with risk or uncertainty, it often is very helpful to determine to what degree changes in an estimate would affect an investment decision; i.e., how *sensitive* a given investment situation is to changes in a particular factor which is not known with certainty. If a particular factor can be varied over a wide range without causing much effect on the investment decision, the decision under consideration is said not to be sensitive to that particular factor. Conversely, if a small change in the relative magnitude of a factor will reverse an investment decision, the decision is highly sensitive to that factor. By knowing the most uncertain factors in a given investment situation, and by understanding how sensitive the venture is to these factors, we can make investment decisions on a more rational basis.

evaluation of risk

It is useful to consider the factors that may affect the risk involved in an investment, so that they may be related to the rate of return that must be met or exceeded for an investment to be justified. Similarly, they may be included as intangible (irreducible) considerations in reaching a final decision.

The factors that affect risk are many and varied. It would be almost impossible to list and discuss all of them. There are four, however, which nearly always are present.

The first factor, which is always present, is the *possible inaccuracy of the figures used in the study*. If exact information is available regarding the items of income and expense, the resulting accuracy should be good. If, on the other hand, little factual information is available, and nearly all the values have to be estimated, the accuracy may be high or low, depending upon the manner in which the estimated values are obtained. Are they sound scientific estimates or merely guesses?

The accuracy of the income figures is difficult to determine. We can usually be guided only by the method by which they were obtained. If they are based upon a considerable amount of past experience or have been determined by adequate market surveys, a fair degree of reliance may be placed on them. On the other hand, if they are merely the result of guesswork, with a considerable element of hope thrown in, they must of course be considered to contain a sizable element of uncertainty. Thus, to a great degree the amount of risk resulting from uncertainty in the income figures must be determined by exercising good judgment. Evaluation of this risk requires mature judgment and experience.

If the income is in the nature of a saving in existing operating costs, there should be less risk involved. It is usually easier to determine what the exact saving will be, since we have considerable experience and past history on which to base the estimates.

In most cases the income figures will contain more error than any other portion of a study, with the possible exception of depreciation costs. There should be no large error in estimates of capital required, except perhaps in the amount allowed above the actual cost of plant and equipment. If we can feel confident that the amount allowed for this purpose is on the high side, the resulting study is apt to be conservative. This element again requires careful study and experience, so we can be certain that we are not underestimating the amount of required capital.

Among the cost figures depreciation is usually the element that should be considered most carefully. It is seldom that the actual physical life of the plant or equipment can be used in determining depreciation. It must always be remembered that only through providing for depreciation is the capital recovered. It is generally better to err on the side of too short a depreciation period (estimated economic life) than to make the period too long. Of course, this process can be carried to an extreme. In determining the amount of risk that results from depreciation estimates, we should be guided by past experience of the company involved as well as that of other companies that have operated similar properties. The important thing is to be certain that the period selected is short enough to assure that the invested capital will be recovered.

The accuracy of the other cost elements will depend to a great degree upon how carefully the estimates have been prepared. If thorough investigations have been made and all of the items considered in detail, it is reasonable to assume that the estimates will not be in great error.

The second factor affecting risk is the *type of business involved*. Some lines of business are notoriously less stable than others. For example, most mining enterprises are more risky than large retail food stores. However, we cannot arbitrarily say that an investment in any retail food store always involves less risk than investment in mining property. Whenever capital is to be invested in an enterprise, the nature and past history of the business should be considered in deciding what risk is present. In this connection it becomes apparent that investment in an enterprise that is just being organized, and thus has no past history, is usually rather uncertain.

A third factor affecting risk is the *type of physical plant and equipment involved*. Some types of structures have rather definite economic lives and secondhand values. Little is known of the physical or economic lives of others, and they have almost no resale value. A good engine lathe generally can be used for many purposes in nearly any fabrication shop. Depreciation on such a lathe can be estimated fairly accurately. Quite different would be a special type of lathe that was built to do only one unusual job. Its value would be dependent almost entirely upon the demand for the special task that it can perform. Thus the type of physical property involved will have a direct bearing upon the accuracy of the depreciation figures and will thereby affect the risk. Where money is to be invested in specialized plant and equipment, this factor should be considered carefully.

The fourth, and very important, factor that must always be considered in evaluating risk is the *length of the assumed study period*. The conditions that have been assumed in regard to income and expense must exist throughout the study period in order for us to obtain the predicted profit. A long study period naturally decreases the probability of all the factors turning out as estimated. Therefore, a long write-off period, even though justified by the probable economic life of the equipment involved, always increases the risk in an investment.

breakeven analysis and uncertainty

Essentially all data employed in an economy study are uncertain simply because they represent estimates of the future. Breakeven analyses are useful when one must make a decision between alternatives which are highly sensitive to a variable or parameter, and that variable is difficult to estimate. Through breakeven analysis, one can solve for the value of that variable or parameter at which the conclusion is a standoff. That value of the variable is known as the *breakeven point*. (The use of breakeven points with respect to production and sales volumes was briefly discussed in Chapter 2.) If one can then estimate whether the actual outcome of that variable will be higher or lower than the breakeven point, the best alternative becomes apparent.

The following are examples of common variables for which breakeven analyses might provide useful insights into the decision problem:

1. *Revenue.* Solve for the annual revenue required to equal (breakeven with) the annual costs.
2. *Rate of return.* Solve for the rate of return at which the two alternatives are equally desirable.
3. *Salvage value.* Solve for future equipment resale value that would result in indifference as to timing of retirement.
4. *Equipment life.* Solve for the use time required for an alternative to be justified.
5. *Hours of operation.* Solve for the hours of utilization per year at which an alternative is justified or at which two alternatives are equally desirable.

The usual breakeven problem involving two alternatives can be most easily approached mathematically by equating the annual costs (or annual worths) of the two alternatives expressed as a function of that variable. Examples below will illustrate both mathematical and graphical solutions to typical breakeven problems.

EXAMPLE 6-1

Suppose that there are two alternative electric motors that provide 100-hp output. An Alpha motor can be purchased for $1,250 and has an efficiency of 74%, an estimated life of 10 years, and estimated maintenance costs of $50 per year. A Beta motor will cost $1,600 and has an efficiency of 92%, a life of 10 years, and annual maintenance costs of $25. Taxes and insurance costs on either motor will be $1\frac{1}{2}\%$ of the investment per year. If capital is worth 8%, how many hours per year would the motors have to be operated at full load for the annual costs to be equal? Assume negligible salvage values for both and that electricity costs $0.012 per kWh.

Solution by Mathematics

Note: 1 hp = 0.746 kW and input = output/efficiency.

If N = number of hours operation per year, for the Alpha motor, components of the annual cost, A.C.$_\alpha$, would be as follows:

Capital recovery cost (depreciation and minimum profit):
$$\$1,250 \times (A/P, 8\%, 10) = \$1,250 \times 0.149 = \$186$$
Operating cost for power:
$$(100 \times 0.746 \times N \times \$0.012)/0.74 = \$1.21N$$
Maintenance cost:
$$\$50$$
Taxes and insurance:
$$\$1,250 \times 0.015 = \$18.75$$

Similarly, for the Beta motor, components of the annual cost, A.C.$_\beta$, in the same order as above would be as follows:

Capital recovery cost (depreciation and minimum profit):
$$\$1,600 \times (A/P, 8\%, 10) = \$1,600 \times 0.149 = \$238$$
Operating cost for power:
$$(100 \times 0.746 \times N \times \$0.012)/0.92 = \$0.973N$$
Maintenance cost:
$$\$25$$
Taxes and insurance:
$$\$1,600 \times \$0.015 = \$24$$

For the breakeven point, A.C.$_\alpha$ = A.C.$_\beta$. Thus

$$\$186.00 + \$1.21N + \$50.00 + \$18.75 = \$238.00 + \$0.973N + \$25.00 + \$24.00$$
$$\$254.75 + \$1.21N = \$287.00 + \$0.973N$$
$$N = 152 \text{ hours } \blacksquare$$

Solution by Graphics

Figure 6-1 shows a curve for the total annual costs of each alternative motor as a function of the number of hours of operation per year. The constant annual costs (*Y*-intercepts) are $254.75 and $287.00 for Alpha and Beta, respectively, and the costs that vary directly with hours of operation per year (slope of lines) are $1.21 and $0.973 for Alpha and Beta, respectively. Of course, the breakeven point is the value of the independent variable at which the annual cost curves for the two alternatives intersect. $\blacksquare$

A graphical solution as shown in Figure 6-1 is particularly useful to give the analyst or decision maker a visual portrayal of the *difference* in the economic desirability for alternatives over a wide range of the variable (hours of operation) under consideration.

There are many situations in which the relationship between the dependent and independent variables is not continuous and cannot therefore be expressed readily in mathematical terms. In other cases the relationship may be complex enough that the time required to develop a mathematical formula would be so great that it would be uneconomical to solve the problem in this manner. In such cases a graphical solution may be used to determine the breakeven point. The following is a case where a graphical solution is used to advantage.

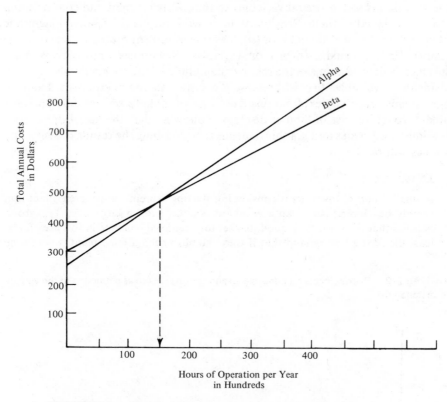

FIGURE 6-1.   Graphical solution of breakeven point for Example 6-1.

Products ordered from a wholesale drug company are collected in large wire-mesh baskets and taken to the shipping department for packing. Two methods of packing are used—one with heavy paper bags and the other with cartons. Figure 6-2 shows the packing costs for the two methods as a function of the volume of the merchandise. It is apparent that the relationship of volume and cost is not a simple one and that for shipments whose volume is less than approximately 3,500 cubic inches the bag method should be used, and that shipments whose volume is larger should be packed in cartons. In this instance, since the sides of the baskets are nearly vertical, the packers are able to determine which packing method to use and the correct size of bag or carton merely by checking the height of the merchandise in the basket against a colored vertical scale on the end of the packing table.

sensitivity analysis

In a great many cases a simple breakeven analysis is not feasible because several factors may vary simultaneously as the single variable under study is varied. In such instances, and in fact in most cases, it is helpful to determine how sensitive the situation is to the several variables so that proper weight and consideration may be assigned to them. Sensitivity, in general, means the relative magnitude of change in the measure of merit (such as rate of return) caused by one or more changes in estimated elements or variables. Sometimes sensitivity is more specifically defined to mean the relative magnitude of the change in one or more elements or variables that will reverse a decision among alternatives. Example 6-2 extends sensitivity explorations from one variable to several different variables through the use of tabular displays. Following that, the use of optimistic–pessimistic estimates and graphical means for describing the results of sensitivity studies will be shown.

EXAMPLE 6-2

A small group of investors is considering starting a premixed-concrete plant in a rapidly developing suburban area about 5 miles from a large city. The group believes that there will be a good market for premixed concrete in this area for at least the next 10 years, and that if they establish such a local plant it would be

FIGURE 6-2.   Packing costs for paper sacks and cartons shown as a function of the volume of merchandise.

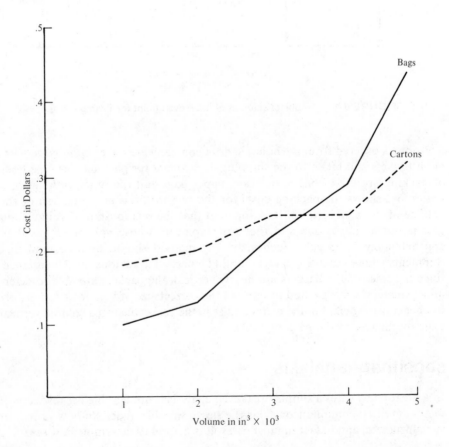

unlikely that another local plant would be established. Existing plants in the adjacent city would, of course, continue to serve this new area. The plant will cost $18,000, and it would have a capacity of 72 cubic yards of concrete per day. To deliver the concrete, four trucks would be required, costing $8,000 each, having an estimated life of 5 years and a trade-in value of $500 each at the end of that time. In addition to the four truck drivers, who would be paid $21.00 per day each, four people would be required to operate the plant and office, at a cost of $67.00 per day. Annual operating and maintenance costs for the plant are estimated at $1,750, for the office at $500, and for each truck at $2,250. Raw-material costs are estimated to be $9.40 per cubic yard of concrete. Payroll taxes, vacations, and other fringe benefits would amount to 12% of the annual payroll. Taxes and insurance on each truck would be $320. The investors would not contribute any labor to the business, but a manager would be employed at an annual salary of $8,000. The investors believe the plant could operate at about 75% of capacity 250 days per year since it is located in an area where the weather is mild throughout the year. Delivered, premixed concrete currently is selling for $16 per cubic yard. They demand that the project be written off within 10 years, and their capital currently is earning about 15% before income taxes. It is desired to find the E.R.R.R. for the expected conditions and to perform sensitivity studies for certain variables.

Solution by E.R.R.R. Method

Annual revenue: $72 \times 250 \times \$16 \times 0.75 = \$216,000$
Annual expenses:
1. Depreciation:
   Plant: $\$18,000(A/F, 15\%, 10)$ $= \$ \quad 890$
   Trucks: $4(\$8,000 - \$500)(A/F, 15\%, 5) = \quad 4,440$
                                                                        $\$ \quad 5,330$
2. Labor:
   Plant and office: $\$67 \times 250$ $= \quad 16,750$
   Truck drivers: $4 \times \$21 \times 250$ $= \quad 21,000$
   Manager $= \quad 8,000$
                                                                        $\$ 45,750$
3. Payroll taxes, etc.: $\$45,750 \times 0.12$ $=$ $5,490$
4. Taxes and insurance:
   Plant $= \quad 360$
   Trucks: $\$320 \times 4$ $= \quad 1,280$
                                                                        $\$ \quad 1,640$
5. Operation and maintenance:
   Plant $= \quad 1,750$
   Office $= \quad 500$
   Trucks: $\$2,250 \times 4$ $= \quad 9,000$
                                                                        $\$ 11,250$
6. Materials: $72 \times 0.75 \times 250 \times \$9.40$ $=$ $\$126,900$
                                                                        ─────────
   TOTAL $\$196,360$

The prospective annual profit would be $\$216,000 - \$196,360 = \$19,640$. This would result in an E.R.R.R. of $\$19,640/\$50,000 = 39.3\%$. ∎

In Example 6-2 there are three variable factors that are of great importance and that must be estimated: *capacity utilization*, the *selling price of the product*, and the *economic life of the plant*. A fourth factor—raw-material costs—is important, but any significant change in this factor also would be equally effective to competitors and probably would be reflected in a corresponding change in the selling price of mixed concrete. The other cost elements should be determinable with considerable accuracy. Therefore, we would like to investigate the effect of variations in the plant utilization, selling price, and economic life. Sensitivity analysis presents a good method for doing this, as will be shown below.

sensitivity to capacity utilization

As a first step we must determine how cost factors would vary, if at all, as capacity utilization is varied. In this case, it is probable that the cost items listed under groups 1, 2, 3, and 4 in the previous tabulation would be virtually unaffected if capacity utilization should vary over a quite wide range—from 50 to 100%, for example. To meet peak demands, the same amount of plant, trucks, and personnel probably would be required. Group 5 costs for operation and maintenance would be affected somewhat. For this type of factor we must try to determine what the variation would be or make a reasonable assumption as to the probable variation. For this case it will be assumed that one half of these costs would be fixed and the other half would vary by a straight line relationship. Certain other factors, such as the cost of materials in this case, will vary in direct proportion to the capacity utilization.

Using these assumptions, Table 6-1 shows how the revenue, costs, profits, and rate of return would change with different capacity utilizations. It will be noted that the rate of return is moderately sensitive to capacity utilization. The plant could be operated at a little less than 65% of capacity, instead of the assumed 75%, and still produce a rate of return equal to the 15% that the investors' capital now is earning. Also, quite clearly, if they should be able to operate above the assumed 75% of capacity, the rate of return would be very good. Thus this type of analysis provides those who must make the decision with a good idea as to how much leeway they would have in capacity utilization and still have an acceptable venture.

sensitivity to selling price

Examination of the sensitivity of the project to the selling price of the concrete reveals the situation shown in Table 6-2. The values in this table assume that the plant would operate at 75% of capacity; thus the costs would remain constant, with only the selling price varying. Here it will be noted that the project is quite sensitive to price. A decrease in price of about $7\frac{1}{2}$% would drop the rate of return to less than 8%. Since a decrease of $7\frac{1}{2}$% is not very

table 6-1

revenues, costs, profits, and rates of return for premixed-concrete plant for various capacity utilizations at selling price of $16

	50% Capacity	60% Capacity	65% Capacity	70% Capacity	80% Capacity
Annual revenue	$144,000	$172,800	$187,200	$201,600	$230,500
Annual expenses:					
Depreciation (sinking fund method)	5,330	5,330	5,330	5,330	5,330
Labor	45,750	45,750	45,750	45,750	45,750
Payroll taxes and similar items	5,490	5,490	5,490	5,490	5,490
Taxes and insurance	1,640	1,640	1,640	1,640	1,640
Operations and maintenance	9,645	10,290	10,600	10,930	11,570
Materials	84,500	101,500	110,000	118,300	135,300
TOTAL	$152,355	$170,000	$178,810	$187,440	$205,080
Annual profit	(−$8,355)	$2,800	$8,390	$14,160	$25,420
E.R.R.R.	−16.7%	5.6%	16.8%	28.3%	50.8%

table 6-2
effect of various selling prices on the profit and rate of return for premixed-concrete plant operating at 75% of capacity

	Selling Price			
	$16.00	$15.50(3%)[a]	$15.20(5%)[a]	$14.80(7½%)[a]
Annual revenue	$216,000	$209,300	$205,000	$200,000
Annual expense	196,360	196,360	196,360	196,360
PROFIT	$ 19,640	$ 12,940	$ 8,640	$ 3,640
E.R.R.R.	39.3%	25.9%	17.3%	7.3%

[a] Percentage values shown in parentheses are reductions in price below $16.

large, the investors would want to make a thorough study of the price structure of concrete in the area of the proposed plant, particularly with respect to the possible effect the increased competition that the new plant would create.

sensitivity to economic life

The effect of the third factor, assumed economic life, can be investigated very readily. If a life of 5 years were assumed for the plant, instead of the assumed value of 10 years, the only factor in the study that would be changed would be the cost of plant depreciation. On a 5-year write-off basis this cost would be $18,000 \times (A/F, 15\%, 5) = \$2,665$ per year, an increase of $1,775. This would reduce the profit, at 75% capacity, by the same amount, to $17,865. The E.R.R.R. then would be reduced by $1,775/\$50,000 = 3.6\%$ to $17,865/\$50,000 = 35.7\%$. Clearly, the venture is quite insensitive to the assumed economic life.

With the added information supplied by the sensitivity analyses that have just been described, those who would have to make the investment decision concerning the proposed concrete plant would be in a much better position to make a decision than if they had only the initial study results, based on an assumed 75% of capacity, available to them. They would know which factors were critical and thus could seek more information about these particular items if desired.

sensitivity to capacity utilization and selling price

It is quite apparent that two or more of the factors might vary simultaneously from the conditions assumed in the original study. For example, what would happen if the price of concrete dropped 5% and at the same time the plant would be operated at only 70% of capacity? Obviously, such combined con-

table 6-3

E.R.R.R. for various combinations of capacity utilization and selling price for premixed-concrete plant

Selling Price	Capacity (%)					
	50	60	65	70	75	80
$16.00	−16.7	5.6	16.8	28.3	39.3	50.8
$15.50 (3%)[a]	−30.1	−7.8	3.4	14.9	25.9	37.4
$15.20 (5%)[a]	−38.7	−16.4	−5.2	6.3	17.3	28.8
$14.80 (7½%)[a]	−48.7	−26.4	−15.2	−3.7	7.3	18.8

[a] Percentage values shown in parentheses are reductions in price below $16.

ditions can be studied. A two-dimensional matrix can be developed to show the results of the varying of two factors, and such a representation often is quite helpful in decision making. Such a matrix showing the E.R.R.R. for all combinations of capacity utilization and selling price considered for the premixed-concrete plant is shown in Table 6-3.

optimistic–pessimistic estimates and graphical ways to describe sensitivity[†]

A very useful method for exploring sensitivity is to change one or more variables to be estimated in a favorable outcome (optimistic) direction and in an unfavorable outcome (pessimistic) direction so as to investigate the effect of these various changes on the economy study result. This represents a simple method for including uncertainty in the analysis.

As an example, consider a new machine for which the optimistic and pessimistic (combined with the most likely or best) estimates are as shown in Table 6-4 (p. 158). Also shown at the end of Table 6-4 are the net annual worths (A.W.s) for all three estimation conditions. Note that the net A.W. for optimistic conditions is very favorable (+ $74,000), while for pessimistic conditions it is quite unfavorable (− $33,100). After having this information the decision maker may be willing to make a go–no go decision on the new machine. However, he should recognize that these are extreme outcomes—the optimistic net A.W. assumes that all estimated variables turn out according to the optimistic estimates, and the pessimistic net A.W. assumes that all estimated variables turn out per the pessimistic estimates. It is reasonable to assume that this will not happen, but that different variables may turn out to have different optimistic versus most

[†] Adapted from J. R. Canada and W. G. Sullivan, "To Understand the Decision Problem, Say It with Pictures," Proceedings of Spring National Conference, American Institute of Industrial Engineers, 1977.

table 6-4

optimistic–most likely–pessimistic estimates and net annual worths
for new machine

	Estimation Condition		
	Optimistic (O)	Most Likely (M)	Pessimistic (P)
Investment	$150,000	$150,000	$150,000
Life	18 years	10 years	8 years
Salvage value	0	0	0
Annual savings	$110,000	$70,000	$50,000
Annual disbursements	$20,000	$43,000	$57,000
Minimum attractive rate of return	8%	8%	8%
Net annual worth (net A.W.)	+ $74,000	+ $4,650	− $33,100

likely versus pessimistic outcomes. One good way to reflect such results is shown in Table 6-5, which shows net annual worths for all combinations of estimated outcomes for three of the variables to be estimated. It could have been done for four or more variables if there were more variables subject to significant variation, but the size of the matrix display would grow enormously. Even the 3 × 3 × 3 matrix shown in Table 6-5 quickly becomes cumbersome or discouraging because of the proliferation of numbers. There are ways around this, as we shall show later.

table 6-5

net annual worths ($) for all combinations of estimated outcomes for annual savings, annual disbursements, and life—new machine

		Annual Disbursements								
		(O)			(M)			(P)		
		Life			Life			Life		
		(O)	(M)	(P)	(O)	(M)	(P)	(O)	(M)	(P)
Annual Savings	(O)	74,000	67,650	63,900	51,000	44,650	40,900	37,000	30,650	26,900
	(M)	34,000	27,650	23,900	11,000	4,650	900	−3,000	−9,350	−13,000
	(P)	14,000	7,650	3,900	−9,000	−15,350	−19,100	−23,000	−29,350	−33,100

Key to estimates:

(O) Optimistic
(M) Most likely
(P) Pessimistic

table 6-6
results in table 6-5 made easier to interpret (net annual worths in $000s)

		(O)			Annual Disbursements (M)			(P)	
		Life			Life			Life	
	(O)	(M)	(P)	(O)	(M)	(P)	(O)	(M)	(P)
Annual Savings (O)	(74)	(68)	(64)	(51)	45	41	37	31	27
Annual Savings (M)	34	28	24	11	5	1	−3	−9	−13
Annual Savings (P)	14	8	4	−9	−15	−19	−23	−29	−33

Key to estimates:

(O) Optimistic
(M) Most likely
(P) Pessimistic

Key to highlight notations:

◯ Net annual worth > $50,000 (4 out of 27)
(combinations)

── Net annual worth < $0 (9 out of 27)
(combinations)

FIGURE 6-3. Results in Table 6-4, using graphical representation (net annual worths rounded to the nearest $5,000).

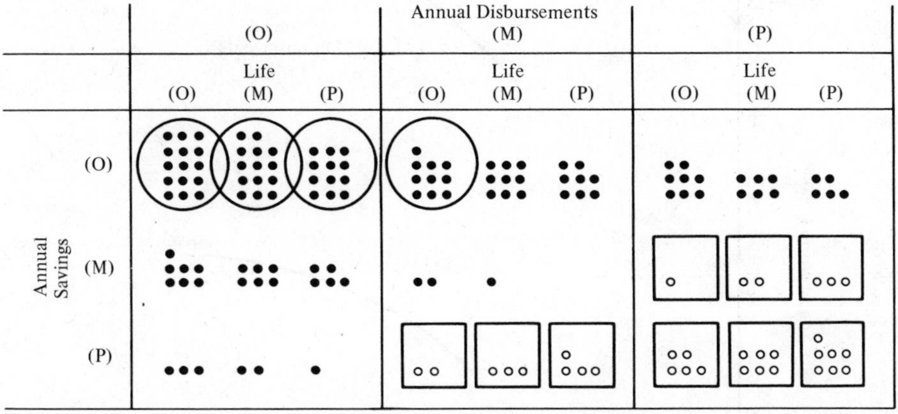

Key to symbols:

• $5,000 of annual worth

○ −$5,000 of annual worth

Key to highlight notations:

◯ Annual worth > $50,000

▢ Annual worth < $0

Key to estimates used:

	Optimistic (O)	Most Likely (M)	Pessimistic (P)
Annual disbursements	$20,000	$43,000	$57,000
Life (year)	10 years	10 years	8 years
Annual savings	$110,000	$70,000	$50,000
Investment		$150,000	
Salvage value		0	
Minimum attractive rate of return		8%	

making matrices easier to interpret

It should be recognized that the net A.W. numbers in Table 6-5 are the results of estimates subject to varying degrees of uncertainty. Hence little information of value would be lost if the numbers were rounded to the nearest thousand (or maybe even the nearest 10,000). Further, it may be thought that management is most interested in the number of combinations of conditions in which the net A.W. is, say, (a) more than $50,000, and (b) negative, or less than $0. Table 6-6 shows how Table 6-5 might be changed to make it easier to read and interpret.

FIGURE 6-4. Two-dimensional graph for displaying the results of Table 6-5 in net annual worths for the most likely annual savings only.

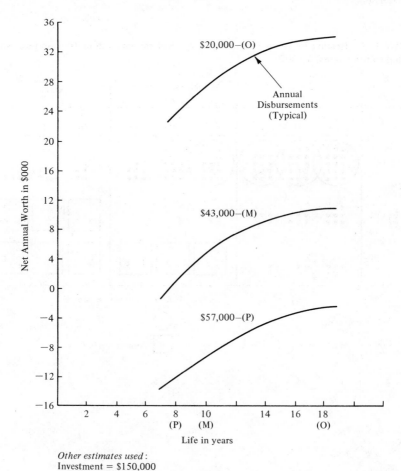

Other estimates used:
Investment = $150,000
Salvage value = 0
Annual savings = $70,000
Minimum attractive rate of return = 8

Figure 6-3 (p. 159) goes even further than Table 6.6 to remove burdensome number reading by portraying the results graphically. Figure 6-3 uses a dot to represent each $5,000 of annual worth, but histogram bars could have been used just as well. Once again, combinations of conditions in which the net annual worths are greater than $50,000 or less than $0 are highlighted, assuming that these would be of particular interest to management.† A key for interpreting these results is shown on the bottom of Figure 6-3. It is included to emphasize the desirability of tables and figures having as much explanatory information as needed for the convenience of the user without cluttering the main messages to be conveyed.

FIGURE 6-5. Example graph of sensitivity to multiple variables, each independently deviating from the most likely estimate.

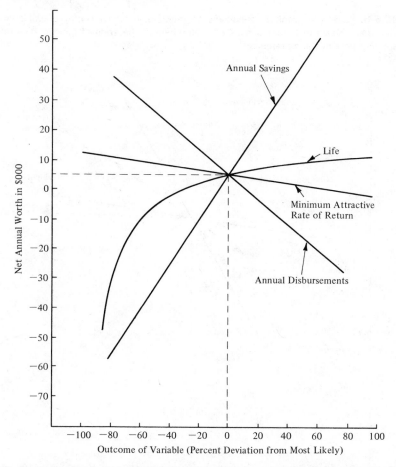

† It should be noted, however, that each combination of conditions in Figure 6-3 is not necessarily equally likely. Therefore, statements such as "There are 9 chances out of 27 that we will lose money on this project" are not appropriate in this example.

graphical sensitivity displays

Figure 6-4 (p. 160) demonstrates use of a two-dimensional graph with multiple lines to show net A.W. as a function of changes in two of the variables given in Table 6-4. The lines connecting the plotted points in Figure 6-4 are curvilinear, reflecting the actual relationship between net annual worth and project life. If it were desired, additional graphs could be constructed using other variables and displayed together with Figure 6-4 for comparison purposes.

An effective way of displaying and examining sensitivity is to graph the measure of merit for independent variation of all variables of interest by expressing variation for each on a common abscissa in terms of percent deviation from its respective or most likely value. This is shown in Figure 6-5 (p. 161) for the

FIGURE 6-6. Example graph of sensitivity to multiple variables (with several changes from what was shown in Figure 6-5). *Note :* Numbers at the ends of most curves reflect optimistic and pessimistic extremes.

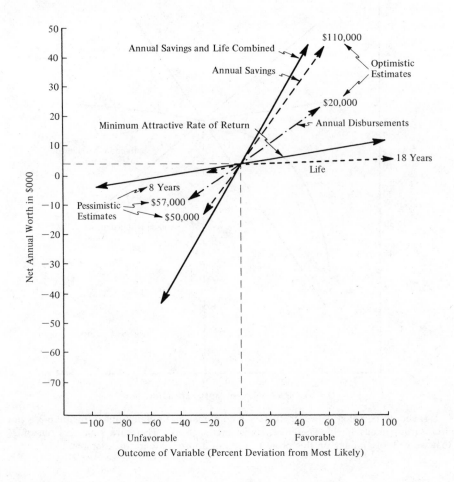

new-machine example, for which the most likely estimates were given in Table 6-4. Figure 6-5 shows, among other things, that the result is relatively sensitive to changes in annual savings, annual disbursements, and decreases in project life. It also shows that the result is relatively insensitive to changes in the M.A.R.R. and to increases in the project life.

Figure 6-5 can be altered in numerous ways that may make it more useful as a management device for exploring sensitivity. Some examples are shown in Figure 6-6. Changes made are: (1) showing all deviations in a favorable direction together and in an unfavorable direction together on the abscissa; (2) showing deviations of two or more variables in the same direction by the same percentage together (done for annual savings and life); and (3) showing sensitivity curves only for a certain range of estimates considered of reasonable concern (done using optimistic and pessimistic estimates as extremes).

Another type of sensitivity test that is often quite valuable is to determine the relative (or absolute) change in one or more variables which will just reverse the decision. Applied to the new-machine example in Table 6-4, this means the relative change that will decrease the net A.W. by $4,650 so that it reaches $0. Figure 6-7 shows this using a table and histogram bars of varying lengths to emphasize facts such as that the indicated economic merit of the machine is (1) most sensitive to changes in the estimated annual receipts, and (2) least sensitive to changes in the salvage value.

It is clear that with even a few variables the number of possible combinations of conditions in a sensitivity analysis can become quite large, and the task of investigating all of them might be quite time-consuming and costly. Ordinarily, a sensitivity analysis involves eliminating from consideration those factors to which the proposal is quite insensitive and reducing the range of conditions for other variables to be studied in accordance with the degree of sensitivity of each. Thus the number of combinations of conditions included in the analysis hopefully can be kept to a manageable size. Digital computers are often quite useful for sensitivity analyses, particularly when evaluating the effect of simultaneous

FIGURE 6-7.   Sensitivity of variable estimates to decision reversal.

	Most Likely Estimate	Estimate Outcome	Change Amount	Change Amount as Percent of Most Likely	
			To Reverse Decision (Decrease A.W. to $0)		
Investment	$150,000	$181,200	+$31,200	+20.8%	
Life	10 Years	7.3 Years	−2.7 Years	−27. %	Least Sensitive
Salvage Value	0	− 67,890	− 67,890	∞	
Annual Savings	70,000	− 65,450	− 4,650	− 5.6%	
Annual Disbursements	43,000	47,650	+ 4,650	+10.8%	Most Sensitive
Minimum Attractive Rate of Return (M.A.R.R.)	8%	12.5%	+ 4.5%	+ 56%	

variations in several variables or elements. Creative displays of relevant information in tabular and in graphical form can be invaluable in promoting insight into, and in communicating about, decision problems.

expected value

The concept of *expected value* often is quite helpful in making decisions when uncertainty is involved. Expected values are based on probability, which is usually considered to be the long-run relative frequency with which an event occurs, or the subjectively judged likelihood that it will occur. The expected value is the product of the probability that a revenue, cost, or other variable will occur and its value (numerical outcome) if it does occur. Thus, if the probability that a cost will occur is 0.4 and the cost if it does occur is $10,000, its expected value is $10,000 × 0.4 = $4,000. If there is more than one possible outcome of the variable, the expected value of that variable is the sum of the products of the outcomes times the respective probabilities of those outcomes. Thus, if the economic life of a project is estimated to be 10 years with 0.4 probability and 15 years with 0.6 probability, the expected project life is 10 × 0.4 + 15 × 0.6 = 4 + 8 = 12 years.†

The expected value concept applies theoretically to long-run conditions in which it is assumed the event is going to occur repeatedly. However, application of the expected value principle is often useful even when investments are not going to be made repeatedly over the long run.

EXAMPLE 6-3

Applying the expected value approach to the premixed-concrete plant discussed in Example 6-2, suppose that the probabilities of attaining various capacity utilizations are

Capacity (%)	Probability
60	0.10
70	0.30
75	0.50
80	0.10

It is desired to determine total annual expected revenue and the expected E.R.R.R. from the project.

† In general, expected value of any outcome, x, can be computed as

$$E(x) = \sum_{\text{all } x} x \cdot P(x)$$

where $E(x)$ = expected value of x
$P(x)$ = probability of x occurring

Solution

Capacity (%)	(A) Probability	(B) Revenue[a]	(A) × (B) Expected Revenue
60	0.10	$172,800	$ 17,280
70	0.30	201,600	60,480
75	0.50	216,000	108,000
80	0.10	230,000	23,000
	TOTAL EXPECTED REVENUE = ($\sum$ A × B) =		$208,760

[a] From Table 6-1.

This revenue corresponds to a capacity factor of about 73.5%. The corresponding total cost would be about $191,800, and the annual profit $16,960, producing an approximate expected E.R.R.R. = $16,960/$50,000 = 33.9%. ∎

evaluation of alternatives that consider risk of loss, using expected values

There are situations, such as flood control projects, in which future losses due to natural or man-made risks can be decreased by increasing the amount of capital that is invested. Drainage ditches or dams, built to control floodwaters, may be constructed in different sizes, costing different amounts. If correctly designed and used, the larger the size, the smaller will be the resulting damage loss when a flood occurs. As we might expect, the most economical size would provide satisfactory protection against most floods, although it could be anticipated that some overloading and damage may occur at infrequent periods.

EXAMPLE 6-4

A drainage ditch in a mountain community in the West where flash floods are experienced has a capacity sufficient to carry 700 cubic feet per second. Engineering studies produce the following data regarding the probability of water flow in any one year versus the probability the flow will be exceeded and the cost of enlarging the ditch:

Water Flow (ft³/sec)	Probability of a Greater Flow Occurring in Any One Year	Investment to Enlarge Ditch to Carry This Flow
700	0.20	—
1,000	0.10	$20,000
1,300	0.05	30,000
1,600	0.02	44,000
1,900	0.01	60,000

Records indicate that the average property damage when serious overflow occurred will amount to $20,000. It is believed that this would be the average damage whenever the storm flow *was greater* than the capacity of the ditch. Reconstruction of the ditch would be financed by 40-year bonds bearing 5% interest. It is thus computed that the annual cost for debt repayment (principal of the bond plus interest) would be approximately 6% of the initial cost, for $(A/P, 5\%, 40) = 0.058$. It is desired to determine the most economic ditch size (water flow capacity).

Solution

The total expected annual cost for the structure and property damage for all alternative ditch sizes would be as follows:

Water Flow (ft^3/sec)	Annual Investment Cost[a]	Expected Annual Property Damage[b]	Total Expected Annual Cost
700	—	$20,000 × 0.20 = $4,000	$4,000
1,000	$20,000 × 6% = $1,200	20,000 × 0.10 = 2,000	3,200
1,300	30,000 × 6% = 1,800	20,000 × 0.05 = 1,000	2,800
1,600	44,000 × 6% = 2,640	20,000 × 0.02 = 400	3,040
1,900	60,000 × 6% = 3,600	20,000 × 0.01 = 200	3,800

[a] These amounts are obtained by multiplying 6% by the investment required to increase water flow capacity to the size indicated.

[b] These amounts are obtained by multiplying $20,000 by the probability of greater water flow occurring.

From these calculations it may be seen that greatest economy would be achieved by enlarging the ditch so that it would carry 1,300 cubic feet per second, with the expectation that a greater flood might occur 1 year out of 20 on the average and cause property damage of $20,000. ∎

It should be noted that when loss of life or limb might result from the occurrence of a risky event, there usually is considerable pressure to disregard pure economy and build such projects in recognition of the intangible values associated with human safety.

The following illustrates the same principles in Example 6-4 except that it is applied to safety alternatives involving electrical circuits, utilizes a fixed charge rate as a computational convenience, and describes the loss itself in terms of probabilities.

EXAMPLE 6-5

There are three alternatives for the protection of electrical circuits, with the following required investments and probabilities of failure:

Alternative	Investment	Probability of Loss in Any Year
A	$ 90,000	0.40
B	100,000	0.10
C	160,000	0.01

If a loss does occur, it will cost $80,000 with probability 0.65, and $120,000 with probability 0.35. The probabilities of loss in any year are independent of the probabilities associated with a loss if it does occur. The annual fixed charge rate to cover the cost of depreciation, interest on investment, and maintenance is 20% of any investment. It is desired to determine which alternative is best based on expected total annual costs.

Solution

The expected cost of a loss, if it occurs, can be calculated as

$$\$80,000(0.65) + \$120,000(0.35) = \$94,000$$

Alternative	Annual Fixed Charges on Investment	Expected Annual Cost of Failure	Total Expected Annual Cost
A	$ 90,000(20%) = $18,000	$94,000(0.40) = $37,600	$55,600
B	100,000(20%) = 20,000	94,000(0.10) = 9,400	29,400
C	160,000(20%) = 32,000	94,000(0.01) = 940	32,940

Thus alternative B is shown to be the best based on total expected annual cost, or long-run average. However, one might rationally choose alternative C so as to reduce significantly the chance of a $80,000 or $120,000 loss occurring in any year in return for a modest increase in the total expected annual cost. ∎

One of the prime problems when expected values are to be computed is the determination of the probabilities. In most cases there is no history of previous cases for the particular venture being considered. Therefore probabilities seldom can be based on historical data and rigorous statistical procedures. In most cases it is necessary that the analyst, or person making the decision, rely on judgment or even intuition in estimating the probabilities. This fact makes some persons hesitate to use the expected-value concept, inasmuch as they cannot see the merit in applying such a technique to improve the evaluation of uncertainty when some of the very basic factors used are not known with certainty. Although this argument has merit, the fact is that because economy studies always deal with future events there must be an extensive amount of estimating involved. Furthermore, even if the probabilities could be based accurately on past history, there rarely is any assurance that the future will repeat the past. In such situations structured methods for assessing "subjective" probabilities are used often in practice.† Also, even if we must estimate the probabilities, the very process of doing so requires us to give some thought to the fact that because of uncertainty the future results can be variable, and to do some thinking as to what the possible outcomes might be. Such forced thinking is likely to produce better results than little or no thinking about such matters.

Commonly known distributions such as binomial, beta, Poisson, and normal are often appropriate and convenient for describing the variability of estimated outcomes. The nature and use of such distributions, sometimes together with simplified techniques for estimating in terms of those distributions, are available in numerous books on probability and statistics and stochastic decision analysis.‡

decision rules for complete uncertainty and decision making

There is no doubt that recognition of uncertainty plays a considerable role in decision making. It also is evident that decision makers often seek methods by which certain logical criteria can be applied, with some consistency, in making investment decisions. However, it is equally clear that most decision makers refuse to be bound, at all times, by the criteria that they previously may have agreed were logical. Such logical criteria could, in most cases, be applied by a computer, but most decision makers believe that they can better take into account any variations from the assumed conditions that may exist and consequently do a better job than a machine can do. Thus, although the development and use of logical criteria for governing investment decisions are helpful, it appears that personal evaluation, judgment, and even whims are going to continue to be important factors in decision making where uncertainties exist. Furthermore, the use of most decision criteria requires value judgments to be made regarding intangible factors, and value judgments can change radically as conditions and time change. Also, although we may show that the decisions of a particular person are not always *logical*, this does not prove that they are *incorrect*. However, decision rules can be helpful, in the same way that sensitivity analyses are helpful; they may provide more guidance to decision makers so that they may better understand the implications of their decisions.

The following is a description of several decision rules that can be applied when there is supposedly complete uncertainty (ignorance) as to the probabilities of various possible outcomes. The example shows some of the usefulness and inadequacies of these decision rules.

† For further information, see W. G. Sullivan and W. W. Claycombe, *Fundamentals of Forecasting* (Reston, Va.: Reston Publishing Company, Inc., 1977), Chap. 6.

‡ See J. R. Canada, *Intermediate Economic Analysis for Management and Engineering* (Englewood Cliffs, N.J.: Prentice-Hall, Inc., 1971), Chap. 10.

EXAMPLE 6-6

Suppose that a contractor can accomplish a given job in 1 year by any of three methods. If method A is used and good weather persists, the profits (expressed in net present worth) will be $175,000. If mixed weather occurs, the profits will be $100,000, and if bad weather occurs, he will lose $20,000. Using method B, the profits will be $15,000 for good weather, $40,000 for mixed weather, and $35,000 for poor weather. With method C, the profit will be $20,000 regardless of the weather. These data are summarized in Table 6-7. Note that the problem is characterized by known quantitative outcomes according to various states of nature but with unknown probabilities of each state of nature occurring.

Solution by Maximin (or Minimax) Rule

The *maximin rule* is quite conservative, for it involves determining the minimum profit associated with each alternative and then selecting that alternative which maximizes (is the highest of) these minimum profits. Similarly, in the case of costs, the *minimax rule* suggests that the decision maker determine the maximum cost associated with each alternative and then select that alternative which minimizes (is the lowest of) those maximum costs.

For this example problem as depicted in Table 6-7, the minimum possible profits are: − $20,000 if method A is chosen; $15,000 if method B is chosen; and $20,000 if method C is chosen. The $20,000 for method C is the maximum of these minimum profits, so C would be the method chosen by this conservative rule. ∎

Solution by Maximax (or Minimin) Rule

The maximax rule is the most optimistic or nonconservative of the decision rules for complete uncertainty. It involves determining the maximum profit associated with each alternative and then selecting that alternative which maximizes these maximum profits. Similarly, in the case of costs, the minimin rule suggests that the decision maker determine the minimum cost outcome for each alternative and then select that alternative which minimizes (is the lowest of) those minimum costs.

For the example problem as depicted in Table 6-7, the maximum possible profit is $175,000 if method A is chosen; $40,000 if method B is chosen; and

Table 6–7
financial profitability data for example 6–6 involving complete uncertainty

Method	State of Nature		
	Good Weather	Mixed Weather	Bad Weather
A	$175,000	$100,000	− $20,000
B	15,000	40,000	35,000
C	20,000	20,000	20,000

$20,000 if method C is chosen. The $175,000 for method A is the maximum of these maximum profits, so A would be the indicated choice method by this optimistic rule. ∎

Solution by Laplace Rule

This rule simply assumes that all possible outcomes (states of nature) are equally likely so that the choice is based on expected outcomes as calculated using equal probabilities for all outcomes. There is a common tendency to implicitly use this assumption in situations where there is no evidence to the contrary, but the assumption (and hence the rule) is of highly questionable merit.

For the example problem depicted in Table 6-7, the expected outcome, E, for each method can be calculated as:

$$E_A = \$175,000(0.333) + \$100,000(0.333) - \$20,000(0.333) = \$85,000$$
$$E_B = \$15,000(0.333) + \$40,000(0.333) + \$35,000(0.333) = \$30,000$$
$$E_C = \$20,000(0.333) + \$20,000(0.333) + \$20,000(0.333) = \$20,000$$

Thus alternative A has the highest expected profit and would be the choice by this optimistic rule. ∎

Solution by Hurwicz Principle or Rule

This rule is intended to reflect any degree of moderation between extreme optimism (as in maximax rule) and extreme conservatism (as in maximin rule) which the decision maker may wish to choose. The rule may be stated explicitly as follows: Select an index of optimism, Z, such that $0 \leq Z \leq 1$. For each alternative, compute the weighted outcome: $Z \times$ (value of profit or cost if most favorable outcome occurs) $+ (1 - Z) \times$ (value of profit or cost if least favorable outcome occurs). Choose the alternative that optimizes the weighted outcome.

Applied to the example depicted in Table 6-7, assuming an index of optimism, Z, of 0.2, the weighted outcome which we will denote as expected outcome for each method is

$$E_A = \$175,000(0.2) - \$20,000(0.8) = \$19,000$$
$$E_B = \$40,000(0.2) + \$15,000(0.8) = \$20,000$$
$$E_C = \$20,000(0.2) + \$20,000(0.8) = \$20,000$$

Thus methods B and C are shown to be equally desirable using the Hurwicz rule and an index of optimism of 0.2. If the index of optimism were lower than 0.2, method C would be the choice. ∎

If the example above were not a problem involving complete uncertainty, but rather the probabilities of the various outcomes or states of nature were reasonably estimable, one would probably want to base the decision at least in part on expected outcomes using those probabilities.

EXAMPLE 6-7

Given the problem as depicted in Example 6-6 and Table 6-7, assume that the probabilities of each type of weather can be estimated as follows: good, 0.3;

mixed, 0.2; and bad, 0.5. Determine the expected profit for each method based on these probabilities.

Solution

$$E_A = \$175,000(0.3) + \$100,000(0.2) - \$20,000(0.5) = \$62,500$$
$$E_B = \$15,000(0.3) + \$40,000(0.2) + \$35,000(0.5) = \$30,000$$
$$E_C = \$20,000(0.3) + \$20,000(0.2) + \$20,000(0.5) = \$20,000$$

Thus method A would be the choice based on expected outcomes with the probabilities as given. ∎

Actually, for the problem in Examples 6-6 and 6-7, many would eliminate method A from consideration because of the possibility of a significant loss being incurred, although the expected profit is higher than for the other alternatives. Rather few business executives want to undertake a venture when it is clear that there is considerable likelihood of a loss being sustained, even though the possible profit may be quite great. Instead, they prefer to select between alternatives that offer varying amounts of profit, but which have relative certainty. Consequently, our unwillingness to gamble, with loss as a possible outcome, often is an important factor in investment decisions, and the extent of such unwillingness will vary from person to person, and even for a given person at different times. Therefore, it is doubtful, even if desirable, that a completely logical basis for decision making can be achieved. Probably the best that can be expected is that decision makers can recognize and justify the bases on which they make decisions and achieve a reasonable degree of consistency in their decisions in the light of the value objectives they have. Perhaps life is sometimes more interesting because decisions cannot always be predicted on a "logical" basis.

miscellaneous methods for considering multiple attributes in economy studies

There are many quantitative and nonquantitative methods and techniques that can be used for analyzing alternatives in which there are multiple attributes (objectives or factors) to be considered.† These methods are applicable to decision making in the private sector as well as the public sector. Detailed presentations are beyond the scope of this book, but the following is a partial listing and brief descriptions and examples of several such methods.

1. *Ordinal scaling*—ranking attributes in order of decreasing importance. For example, the three relevant attributes for a decision problem might be safety, cost, and completion (speed). After due consideration, decision maker(s) might

† See, for example, J. R. Canada and J. A. White, *Capital Investment Decision Analysis for Management and Engineering* (Englewood Cliffs, N.J.: Prentice-Hall, Inc., in press).

decide that cost is more important than either safety or completion, and that safety is more important than completion. Thus the rank ordering would be

1. Cost.
2. Safety.
3. Completion.

This information could then be used for further subjective or objective consideration of alternatives, as shown next.

2. *Weighting attributes*—quantifying relative importance of attributes on a dimensionless scale from, say, 0 to 1 or 0 to 100. For example, the three attributes in method 1 might be weighted as follows on a 0 to 100 scale, with 100 being assigned to the most important attribute:

$$
\begin{array}{lc}
\text{Cost} & 100 \\
\text{Safety} & 60 \\
\text{Completion} & \underline{40} \\
\Sigma = 200
\end{array}
$$

The relative weights above can be used as they are. However, it is commonly thought helpful to "normalize" them by dividing each by the total and multiplying times 100 so that each weight will be points out of 100, which are the terms in which people commonly think. Thus we can "normalize" the weight above as follows:

Attribute	Weight	Normalized Weight
Cost	100	(100/200)100 = 50
Safety	60	(60/200)100 = 30
Completion	40	(40/200)100 = 20
	$\Sigma = 200$	$\Sigma = 100$

3. *Weighted evaluation of alternatives*—quantifying how well each alternative (or investment opportunity) meets each attribute on a dimensionless scale and then summing the product of the "evaluations" and their respective "weightings" (from method 2) for each alternative. For example, suppose that a decision maker is comparing alternative bridge designs A and B according to the three weighted attributes given above. Upon due consideration, he arrives at the following mostly subjective evaluation ratings of how well each alternative meets each attribute (on an arbitrary scale of 0 to 10).

	Alternative	
Attribute	A	B
Cost	7	9
Safety	8	6
Completion	10	9

table 6-8
calculation of weighted evaluation of alternatives

Attribute	Normalized Attribute Weight	Alternative A		Alternative B	
		Evaluation Rating	Weighted Evaluation [a]	Evaluation Rating	Weighted Evaluation [a]
Cost	50	7	35	9	45
Safety	30	8	24	6	18
Completion	20	10	20	9	18
			$\Sigma = 79$		$\Sigma = 81$

[a] Weighted evaluation = normalized attribute weight $\times \dfrac{\text{Evaluation rating}}{10}$.

Table 6-8 shows the calculations needed to determine the weighted evaluation for each alternative. The footnote at the bottom of Table 6-8 shows that the evaluation rating was divided by 10. This was done merely to keep the total weighted evaluation in terms of points out of 100. Thus alternative B is shown to be slightly better than A by a score of 81 to 79.

4. *Alternatives–objectives score card*—displaying a matrix of alternatives vs. objectives (attributes) together with numbers and/or other symbols to represent how well each alternative meets each objective. As an example, suppose that there are four alternative bridge designs to be compared on the basis of the three attributes used in the example above. Figure 6-8 shows a typical "score card" display. Using such a display, the decision maker should be aided in making his selection according to his subjective (usually nonquantified) feelings about the relative importance of the various attributes and the corresponding

FIGURE 6-8.   Example alternatives—objectives score card.

Objectives (Attributes)	Bridge Design Alternatives			
	A	B	C	D
Cost (P.W. in millions of $)	⬭13	10	▢9	11
Safety (category or rank)	Good	Fair	◯Poor	▢Excellent
Completion (months)	▢15	▢18	◯20	19

Key: ▢ Best alternative for attribute factor

◯ Worst alternative for attribute factor

measures and/or relative indications of desirability for each alternative. No further quantification is necessary before a choice may be made.

cost-effectiveness analysis

The requirements for cost-effectiveness analysis are essentially the same as for any other type of economic analysis. That is, the following three requirements should be satisfied:

1. Alternative means for meeting the goals must exist.
2. The alternatives to be evaluated must have common goals or purposes.
3. Measures of cost as well as effectiveness must be estimable.

Effectiveness can, of course, be expressed in whatever terms or scales are appropriate rather than having to be reduced to dollars. Decision criteria using this approach involve subjective or objective trade-offs between various levels of cost and the evaluation of other attributes for each alternative to judge the optimal combination.

The following is a description of the selection procedure that could be employed when using the cost-effectiveness approach for evaluating mutually exclusive alternatives, and there are several attributes to be considered. As an example, we will use the information on the alternative bridge designs given in Figure 6-8. The alternatives can first be ranked in order of their capability to satisfy the most important attribute or measure of effectiveness. For the bridge design problem, let us again assume that the most important attribute is cost and that any other attributes (such as safety and completion) would be ranked in secondary positions. Often this procedure will result in elimination of the least promising candidates. If the cost and effectiveness in meeting noncost attributes for the top contender are both superior to the respective values for other remaining candidates, the choice is obvious. However, it is more common that the costs as well as other attributes of effectiveness for the various candidates differ significantly in conflicting directions, so selection must be made or based on judgmental trade-offs using the incremental approach. That is, one should start with the lowest-cost alternative as a base and judgmentally determine if the incremental cost is justified by the incremental effectiveness for each successively higher-cost alternative.

Applied to the example above, for which data are given in Figure 6-8, the lowest-cost alternative is C and the next-lowest alternative is B. For the extra $1 million required for B, safety would be improved from "poor" to "fair" and completion time would improve from 20 months to 18 months. Let us suppose that the decision maker decides B is thus justified.

Next, B would be compared to the next-lowest-cost alternative, D. For the extra $11 − $10 = $1 million required, safety would be improved from "fair" to "excellent," but completion time would be increased from 18 months to 19 months. Based on this information, let us suppose that the decision maker decides that D is justified compared to B, so D is now the leading contender.

As a final incremental comparison, D should be compared to the next-lowest-cost alternative, A. For the extra $13 − $11 = $2 million required, safety is made worse (from "excellent" to "good"). However, this is offset somewhat by the fact that the completion time would be improved by 19 − 15 = 4 months. Let us assume that the decision maker decides that the extra cost for A cannot be justified. Hence the best final choice would be alternative D.

The example above assumed that there were no attributes important to the choice between the alternatives other than the three indicated. In normal practice, there would be other intangible or irreducible factors that should be considered in addition to these quantifiable measures before the selection decision is made.

The above methods, even though rather simple to apply, usually involve some important assumptions that affect the validity of results significantly. For example, most of the methods assume that the important attributes can be identified and quantified or ranked independently of one another so that the results are additive for any given alternative. Even though such assumptions are not fully valid in many practical applications, such methods as described above are very valuable as guides and "thinking forcers" for decision makers considering multiple attributes.

problems

6-1. Why should the effects of uncertainty be considered in engineering economy studies? What are some likely sources of risk and uncertainty in these studies?

6-2. Consider these two alternatives:

	Alternative 1	Alternative 2
First cost	$5,000	$8,000
Annual receipts	$1,600	$2,000
Annual disbursements	$400	$500
Estimated salvage value	$1,000	$1,200
Life	8 years	10 years

Suppose that the salvage value of alternative 1 is known with certainty. By how much would the estimate of salvage value for alternative 2 have to vary so that the initial decision based on the data above would be reversed? The minimum attractive rate of return is 15%.

6-3. Referring to Problem 6-2, suppose there is concern over the accuracy of estimated annual receipts for alternative 2. Is this concern justified in terms of the sensitivity of the initial decision to possible changes in this quantity? How sensitive is the initial decision to estimates of the life of alternative A?

6-4. Two electric motors are being considered to power an industrial hoist. Each is capable of providing 100 horsepower. Pertinent data for each motor are as follows:

	Motor	
	D-R	Westhouse
Investment	$2,500	$3,200
Electrical efficiency	0.74	0.92
Maintenance/year	$40	$60
Life	10 years	10 years
M.A.R.R.	8%	8%

If the expected usage of the hoist is 500 hours per year, what would the cost of electrical energy have to be (in cents/kWh) before the D-R motor is favored over the Westhouse motor? [*Note:* 1 horsepower(hp) = 0.746 kilowatt.]

6-5. In a certain building project, the amount of concrete to be poured during the next week is uncertain. The foreman has estimated the following probabilities of concrete poured.

Amount (cubic yards)	Probability
1,000	0.1
1,200	0.3
1,300	0.3
1,500	0.2
2,000	0.1

Determine the expected value (amount) of concrete to be poured next week.

6-6. A bridge is to be constructed now as part of a new road. Engineers have determined that traffic density on the new road will justify a two-lane road and bridge at the present time. Because of uncertainty regarding future use of the road, the time at which an extra two lanes will be required is currently being studied. The estimated probabilities of having to widen the bridge to four lanes at various times in the future are as follows:

Widen Bridge in:	Probability
3 years	0.1
4 years	0.2
5 years	0.3
6 years	0.4

The two-lane bridge will cost $200,000 and the four-lane bridge, if built at one time, will cost, $350,000. The cost of widening a two-lane bridge will be an extra $200,000 plus $20,000 for every year that construction is delayed. If money can earn 10% interest, what would you recommend?

6-7. In Problem 6-6, perform an analysis to determine how sensitive the choice of a four-lane bridge built at one time versus a four-lane bridge that is constructed in two stages is to the interest rate. Will an interest rate of 12% reverse the initial decision?

6-8. In reference to Problem 6-6, suppose that instead of specifying probabilities for times at which the four-lane bridge will be required, these estimates have been made:

Pessimistic estimate	4 years
Most likely estimate	5 years
Optimistic estimate	7 years

In view of these estimates, what would you recommend? What difficulty do you have in interpreting your results? List some advantages and disadvantages of this method of preparing estimates.

6-9. The owner of a ski resort is considering installing a new ski lift, which will cost $90,000. Costs, other than depreciation, for operating and maintaining the lift are estimated to be $150 per day when operating. The Weather Service estimates that there is a 60% probability of 80 days of skiing weather per year, 30% probability of 100 days, and 10% probability of 120 days per year. The operators of the resort estimate that during the first 80 days of adequate snow in a season, an average of 500 people will use the lift each day, at a fee of $1 each. If 20 additional days are available, the lift will be used by only 400 people per day during the extra period; and if 20 more days of skiing are available, only 300 people per day will use the lift during those days.

The owners desire to recover any invested capital within 5 years, and want at least a 25% rate of return. Should the lift be installed?

6-10. To manufacture a new product, equipment costing $100,000 must be installed. It is estimated that it would have a life of 7 years and a salvage value of $20,000 at the end of that time. Costs for producing the product are:

Direct labor	$6 per unit
Direct materials	$3 per unit
Overhead	$10,000 + $2 per unit

The product will sell for $20 per unit, but there is considerable uncertainty as to the number of units that can be sold. The company's capital is worth 10%, before income taxes. Determine the rate of return (I.R.R.) for sales volumes of (a) 3,000 units, and (b) 5,000 units.

6-11. A dam is being planned for a river that is subject to frequent flooding. From past experience, the probabilities that water flow will exceed the design capacity of the dam, plus relevant cost information, are as follows:

Design of Dam	Probability of Greater Flow	Required Investment
A	0.100	$180,000
B	0.050	195,000
C	0.025	208,000
D	0.015	214,000
E	0.006	224,000

Estimated damages that occur if water flows exceed design capacity are $150,000, $160,000, $175,000, $190,000, and $210,000 for design A, B, C, D, and E, respectively. The life of the dam is expected to be 50 years, with negligible salvage value. For an interest rate of 6%, determine which design should be implemented. What nonmonetary considerations might be important to the best selection?

6-12. A diesel generator is needed to provide auxiliary power in the event that the primary source of power is interrupted. At any given time, there is a 0.1%

probability that the generator will be needed. Various generator designs are available, and more expensive generators tend to have higher reliabilities should they be called on to produce power. Estimates of reliabilities, investment costs, maintenance costs, and damages resulting from a complete power failure (i.e., the standby generator fails to operate) are given below for three alternatives.

Alternative	First Cost	Operating and Maintenance Costs/Year	Reliability	Cost of Power Failure	Salvage Value
R	$200,000	$5,000	0.96	$400,000	$40,000
S	170,000	7,000	0.95	400,000	25,000
T	214,000	4,000	0.98	400,000	38,000

If the life of each generator is 10 years and M.A.R.R. = 12%, which generator should be chosen?

6-13. Every time an automatic welding machine fails, the cost in idle labor and repairs is $1,000. The outage time averages 4 working hours and the plant works 2,000 hours a year. Suppose that the machine could fail a maximum of 250 times in a given year.

The probabilities of failure during a year are assessed from a certain trade association to be: no failures, 0.050; 1, 0.113; 2, 0.209; 3, 0.333; 4, 0.201; 5, 0.080; 6, 0.009; 7, 0.003; 8, 0.001; 9, 0.0008; 10, 0.00008.

A standby machine with an expected life of 10 years can be procured for $10,000 with $2,000 salvage value. The annual disbursements to keep it ready to run are $500. The probability of its breaking down during a standby run are nil. The minimum required rate of return is 15%. Make a recommendation regarding whether to purchase the standby machine.

6-14. There is a 10% chance that bad weather will delay construction work on a building beyond the penalty date and cause a penalty to a building contractor of $75,000. Should the contractor spend $15,000 for special all-weather equipment that will reduce the chance of a penalty to 4%? The building is large and will require 2 years to finish. Assume that money is worth 15% before taxes and that nonunion labor is completing this job.

6-15. Consider the matrix of profits below for five mutually exclusive alternatives (only one of the five will be chosen).

Alternative	State of Nature (Degree of Tax Reform)		
	Drastic	Moderate	None
A	12	18	18
B	14	14	15
C	14	26	15
D	10	22	14
E	18	12	10

(a) Which alternative is best when using the maximin rule?

(b) Which is best by the Laplace rule?

(c) Determine the best alternative by the Hurwicz principle when $Z = 0.7$.

6-16. For the following matrix of costs, select the best course of action with each of these decision rules: (a) minimax; (b) Laplace; (c) minimin.

Mutually Exclusive Alternative	State of Nature			
	θ_1	θ_2	θ_3	θ_4
S	18	18	10	14
T	14	14	14	14
U	5	26	10	14
V	14	22	10	10
W	10	12	12	10

6-17. When a new product is placed on the market, its sales are highly dependent on the reactions of producers of competitive products. In deciding between the manufacture of two alternative products, a producer estimates that the relationship between possible annual profits and the competitor's reactions are as follows:

Product	Competitor's Reaction		
	Aggressive	Normal	Minimal
A	$100,000	$300,000	$500,000
B	− 200,000	500,000	750,000

(a) Assuming competitors' reactions cannot be predicted, which product would be selected if the maximin rule is applied?

(b) Under the same assumptions, which product would be selected if the maximax rule is used?

(c) Assume that the probabilities of the competition reacting in each of the indicated manners are thought to be equal. Which product would produce the maximum expected profits?

6-18. Develop and demonstrate a procedure for considering multiple attributes associated with choosing among job offers (e.g., consider your first job after graduating from college).

appendix 6-A: using the computer program for evaluating investment opportunities

Earlier in this chapter it was mentioned that computers and computer programs can be helpful in analyzing the effects of uncertainties in investment decision making. A number of useful programs are being developed and are being used increasingly. A brief discussion of one such program, entitled PANIC, will be given here.† It contains some terms regarding depreciation and income taxes, which are described in Chapters 9 and 10 of this text.

Essentially this program determines the expected internal rate of return, on an after-tax basis if desired, and provides the analyst or decision maker with the opportunity to specify the manner in which the following factors will vary throughout the life of the project:

1. Earnings.
2. Salvage value.

The user of the program also must specify the following:

3. The depreciable investment.
4. Average service life expected.
5. Maximum life believed possible.
6. Minimum salvage value.
7. Percent of borrowed capital to be used.
8. Total invested capital.
9. Income tax rate.
10. Method of depreciation (straight line, SYD, etc.).
11. Interest payments on borrowed capital.
12. Reinvestment interest rate (whether at a specified rate or at the internal rate).
13. Obsolescence factor (optional). (This causes a built-in probability equation to be used to generate probabilities for the project continuing through any given year, and thus produces expected rates of return. Basically, the user estimates whether the obsolescence will be average, greater than average, or less than average.)

For most of the factors the user can specify individual values for each year of life, or he can select from several choices. For example, concerning salvage or trade-in value, four *modes* are offered:

Mode 1:   Individual values for each year are supplied by the user.
Mode 2:   Salvage value declines in a straight line.
Mode 3:   Salvage value declines exponentially: $S(n) = e^{-bn}$.
Mode 4:   Salvage value is book value, as determined by the depreciation method used.

The results obtained by the use of the PANIC program are illustrated in Figure 6-9. Looking at Figure 6-9, the reader will note that a depreciable investment of $169,360 is involved, with an estimated average life of 8 years and a maximum life of 12 years, and 25% of the required capital is to be borrowed. The *modes of operation* used (denoted 242312) are as follows:

† An extensive presentation of this program is contained in *Bulletin E20-8085*, "General Information Manual—Capital Investments," obtainable from the Technical Publications Department of the International Business Machines Corporation, White Plains, N.Y. The material used here is copyrighted by and used through the courtesy of the International Business Machines Corporation.

2—Uniform earnings.
4—SYD depreciation.
2—Uniform annual payments of principal and interest.
3—Exponential decline of salvage value.
1—Reinvestment rate is the internal rate.
2—Probabilities computed; obsolescence factor 0.50 (average).

The significant results of the study are contained in the last two columns and in the "expected return" figure. It will be noted that there is a low probability of 0.011 that the project might have an economic life of only 1 year. If this should occur, the rate of return would be −5.6%. The probabilities that the economic life will be 3, 4, 5, and so on, years increase until a maximum probability of 0.134 is indicated for a life of 8 years. For this life the rate of return would be 16.2%. However, there is an equal probability that, for the conditions assumed, the economic life will be 9 years rather than 8 years. If this occurs, the rate of return would be 17.5%. Similarly, there are quite good probabilities that the life might be 10, 11, or even 12 years. When all these probabilities and the corresponding rates of return are considered, the *expected* rate of return for the project is determined to be 14.5%. Note that this is less than the 16.2% rate for

FIGURE 6-9. Printout resulting from the IBM PANIC investment-analysis program. (Courtesy International Business Machines Corporation.)

```
                          CAPITAL INVESTMENT PROGRAM

   DEPRECIABLE INVESTMENT        $   169360    AVERAGE LIFE      8    MAXIMUM LIFE    12
   MINIMUM SALVAGE VALUE         $    16940    % BORROWED       25
   TOTAL INVESTED CAPITAL        $   177160    INCOME TAX %     52
   MODES OF OPERATION                242312
   EARNINGS FACTORS                          $    59570
   DEPRECIATION FACTORS                 10          10
   INTEREST PAYMENT FACTORS              8         500
   SALVAGE VALUE FACTORS               032
   REINVESTMENT INTEREST
   OBSOLESCENCE/DETERIORATION          .50
```

YEAR	EARNINGS	DEPRECIATION	INTEREST	CASH FLOW	SALVAGE VALUE	RATE OF RETURN	PROBABILITY FACTOR
1	$ 59570	$ 27741	$ 2215	$ 44171	$ 122983	−5.6%	.011
2	59570	24967	1983	42608	89306	.0	.021
3	59570	22193	1739	41039	64851	3.7	.036
4	59570	19419	1484	39464	47092	7.5	.056
5	59570	16645	1215	37881	34197	10.0	.080
6	59570	13871	933	36292	24833	12.5	.105
7	59570	11096	637	34695	18033	14.3	.124
8	59570	8322	327	33091	16940	16.2	.134
9	59570	5548		31479	16940	17.5	.134
10	59570	2774		30036	16940	18.1	.122
11	59570			28594	16940	18.7	.100
12	59570			28594	16940	19.4	.077

```
   EXPECTED RETURN          14.5 %
```

the estimated life of 8 years, because of the possibility that the project might not last 8 years.

The availability of the probabilities and the individual rates of return for each possible length of life, and the expected rate of return, provide considerably more useful information than would be available from an ordinary internal rate of return study based on an expected life of 8 years. But probably the greatest advantage of such a program is the fact that we can quickly complete another analysis, or as many as desired, for a different set of conditions (modes), thus being able to vary a number of estimated, and uncertain, variables.

It should be pointed out, however, that the results achieved by such a program can be no better in quality than the data fed into the program, and most of the values must be estimated by a human being. Thus we should not be misled by the fact that a somewhat sophisticated computer program produces a lot of information that otherwise would not be available.† However, if used properly, such programs can provide much useful information for the decision maker.

---

† It should be noted that the PANIC program will not throw out those cases that have no solutions or that have multiple solutions. Also, it does not take capital gains and losses into account.

chapter 7

selections
among alternatives

Most engineering and business endeavors can be carried out by more than one method or alternative. Such alternatives can be classified as *mutually exclusive*, meaning the choice of one excludes the choice of any other alternative. Typically, the several alternatives available require the investment of different amounts of capital, and the out-of-pocket costs (disbursements) usually will be different. Sometimes the several alternatives may produce different revenues, and frequently the physical lives of the assets, and even the economic lives of the projects, will be different. Because in such situations different levels of investment produce different economic outcomes, we must make an economy study to determine how much should be invested. Economy studies of mutually exclusive alternatives are the principal topic of this chapter. Studies of independent alternatives (opportunities) as well as sets of mutually exclusive, independent, and contingent projects will be considered briefly in the last section of this chapter.

a basic philosophy for studies of alternatives

Alternative investment studies are simplified if we assume that it already has been determined that the project is economically desirable; there is some level of investment that provides an acceptable return. This would be determined by the basic type of economy study discussed in Chapter 5. The decision regarding several investment alternatives, or levels, is further simplified if we remember that the purpose of investment is to obtain the greatest possible return from *each* dollar of capital. Theoretically, we should consider each unit of capital separately before investing it. In actual practice, there are usually only a limited number of choices. The problem of deciding which alternative should be used is made easier if we adopt a simple policy, which can be stated as follows: *The alternative that requires the minimum investment of capital and will produce satisfactory functional results will always be used unless there are definite reasons why an alternative requiring a larger investment should be adopted.* Under this policy we would consider the alternative that requires the least investment of capital to be the basic plan. All other alternatives would then be compared with it. Additional amounts of capital would not be invested unless some distinctively advantageous results, such as a sufficiently high rate of return on the added capital, would be obtained.

The investment of additional amounts of capital in various alternatives usually results in increased capacity, increased revenue, decreased operating expenses, or increased life. We expect through these means to obtain greater profit. Any change in profit that occurs is usually the direct result of the additional capital that was invested. Thus we may measure the effectiveness of the investment of additional increments of capital beyond the minimum that would be required to obtain the desired results. If the additional return obtained by investing an additional amount of capital is better than could be obtained from investment of the same additional capital elsewhere, the investment probably should be made. If such is not the case, we obviously would not invest more than the minimum amount required.

The philosophy above can be reduced to the following guideline principles when comparing mutually exclusive projects by any rate of return (R.R.) method. Here the rate of return method can be either the I.R.R. or E.R.R. or E.R.R.R.

1. Each increment of investment capital must justify itself (by sufficient R.R. on that increment).
2. Compare a higher investment alternative against a lower investment alternative only if that lower investment alternative is justified.
3. Choose the alternative project that requires the highest investment for which each increment of investment capital is justified.

This choice criterion assumes that the firm wants to invest any capital needed as long as a sufficient R.R. is expected to be earned on each increment of capital. In general, a sufficient R.R. is any R.R. that is greater than the minimum attractive rate of return. (M.A.R.R.)

When alternatives are compared by using equivalent worth methods (such as A.W., P.W., and F.W.), the indicated choices will be completely consistent with the principles above. Thus the preferred alternative will be the same as for R.R. methods, as will be shown in some of the following examples.

alternatives having identical (or no known) revenues and lives

A good many economy studies are made of situations in which the revenues from the various alternatives will be identical, or may be assumed to be identical, and the economic lives are also the same. Such conditions lead to the simplest type of alternative investment study. Such a case may be illustrated by the following example, which will be analyzed by each of the six economy study methods introduced in Chapter 5.

EXAMPLE 7-1

A company is going to install a new plastic-molding press. Four different presses are available. The essential differences as to cost and operating expenses are as follows:

| | Press | | | |
	A	B	C	D
Investment (installed)	$6,000	$7,600	$12,400	$13,000
Economic life	5 years	5 years	5 years	5 years
Annual disbursements				
Power	$ 680	$ 680	$1,200	$1,260
Labor	6,600	6,000	4,200	3,700
Maintenance	400	450	650	500
Property taxes and insurance	120	152	248	260
TOTAL ANNUAL DISBURSEMENTS	$7,800	$7,282	$6,298	$5,720

Each press will produce the same number of units. However, because of different degrees of mechanization, some require different amounts and classes of labor and have different operation and maintenance costs. None is expected to have a salvage value. Any capital invested in this company is expected to earn at least 10% before taxes. Which press should be chosen?

Solution of Example 7-1 by Annual Cost (A.C.) Method

Table 7-1 shows the analysis by the A.C. method. The economic criterion is to choose that alternative with the minimum A.C., which is press D. However, it should be noted that press B is very close, showing a total A.C. of only $9,287 — $9,149 = $138 more than for press D. █

table 7-1
comparison of four molding presses by A.C. method

| | Press | | | |
	A	B	C	D
Annual expenses:				
Disbursements	$7,800	$7,282	$6,298	$5,720
Capital recovery (depreciation + interest):				
= (investment) × $(A/P, 10\%, 5)$	1,583	2,005	3,271	3,429
TOTAL A.C.	$9,383	$9,287	$9,569	$9,149 █

table 7-2

comparison of four molding presses by P.W.-C. method

	Press			
	A	B	C	D
Present worth of:				
Investment	$ 6,000	$ 7,600	$12,400	$13,000
Disbursements:				
(annual disbursements)				
$\times (P/A, 10\%, 5)$	29,568	27,605	23,874	21,683
TOTAL P.W.-C.	$35,568	$35,205	$36,274	$34,683 ∎

When alternatives for which revenues are not know are compared using the present worth method, that method is more commonly and descriptively called the present cost (P.C.) or present worth–cost (P.W.-C.) method. Just as for the A.C. method, the alternative that has the minimum P.C. is assumed to be the most desirable.

Solution of Example 7-1 by Present Worth–Cost (P.W.-C.) Method

Table 7-2 shows the analysis by the P.W.-C. method. The economic criterion is to choose that alternative with the minimum P.W.-C., which of course, consistent with the other methods, is press D. Again, press B is shown to be a close second choice. ∎

Solution of Example 7-1 by Future Worth–Cost (F.W.-C.) Method

Table 7-3 shows the analysis by the F.W.-C. method. Once again, press D is shown to be the least costly, and the order of desirability is the same as for the A.W. and F.W. methods. ∎

table 7-3

comparison of four molding presses by F.W.-C. method

	Press			
	A	B	C	D
Future worth of:				
Investment:				
(investment) $\times (F/P, 10\%, 5)$	$ 9,663	$12,240	$19,970	$20,937
Disbursements:				
(annual disbursements)				
$\times (F/A, 10\%, 5)$	47,620	44,457	38,450	34,921
TOTAL F.W.-C.	$57,283	$56,697	$58,420	$55,858

Solution of Example 7-1 by Internal Rate of Return (I.R.R.) Method

Table 7-4 provides a tabulation of cash flows and a calculated I.R.R. for each increment of investment considered. *In this table and in subsequent economy studies of alternatives by a rate of return (R.R.) method, the symbol "Δ" is used to mean "incremental" or "change in," and the letters on the end of arrows indicate the projects for which the increment is considered.* Note that the alternatives are ordered according to increasing amounts of investment to facilitate step-by-step consideration of each increment of investment.

The first increment subject to analysis is the $7,600 − $6,000 = $1,600 extra investment required for press B compared to press A. For this increment, annual disbursements are reduced or saved by $7,800 − $7,282 = $518. The I.R.R. on the incremental investment is the interest rate at which the present worth of the incremental net cash flows is zero. Thus

$$-\$1,600 + \$518(P/A, i', 5) = 0$$

$$(P/A, i', 5) = \$1,600/\$518 = 3.10$$

Interpolating between the tabled factors for the two closest interest rates, one can find that i' = I.R.R. = 18.5%. Since 18.5% > 10% M.A.R.R., the increment A → B is justified.

The next increment subject to analysis is the $12,400 − $7,600 = $4,800 extra investment required for press C compared to press B. This increment results in annual disbursement reduction of $7,282 − $6,298 = $984. The I.R.R. on increment B → C can then be determined by finding the i' at which

$$-\$4,800 + \$984(P/A, i', 5) = 0$$

Thus i' = I.R.R. can be found to be approximately 0.1%. [*Note:* $(P/A, 0\%, 5)$ = 5.00.] Since 0.1% < 10%, the increment B → C is *not* justified, and hence we can say that press C itself is not justified.

The next increment that should be analyzed is B → D, and not C → D. This is because press C has already been shown to be unjustified, and hence it can no

table 7-4
comparison of four molding presses by I.R.R. method

	Press			
	A	B	C	D
Investment	$6,000	$7,600	$12,400	$13,000
Annual disbursements	$7,800	$7,282	$6,298	$5,720
Life	5 years	5 years	5 years	5 years
Increment considered:		A → B	B → C	B → D
Δ Investment		$1,600	$4,800	$5,400
Δ Annual disbursements (savings)		$518	$984	$1,564
I.R.R. on Δ investment		18.5%	0.1%	13.8%
Is increment justified?		Yes	No	Yes

longer be a valid basis for comparison with other alternatives.† For $B \rightarrow D$, the incremental investment is $13,000 − $7,600 = $5,400 and the incremental savings in annual disbursements is $7,282 − $5,720 = $1,564. The I.R.R. on increment $B \rightarrow D$ can then be determined by finding the i' at which

$$-\$5,400 + 1,564(P/A, i', 5) = 0$$

Thus i' can be found to be 13.8%. Since 13.8% > 10%, the increment $B \rightarrow D$ is justified.

Based on the preceding analysis, press D would be the choice because it is the highest investment for which each increment of investment capital is justified. In choosing press D, one is actually accepting the first $6,000 investment as necessary without justification, the $1,600 increment $A \rightarrow B$ earning 18.5%, and the $5,400 increment $B \rightarrow D$ earning 13.8%. Figure 7-1 depicts these results for all separable increments that comprise the total investment of $13,000 in press D. Note that this choice is reached without calculation of the I.R.R. on the total investment for press D or *any* of the individual presses. ▌

The underlying rationale for the type of analysis above is that the firm wants to invest capital if and only if it is necessary or will earn the minimum required return, which in this case is 10%. Thus the firm supposedly has opportunities to invest any capital not used for one of the presses in comparable risk projects elsewhere and earn 10%.

An error commonly made in the type of analysis above is to choose the alternative either with the highest overall R.R. or with the highest R.R. on an incremental investment. Neither criterion is correct generally. For example, in the problem above, one might choose press B rather than press D because the I.R.R. for increment $A \rightarrow B$ is 18.5% and for increment $B \rightarrow D$ is 13.8%. However, the choice of press B means that one is forgoing the opportunity to invest

FIGURE 7-1. Representation of increments and I.R.R. on increments considered in justifying press D in Example 7-1. (a) Analysis breakdown, and (b) final choice.

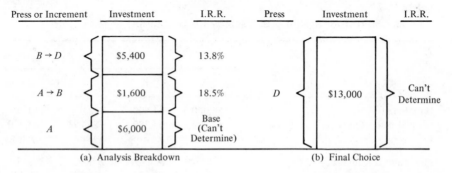

(a) Analysis Breakdown   (b) Final Choice

† If the I.R.R. were computed for increment $C \rightarrow D$, it would be in excess of 80%. This only means that press D is *very* attractive compared to press C, but since press C is not justified, this is not a valid basis for comparison.

table 7-5
comparison of four molding presses by E.R.R. method

	Press			
	A	B	C	D
Investment	$ 6,000	$ 7,600	$12,400	$13,000
F.W. of annual disbursements:				
(annual disbursements)				
× (F/A, 10%, 5)	$47,620	$44,457	$38,450	$34,921
Increment considered		A → B	B → C	B → D
Δ Investment		$1,600	$4,800	$5,400
Δ F.W. of annual disbursements				
(savings)		$3,163	$6,007	$9,536
E.R.R. on Δ investment		14.6%	4.6%	12.0%
Is increment justified?		Yes	No	Yes

an extra $5,400 to obtain D (i.e., investment in B → D) and thereby make 13.8% on that $5,400. If only 10% is the return obtainable from alternative use of the money, clearly one should invest in D rather than B.

Solution of Example 7-1 by External Rate of Return (E.R.R.) Method

The rationale and criteria for using the E.R.R. method to compare alternatives are the same as for the I.R.R. method. The only difference is in the calculation methodology. Table 7-5 provides a tabulation of the calculations and answers for each increment of investment considered. The annual disbursements are compounded to the end of the 5-year life at e = M.A.R.R. = 10%.

To illustrate the procedure for finding the E.R.R. on a Δ investment, consider increment A → B. The formula is

$$1,600(F/P, i'\%, 5) = \$518(F/A, 10\%, 5)$$

$$(F/P, i'\%, 5) = [\$518(6.1051)]/1,600 = 1.977$$

Linear interpolation between 12% and 15% results in i' = 14.6%, which is the E.R.R. Using the M.A.R.R. = 10%, increment A → B is justified. Using similar calculations, increment B → C, earning 4.6%, is not justified and increment B → D, earning 12.0%, is justified. Hence press D is again, as anticipated, the choice. ∎

Solution of Example 7-1 by Explicit Reinvestment Rate of Return (E.R.R.R.) Method

The rationale and criteria for using the E.R.R.R. method to compare alternatives are the same as for the I.R.R. method. The only difference is in the calculation methodology. Table 7-6 provides a tabulation of the calculation and answers for each increment of investment considered. Keeping in mind that the minimum

table 7-6
comparison of four molding presses by E.R.R.R. method

	Press			
	A	B	C	D
Investment	$6,000	$7,600	$12,400	$13,000
Annual expenses				
Disbursements	7,800	7,282	6,298	5,720
Depreciation:				
(investment) × $(A/F, 10\%, 5)$	983	1,245	2,031	2,129
TOTAL ANNUAL EXPENSES	$8,783	$8,527	$8,329	$7,849
Increment considered		A → B	B → C	B → D
Δ Investment		$1,600	$4,800	$5,400
Δ Total annual expenses		$256	$198	$678
E.R.R.R. on Δ investment		16%	4.1%	12.5%
Is increment justified?		Yes	No	Yes

required R.R. is 10%, increment A → B earning an E.R.R.R. of 16% is justified, increment B → C earning 4.1% is not justified, and increment B → D earning 12.5% is justified. Hence press D would be the choice, which of course is the same as when using the E.R.R. method.

To find the E.R.R.R. on the incremental investment for any two alternatives, one need only divide the incremental total annual expenses (including depreciation using the sinking fund method) by the incremental investment for those two alternatives. As an example, for increment A → B in Table 7-6, the incremental total annual expense is $8,783 − $8,527 = $256, and the incremental investment is $7,600 − $6,000 = $1,600. Hence the E.R.R.R. on increment A → B is $256/$1,600 = 16%. One can similarly calculate the E.R.R.R. on all other applicable incremental investments. ∎

Summary Comparison of Economy Study Methods Based on Example 7–1 Results

Near the end of Chapter 5, the "Summary Comparison ..." section stated that all six methods will give a valid answer for the reinvestment assumption inherent in that method. It further stated that the equivalent annual, present, and future worth methods give the same ranking of alternatives, because all assume that funds can be reinvested at the same rate, normally the M.A.R.R. The equivalent worth results for Example 7-1 were as follows:

	Press			
Method	A	B	C	D
Annual cost	$ 9,383	$ 9,287	$ 9,569	$ 9,149
Present worth–cost	35,568	35,205	36,274	34,683
Future worth–cost	57,283	56,697	58,420	55,858

In general, the results for any two alternatives, say A and B, is a constant ratio as follows:

$$\frac{A.C._A}{A.C._B} = \frac{P.W.\text{-}C._A}{P.W.\text{-}C._B} = \frac{F.W.\text{-}C._A}{F.W.\text{-}C._B}$$

Further, for any alternative,

$$A.C. = P.W.\text{-}C.(A/P, i\%, N) = F.W.\text{-}C.(A/F, i, N)$$

and

$$P.W.\text{-}C. = F.W.\text{-}C.(P/F, i\%, N)$$

The rate of return methods will always give consistent answers regarding which investments (or incremental investments) will earn more or less than the M.A.R.R.[†] However, the calculated rates of return will differ somewhat between the methods because of the differing reinvestment assumptions. The results for Example 7-1, in which M.A.R.R. = 10%, were as follows:

Method	Increment Considered (%)		
	A → B	B → C	B → D
I.R.R.	18.5	0.1	13.8
E.R.R.	14.6	4.6	12.0
E.R.R.R.	16.0	4.1	12.5

alternatives having different lives

In many cases an alternative that requires a larger investment of capital than another will have not only higher revenues and/or lower out-of-pocket costs but also a longer economic life. This complicates somewhat the analysis of alternative investments. In order to make economy studies of such cases, we must adopt some procedure that will put the alternatives on a comparable basis. Two types of assumptions commonly are employed, which we shall call the (1) repeatability assumption and the (2) coterminated assumption.

The *repeatability assumption* involves two main subassumptions as follows:

(a) The period of needed service for which the alternatives are being compared is either indefinitely long or a length of time equal to a common multiple of the lives of the alternatives.

(b) What is estimated to happen in the first life cycle will happen in all succeeding life cycles, if any, for each alternative.

The repeatability assumption is usually made in economic analyses by default (i.e., because there is no good basis for estimates to the contrary). They are implicitly contained in all examples and problems illustrating all methods of economic evaluation herein unless otherwise indicated.

[†] Statement is based on the usual assumption for E.R.R. and E.R.R.R. methods that $e\% = $ M.A.R.R.

Whenever alternatives to be compared have different lives and one or both of the conditions is not appropriate, it is necessary to enumerate what receipts and what disbursements are expected to happen at what points in time for each alternative for as long as service will be needed or the irregularity is expected to exist. This enumerated information can then be used to make the economy study.

The *coterminated assumption* involves using a finite study period for all alternatives. This time span may be the period of needed service or any arbitrarily specified length of time such as (a) the life of the shorter-lived alternative, or (b)

FIGURE 7-2. Repeatability versus coterminated assumptions.

Repeatability Assumption: (For lowest common multiple of lives, 50 years)

M–1	M–2	M–3	M–4	M–5

0   10   20   30   40   50   years

N–1	N–2

0   25   50   years

Coterminated Assumption

(A) At the life of the shorter–lived alternative:

M–1

0

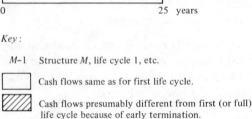

0   10   years

(B) At the life of the longer–lived alternative:

M–1	M–2	M–3

0   10   20   25   years

N–1

0   25   years

Key:

M–1   Structure M, life cycle 1, etc.

☐   Cash flows same as for first life cycle.

▨   Cash flows presumably different from first (or full) life cycle because of early termination.

the life of the longer-lived alternative, or (c) what is sometimes descriptively called the "organization's planning horizon."

Figure 7-2 shows the effect of these two assumptions for structures M and N, which have lives of 10 and 25, years, respectively, and which will be the subject of Example 7-2. The objective is to choose either M or N because here the "do nothing" option does not exist.

comparisons using the repeatability assumptions

Consider the following example for which economic analyses will be made using different economy study methods and the repeatability assumption.

EXAMPLE 7-2

	Structure M	Structure N
First cost	$12,000	$40,000
Economic life	10 years	25 years
Salvage value at end of life	0	$10,000
Annual disbursements for O + M	$2,200	$1,000

Determine which structure is better if the M.A.R.R. is 5% using the repeatability assumption.

Solution of Example 7-2 by A.C. Method

	Structure M	Structure N
Annual costs:		
Disbursements	$2,200	$1,000
C.R. cost:		
$12,000(A/P, 5%, 10)	1,554	
($40,000 − $10,000)(A/P, 5%, 25) + $10,000(5%)		2,630
TOTAL A.C.	$3,754	$3,630

Since $3,630 is less than $3,754, structure N is shown to be the better economic choice. ∎

Solution of Example 7-2 by P.W.-C. Method

The cardinal principle in comparing alternatives by the P.W.-C. method is that all alternatives should be compared over the same length of time (study period). By use of the repeatability assumption, this is most conveniently the lowest common multiple of the lives. In this case 50 years of service can be provided by building structure M and replacing it four times—a total of five structures. The same length of service can be obtained by building structure N and replacing it once—a total of two such structures.

Present Worth of Costs	Structure M
Original investment	$12,000
First replacement: $12,000($P/F$, 5%, 10)	7,350
Second replacement: $12,000($P/F$, 5%, 20)	4,520
Third replacement: $12,000($P/F$, 5%, 30)	2,770
Fourth replacement: $12,000($P/F$, 5%, 40)	1,700
Annual disbursements: $2,200($P/A$, 5%, 50)	40,170
TOTAL P.W.-C.	$68,510

Present Worth of Costs	Structure N
Original investment	$40,000
First replacement: ($40,000 − $10,000)($P/F$, 5%, 25)	8,850
Annual disbursement: $1,000($P/A$, 5%, 50)	18,260
Less salvage of first replacement: $10,000($P/F$, 5%, 50)	− 870
TOTAL P.W.-C.	$66,240

Thus the P.W.-C. for structure N is lower, indicating once again that N is better. ▮

Solution of Example 7-2 by I.R.R. Method

The I.R.R. on the incremental investment for any two alternatives can be found by any of the following approaches†:

1. Finding the rate at which the present worth of the net cash flow of the differences between the two alternatives is equal to zero, or
2. Finding the rate at which the present worths (or present worth–costs) of the two alternatives are equal, or
3. Finding the rate at which the annual worths (or annual costs) of the two alternatives are equal.

The first approach (above) was used in solving Example 7-1 by the I.R.R. method. We will illustrate the second approach for this example.

The P.W.-C.'s for the two structures as calculated in the last solution were very close. This means the interest rate used, 5%, is close to the I.R.R. answer. If an interest rate of 6% were used, the present worth of costs for structures M and N would be $60,000 and $62,210, respectively. Since we desire to find the interest rate at which the present worths for the alternatives are equal (difference in present worths = 0), the results can be summarized, and the answer interpolated as follows:

† When the cash flow difference between alternatives changes sign more than once, the I.R.R. method can result in two or more answers, as demonstrated in Appendix 5-B.

Structure	M	N	Difference (N − M)
Total P.W.-C. at 5%	$68,510	$66,240	− $2,270
Total P.W.-C. at i'%			0
Total P.W.-C. at 6%	60,290	62,210	1,920

$$i' = 5\% + \left(\frac{2,270}{2,270 + 1,920}\right)(6\% - 5\%) = 5.5\%$$

Thus the I.R.R. on the incremental investment required for structure N relative to structure M is 5.5%. Since 5.5% is greater than the 5% M.A.R.R., the extra investment is justified, and structure N would be chosen based on this analysis. If i' had been less than 5%, structure M would have been chosen. ∎

Solution of Example 7-2 by E.R.R.R. Method

	Structure M	Structure N
Annual expenses:		
Disbursements	$2,200	$1,000
Depreciation:		
$12,000(A/F, 5\%, 10)$	954	
$(\$40,000 - \$10,000)(A/F, 5\%, 25)$		630
TOTAL ANNUAL EXPENSES	$3,154	$1,630

Δ Annual expenses (annual savings): $3,154 − $1,630 = $1,524
Δ Investment: $40,000 − $12,000 = $28,000
E.R.R.R. on Δ Investment: $1,524/$28,000 = 5.4%

Since 5.4% > 5%, structure N is again shown to be the more desirable alternative. ∎

comparisons using coterminated assumption

If the period of needed service is less than a common multiple of the lives, that should be the study period used for economic study purposes. Even if the period of needed service is not known, it is often thought convenient to use some arbitrarily specified study period for both alternatives. In such cases, a some-times-important concern is the salvage value to be assigned to any alternative that will not have used up its normal economic life as of the end of the study period.

A cotermination point commonly used is the life of the shortest-lived alternative. The following example is illustrative.

EXAMPLE 7-3

Suppose that we are faced with the same alternatives as in Example 7-2, as follows:

	Structure M	Structure N
First cost	$12,000	$40,000
Economic life	10 years	25 years
Salvage value at end of life	0	$10,000
Annual disbursements	$2,200	$1,000
M.A.R.R.	5%	5%

It is desired to compare the economics of the alternatives by terminating the study period at the end of 10 years and assuming a salvage (remaining) value for structure N at that time of, say, $21,100.†

Solution of Example 7-3 by A.C. Method

	Structure M	Structure N
Annual costs:		
Disbursements	$2,200	$1,000
C.R. cost:		
$12,000(A/P, 5%, 10)$	1,554	
($40,000 − $21,100)(A/P, 5\%, 10) + $21,100(5\%)$		3,503
TOTAL A.C.	$3,754	$4,503

Structure M now has the lower annual cost, thus reversing the indicated decision for the same problem using the repeatability assumption. ∎

Solution of Example 7-3 by E.R.R.R. Method

	Structure M	Structure N
Annual expenses:		
Disbursements	$2,200	$1,000
Depreciation:		
$12,000(A/F, 5%, 10)$	955	
($40,000 − $21,000)(A/F, 5\%, 10)$		1,503
TOTAL	$3,155	$2,503

Δ Annual expenses (savings): $3,155 − $2,503 = $652
Δ Investment: $40,000 − $12,000 = $28,000
E.R.R.R. on Δ investment: $652/$28,000 = 2.3%

Since 2.3% < 5%, structure M is again the economic choice. ∎

It should be expected that different assumptions for determining the salvage value of the longer lived alternative could result in different decisions. This

† The $21,100 happens to be the book value at the end of the tenth year, assuming the sum-of-years'-digits method of depreciation. Computation of book values will be explained in Chapter 9.

situation merely points up the importance of basing the salvage value determination on the best available information or estimate as to how the salvage value will, in fact, vary over time.

comparisons with nonrepeating cash flows and using coterminated assumption

Of course, in many situations it is not reasonable to expect cash flows to repeat for life cycles after the first. In such situations the P.W. or F.W. method is usually easiest to use for comparison of alternatives after enumerating the magnitude and occurrence of expected cash flows.

EXAMPLE 7-4

Suppose that we are faced with the same alternatives as in Example 7-2 except now we want to terminate the study period at the end of 25 years, either because the firm thinks the services of a structure will be needed only for that time, or because it arbitrarily wants to use that length of study period. Further, the firm estimates that the same first cost and annual disbursements for structure M will not be repeated after the first life cycle. Rather, it estimates that the first cost will be $18,000 for the first replacement and $20,000 for the second replacement; and

FIGURE 7-3. Cash flow diagram for Example 7-4.

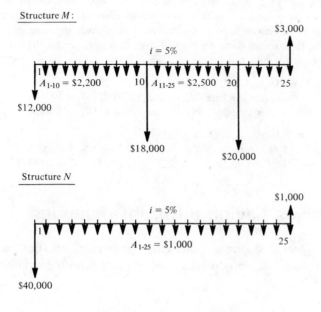

that the annual disbursements will be \$2,500 for both the first and second replacements. Since the third replacement will be terminated after only 5 years of service, it is estimated to have a salvage value then of \$3,000. Estimates for structure N are the same as given for Example 7-2, and the M.A.R.R. = 5%. Figure 7-3 shows a cash flow diagram for both alternatives.

Solution of Example 7-4 by P.W.-C. Method

Structure M:

Original investment	\$12,000
First replacement: \$18,000(P/F, 5%, 10)	11,050
Second replacement: \$20,000(P/F, 5%, 20)	7,538
Annual disbursements:	
Years 1–10: \$2,200(P/A, 5%, 10)	16,988
Years 11–25: \$2,500(P/A, 5%, 15)(P/F, 5%, 10)	15,930
Less: salvage on second replacement: \$3,000(P/F, 5%, 25)	− 886
TOTAL P.W.-C.	\$62,620

Structure N:

Original Investment	\$40,000
Annual disbursements: \$1,000(P/A, 5%, 25)	14,094
Less: salvage on original investment: \$1,000(P/F, 5%, 25)	− 295
TOTAL P.W.-C.	\$53,799

Thus, with the changed conditions, structure N appears to be considerably more economical than structure M. ∎

Solution of Example 7-4 by A.C. and F.W.-C. Methods

The easiest way to compare the annual costs and the future worth of costs for the alternatives in Example 7-4 is to determine these quantities directly from the equivalent present worth–cost results obtained above. Because the study period is 25 years, the annual costs are spread over that period and the future worths of costs occur at the end of that period. Thus

A.C. = P.W.-C. × (A/P, 5%, 25)
   For structure M: = \$62,620(0.0710) = \$4,446
   For structure N: = \$53,799(0.0710) = \$3,820

F.W.-C. = P.W.-C. × (F/P, 5%, 25)
   For structure M: = \$62,620(3.3864) = \$212,056
   For structure N: = \$53,799(3.3864) = \$182,185 ∎

alternatives having different revenues

It is to be expected that many alternative investment situations will result in different revenues being produced by two or more investment possibilities. Any of the six main methods of comparison presented in Chapter 5 again may generally be used for such economy studies without difficulty, if their respective reinvestment assumptions are applicable.

In general, a firm should be willing to invest additional increments of capital as long as each increment is justified by a sufficient return. This criterion is made explicit when using a rate of return method and implicitly exists when using an equivalent worth method.

EXAMPLE 7-5

The following are estimated data for two investment alternatives, A and B, for which revenues as well as costs are known, and which have different lives. If the minimum attractive rate of return is 10%, show which project is more desirable using each of the four basic methods.

	A	B
Investment	$3,500	$5,000
Annual revenue	$1,900	$2,500
Annual disbursements	$645	$1,383
Estimated life	4 years	8 years
Net salvage value	0	0

Solution of Example 7-5 by A.W. Method

	A	B
Annual revenue	$1,900	$2,500
Annual expenses:		
Disbursements	645	1,383
C.R. cost:		
$3,500(A/P, 10\%, 4)	1,104	
$5,000(A/P, 10\%, 8)		937
TOTAL ANNUAL EXPENSES	$1,749	$2,320
Net A.W. (revenue − expenses)	$151	$180

Since alternative B has a higher net A.W., it is shown to be the better economic choice. ∎

Solution of Example 7-5 by P.W. Method

The expected lives of the alternatives are 4 and 8 years, respectively. The lowest common multiple is 8 years, which will be taken to be the length of the study period.

	A	B
Annual revenue:		
$1,900(P/A, 10\%, 8)$	$10,136	
$2,500(P/A, 10\%, 8)$		$13,337
TOTAL REVENUE	$10,136	$13,337
Annual disbursements:		
$645(P/A, 10\%, 8)$	3,441	
$1,383(P/A, 10\%, 8)$		7,378
Original investment	3,500	5,000
First replacement: $3,500(P/F, 10\%, 4)$	2,390	
TOTAL COSTS	$9,331	$12,378
Net P.W. (revenue − costs)	$805	$959

Since alternative B has a higher net P.W., it again is shown to be the better economic choice. ∎

Solution of Example 7-5 by I.R.R. Method

The first step is to find the I.R.R. for the lowest investment project, alternative A†: Using the present worth of net cash flow relationship:

$$- \$3,500 + (\$1,900 - \$645)(P/A, i'\%, 4) = 0$$

$$(P/A, i'\%, 4) = \frac{\$3,500}{\$1,900 - \$645} = 2.79$$

Interpolating in interest tables between 15% and 20%:

$$i'\% = 15\% + \left(\frac{2.86 - 2.79}{2.86 - 2.59}\right)(20\% - 15\%) = 16.5\%$$

Since $16.5\% > 10\%$, alternative A is justified.

The next step is to find the I.R.R. for the incremental investment required for B instead of A. Instead of using our option of finding either the i' at which the P.W. of the net cash flow for the difference between the alternatives = 0, or the i' at which the P.W. of the two alternatives are equal, we will find the i' at which the A.W. of the two alternatives are equal. Thus

$$i' = ?$$

$$- \$3,500(A/P, i'\%, 4) + \$1,900 - \$645 = - \$5,000(A/P, i'\%, 8) + \$2,500 - \$1,383$$

at $i' = 10\%$:

$$- \$3,500(0.3155) + \$1,255 \overset{?}{=} - \$5,000(0.1874) + \$1,117$$
$$\$150 \neq \$181$$

at $i' = 15\%$:

$$- \$3,500(0.3503) + \$1,255 \overset{?}{=} - \$5,000(0.2229) + \$1,117$$
$$\$25 \neq \$2$$

† The lowest investment project constitutes the first increment, going from nothing to that investment.

Summarizing the results and linearly interpolating, we obtain:

	A	B	Difference (B − A)
At $i = 10\%$	\$150	\$181	\$31
$i' = X\%$			0
At $i = 15\%$	25	2	− 23

$$i = 10\% + \left(\frac{31}{31 + 23}\right)(15\% - 10\%) = 12.9\%$$

Thus the I.R.R. on the incremental investment is 12.9%, which is greater than the M.A.R.R. of 10%. Hence the incremental investment is justified, and alternative B is the better economic choice.

Figure 7-4 illustrates graphically how the I.R.R. on the incremental investment is determined. The graph shows the A.W. to vary as a nonlinear function of i, rather than assuming linearity between $i = 10\%$ and 15% as was done for the mathematical interpolation above.

Note that the choice of alternative B above was made *without* computing the I.R.R. on the total investment in alternative B. Actually, such a computation is unnecessary in this case, for if each increment of investment comprising a project is justified, then the total investment is also justified.† In this case, the \$3,500 for

FIGURE 7-4.   Finding I.R.R. on incremental investment in Example 7-5.

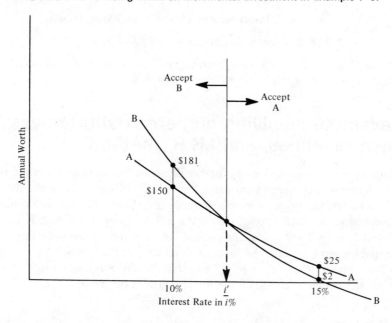

† In the case of alternatives for which only costs are known, this assumes that the alternative with the lowest first cost is found acceptable.

alternative A will earn 16.5% and the $1,500 increment for alternative B compared with A will earn 12.9%. Thus the I.R.R. for the total $5,000 investment in alternative B is a weighted average of 16.5% and 12.9%. ∎

Solution of Example 7-5 by E.R.R.R. Method

	A	B
Annual revenue	$1,900	$2,500
Annual expenses:		
Disbursements	645	1,383
Depreciation:		
$3,500(A/F, 10\%, 4)$	754	
$5,000(A/F, 10\%, 8)$		437
TOTAL EXPENSES	$1,399	$1,820
Annual profit	$501	$680

The E.R.R.R. for the lowest investment project, alternative A, is

$$\text{E.R.R.R.} = \$501/\$3,500 = 14.3\%$$

Since 14.3% > 10%, alternative A is justified.

The E.R.R.R. for the incremental investment required for B instead of A, (A → B), is computed as follows:

$$\Delta \text{ Annual profit} = \$680 - \$501 = \$179$$

$$\Delta \text{ Investment} = \$5,000 - \$3,500 = \$1,500$$

$$\text{E.R.R.R. on } \Delta \text{ investment} = \$179/\$1,500 = 11.9\%$$

Since 11.9% > 10%, the incremental investment is justified, and alternative B is again shown to be the better economic choice. ∎

an example involving numerous alternatives, known revenues, and I.R.R. method

The following example is given to further illustrate the principle that the return on each increment of investment capital should be justified. To make the computations easier, the projects in this example each have a salvage value equal to the investment. In such special cases, the I.R.R. (and the E.R.R.R.) can be calculated directly by dividing the annual net cash inflow (revenue or savings) by the investment amount. The reader will recall that the symbol "Δ" is used to mean "incremental" or "change in," and that the letters on the ends of an arrow designate the projects for which the increment is considered.

EXAMPLE 7-6

Given the following alternative projects A through F:

	A	B	C	D	E	F
Investment	$1,000	$1,500	$2,500	$4,000	$5,000	$7,000
Annual savings in cash disbursements	150	375	500	925	1,125	1,425
Salvage value	1,000	1,500	2,500	4,000	5,000	7,000

If the company is willing to invest any capital which will earn at least 18% M.A.R.R., find which mutually exclusive alternative, if any, should be chosen using the I.R.R. method.

Solution

Increment Considered	A	B	B → C	B → D	D → E	E → F
Δ Investment	$1,000	$1,500	$1,000	$2,500	$1,000	$2,000
Δ Annual savings	$150	$375	$125	$550	$200	$300
I.R.R. on Δ investment	15%	25%	12.5%	22%	20%	15%
Is increment justified?	No	Yes	No	Yes	Yes	No

By the analysis above, alternative E would be chosen because it is the alternative requiring the highest investment for which each increment of investment capital is justified. Note that this analysis was performed without even considering the rate of return on the total investment for each of the alternatives. It is instructive to note that the I.R.R. on the total investment for each alternative is:

Alternative Project	A	B	C	D	E	F
I.R.R.	15%	25%	20%	23%	22.5%	20%

Note that alternative B has the highest overall I.R.R. and that alternative F has an overall I.R.R. which is greater than the minimum of 18%. Nevertheless, alternative E would be chosen on the rationale that the firm wants to invest any increment of capital when and only when the increment will earn at least the M.A.R.R. ∎

the capitalized worth method

One special variation of the present worth method involves the determination of the worth of all receipts and/or disbursements over an infinitely long length of time. This is known as the *capitalized worth* (C.W.) *method*. If disbursements only are considered, results obtained by this method can be more appropriately expressed as *capitalized cost*. This is a convenient basis for comparison when the period of needed service is indefinitely long or when the common multiple of the lives is very long and the repeatability assumption is applicable.

The capitalized worth of a perpetual series of end-of-period uniform payments A, with interest at $i\%$ per period, is $A(P/A, i\%, \infty)$. From the interest formulas, it can be seen that $(P/A, i\%, N) \rightarrow 1/i$ as N becomes very large. Thus capitalized worth $= A/i$ for such a series, as was given in Equation 4-22.

The annual worth of a series of payments of amount $\$X$ at the end of each kth period with interest at $i\%$ per period is $\$X(A/F, i\%, k)$. The capitalized worth of such a series can thus be calculated as $\$X(A/F, i\%, k)/i$, as was given in Equation 4-23.

EXAMPLE 7-7

Compare structures M and N given in Example 7-2 by the capitalized worth (cost) method.

Solution

	Capitalized Cost	
	Structure M	Structure N
First cost	$12,000	$40,000
Replacements:		
$12,000(A/F, 5\%, 10)/0.05	19,080	
($40,000 − $10,000)(A/F, 5\%, 25)/0.05		12,600
Annual disbursements:		
$2,200/0.05	44,000	
$1,000/0.05		20,000
TOTAL CAPITALIZED COST	$75,080	$72,600

Thus structure N is the indicated better alternative, which is, of course, consistent with the results for Example 7-2 by the other methods of comparison. As an aside, since we have previously calculated the equivalent annual costs for these two alternatives, the easiest way to have determined the capitalized costs was to use the relation

$$\text{C.C.} = \frac{\text{A.C.}}{i}$$

For structure M: $3,754/0.05 = $75,080

For structure N: $3,630/0.05 = $72,600 ∎

comparison of alternatives by the payout period method

Some people persist in comparing alternatives by the payout (payback) period method. In many cases this procedure gives misleading or erroneous results, so its use should be avoided except as a supplement to analysis by a correct method.

If the life of each alternative is the same and the risks are comparable, the method does rank the alternatives relative to each other correctly. Otherwise, the method is useful primarily as an indicator of relative risk. The higher the payout period, the greater the opportunity for events to not turn out as originally estimated, and thus the greater the risk. The following is a typical case involving three alternatives:

	A	B	C
Required investment	$5,000	$5,000	$6,000
Estimated useful life (N)	8 years	5 years	13 years
Available for payout (annual revenue minus out-of-pocket costs)	$2,250	$2,500	$3,000
Payout period = investment/available for payout	2.2 years	2.0 years	2.0 years

By this payout analysis, alternative B would be selected, since it has a shorter payout period than alternative A and requires less capital than C. If, however, the three alternatives are compared on the basis of the rate of return (E.R.R.R. method), with proper consideration given to the estimated lives, the results are as follows:

	A	B	C
Annual revenue minus out-of-pocket costs	$2,250	$2,500	$3,000
Less: annual depreciation:			
Investment (A/F, 8%, N)	470	853	279
Annual profit	$1,780	$1,647	$2,721
E.R.R.R. = annual profit/investment	$35\frac{1}{2}$%	33%	45%

By this analysis, alternative B is found to be the least desirable of the three, and, if capital is available, alternative C is more desirable than either A or B.

Because the payout period method does not take expected life into account, its use is not recommended for comparing alternatives having different lives. Rather than indicate overall profitability or desirability, the payout method shows only the number of years required to recover investment capital, which is sometimes a useful measure of risk.

an example of increasing investment affecting the rate of return

An excellent example of how increased investment can affect the rate of return is found in a study that was made by W. C. Clark and J. L. Kingston† a number

† W. C. Clark and J. L. Kingston, *The Skyscraper* (New York: American Institute of Steel Construction).

of years ago. This study, which grew out of experience related to the Empire State Building, whose height was uneconomical at the time it was constructed and caused great financial loss to the owners, is particularly interesting because it shows all the possible effects that can result from increased investment.

The study was made for a theoretical office building of different heights, and corresponding investments, to be built on a particular plot of ground in New York City. The heights of the buildings considered were 8, 15, 22, 30, 37, 50, 63, and 75 stories. Figures 7-5 and 7-6 show the results of the study. The bar chart in Figure 7-5 shows the rate of return that would be obtained from each *increment* of investment required to increase the height of the building. The curve of Figure 7-6 shows the rate of return on the *total* investment for the various heights. It will be noted that a building of only 8 stories would produce a rate of return of only 4.2%—insufficient to justify investment. The increment of capital required to increase the height from 8 to 15 stories would earn a return of about 23%, and would raise the return from the total investment to about 6.4%, possibly making such a building justified. As the increments of height were increased, there was a general falling off of the incremental rate of return, accompanied by an increase in the rate of return from the total investment, up to a height of 63 stories. Beyond this height the total rate of return also decreased, and it dropped to zero for a building of 130 stories.

FIGURE 7-5. Return yielded by incremental investments required to add various increments to the height of a proposed skyscraper.

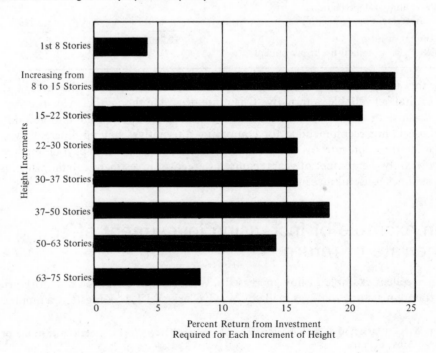

Percent Return from Investment
Required for Each Increment of Height

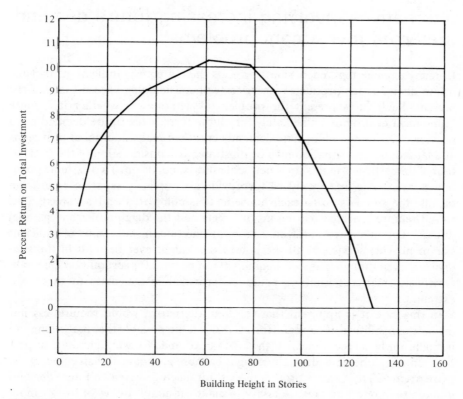

FIGURE 7-6. Return on total investment required for building proposed skyscrapers of various heights.

This study points out a number of the important conditions that may arise in alternative investment studies. First, the investment of an amount beyond the minimum may be necessary to make a project sufficiently effective to provide enough return to be financially sound. Second, it is necessary to know how far the investing of additional capital should be carried. The amount of investment that will bring a maximum rate of return upon the total investment has appeal as being the best. However, from a practical viewpoint, if no better opportunity for investment of additional amounts of capital exists than is offered by the increase above the point of maximum total rate of return, it is logical to invest the additional increment and obtain this contentment return. This process could continue as long as the return from each additional increment is greater than the contentment (minimum attractive) rate. Third, the risks may increase with each increment of investment, owing to decreased flexibility of the invested capital as more and more is tied up in a single property.

alternatives involving increasing future demand (deferred investment problems)

In many engineering and business projects there are clear indications that the future demands for products or services will considerably exceed those of the present. Such cases present the problem of determining whether it is more economical to provide immediately for all the foreseeable future demand or to provide only for the immediate demands and then make additional provisions at a later date. The situation may be illustrated by consideration of the installation of domestic gas mains for a new real estate subdivision. Two extreme possibilities exist. The first would be to provide a completely separate main from outside the subdivision to each house as it is completed and occupied. This would require repeated tearing up of streets and the digging of many parallel ditches. The other extreme would be to at one time install mains of sufficient size to provide service to all the houses that might ever be built in the subdivision, even though many of them might not be built for several years, if at all. Obviously, many alternatives could be provided intermediate to these two extremes.

In this case it is apparent that the first alternative would require less immediate expenditure of capital. However, it is quite possible that in the long run it might not be as economical as the second extreme. We would at once suspect that some intermediate alternative might be more economical than either of the two extremes. Thus, where provisions exist for meeting increased future demand in more than one way, it is necessary to make economy studies of two or more alternative methods to determine which will be the more economical.

Such studies ordinarily involve relatively long-range demands, usually from 10 to 50 or more years. They, therefore, are encountered most frequently in connection with public works—such as roads, bridges, water reservoirs, or sewers—or in connection with public utilities—such as power plants, gas and water mains, or telephone central offices and cable installations. It will be noted that all such projects involve relatively stable services for which the demand is likely to continue, and the necessity for meeting the future demands, whatever they may be. In some cases there is a limitation of site location. For example, in locating the central office of a telephone company a difference of a block or two may mean thousands of dollars in cable costs. Unless the correct site is determined and acquired well in advance, it may not be available when the actual demand occurs.

There are two factors that make economy studies involving variable future demand different from those alternative investment studies that have been discussed previously in this chapter. The first is that the alternatives may involve different amounts of invested capital at different times. This is illustrated in Figure 7-7. Here the demand is assumed to increase from an initial value to a larger amount at a future date. This demand could be met in two alternative ways. Alternative A would be to immediately provide capacity to meet all the

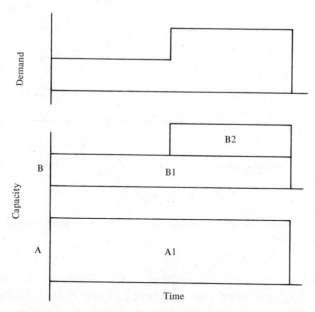

FIGURE 7-7.   Two methods of providing capacity to meet a demand that will increase at a future date.

anticipated future demands. Alternative B, on the other hand, would immediately provide capacity B1 sufficient only to meet the demand of the first period of years. At a later date capacity B2 would be added to enable the demand of the later years to be met. Quite clearly these two alternatives involve different investments of capital initially, and alternative B involves different amounts of capital investment at different times. It also is evident that the revenues, costs, and profits probably would not be uniform throughout the investment period, regardless of which alternative is selected.

The second factor that often must be dealt with in future demand studies is that illustrated in Figure 7-8. If, in the previous example, alternative B were to consist of two units, B1 and B2, each having a life of n years, it is clear that at the end of the needed life the installation B2 would have $m - n$ years of life still available. This situation, obviously, presents the same problem that was discussed earlier in this chapter in connection with alternatives that have different lives. One way of dealing with this extra life (sometimes called *hangover life*) is to assign an anticipated salvage value at the end of the needed life, as is commonly done with the coterminated assumption. However, another procedure sometimes exists in the case of future demand studies. Because it is recognized that the deferred capacity (unit B2 in this case) will be needed for only $n - q$ years (Figure 7-8), in some cases it is possible to build the second unit as an "inferior" unit that will, in fact, have a shorter life than the unit that is provided initially. This is what was portrayed for unit B2 in Figure 7-7.

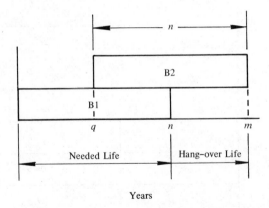

FIGURE 7-8. "Hang-over life" resulting from a deferred capacity provision which has life beyond that needed.

When these two factors are handled properly, alternative investment studies involving increasing future demand are little different from ordinary alternative investment studies. Generally, the P.W. method is most convenient to use in these cases.

an example with variable demands and coterminated asset lives

The following is a typical example of an alternative investment situation involving variable future demand and nonuniform investments and costs. The solution will be shown by the present worth–cost method. It will then be pointed out why the internal rate of return procedure cannot be applied without modification in this particular case—a not-uncommon situation when hangover lives are involved.

EXAMPLE 7-8

In 1980 a company is deciding to build a new power plant. It believes that the plant would be needed for a number of years, but no absolute need is seen after 10 years. Two plans were being considered. Plan A is to build immediately a plant of sufficient size to meet all the needs of the next 10 years—1981 through 1990. Such a plant would be of 5,000-kilowatts capacity and cost $4,000,000. The life of this plant is estimated to be at least 10 years. Plan B would be to construct a plant of 3,000-kilowatts capacity immediately to meet the demands of the first 6 years and at the end of 6 years to enlarge this plant by an additional 2,000 kilowatts to meet the needs of the last 4 years. The 3,000-kilowatt plant would cost $3,100,000, and the additional 2,000 kilowatts would cost $1,800,000. Annual property taxes and insurance would amount to 2% of the first cost in each case.

Although the 2,000-kilowatt addition would have an economic life of at least 10 years if needed, it is believed that it could be disposed of for $300,000 at the end of 1990 if it should not be needed beyond that date. Annual operating and maintenance costs for the two plans were estimated to be as follows:

	Plan A	Plan B
First and second years	$480,000	$450,000
Third and fourth years	540,000	520,000
Fifth and sixth years	610,000	600,000
Seventh and eighth years	785,000	800,000
Ninth and tenth years	960,000	970,000

The minimum attractive rate of return is 7%.

Solution by P.W.-C. Method

For Plan A:

		P.W.-C.
Investment in 5,000-kW plant		$4,000,000
Taxes and insurance = $4,000,000 \times 0.02 \times (P/A, 7\%, 10)$		562,000

Operation and maintenance costs:

First and second years:
$480,000(P/A, 7\%, 2)$ $ 868,000

Third and fourth years:
$540,000 \times (P/A, 7\%, 2)(P/F, 7\%, 2)$ 853,000

Fifth and sixth years:
$610,000 \times (P/A, 7\%, 2)(P/F, 7\%, 4)$ 841,000

Seventh and eighth years:
$785,000 \times (P/A, 7\%, 2)(P/F, 7\%, 6)$ 946,000

Ninth and tenth years:
$960,000 \times (P/A, , 7\%, 2)(P/F, 7\%, 8)$ 1,010,000

TOTAL $4,518,000

Total P.W.-C. $9,080,000

For Plan B:

		P.W.-C.
Investment in 3,000-kW plant		$3,100,000

Investment in 2,000-kW plant built 6 years hence:
$1,800,000 \times (P/F, 7\%, 6)$ $1,200,000

Less present worth salvage:
$300,000 \times (P/F, 7\%, 10)$ − 152,000

Net present worth–cost $1,048,000

Taxes and insurance:

First 6 years:
$3,100,000 \times 0.02 \times (P/A, 7\%, 6)$ 296,000

Last 4 years:
$4,900,000 \times 0.02$
$\times (P/A, 7\%, 4)(P/F, 7\%, 6)$ 221,000

TOTAL $ 517,000

Operation and maintenance costs:
First and second years:
$450,000(P/A, 7%, 2) $ 814,000
Third and fourth years:
$520,000(P/A, 7%, 2)(P/F, 7%, 2)   821,000
Fifth and sixth years:
$600,000(P/A, 7%, 2)(P/F, 7%, 4)   827,000
Seventh and eighth years:
$800,000(P/A, 7%, 2)(P/F, 7%, 6)   964,000
Ninth and tenth years:
$970,000(P/A, 7%, 2)(P/F, 7%, 8)   1,020,000

| | TOTAL | $4,446,000 |

| Total P.W.-C. | $9,111,000 |

On the basis of present worth–cost, plan A is slightly more economical and would be used. However, from a practical viewpoint, this is a case where we might very well give considerable additional consideration to plan B, inasmuch as the extra cost is not great ($31,000) and it permits a portion of the investment to be deferred. Furthermore, if the service should be required after 1990, the 2,000-kilowatt plant presumably would be usable. ∎

Solution by I.R.R. Method

If we were to attempt a comparison of the two plans by the internal rate of return method, the comparison of the annual cash flows would be as follows (initial costs of plants plus taxes and insurance and operations and maintenance, less salvage value):

Year	Plan A	Plan B	Increment (A − B), or (B ÷ A)
0	− $4,000,000	− $3,100,000	− $ 900,000
1	− 560,000	− 512,000	− 48,000
2	− 560,000	− 512,000	− 48,000
3	− 620,000	− 582,000	− 38,000
4	− 620,000	− 582,000	− 38,000
5	− 690,000	− 662,000	− 28,000
6	− 690,000	− 2,462,000	+ 1,772,000
7	− 865,000	− 898,000	+ 33,000
8	− 865,000	− 898,000	+ 33,000
9	− 1,040,000	− 1,068,000	+ 28,000
10	− 1,040,000	− 768,000	− 272,000

The set of cash flows shown in the last column in the preceding tabulation would result from the greater initial investment required for plan A. For determination of the internal rate of return, the present worth of these cash flows would have to be equated to zero. However, it will be noted that the final cash flow is negative, so that the cash flow reverses sign more than once—a condition that may prevent the calculation of a correct internal rate of return. This problem and suggested solution methodology is described in Appendix 5-B.

An adjustment that can be made to the "Increment $(A - B)$" cash flows to allow one to calculate the I.R.R. is to make equivalent moves in one or more cash flows so that there is only one reversal in sign over time. In general, the less the cash flow amount(s) so treated and the shorter the time over which it is (they are) moved, the less will be the influence on the final calculated I.R.R. due to such adjustments. The usual interest rate at which equivalence adjustments are made is normally the M.A.R.R. For this problem, one possible adjustment is to move the $-\$272,000$ from year 10 to year 5 at $i = 7\%$. The equivalent value at year 5 is

$$P_5 = F_{10}(P/F, 7\%, 5) = -\$272,000(0.7130) = -\$194,000$$

The $-\$194,000$ added to the $-\$28,000$ already at year 5 results in a net $-\$222,000$ The cash flows, which now exhibit only one reversal in sign, are summarized in the second column of Table 7-7 (p. 214). Note that there is no remaining difference in cash flow for year 10. Table 7-7 also contains the present worth factors and calculations to facilitate the determination of the I.R.R. on the incremental investment. This I.R.R. is shown to be 7.6%, which, when compared to the minimum attractive return of 7%, again indicates that the incremental investment in A is justified. ∎

determination of the breakeven deferment period

As was pointed out previously, there is always some uncertainty as to what future requirements will be and when they will occur. Provision of a certain amount of capacity will establish a limit to the amount of service that can be provided in the future. It often is helpful to know at what future date a deferred investment will be needed so that an alternative permitting deferred investment will break even with one that provides immediately for all future demands. Where only the costs of acquiring the assets by the two alternatives need to be considered, or where the annual costs through the entire life are not affected by the date of acquisition of the deferred asset, the breakeven point may be determined very easily and may be helpful in arriving at a decision between alternatives.

EXAMPLE 7-9

In the planning of a two-story municipal office building the architect has submitted two designs. The first provides foundation and structural details so that two additional stories can be added at a later date without modifications to the original structure. This building would cost $1,400,000. The second design, without such provisions, would cost only $1,250,000. If the first plan is adopted, it is estimated that an additional two stories could be added at a later date at a cost of $850,000. If the second plan is adopted, however, considerable strengthening and reconstruction would be required, which would add $300,000 to the cost of a two-story addition.

table 7-7
I.R.R. analysis on incremental investment in example 7-8

Year End, N,	(1) Cash Flow Increment (A − B)	After Cash Flow Adjusted to Remove Second Reversal of Sign			
		(2) $(P/F, 7\%, N)$	(3) = (1) × (2) Present Worth at $i = 7\%$	(4) $(P/F, 8\%, N)$	(5) = (1) × (4) Present Worth at $i = 8\%$
0	− $900,000	1.0000	− $900,000	1.0000	− $900,000
1	− 48,000	0.9346	− 44,900	0.9259	− 44,500
2	− 48,000	0.8734	− 42,000	0.8573	− 41,100
3	− 38,000	0.8163	− 31,000	0.7938	− 30,200
4	− 38,000	0.7629	− 29,000	0.7350	− 27,900
5	− 222,000	0.7130	− 158,000	0.6808	− 151,000
6	+ 1,772,000	0.6663	+ 1,181,100	0.6302	+ 1,118,000
7	+ 33,000	0.6227	+ 20,500	0.5835	+ 19,250
8	+ 33,000	0.5820	+ 19,200	0.5403	+ 17,850
9	+ 28,000	0.5439	+ 15,200	0.5002	+ 14,000
10	0		$\Sigma = +\$31,100$		$\Sigma = -\$25,600$

$$i' = \text{I.R.R.} = 7\% + \left(\frac{\$31,100}{31,100 + 25,600}\right)(8\% - 7\%) = 7.6\%$$

Assuming that the building is expected to be needed for 75 years, by what time would the additional two stories have to be built to make the adoption of the first design justified? Assume interest at 6%.

Solution

The breakeven deferment period is determined as follows:

	Provide Now	No Provision
Present worth–cost:		
First unit	$1,400,000	$1,250,000
Second unit	$850,000 $\times$ $(P/F, 6\%, N)$	$1,150,000 $\times$ $(P/F, 6\%, N)$

Equating total present worth–costs:
$$\$1,400,000 + \$850,000 \times (P/F, 6\%, N) = \$1,250,000 + \$1,150,000 \times (P/F, 6\%, N)$$

Solving, we have

$$(P/F, 6\%, N) = 0.5$$

From the 6% table in Appendix E, $N = 12$ years (approximately). Thus, if the additional space will be required in less than 12 years, it would be more economical to make immediate provision in the foundation and structural details. If the addition would not be likely to be needed until after 12 years, greater economy would be achieved by making no such provisions in the first structure. ∎

other factors affecting deferred investment studies

Because studies of the type discussed in this chapter involve future demand, often over a considerable period of time, it is obvious that considerable attention must be given to possible effects that the future may have.

The type of project being considered should be kept in mind in evaluation of the probable accuracy of the future-demand estimates. For some types of industries, such estimates can be made with considerable accuracy. For other types the inaccuracy is notorious.

It has been pointed out that the fact that some of the assets are not needed immediately tends to favor deferring a portion of the investment. Another factor having the same effect is that such an alternative requires less immediate financing. Even though alternatives are compared on the basis of the present worth of the costs, ordinarily no capital will be set aside immediately to pay for a deferred asset. Capital is scarce in most firms. Therefore, firms usually favor an alternative that will reduce current capital requirements.

On the other hand, alternatives requiring immediate investment sometimes are justified on the basis that increasing demand will soon make them economical.

If there is a good chance of demand increasing rapidly, this is an important factor to be considered. Delay in providing an asset might make it impossible to meet rapidly increasing demand.

There are some cases in which demand depends upon a service or product being available. For example, the demand for gasoline stations on an interstate highway that does not exist is zero. On the other hand, if a long highway without intermediate entry and exit roads were built without service stations, travel on it would not be great. The proper combination of highway and service stations would have a great effect on the services from each. Thus it sometimes is necessary to provide a service before any realistic evaluation of the demand can be made. When the service is available, the demand often can be developed.

Occasionally, the possibility of material or equipment shortages in the future may be an important factor in favoring immediate investment. Over short periods, particularly in times of national crisis, material shortages may sometimes be anticipated. In such cases we might wish to favor immediate provision over deferred provision. If study periods are more than a few years, however, the probability of foreseeing material shortages is rather low.

comparing independent projects (opportunities)

All previous examples of the comparison of projects in this chapter assume that the alternatives are mutually exclusive, that is, that the choice of one project excludes the choice of any other project so that at most one project under consideration will be chosen. Any of the six basic economy study methods can also be used for the comparison of independent projects. By independent is meant that the choice of one does not affect the choice of any other such that any number of projects may be chosen as long as sufficient capital is available. Independent project alternatives are often descriptively called *opportunities*.

EXAMPLE 7-10

Given the following independent projects, determine which should be chosen using the annual worth method, if the minimum required rate of return is 10% and if there is no constraint on total investment funds available.

Project	Investment, P	Life (years)	Salvage Value, F	Net Annual Cash Flow
X	$10,000	5	$10,000	+ $2,300
Y	12,000	5	0	+ 2,800
Z	15,000	5	0	+ 4,067

Solution

Project	(1) Net Annual Cash Flow	(2) C.R. (= Annual Cost of Depreciation + Minimum Profit): $(P - F)(A/P, 10\%, 5) + F(10\%)$	(3) = (1) − (2) Net Annual Worth
X	+ $2,300	$1,000	+ $1,300
Y	+ 2,800	3,166	− 366
Z	+ 4,067	3,957	+ 110

Thus projects X and Z, having positive net annual worths, would be satisfactory for investment, but project Y would not be satisfactory. The same indication of satisfactory projects and unsatisfactory project would be obtained using other economy study methods. ∎

consideration of sets of mutually exclusive, independent, and contingent projects

It is helpful to think of sets of projects in three major groups as follows:

1. *Mutually exclusive*—which means that at most one project out of the group can be chosen.
2. *Independent*—which means that the choice of a project is independent of the choice of any other project in the group.
3. *Contingent*—which means that the choice of a project is conditional on the choice of one or more other projects.

It is common for decision makers to be faced with sets of mutually exclusive, independent, and/or contingent investment projects. For example, a contractor might be considering investing in a dump truck, and/or a power shovel, and/or an office building. For each of these types of investments, there may be two or more mutually exclusive alternatives, that is, brands of dump trucks, and types of power shovels, and designs of office buildings, respectively. While the economic choice of an office building is probably independent of that of either dump trucks or power shovels, the choice of any type of power shovel may be contingent (conditional) on the decision to purchase a dump truck.

In order to provide a simple method of handling the various types of projects as well as to provide some insight into mathematical programming formulations of this type of decision problem, a general approach is recommended herein. This approach requires that all investment projects be listed and that all the feasible combinations of projects be enumerated. Such combinations will be called *mutually exclusive combinations*. Each combination of projects is mutually exclusive since each is unique and the acceptance of one combination of investment projects precludes the acceptance of any of the other combinations. The cash flow of each combination is determined simply by adding, period by

table 7-8

combinations of three mutually exclusive projects[a]

Mutually Exclusive Combination	Project			Explanation
	X_A	X_B	X_C	
1	0	0	0	Accept none
2	1	0	0	Accept A
3	0	1	0	Accept B
4	0	0	1	Accept C

[a] For each investment project there is a binary variable X_j that will have the value 0 or 1 indicating that project j is rejected (0), or accepted (1). Each row of binary numbers represents an investment alternative (mutually exclusive combination). This convention is used throughout this chapter.

period, the cash flows of each project contained in the mutually exclusive combination being considered.

For example, suppose that we have three projects, A, B, and C. If the projects themselves are all mutually exclusive, then the four possible mutually exclusive combinations are given in binary form in Table 7-8. If, by chance, the firm felt that one of the projects must be chosen (i.e., it is not permissible to turn down all alternatives) then mutually exclusive combination 1 would be eliminated from consideration.

If the three projects were independent, there are eight mutually exclusive combinations, as shown in Table 7-9.

To illustrate one of many possible instances of contingent projects, suppose that A is contingent on the acceptance of both B and C, and that C is contingent on the acceptance of B. Now there are four mutually exclusive combinations, as shown in Table 7-10.

table 7-9

mutually exclusive combinations of three independent projects

Mutually Exclusive Combination	Project			Explanation
	X_A	X_B	X_C	
1	0	0	0	Accept none
2	1	0	0	Accept A
3	0	1	0	Accept B
4	0	0	1	Accept C
5	1	1	0	Accept A and B
6	1	0	1	Accept A and C
7	0	1	1	Accept B and C
8	1	1	1	Accept A, B, and C

table 7-10
mutually exclusive combinations
of three projects with
particular contingence

Mutually Exclusive Combination	Project		
	X_A	X_B	X_C
1	0	0	0
2	0	1	0
3	0	1	1
4	1	1	1

Suppose that one is considering two independent sets of mutually exclusive projects. That is, projects A1 and A2 are mutually exclusive while projects B1 and B2 are mutually exclusive. However, the selection of any proposal from the set of proposals A1 and A2 is independent of the selection of any proposal from the set of proposals B1 and B2. For example, the decision problem may be to select at most one dump truck out of two brands being considered, and to select at most one office building out of two designs being considered. Table 7-11 shows all mutually exclusive combinations.

Example 7-11 provides an illustration involving use of mutually exclusive combinations to select an optimal set of projects under capital constraints.

table 7-11
mutually exclusive combinations for two
independent sets of mutually exclusive
projects

Mutually Exclusive Combination	Projects			
	X_{A1}	X_{A2}	X_{B1}	X_{B2}
1	0	0	0	0
2	1	0	0	0
3	0	1	0	0
4	0	0	1	0
5	0	0	0	1
6	1	0	1	0
7	1	0	0	1
8	0	1	1	0
9	0	1	0	1

EXAMPLE 7-11

The following are prospective projects, their interrelationships, and respective cash flows for an organization for the coming budgeting period. Using the net P.W. method and M.A.R.R. = 10%, determine what combination of projects is best if the capital to be invested is (a) unlimited, and (b) limited to $48,000.

Project B1 ⎰
Project B2 ⎱ mutually exclusive

Project C1 ⎰ mutually exclusive and
Project C2 ⎱ dependent on the acceptance of B2

Project D   contingent on the acceptance of C1

Project	Cash Flow ($000s) for End of Year					Net P.W. ($000s) at M.A.R.R. = 10%
	0	1	2	3	4	
B1	−50	20	20	20	20	+13.4
B2	−30	12	12	12	12	+ 8.0
C1	−14	4	4	4	4	− 1.3
C2	−15	5	5	5	5	+ 0.9
D	−10	6	6	6	6	+ 9.0

Solution

The net P.W. for each project by itself is shown on the right-hand column of the matrix above. As a sample calculation, the net P.W. for project B1 is

$$- \$50,000 + \$20,000(P/A, 10\%, 4) = + \$13,400$$

The mutually exclusive combinations are as follows:

Mutually Exclusive Combination	Project				
	B1	B2	C1	C2	D
1	0	0	0	0	0
2	1	0	0	0	0
3	0	1	0	0	0
4	0	1	1	0	0
5	0	1	0	1	0
6	0	1	1	0	1

The combined cash flows and the net P.W. for each mutually exclusive combination are as follows:

Mutually Exclusive Combination	Cash Flow ($000s) for End of Year						Invested Capital ($000s)	Net P.W. ($000s) at M.A.R.R. = 10%
	0	1	2	3	4	5		
1	0	0	0	0	0	0	0	0
2	−50	20	20	20	20	20	50	13.4
3	−30	12	12	12	12	12	30	8.0
4	−44	16	16	16	16	16	44	6.7
5	−45	17	17	17	17	17	45	9.0
6	−54	22	22	22	22	22	54	16.0

Examination of the right-hand column reveals that mutually exclusive combination 6 has the highest net P.W. if capital available (in year 0) is unlimited, as specified in part (a). If, however, capital available is limited to $48,000, as specified in part (b), both mutually exclusive combinations 2 and 6 are not feasible. Of the remaining mutually exclusive combinations, 5 is best, which means that projects B2 and C2 would be selected for a net P.W. = $9,000. ∎

For problems that involve a relatively small number of projects, the general technique just presented for arranging various types of projects into mutually exclusive combinations is computationally practical. However, for larger numbers of projects the number of mutually exclusive combinations becomes quite large, and therefore this approach becomes computationally cumbersome. Linear (integer) programming provides a practical means to accomplish the same result and will be described in Chapter 17.

problems

7-1. A real estate operator has a 30-year lease on a plot of land. He gets estimates on the costs and income of various types of structures on the piece of land as follows.

	Cost of Structure	Receipts Less Disbursements[a]
Apartment house	$300,000	$69,000/year
Theater	200,000	40,000/year
Department store	250,000	55,000/year
Office building	400,000	76,000/year

[a] Disbursements do not include depreciation, of course.

Each structure is expected to have a salvage value equal to 20% of its initial cost. If the investor requires a minimum attractive rate of return of at least 12% before taxes on all his investments, which structure (if any) should she build?

(a) Use the annual worth method to determine which type of structure (if any) should be chosen.

(b) Use the I.R.R. method to investigate incremental differences between alternatives and to recommend which structure should be selected.

7-2. Work Problem 7-1 by using the E.R.R. and E.R.R.R. methods when the reinvestment rate is 12%. Calculate the present worth of each alternative at 12%. Why do all methods lead to the same choice?

7-3. The Consolidated Oil Company must install antipollution equipment in a new refinery to meet Federal clean air legislation. Four types of equipment are being considered, which will have investment and annual operating costs as follows:

	Equipment			
	A	B	C	D
Investment	$600,000	$760,000	$1,240,000	$1,600,000
Power	68,000	68,000	120,000	126,000
Labor	40,000	45,000	65,000	50,000
Maintenance	660,000	600,000	420,000	370,000
Taxes and insurance	12,000	15,000	25,000	28,000

Assuming an economic life of 10 years for each type, no salvage value, and that the company wants a before-tax minimum return of 20% on its capital, calculate the future worth of each alternative to determine which one should be purchased.

7-4. Work Problem 7-3 with the I.R.R. and E.R.R. methods when the reinvestment rate (e) equals 20%.

7-5. A large sailboat is being built in a coastal town. In specifying air conditioning equipment and insulation for the living quarters of the vessel, engineers have found various combinations of compressors and insulation thickness to be feasible. A small air-conditioner compressor causes the need for more insulation but has slightly lower operating and maintenance costs compared to a larger compressor. Three combinations are being considered that have the following investment and annual costs.

	Combination		
	A	B	C
First cost of compressor	$7,000	$ 4,200	$ 5,500
First cost of insulation	7,000	14,000	10,000
Out-of-pocket costs/year	950	300	600

Each system is expected to have a 20-year life, with no salvage value. The before-tax minimum attractive rate of return is 20%. Compare the alternatives with the annual worth method and the E.R.R.R. method when $e = 20\%$.

7-6. Work Problem 7-5 by using the present worth method and the I.R.R. method. Which alternative should be recommended?

7-7. A construction company is going to purchase several heavy-duty trucks. Its M.A.R.R. before taxes is 18%. It is considering two makes, and the following relevant data are available.

	Wiltsbilt	Big Mack
Cost	$10,000	$15,000
Life (estimated by manufacturer)	3 years	5 years
Salvage value at end of life	$2,000	$3,000
Annual out-of-pocket costs	$4,000	$3,000

(a) Which type of truck should be selected when the repeatability assumption is appropriate?

(b) Which type of truck would you recommend if the study period is limited to 3 years (coterminated assumption) and it is estimated that a Big Mack truck will have a salvage value of $5,600 at that time?

7-8. In building the landing strip at a small municipal airport, which is not used by commercial planes, one method of construction will cost $500,000 and have an estimated life, with proper maintenance of 40 years. The annual cost of such maintenance is estimated to be $10,000 per year.

An alternative type of construction would cost only $275,000, but at the end of 10 years it is estimated that another $250,000 will have to be spent for resurfacing and other major repair work. It would then last another 10 years. This resurfacing would involve closing the field for a month and would result in a revenue loss of $20,000 and considerable inconvenience. If this alternative is used, it is estimated that the annual maintenance will cost $5,000. If capital costs the municipality 6%, which alternative would you recommend if the repeatability assumption is used?

7-9. A small branch office of a major retailing firm is planning to purchase a mini-computer for simple regression analysis of sales data. Three machine companies have supplied cost, service life, and salvage data (shown below). The minimum attractive rate of return is 12%.

	Machine		
	A	B	C
Initial cost	$800	$1,400	$900
Annual service contract cost	$180	$150	$170
Life	5 years	8 years	5 years
Salvage value	$200	$600	$100

If the expected annual saving in labor is $500 regardless of which machine is purchased, what recommendation would you make to the boss? Assume that the coterminated life assumption is valid and that after 5 years the market value of machine B is expected to be $825.

7-10. A company is planning to purchase a small executive airplane. It has about decided to buy an "Airbird" model, costing $300,000, and has based its decision on an assumed economic life of 6 years. A company vice president favors an "Eaglejet" plane, and has presented data that shows rather convincingly that its annual out-of-pocket operating costs would be $42,000 per year less, and it would have an equally long economic life. This plane, however, would cost considerably more to purchase. It is believed that either plane would have a salvage value of about 30% of first cost at the end of 6 years. If the company's capital is worth 15% and income tax effects are neglected, how much can it afford to pay for the more expensive plane?

7-11. In the Rawhide Co., Inc. (cosmetics manufacturers), plant decisions regarding approval of proposals for plant investment are based upon a stipulated minimum attractive rate of return of 20% before income taxes. The following five packaging devices were compared assuming a 10-year life and zero salvage value for each. Which one (if any) should be selected? Make any additional calculations you think are needed.

	Packaging Equipment				
	A	B	C	D	E
Investment	$30,000	$50,000	$55,000	$60,000	$70,000
Net annual return	$11,000	$14,100	$16,300	$16,800	$19,200
Rate of return (I.R.R.)	35.5%	25.2%	26.9%	25.0%	24.3%
Rate of return (I.R.R.) on increment of investment (compared to alternative with next lower investment)	—	7.4% (A → B)	42.7% (B → C)	0.0% (C → D)	20.2% (D → E)

7-12. You have been asked to evaluate the economic implications of various methods for cooling condenser effluents from a 350-megawatt steam-electric plant. In this regard, cooling ponds and once-through cooling systems have been eliminated from consideration because of their adverse ecological effects. It has been decided to use cooling towers to dissipate waste heat to the atmosphere. There are two basic types of cooling towers: wet and dry. Furthermore, heat may be removed from condenser water by (1) forcing (mechanically) air through the tower or (2) allowing heat transfer to occur by making use of natural draft. Consequently, there are four basic cooling tower designs that could be considered.

Assuming that the cost of capital to the utility company is 10%, your job is to recommend the best alternative (i.e., the least expensive during the service life) in view of the data below. Further assume that each alternative is capable of satisfactorily removing waste heat from the condensers of a 350-megawatt power plant. What *noneconomic* factors can you identify that might also play a role in the decision-making process?

alternative types of cooling towers for a 350-megawatt
fossil-fired power plant operating at full capacity[a]

	Alternative			
	Wet Tower Mech. Draft	Wet Tower Natural Draft	Dry Tower Mech. Draft	Dry Tower Natural Draft
Initial cost	$3 million	$4.2 million	$3.6 million	$4.9 million
Power for I.D fans	40 200-hp induced-draft fans	None	20 200-hp I.D. fans	None
Power for pumps	20 150-hp pumps	20 150-hp pumps	40 100-hp pumps	40 100-hp pumps
Mechanical maintenance/year	$0.15 million	$0.10 million	$0.17 million	$0.12 million
Service life	30 years	30 years	30 years	30 years
Salvage value	0	0	0	0

[a] 100 hp = 74.6 kW; cost of power to plant is 0.7 cent per kWh; induced-draft fans and pumps operate around the clock for 365 days/year (continuously). Assume that electric motors for pumps and fans are 100% efficient.

7-13. The demand for a certain product is expected to be constant for 4 years and then to increase sharply and continue at the higher level for the foreseeable future. Two alternative methods are available for providing the equipment required to produce the product. Method A is to provide one installation, A1, at a cost of $50,000, which will have a life of 10 years and will have a capacity sufficient to meet the requirements during the first 4 years. At the end of 4 years a second unit, A2, costing $40,000, would be added, which would have a life of 6 years and would supply sufficient output to meet the increased demand for the remainder of the 10-year period. With this alternative the relevant annual out-of-pocket costs would be $2,000 during years 1–4, and $4,000 during years 5–10.

Alternative B is to provide a larger installation at a cost of $75,000 which would have a 10-year life and would have sufficient capacity to provide all the needs for the 10-year period. With this installation the relevant out-of-pocket costs would be $3,000 per year during years 1–4, and $3,500 per year during years 5–10.

Assuming that no installation would have any salvage value at the end of life, and that capital is worth 15%, which installation should be made?

7-14. An electric utility company must install an underground power cable in a new housing subdivision. To excavate and backfill the ditch costs $1.50 per linear foot if it is done before the pavement is installed. The cable costs $2.10 per foot. Annual taxes on all installations is approximately 1% of the cost of the cable.

One street of this subdivision, 900 feet in length, will serve as an access to another subdivision, which, according to present plans, will be started in about 5 years. If power cable for this second subdivision is not installed at the present time, when it is installed later the pavement will have to be removed and replaced, which will double the cost of the ditch but will not affect the tax rate. In addition, it is expected that all costs will increase by at least 6% per year during the next 20 years. If capital is worth 10% to the company, should the cable for the future subdivisions be installed now? Assume that the life of all installations is 25 years.

7-15. In Problem 7-14, determine the time at which the cable for the future subdivision would have to be installed in order for the deferred and immediate installations to be equally economical.

7-16. An oil company must purchase a new piece of drilling equipment. One type A, will cost only $60,000, but it is powered from an external source, necessitating building and moving temporary power lines. It is estimated that this relocation of power lines will cost $5,000 per year during the estimated 15-year life of the drilling equipment. Power would cost $18,000 per year, maintenance $2,000 per year, and labor $25,000 per year.

A self-contained diesel-electric drilling unit, B, will cost $85,000 and have an estimated life of 10 years and no salvage value. Estimated fuel costs would be $12,000 per year, maintenance $3,500 per year, and labor $27,500 per year. Annual taxes and insurance for either type of equipment will be 2% of first cost. Average earnings of capital have been about 10% per year. Which equipment would you recommend?

7-17. A state highway department is planning a new highway to a newly-opened recreation area. At present a four-lane road will be built, but it is expected that at a later date a second four-lane road will be added, so as to provide a divided highway.

If an eight-lane bridge is built now, it will cost $1,640,000. A single four-lane bridge can be built now for $1,000,000. It is estimated that a second four-lane bridge at a later date will cost $1,200,000. Annual upkeep on each four-lane bridge would be $18,300, and for the eight-lane bridge would be $30,000. Money to build the bridges will cost the state 4%.

Assume a 50-year functional life for the project and determine the earliest time at which the additional four lanes would be required to make it economical to build the eight-lane bridge immediately.

7-18. Estimates for a proposed development are as follows:

Plan A has a first cost of $50,000, a life of 25 years, a $5,000 salvage value, and annual maintenance of $1,200.

Plan B has a first cost of $100,000, a life of 50 years, no salvage value, and annual maintenance of $6,000 for the first 15 years and $1,000 per year for years 16–50.

Assuming interest at 8%, compare the two plans by use of the capitalized worth (cost) method.

7-19. In the design of a certain system, two alternatives are under consideration. These alternatives are as follows.

	Plan A	Plan B
First cost	$50,000	$120,000
Life	20 years	40 years
Salvage value	$10,000	$20,000
Annual disbursements	$9,000	$6,000

If *perpetual service life* is assumed, which of these alternatives do you recommend? Assume the interest rate to be 6%.

7-20. Refer back to Problem 5-14. Calculate the capitalized worth of motor R and motor S at a minimum attractive rate of return of 20%. Compare present worths of Problem 5-14 with the calculated capitalized worth. Why is there a difference in numerical values for each type of motor?

7-21. Your company has $20,000 in "surplus" funds which it wishes to invest in new revenue-producing projects. There have been *three* independent sets of mutually exclusive proposals developed. The service life of each is 5 years and all salvage values are zero. You have been asked to perform an analysis to maximize total present worth of investment funds available (i.e., $20,000). If the cost of capital is 12%, which combination of proposals would you recommend?

	Proposal	First Cost	Cash Profit/Year
Mutually exclusive	A1	− $ 5,000	+ $1,500
	A2	− 7,000	+ 1,800
Mutually exclusive	B1	− 12,000	+ 2,000
	B2	− 18,000	+ 4,000
Mutually exclusive	C1	− 14,000	+ 4,000
	C2	− 18,000	+ 4,500

7-22. Various methods are proposed to reduce the cost of an operation. The investment costs and savings for each alternative are listed below, The lives of all alternatives are 5 years, and salvage values in the fifth year are 50% of investment costs.

Alternative	Investment	Annual Cash Savings
A	$ 50,000	$15,000
B	90,000	26,500
C	75,000	23,000
D	150,000	32,000

Suppose that there is a $120,000 budget constraint on investment funds. If the company requires a minimum return of 20% on all money invested, use the present worth method and determine which alternative should be recommended.

7-23. (a) List all mutually exclusive alternatives when these conditions exist among the proposals below:

Proposal B1 ⎫ independent (one or both can be chosen)
Proposal B2 ⎭

Proposal C1 ⎫ mutually exclusive and dependent on the acceptance of B2
Proposal C2 ⎭

Proposal D   dependent (contingent) on the acceptance of B1

(b) In view of the cash flow pattern of each proposal, which alternative would you select if M.A.R.R. = 12% (before taxes)?

	End-of-Year Cash Flow ($000s), Year:				
	0	1	2	3	4
B1	−50	20	20	20	20
B2	−30	12	12	12	12
C1	−14	4	4	4	4
C2	−15	5	5	5	5
D	−10	6	6	6	6

7-24. A firm is considering the development of several new products. The products under consideration are listed below and products in each group are mutually exclusive.

Group	Product	Development Cost	Annual Net Cash Income
A	A1	$ 500,000	$ 90,000
	A2	650,000	110,000
	A3	700,000	115,000
B	B1	600,000	105,000
	B2	675,000	112,000
C	C1	800,000	150,000
	C2	1,000,000	175,000

At most one product from each group will be selected. The firm has a minimum attractive rate of return of 10% and a budget limitation on development costs of $2,100,000. The life of all products is assumed to be 10 years, with no salvage value.

(a) List all mutually exclusive combinations.

(b) Using the net present worth criterion, which alternative should be selected?

chapter 8

estimating, inflation, and costs

introduction to estimating

Economic analyses, by nature, are concerned with present and future consequences. The central thrust of economic analyses is to use information regarding the present situation, as well as our appraisal of the future conditions, to make sound decisions. It may be relatively easy to determine the present situation, but it is usually quite difficult to look into the future.

The purpose of the estimating function of a firm is not to produce exact data but to obtain numbers having a high probability of falling within an acceptable range. Neither a preliminary estimate nor a final estimate is expected to be exact; rather, it should adequately suit the need at a reasonable cost. Preparation of precise estimates would be excessively time consuming and expensive, and even if they were possible to obtain they would still be subject to estimation errors.

Deviations between estimated and actual outcomes result from innumerable factors. Two of the most important are human errors in making estimates and unpredictable changes in circumstances. There are so many possible approaches to estimating a given quantity that the determination of which approach(es) should be used in what level of detail is a significant problem in itself.

It is useful to think of future events as resulting in part from factors that have determined these events in the past, and resulting in part from factors that are new and different. To the extent that, say, future production costs are the product of the same factors that have determined past production costs, the analyst may use explicit prediction techniques to extend past experience into forecasts of the future. However, to the extent that future production costs depend on factors that were not active in the past, the past data fail and one must rely on managerial experience and judgment. Forecasting, perhaps more than any other aspect of decision making, involves the combined skill and experience of both the analyst and management.

229

sources of estimates

An estimate can vary from an instant, top-of-the-head guess to a very detailed and hopefully accurate prognostication of the future. The level of detail and accuracy of an estimate should depend upon:

1. The estimability of that which is to be estimated.
2. Methods or techniques employed.
3. Qualifications of estimator(s).
4. Time and effort available and justified by the importance of the study.
5. Sensitivity of study results to the particular estimate.

Regardless of how estimates are made, individuals who use them should have specific recognition that the estimate will be in error to some extent. Even the use of sophisticated estimation techniques will not, in itself, eliminate error, although it will hopefully minimize estimation errors, or will at least provide better recognition of the anticipated degree of error.

The variety of sources from which information useful in estimating can be obtained is too great for complete enumeration. The following four major sources of information, ordered roughly according to decreasing importance, are described in subsequent sections.

1. Accounting records.
2. Other sources within the firm.
3. Sources outside the firm.
4. Research and development.

1. *Accounting records.* It should be emphasized that although data available from the records of the accounting function are a prime source of information for economic analyses, such data are very often not suitable for direct, unadjusted use.

A very brief and simplified description of the accounting process was given in Chapter 1. In its most basic sense accounting consists of a series of procedures for keeping a detailed record of monetary transactions between established categories of assets, each of which has an accepted interpretation useful for its own purposes. The data generated by the accounting function are often inherently misleading for economic analyses, not only because they are based on past results, but also because of the following limitations:

(a) The accounting system is rigidly categorized. Categories of various types of assets, liabilities, net worth, income, and expenses for a given firm may be perfectly appropriate for operating decisions and financial summaries, but rarely are they fully appropriate to the needs of economic analyses and decision making involving long-term considerations.

(b) Standard accounting conventions cause misstatements of some types of financial information to be "built into" the system. These misstatements tend to be based on the philosophy that management should avoid overstating the value of its assets or understating the value of its liabilities and should therefore assess them very conservatively. This leads to such practices as (1) not changing the stated value of one's resources as they appreciate due to rising market prices, and (2) depreciating assets over a much shorter life than actually expected. As a result of such accounting practices, the analyst should always be careful about treating such resources as cheaply (or, sometimes, as expensively!) as they might be represented in the accounting records.

(c) Accounting data have illusory precision and implied authoritativeness. Although it is customary to present data to the nearest dollar or the nearest cent, the records are not nearly that accurate in general.

In summary, accounting records are a good source of historical data, but have severe limitations when used in making estimates for economic analyses. Moreover, accounting records rarely contain direct statements of incremental costs or opportunity costs, both of which are essential in most economic analyses.

2. *Other sources within the firm.* The usual firm has a large number of people and records that may be excellent sources of estimates or information from which estimates can be made. Examples of functions within firms that keep records useful to economic analyses are sales, production, inventory, quality, purchasing, industrial engineering, and personnel. Professional colleagues, supervisors, and workers in production/service areas can provide insights or suggest sources that can be obtained readily.

3. *Sources outside the firm.* There are numerous sources outside the firm that provide information helpful for estimating. The main problem is in determining those that are most beneficial for particular needs. The following is a listing of commonly utilized outside sources:

(a) *Published information,* such as technical directories, trade journals, U.S. government publications, and comprehensive reference books, offer a wealth of information to the knowledgeable or persistent searcher.

(b) *Personal contacts* are excellent potential sources. Vendors, salespeople, professional acquaintances, customers, banks, government agencies, chambers of commerce, and even competitors are often willing to furnish needed information on the basis of a serious and tactful request.

(c) *Cost indexes and production indexes* are probably the most valuable estimating sources outside the firm. Cost indexes provide a means for converting past costs to present costs through the use of dimensionless numbers, called *indexes,* to reflect relative costs at two or more points in time. Some are based on national averages, others are very specialized. For example, the Bureau of Labor Statistics of the U.S. Department of Labor publishes many data on price changes for a wide variety of products in addition to earnings of workers in practically all industries. Some of these data are components of many cost indexes, and others are useful for the construction of highly specialized indexes.

4. *Research and development* (*R&D*). If the information is not published and cannot be obtained by consulting someone who knows, the only alternative may be to undertake R&D to generate it. Classic examples are developing a pilot plant and undertaking a test market program. These activities are usually expensive and may not always be successful; thus this final step is taken only in connection with very important decisions, and when the sources mentioned above are known to be inadequate.

estimates needed for typical economic analyses

Some judgment is required in determining how far to investigate individual variables to be estimated. This judgment should weigh which variables are dominant and deserve more study and which variables, even if drastically misjudged, will not produce significant changes in the overall estimate.

Perhaps the most serious sources of error in estimating result from overlooking important types of costs. A tabular form or checklist is a good means of preventing such oversights, but it is no better than the completeness of the checklist for the particular situation being estimated. Technical familiarity with the project is essential in assuring completeness as well as reasonable accuracy of project estimates.

The following is a brief listing of the types of estimates that are typically needed in economic analyses, together with some discussion of how those estimates might be obtained.

1. *First* (*investment*) *costs*, which may consist of two types:

   (a) *Fixed capital investment*, such as for design and engineering, land purchase and improvement, buildings, equipment, installation, promotional and legal fees, and startup costs.

   (b) *Working capital*, such as for inventories, accounts receivable, cash for wages, materials, and other accounts payable. Working capital is a revolving fund needed to get a project started and to meet subsequent obligations. Normally, it is assumed that working capital can be recovered in total (i.e., 100% salvage value) by the end of the life of a project.

2. *Labor costs* are a function of skill level, labor supply, and time required. Standards for the normal amount of output per labor hour have been developed for many classes of work. Standard times combined with expected wage rates, adjusted for productivity patterns, if any, provide a reasonable estimate of labor costs for repetitive jobs. Labor costs for specialized work can be predicted from bid estimates or quotes by agencies offering the service. It should be remembered that labor costs should consider fringe benefits as well as direct wages.

3. *Maintenance costs* are the ordinary costs required for the upkeep of property and minor changes required for more efficient use. Maintenance costs tend to increase with the age of an asset because more upkeep is required later in life.

4. *Property taxes and insurance* are usually expressed as an annual percentage of first cost in economic comparisons.

5. *Quality and scrap costs* depend upon the types of products and associated quality standards as well as upon abilities of the work force, learning time, and rework possibilities.

6. *Overhead costs* are by definition those costs which cannot be conveniently and practicably charged to particular products or services, and thus are normally prorated among the products or centers on some arbitrary basis. One should guard against using these arbitrary allocations in economic analyses, for the differences in overhead costs brought about by various alternatives being considered are rarely described by these rates. In general, one should consider each individual cost element included in the overhead and estimate how much, if any, each cost element is affected by each alternative.

7. *Revenues* are normally very difficult to predict for all except the most stable markets because of the buying whims of the public.

8. *Economic lives* are often very difficult to estimate and crucial to the analysis result.

9. *Salvage values* are often a function of the economic life and, in the case of long lives, are relatively unimportant to the analysis result.

how estimates are accomplished

Estimates can be prepared in limitless ways, some of which can be categorized as follows:

1. A *conference* of various people who are thought to have good information or bases for estimating the quantity in question. A special version of this is the *Delphi method*, which involves cycles of questioning and feedback in which the opinions of individual participants are kept anonymous.

2. *Comparison* with similar situations or designs about which we have more information and from which we can extrapolate estimates for alternatives under consideration.

3. *Quantitative techniques*, which do not always have standardized names. Some examples are given here with the names used being generally suggestive of the approaches.

(a) *Unit technique*—involves utilizing an assumed or estimated "per unit factor" that can be estimated effectively. Examples are

Capital cost of plant per kilowatt of capacity.
Fuel cost per kilowatt-hour generated.
Capital cost per installed telephone.
Revenue per long-distance call.
Temperature loss per 1,000 feet of steam pipe.
Operating cost per mile.
Maintenance cost per hour.

Such factors, when multiplied by the appropriate unit, give a total estimate of cost or savings.

There are limitless possibilities for breaking estimates into units that can be estimated readily. Examples are

In different units, such as dollars per week, to convert to dollars per year.

A proportion, rather than a number, such as % defective, to convert to number of defects.

A number, rather than a proportion, such as number defective and number produced, to convert to % defective.

A rate, rather than a number, such as miles per gallon, to convert to gallons consumed.

A number, rather than a rate, such as miles and hours traveled, to convert to average speed.

Using an adjustment factor to increase or decrease a known or estimated number, such as defectives reported, to convert to total defectives.

As a simple example, suppose that we need a preliminary estimate of the cost of a particular house. Using a unit factor of, say, $25 per square foot and knowing that the house is approximately 2,000 square feet, the estimated cost would be $25 × 2,000 = $50,000.

While the unit technique is very useful for preliminary estimating purposes, one can be dangerously misled by such average values. In general, more detailed methods can be expected to result in greater estimation accuracy.

(b) *Segmenting technique*—involves decomposing an uncertain quantity into parts that can be separately estimated and then added together. As an example, suppose that we desire to estimate sales of a product X in the Dakotas. The simplest possible segmenting would be to estimate sales separately in North Dakota and South Dakota and then to add the two together.

(c) *Factor technique*—is an extension of the unit method and the segmenting method in which one sums the product of several quantities or components and adds these to any components estimated directly. That is,

$$C = \sum_e C_e + \sum_i f_i \times U_i \qquad (8\text{-}1)$$

where C = value (cost, price, etc.) being estimated

C_e = cost of selected component e that is estimated directly

f_i = cost per unit of component i

U_i = number of units of component i

As a simple example, suppose that we need a slightly refined estimate of the cost of a house consisting of 2,000 square feet, two porches, and a garage. Using unit factors of $20/square feet, $3,000/porch, and $5,000/garage, we can calculate the estimate as

$$\$20 \times 2{,}000 + \$3{,}000 \times 2 + \$5{,}000 = \$51{,}000$$

(d) *Power-sizing technique*—is a sophistication of the unit method and is frequently used for costing industrial plant and equipment in which it is recognized that cost varies as some power of the change in capacity or size. That is,

$$(C_A \div C_B) = (S_A \div S_B)^X \qquad (8\text{-}2)$$

where C_A = cost for plant A $\Big\}$(both in \$ as of point in time for which
$\quad\;\; C_B$ = cost for plant B $\quad$ estimate is desired)
$\quad\;\; S_A$ = size of plant A $\Big\}$(both in same physical units)
$\quad\;\; S_B$ = size of plant B
$\quad\;\; X$ = cost-capacity factor to reflect economies of scale†

As an example, suppose that it is desired to make a preliminary estimate of the cost of a 600-MW fossil power plant if built now. It is known that a 200-MW plant cost \$50 million in 1960 when the appropriate cost index was 800, and that cost index is now 1,200. The power-sizing model estimate, with $X = 0.79$, is

cost now of 200-MW plant: \$50 million $\times$ (1,200 $\div$ 800) = \$75 million
  (call it C_B)

cost now of 600-MW plant: $C_A \div$ \$75 million = (600 $\div$ 200)$^{0.79}$
  (call it C_A)

$$C_A = \$75 \text{ million} \times 2.38 = \$178.6 \text{ million}$$

(e) *Miscellaneous Statistical and Mathematical Modeling Techniques*—can be used for estimating or forecasting the future. Typical of the numerous mathematical modeling techniques that allow one to break down difficult problems to make more and/or reliable estimates, but which will not be described herein, are econometric models, demographic (population characteristic) models, network models, stochastic process models, mathematical programming models, input–output tables, regression models, and exponential smoothing models.

communication problems in estimating

When uncertainty is involved in estimates and information for an investment proposal has to flow from an analyst to a decision maker, or from a subordinate to his boss, the following consequences can occur and should be guarded against:

1. Misinterpretations and ambiguities occur because judgments about uncertainty are difficult to transmit and do not flow easily from one person to another.

† May be calculated/estimated from experience. See p. 137 of W. R. Park, *Cost Engineering Analysis* (New York: John Wiley & Sons, Inc., 1973), for typical factors. For example, $X = 0.68$ for nuclear generating plants and 0.79 for fossil-fuel-generating plants.

2. Biases and distortions occur because:
   (a) The experts are not always sure just what top management wants.
   (b) The performance measurement and control system used by top management often unwittingly stifles any hope of getting an unbiased estimate from a subordinate.
   (c) The capacity for risk taking differs among individuals.
3. The estimating procedures used by analytic experts may be so unsystematic and muddled that the information reported is misleading.

consideration of inflation

Until now we apparently have assumed that prices for goods and services are relatively unchanged over substantial periods of time, or that the effect of such changes are the same on any alternatives considered. Unfortunately, this is not generally a realistic assumption. *Inflation*, which is the phenomenon of rising prices bringing about a reduction in the purchasing power of a given unit of money, is a fact of life and can significantly affect the economic comparison of alternatives.

Annual rates of inflation (often referred to as escalation) vary widely for different types of goods and services and for different times. For example, the U.S. government-sponsored Consumer Price Index rose less than 2% per year during the 1950s, but increased to approximately 6% per year during the mid-1970s. On the other hand, the cost of new housing construction was rising more than 10% per year during the mid-1970s. While it appears that such inflation will continue for the long-term future, it is possible that its opposite, *deflation*, can occur as was true during the recession of the 1930s.

If all cash flows in an economic comparison of alternatives are inflating at the same rate, inflation can be disregarded in before-tax studies. In cases where all incomes and all expenses are not inflating at the same rate, inflation causes differences in economic attractiveness among alternatives and should be taken into account. A typical case is when a project is financed by means of borrowing at a fixed rate of interest. The effect of inflation is to reduce the cost of borrowing, and this may affect the final decision. Conversely, in the same situation the lender fails to realize his desired return on the capital loaned out. The following is an example.

EXAMPLE 8-1

An investor lends $10,000 today to be repaid in a lump sum at the end of 10 years with interest at 8% compounded annually. What is his real rate of return, assuming that inflation is 5% compounded annually?

Solution

In 10 years, the investor will receive his original $10,000 plus interest that has accumulated:

$$F = \$10,000(F/P, 8\%, 10) = \$21,589$$

The purchasing power of a dollar bill, however, has been reduced or eroded, at 5% annual inflation, to

$$P = \$1(P/F, 5\%, 10) = \$0.6139$$

Thus the $21,589 is only worth, in today's purchasing power,

$$\$21,589 \times 0.6139 = \$13,253$$

The $13,253 of today's purchasing power returned for the use of $10,000 represents a real rate of return that may be calculated by finding the $i'\%$, at which

$$\$10,000 = \$13,253 \times (P/F, i'\%, 10) \quad \text{or} \quad i'\% \cong 3\%$$

The real rate of return can be approximated much more expeditiously by simply subtracting the inflation rate, f, from the 8% being charged. Thus the real rate $\cong 8\% - 5\% \cong 3\%$. ∎

We now describe general principles and methods for considering inflation in economic analyses. Let us define two distinct kinds of dollars (or other monetary units such as pesos or rubles) typically encountered in economic analyses if done properly:

1. *Actual* dollars—the actual number of dollars as of the point in time they occur and the usual kind of dollars in terms of which people think. Sometimes called *then-current dollars*, they will be denoted as A$ whenever a distinction needs to be made in this book. As an example, individuals typically anticipate their salaries 2 years hence in terms of actual dollars.
2. *Real* dollars—dollars of the same purchasing power as of some point in time, regardless of when the corresponding actual dollars occur. Sometimes called *constant-worth dollars*, they will be denoted as R$ whenever a distinction needs to be made in this book. As an example, costs of constructing power plants are often estimated in real dollars to provide a consistent basis for comparison.

Actual dollars as of any time, n, can be converted into real dollars of purchasing power at any time, k, by the relation

$$R\$ = A\$ \cdot \left(\frac{1}{1 + f}\right)^{n-k} = A\$ \cdot (P/F, f\%, n - k) \tag{8-3}$$

where f is the inflation rate per period over the n periods. It is most common to express real dollars as dollars as of the time estimates are made or as of the beginning of the study period under consideration.

As an example, suppose that your salary for each of the next 4 years (expressed in A$) is expected to be as follows:

Year	Salary (A$)
1	$15,000
2	16,500
3	17,150
4	18,165

Your expected salary could just as well have been stated in dollars of purchasing power as of some base point in time (i.e., R$). If you assume that the base point in time is year 1, your R$ salary expected for the next 4 years could be expressed as follows:

Year	Salary (R$)
1	$15,000
2	15,000
3	15,000
4	15,000

If the inflation rate, f, is 10% per year, both ways of expressing your expected salary are exactly equivalent. These numbers illustrate a situation that reflects the experience of many people in recent years; that is, even though salaries or wages have been increasing, the purchasing power (R$) of those salaries has not increased correspondingly. Indeed, many people have experienced declining purchasing power because the rate of salary increases has not been as great as the rate of inflationary increases.

real interest rate, inflation, and combined interest rate

Let us define several types of rates necessary to account for the effects of inflation and show how they are used:

1. *Real interest rate*—the increase in *real* purchasing power expressed as a percent per period. This is the interest rate at which R$ outflow is equivalent to R$ inflow and is sometimes known as "monetary return" or "real monetary rate." It is denoted as i.
2. *Inflation rate*—the increase in price of given goods or services as a percent per period. It is denoted as f.
3. *Combined interest rate*—the increase in future sums to cover real interest and inflation, expressed as a percent per period. This is the interest rate at which A$ outflow is equivalent to A$ inflow. It is often denoted as i, but will be denoted as i_c, where it needs to be distinguished from real interest rate. *Most M.A.R.R's used by industry include anticipated inflation and thus are combined interest rates.*

Because the real interest rate or monetary return and the inflation rate have a multiplicative or compounding effect,

$$i_c = (1 + i)(1 + f) - 1$$

$$= i + f + i \times f \qquad \text{(8-4)}$$

$$i_c \cong i + f \qquad \text{(8-5)}$$

The approximation is satisfactory where i and f are not large relative to the accuracy desired.

what interest rate to use in economy studies

In general, the interest rate appropriate for equivalence calculations in economy studies depends upon the type of cash flow estimates as follows:

Approach	If Cash Flows Are in Terms of:	Then the Interest Rate to Use Is:
(a)	Real $	Real interest rate, i
(b)	Actual $	Combined interest rate, i_c

This table should make intuitive sense as follows. If one is estimating in terms of real (uninflated) $, then use the real (uninflated) interest rate. Similarly, if one is estimating in terms of actual (inflated) $, then use the combined (inflated) interest rate. Thus one can make economic analyses using either R$ or A$ with equal validity provided that the appropriate interest rate is used for equivalence calculations. *For cases in which different cash flows are subject to inflation at different rates, all cash flows should be converted into real $, and hence a real interest rate should be used for further equivalence calculations.*

EXAMPLE 8-2

To illustrate the relationship of the two approaches to consideration of inflation, consider a project requiring an investment of $20,000 which is expected to return, in terms of actual dollars, $6,000 at the end of the first year, $8,000 at the end of the second year, and $12,000 at the end of the third year. The rate of inflation is 5% per year, and the real monetary interest rate is 10% per year. (Thus, from Equation 8-4, the combined interest rate is 15.5%.) Table 8-1 shows how the net present worth of the project would be calculated to be − $1,060 using the second approach (b), above.

If the first approach, (a), for considering inflation is used, the outcomes should be estimated in terms of real dollars.

For the example above, with the inflation rate of 5%, if the estimator is precisely consistent, his projections in terms of real dollars (as of time of investment) should be $20,000, $5,720, $7,240, and $10,360 for each year, respectively.

table 8-1

calculation of net present worth with estimates
in actual dollars

Year, N	Outcome (A$)	Discount Factor for Real Interest and Inflation ($P/F, 15.5\%, N$)	Present Worth (rounded)
0	− $20,000	1.000	− $20,000
1	6,000	0.867	5,200
2	8,000	0.745	5,980
3	12,000	0.647	7,760
			$\Sigma \cong -\$\ 1,060$

Table 8-2 shows how, using a real interest rate of 10%, the net present worth of the project would be calculated to be the same − $1,060 as when using the first approach.

It is worthy of note that had the outcomes in actual dollars been discounted by only the real interest rate of 10%, the net present worth would have been calculated to be $1,080, indicating a favorable project. This is in contrast with the − $1,060 net present worth (indicating an unfavorable project) calculated when the correct interest rates corresponding to the types of dollars under consideration were utilized.

After-tax evaluations incorporating actual dollar estimates present some difficulty because interest and depreciation charges are necessarily based on past commitments and generally are unresponsive to inflation. (Taxes are discussed in Chapter 10.) Because interest and depreciation deductions are not responsive to inflation like other cash flows, the taxable income tends to be higher and the after-tax rate of return is consequently lower than if the deductions were based on real dollars. Fortunately, most alternatives in engineering economic comparisons are affected by inflation so nearly the same that inflation is seldom a deciding factor.

table 8-2

calculation of net present worth estimates in real dollars

Year, N,	Outcome (R$)	Discount Factor for Real Interest Only ($P/F, 10\%, N$)	Present Worth (rounded)
0	− $20,000	1.000	− $20,000
1	5,720	0.909	5,200
2	7,240	0.826	5,980
3	10,360	0.751	7,760
			$\Sigma \cong -\$\ 1,060$

fixed and increment costs

In many situations we are concerned with the economic results of changes from existing operating conditions. Sometimes such changes involve the level of operations; in other cases a temporary change in procedures may be involved, but there may be accompanying longer-range effects. Such studies constitute virtually a distinct class of economy studies, inasmuch as they involve *fixed* and *increment costs*, which must be recognized and dealt with properly. Fixed and increment costs will be discussed in depth herein. The concept of *sunk costs* will be treated at the end of the chapter.

Incorrect handling of fixed and increment costs can result in widely different answers and decisions. An illustration may be found in a very simple example. Four college students wish to go home for Christmas vacation, a distance of 400 miles each way. One has an automobile and agrees to take the other three if they will pay the cost of driving the car. When they return from the trip the owner presents each of them with a bill for $34.14, stating that he has kept careful records of the cost of operating his car and has found that, based on his average yearly mileage of 15,000 miles, his cost per mile is $0.128. The three students declare that they feel the charge is too high and ask to see his cost figures. He shows them the following list:

Item	Cost per Mile
Gasoline	$0.040
Oil and lubrication	0.007
Tires	0.009
Depreciation	0.050
Insurance and taxes	0.008
Repairs	0.010
Garage	0.004
TOTAL	$0.128

The three riders, on the other hand, claim that only the costs for gasoline, oil and lubrication, tires, and repairs are a function of mileage driven and thus could be caused by the trip. Since these four costs total only $0.066 per mile, and thus $52.80 for the 800-mile trip, the share for each student would be $52.80/3 = $17.60. Obviously, the opposing views are substantially different. The question is: Which, if either, is correct?

Actually, the correct answer to this apparently simple problem is not a simple one, as will be pointed out later, and more complex problems involving the same basic factors frequently arise.

Ordinarily, capital is invested for the purpose of making provision for future operations. Such investments result in costs being incurred, many of which, such as taxes on property, maintenance costs of buildings and equipment, and even depreciation, may be almost constant regardless of considerable variations in the level of future operations. However, at any time subsequent to the investment these resulting costs are due to past decisions. For economy study purposes, recognition of this category of costs is very important.

Fixed costs are those that will continue, unchanged, whether or not a given change in operations or policy is adopted. In other words, a proposed change will have no effect upon fixed costs. It should be kept in mind, however, that when we use the term *fixed costs* we mean *fixed* only with respect to the particular proposal or change being considered and the span of time involved. A cost that would not be affected by a certain change in operations might be altered greatly by some other change. For example, suppose that Jennifer Jones drives to her bank and parks her car, putting a dime in the parking meter to gain the privilege of parking 1 hour. She concludes her business at the bank in 15 minutes and passing a boutique, decides to buy a gift for a friend. This added transaction requires only 10 minutes, and she returns to her car and drives away, leaving time remaining on the parking meter. In this case the decision to purchase the gift involved no increased cost for parking. As far as this transaction was concerned, the cost for parking was fixed. On the other hand, had Miss Jones, upon leaving the bank, stopped at her club, engaged in a card game, forgotten about the time, and returned to her car after the hour had expired and found a citation for violation of parking regulations on her car, her cost for parking would not have been fixed.

Similarly, a cost that may be fixed for a certain interval of time, regardless of how operations vary, may change if the revised operating procedure becomes permanent. For instance, a contractor who is to make a cut through a large hill, in connection with building a new highway, makes a subcontract to sell the dirt to a company for $1 per load to be used to fill in some swampland. The $1 charge exactly covers the cost of hauling the dirt. As long as he obtains the dirt from the cut being made in the hill, the contractor may correctly assume that the dirt costs him nothing. However, if after the cut is completed he is still obligated to supply additional dirt to the company at the same price, but no longer has any free source of supply, his cost situation would not be fixed. Thus temporarily he could correctly consider the cost of the dirt to be zero; on a longer basis he could not.

The importance of short-range versus long-range viewpoints will be discussed later. In even such a simple situation as the students' Christmas-vacation trip discussed previously, possible long-range effects can be important.

It should also be noted that *fixed costs* do not necessarily have any relationship to what are often referred to as *fixed charges* in ordinary accounting procedure. For example, bond interest is usually classified as a fixed charge. However, a change in operations might conceivably result in increased or decreased total bond interest. Similarly, certain expenses which for accounting purposes ordinarily are thought of as variable, such as labor cost, may remain constant when a proposed change is made; thus they then would be considered as fixed costs.

It is important to remember that "fixed costs" usually are not completely fixed, particularly if the magnitude of activity change is very great. Over small

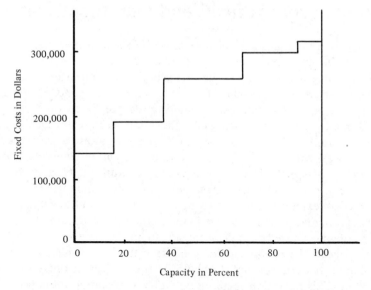

Figure 8-1. Variation of "fixed" costs with volume for a company.

ranges of change, frequently sufficient to cover a proposed activity, they may be substantially, if not completely, constant. Figure 8-1, for example, shows how fixed costs varied in one company as volume varied from zero to 100% of capacity. It may be seen, for example, that if a proposed change in activity involved increasing the capacity utilization from 15 to 20%, there would be an increase in the fixed costs. On the other hand, these costs would be unaffected by a change in capacity utilization from 75 to 85%. Thus the term *fixed costs* must be considered in relation to a specific proposal.

Increment costs are those that arise as the result of a change in operations or policy. Thus they represent the actual increases or decreases in cost resulting from the change.† They are the costs that must be considered in determining the true cost of carrying out the added operations. Like fixed costs, they are applicable only to the specific change from which they result. They may be either positive or negative in amount. Like fixed costs, they should be determined by fact.

From the nature of fixed and increment costs it follows that economy studies of this type generally involve relatively short-range problems. However, this is not necessarily the case, and in many instances what appear to be only short-range situations can have seriously longer-range implications, sometimes leading to quite disastrous results if they are not given proper consideration.

† Such costs sometimes are called *differential costs*. However, because of the mathematical connotation of the word *differential*, which means an extremely small change, this term is not as appropriate as *increment costs*, and thus is less frequently used. The term *variable costs* is sometimes used to mean the same as increment costs. However, this has the drawback of confusion with what are often referred to as *variable costs* for ordinary accounting purposes.

unit costs versus fixed and increment costs

It is quite probable that more mistakes have been made in economic decisions because of the improper use of standard or unit costs based on accounting records than due to any other single cause. Here the relationship of fixed costs, increment costs, and unit costs is of the utmost importance. As was pointed out in Chapter 1, all standard costs and unit costs are based on some level of activity, and a number of arbitrary cost allocations are used in their determination, such as in the allocation of overhead. If operations are to be carried out at a different level, the unit costs are bound to be inaccurate. Since increment cost studies invariably involve changes from existing conditions, it is at once apparent that the use of standard or unit costs may lead to considerable error.

An excellent example of how a wrong decision may result from improper use of unit costs in an economy study is found in the following case. A large manufacturing plant consisted of 11 departments. In one of these automobile batteries were produced. The equipment for producing the hard rubber covers for the cells of these batteries occupied about 100 square feet in one corner of this department. This operation required only one workman 4 hours per shift and was under the supervision of a person who also supervised a number of other operations. The daily production of cell covers was 576. The total cost of operation according to accounting records was as follows:

Labor	$12.00
Material	8.64
Overhead	8.20
TOTAL	$28.84

Unit cost = $28.84/576 = $0.05 per cover

The cell cover operation was the cause of considerable noise, and the percentage of defects was high. It required considerable attention by the foreman and was a general source of grief.

A new foreman was appointed for this department, and he soon became aware of the troubles connected with the making of the cell covers. He happened to have a friend who was the manager of another company that specialized in the production of hard rubber and plastic products. In discussing their problems together, the new foreman mentioned his troubles with the cell cover operation and the fact that it cost $0.05 to produce each cover. His friend thought the price seemed rather high and later made an offer to produce the cell covers from the existing molds for $0.035 each. The department manager computed that the saving involved through having the cell covers manufactured outside the plant would be as follows:

Present cost for 576 cell covers	$28.84
Cost if made outside (576 × $0.035)	20.16
DAILY SAVING	$ 8.68

This appeared to him to be a very worthwhile saving, inasmuch as it could be obtained with no investment of additional capital. As a result the change was made, and the cell covers were made by the outside company.

After the change had been in effect a little over a month, the cost department made a check upon the savings that actually resulted from the change. The investigation showed that the actual saving in overhead was only $0.30 per day, which represented a saving in insurance. The molding press had to be kept available, since it served as a stand-by for a second press that was required for another operation. It was further found that a portion of the material that had been used for the cell covers was a waste product of another operation in the factory and that the actual saving in material was only $6.10 instead of $8.64. This waste product had been included in the unit-cost determination at $2.54 per day, because it would have cost this much to produce this material if it had not been available as scrap. As a result the company was now paying for the 576 cell covers the following:

Remaining overhead	$ 7.90
Waste material	2.54
To outside company	20.16
TOTAL	$30.60

Instead of a saving of $8.68 per day there was actually a loss of $30.60 − $28.84, or $1.76. Arrangements were made immediately to have the molds returned to the plant, and cell covers were again produced within the factory.

A little thought makes it apparent why the use of unit costs failed to give correct results. The unit cost of $0.05 per cover was based upon the assumption that 576 covers *were to be produced* each day. If the covers were not produced, the unit cost had no meaning, since it did not apply to this set of conditions. We might even do a bit of computation and say that the unit cost of *not producing* 576 cell covers each day would be

$$\frac{\$7.90 + \$2.54}{576} = \$0.018 \text{ per cell cover}$$

It becomes apparent from the foregoing example that serious errors may result from the use of unit costs in economy studies. For economy studies of this type we must determine the increment cost of producing the products. The actual (increment) costs that were incurred when the cell covers were produced, but that would not exist if they were not manufactured, would be as follows:

Increment labor	$12.00
Increment material	6.10
Increment overhead (insurance)	0.30
TOTAL INCREMENT COST	$18.40

Dividing $18.40 by 576 gives a unit increment cost of $0.032 for producing the cell covers. In other words, the actual cost of producing each cell cover, *compared to not making any cell covers*, was slightly less than $0.032. This result makes it obvious that an actual loss would result from paying $0.035 each for cell covers produced by an outside company. The department foreman had made the common mistake of using unit costs for a purpose for which they were never intended and of failing to determine the facts that would reveal the increment costs for the alternatives of producing or not producing the cell covers.

two methods for increment cost studies

The example of the molded cell covers illustrates two basic methods that can be used for making increment cost studies. The first is to consider only the increment costs, revenues, or savings. For example, for the cell covers the increments costs were found to be $18.40. By comparing this amount with $20.16, the cost of the alternative of having the cell covers produced by an outside company, we could easily conclude that the existing program for producing them was more economical. This method of analysis, considering only the increment values, usually involves fewer data and sets forth the pertinent facts very clearly. We must, however, be certain that all increment costs are correctly separated from the fixed costs.

The second method is to compare the total of all costs for the two alternatives being considered. Thus, for the cell covers the total cost of making them had been $28.84. The total cost of having them manufactured by an outside company, including those portions of the existing costs that would continue, was $30.60. This procedure requires a careful determination of the fixed costs. It may thus be seen that the first method deals, essentially, with increment costs, and the second deals with the total costs.

capacity factor, load factor, and diversity factor

Three terms commonly are employed when we discuss and measure the utilization of installed capacity, and the fixed and increment costs that result from the degree of its utilization. The first is *capacity factor*. This is defined as the following ratio:

$$\text{capacity factor} = \frac{\text{average actual demand}}{\text{available capacity}} \tag{8-6}$$

Whenever a capacity for production or service exists, the capacity factor at which the equipment is utilized has a vital effect on the economy of the project, since effective use of capacity usually means effective use of capital. Most of the early studies of the effect of capacity factor were made in the utilities industries, where the investment costs are high relative to total production costs. However, inasmuch as the effects are the result of fixed investment, they are just as important to nonutility companies and are now receiving more attention.

As a simple example of the effect of capacity factor, consider the case of a six-passenger station wagon. If it is used by only two persons, it is operated at a capacity of only $33\frac{1}{3}\%$. For such conditions a two-passenger Volkswagen, operated at 100% of capacity, would be expected to be a much more economical means of transportation. On the other hand, if six persons were to be transported, it probably would be considerably more economical to use one six-passenger station wagon than three Volkswagens. In this simple example the effects of capacity upon first cost and its utilization upon operating cost are apparent.

The second term used to measure capacity utilization is *load factor*. This is defined as the following ratio:

$$\text{load factor} = \frac{\text{average actual demand}}{\text{maximum demand}} \tag{8-7}$$

Obviously, if the available capacity were also the maximum demand, capacity factor and load factor would be identical. However, in many instances, particularly where *systems* of facilities are involved, it is very difficult to specify the exact available capacity. By combining the units in different ways and by using them for different purposes, we can vary the capacity. Therefore, when a single unit of capacity, or one having a well-defined capacity, is involved, load factor may be used as a measure of utilization. For systems, capacity factor more commonly is employed.

When we use load or capacity factors, it often is necessary to specify the duration of the maximum demand or capacity load, inasmuch as units and systems may be capable of sustaining overloads for short periods. Thus electric power companies frequently use the 15-minute maximum demand as the basis for determining the load factor of their individual customers.

Diversity factor measures a condition that is very important to public utilities, but it also is of considerable importance in many nonutility industries. Its significance may be explained by a simple example. Assume that a utility company has only two customers. One is a residence that uses the power for lighting. The other is a small factory that uses power for the operation of machinery between the hours of 8 A.M. and 5 P.M. Since the residence uses power only for lighting, it is unlikely that all the lights would ever be turned on between the hours of 8 and 5 during the daytime. Similarly, the factory would not use any appreciable amount of power during the evening, when the lights in the residence would be turned on. If the total connected load in the residence is 3 kilowatts and that of the factory 8 kilowatts, the total possible load that can be thrown on the line is the sum of the two loads, or 11 kilowatts. However, because of the diversity of the load, the maximum demand is actually 9 kilowatts. It is thus seen that the maximum demand may be considerably less than the maximum possible demand because the two or more customers have different demand characteristics.

Ordinarily, a utility has many customers of various types, providing power for residences, stores, factories, street lights, street railways, and so on. No two of these have the same load characteristics. The utility is able to supply all of these diversified customers satisfactorily with a generating capacity considerably less than the sum of all the maximum demands of each consumer. Great effort is made to obtain customers who will use power during periods when others are using very little. Attractive rates are usually offered to those who will buy power during off-peak periods and not consume any during the peak hours.

Diversity factor is defined as the following ratio:

$$\text{diversity factor} = \frac{\text{sum of the individual users' maximum demands}}{\text{maximum demand actually experienced}} \quad (8\text{-}8)$$

Thus a diversity factor of 4 would indicate that the utility would need to have generating equipment of only one-fourth the capacity of the sum of the individual customers' maximum demands.

The discussion of diversity factor has been in terms of an electric utility. Of course, diversity is just as important to other utilities. Water and gas companies must pay attention to the diversity of demand by their consumers. These utilities do have one advantage in that they can provide some storage of their product during off-peak hours to meet the peak loads. Electric companies are able to do this only in rare cases.

A system with a high load factor may have a high diversity factor—but not necessarily. The effect may be just the opposite. High load factor comes from continuous use of connected equipment. If all the customers on a utility system used all their connected equipment all the time, the company would have a load factor of 100%, but the diversity factor would be unity. This is obviously an extreme case.

Generally, any given customer places a demand that shows "peaks and valleys" variation over time. The providers of the service generally desire to have many customers of different types, so their respective maximum demands will occur at different times and thus result in a "smoothed out" total demand. Thus the providers of service generally desire to obtain a high diversity.

increment cost pricing

Numerous uses are made of increment cost theory in business. Among these is the basing of selling prices on increment costs. Management frequently is faced with the existence of surplus or scrap materials or surplus facilities that might be used, at relatively low increment cost, for producing new products or by-products. Inasmuch as the fixed costs of the facilities or operations already are being paid, through charges to the existing products or operations, the only costs for the new products will be the increment costs incurred. On this basis the selling price for the new product can be considerably less than if fixed costs also had to be assigned to it.

An example of this type of application of fixed and increment cost principles is found in the case of a company manufacturing blowers of various sizes.

During a business recession the sales of large blowers were very low. As a result, the company was faced with the necessity of operating at very low output and of laying off a number of its employees. A study of the situation revealed that it had all the facilities necessary for the production of small kitchen ventilating fans. It was found that there would be no fixed costs involved and that the men could work on this product at odd times when they were not employed on the production of blowers. The cost of producing each fan under these conditions was

Increment material	$4.80
Increment labor	1.08
TOTAL INCREMENT COST	$5.88

Other fans of this type were sold at about $18.00 retail, and the dealers were given a 40% discount from this price. This company placed its fans on the market at a retail price of $15.00 and allowed the dealers a discount of 50%. With this pricing, the dealers were able to make more profit on a $15.00 fan than they had previously obtained from an $18.00 sale. The manufacturer obtained $7.50 from each fan. The profit to the manufacturer was

Selling price to retailers	$7.50
Total increment cost	5.88
INCREMENT PROFIT PER FAN	$1.62

As the sales of the fans at $15.00 were considerably greater than had been anticipated, the company was able to show a profit throughout the entire recession. At the same time, the working force was kept at nearly full employment. Thus, by taking advantage of the fact that the additional product could be produced for only the amount of the increment costs, the company was able to avoid a serious loss of profits and to maintain employment. Such a policy is of obvious advantage for short and intermediate purposes. However, in the long run the firm must cover all its costs as well as show sufficient profits to justify continued investment of the owners' capital.

the economy of shutting down plants

The principle of fixed and increment costs is of prime importance in decisions concerning shutting down plants when excess capacity exists. A small manufacturer of specialized chemicals had products used principally by one industry that was experiencing a period of reduced activity. During the previous year the company's net sales were $141,200. Operating costs had been approximately as follows:

Raw materials	$ 42,500
Labor	27,600
Taxes, depreciation, and maintenance	18,200
Managerial and sales expense	14,000
TOTAL	$102,300

With these conditions, the net profit was as follows:

Income from sales	$141,200
Total expenses	102,300
PROFIT	$ 38,900

The company estimated that its sales for the coming year would be only $46,000. Examination resulted in an estimate that its total expenses for operating under these conditions would be as follows:

Raw materials	$12,600
Labor	19,800
Taxes, depreciation, and maintenance	16,000
Managerial and sales expense	12,000
TOTAL EXPENSES	$60,400

The loss that would be incurred by operating under these conditions would be calculated as follows:

Total expenses	$60,400
Income from sales	46,000
TOTAL LOSS	$14,400

An alternative to operating the plant under these conditions was to close the manufacturing plant and maintain only the sales offices. Its products would be manufactured by a large general chemical plant. It was estimated that it could buy its products from the large producer at prices that would enable it to resell them to its customers and break even on the individual transactions; that is, there would be no profit or loss from not producing the chemicals. Under these conditions its expenses for the year would be the following fixed portion of the total expenses.

Taxes, depreciation, and maintenance	$12,800
Managerial and sales expense	12,000
TOTAL LOSS	$24,800

After these figures were assembled, it was at once apparent that from a purely financial viewpoint it was much better policy for the company to operate its chemical plant and produce its own goods even though a loss would follow. The total loss would be only $14,400 from this method of operating, whereas there would be a total loss of $24,800 if the plant were closed.

The analysis above was made on the basis of *total* expenses. Now let us illustrate how to arrive at the same decision using the *increment* viewpoint. The increment cost of producing in the coming year would be

Raw materials	$12,600
Labor	$19,800
Taxes, depreciation, and maintenance: $16,000 − $12,800	$ 3,200
Managerial and sales: $12,000 − $12,000	0
TOTAL INCREMENT EXPENSES	$35,600

The increment gain that can be obtained by producing is

Income from sales	$46,000
Increment expenses	− 35,600
NET INCREMENT GAIN	$10,400

If the net increment gain from operating the plant is anything greater than zero, this indicates that the plant should be operated based on the analysis results. The increment gain is the amount by which the total loss will be *reduced* by virtue of operating the plant rather than not operating the plant. Since the increment gain is $10,400, the plant should be operated.

Thus, if the plant operates, the total loss would be reduced as follows:

Total loss if shut down	$24,800
Net increment gain if operated	− 10,400
TOTAL LOSS IF OPERATED	$14,400

Of course, this total loss of $14,400 is the same as the $14,400 total loss calculated before the problem was approached from an increment viewpoint.

In addition to the purely monetary side of the question, the costs of rehiring and training new workers that would follow later, the possible loss of customers to the larger company from which it was going to purchase its products, and the effect of unemployment on its workers were factors that had to be considered but upon which a monetary value could not easily be placed. It was decided that the plant should be operated at reduced capacity and considerable effort made to develop additional products and new markets for its existing products.

balancing outputs between plants

A modification of the problem involved in the possible shutting down of plants frequently occurs when more than one plant of a company produces the same product, and no plant is operating at capacity. If, as usually is the case, operating costs are not identical in all plants, it is necessary to determine the proper amounts to produce in each plant to assure maximum overall economy.

The general economic principle to follow in such instances is that total costs are minimized if the firm produces as much as possible at that plant(s) with the lowest variable (increment) costs per unit. The following example illustrates such a situation, and in it an instance of arbitrary cost allocation will be found.

EXAMPLE 8-3

A company having its main factory in Illinois produces a by-product in a separate factory adjacent to its main plant. It also produces this same product in a California plant, where it is the sole product. The Illinois plant has a capacity of 15,000 tons per year and is currently producing at the rate of 10,000 tons per year. At this output the unit cost of production is $110 per ton. Included in this unit cost is a charge for material A, which is a waste product of the main Illinois plant and for which the by-product plant pays $5 per ton used. Any excess of this waste product not used in the by-product plant is sold for $2 per ton. The by-product plant is currently using 2,000 tons per year of this waste product. The fixed expenses of the Illinois by-product plant are $850,000 per year.

At the California plant the capacity is 11,000 tons per year, with a current output of 7,000 tons. At this output the unit cost is $94.75 per ton with the fixed expenses being $500,000 per year. At this plant all materials have to be bought at regular market prices. Would any change in the output of either plant be desirable?

Solution

Since neither plant can produce the total output, both will have to continue in operation. Therefore, the fixed costs of each plant will continue, and the decision should be based on the increment costs.

In order to determine the true increment costs at the Illinois plant, the $3.00 per ton bookkeeping profit being made on waste material A must be taken into account. This bookkeeping profit clearly is the result of an arbitrary $5 price that has been assigned to this material when it is "sold" to the by-product plant. Such arbitrary cost or price allocations are not uncommon in business, and some care must be exercised in making economy studies to prevent the true facts from being obscured by them. In this instance the true unit variable cost for the by-product at the Illinois plant would be as follows:

Total annual cost = $110 × 10,000	$1,100,000
Less profit on material A = $3.00 × 2,000	6,000
Actual total annual cost	$1,094,000
Less fixed costs	850,000
Variable costs	$ 244,000

Unit variable cost = $244,000/10,000 = $24.40 per ton

For the California plant the unit variable cost can be calculated as:

Total annual cost = $94.75 × 7,000	$663,250
Less fixed costs	500,000
Total variable costs	$163,250

Unit variable costs = $163,250/7,000 = $23.32 per ton

Therefore, if there are no difficulties regarding extra shipping costs, and so on, it appears that the responsibility for 4,000 tons of output should be shifted so that the California plant operates at capacity and only 6,000 tons are produced by the Illinois plant. The savings will be $24.40 − $23.32 = $1.08 per ton. ∎

utilization of excess capacity by dumping

Another method is often resorted to by companies with a problem of excess plant capacity, particularly those that export a portion of their products. In a number of industries there is virtually an absolute maximum number of units of production that can be used by the domestic market during a given period. Reduction of the selling price would bring few, if any, additional sales. The only outlet for additional products lies in foreign markets. In such situations, sales can often be made in foreign markets only if the selling price is considerably reduced so that the product can compete with foreign-produced goods. This price at which the product must be sold on the foreign market may be considerably less than is obtained from the same product on the domestic market. When goods are produced and sold in this manner, the procedure is sometimes known as *foreign dumping.*

The economy of dumping is based upon the fact that fixed costs are present in nearly all enterprises. When a plant is operating at reduced output, a moderate increase in output will not affect the fixed costs. As a result, these additional increments of production will actually cost less to produce than the others. This situation is illustrated best by an example.

EXAMPLE 8-4

Consider the case of the ABC Company, which has plant capacity for the production of 100,000 units annually. Because of depressed business conditions, only 60,000 units can be sold in the domestic market at a price of $2.27 each. The total unit cost for these products, on the basis of 60,000 annual production, is $2.34. Under these conditions there is obviously a loss of $0.07 on each unit, or a total loss of $4,200 on 60,000 units. Nevertheless, the company would probably produce the 60,000 units for the domestic market because if they shut down, the total loss probably would be more than the $4,200.

The costs of producing 60,000 units, and other quantities between 60,000 and the plant capacity of 100,000, are shown in Table 8-3. In the table it will be noted that none of the costs are actually fixed costs in the strict sense that they are not affected by changes of output. However, factory overhead and selling expense are almost fixed, varying only slightly with changes of production. Material expense is not exactly proportional to output, since increased production makes it possible to obtain some economy in purchasing materials in larger quantities.

table 8-3
variation in unit costs with output

(1) Output (units)	(2) Material	(3) Labor	(4) Factory Overhead	(5) Selling Expense	(6) Total	(7) Unit Cost	(8) Cost per Unit of Last Output Increment
60,000	$31,200	$ 75,600	$22,800	$10,800	$140,400	$2.34	—
70,000	35,700	88,200	24,000	10,950	158,850	2.27	$1.845
80,000	40,000	100,800	25,200	11,200	177,200	2.22	1.835
90,000	44,100	113,400	26,400	11,350	195,250	2.17	1.805
100,000	48,000	126,000	27,600	11,400	213,000	2.13	1.775

This is a situation that usually occurs in economy studies of this type. There are actually very few fixed costs in the strict sense of the term. They usually vary slightly. Similarly, there may be other costs that do not vary quite in proportion to production but are more nearly incremental than fixed. In this case only labor expenses vary in direct proportion to output.

From column 6 of Table 8-3, we may see that the total cost of producing 60,000 units is $140,400. The total cost for producing 70,000 units is $158,850, an increase of $18,450. Thus the cost of producing each of these additional 10,000 units is only $1.845, whereas each of the first 60,000 units cost $2.34 to produce. If 60,000 units are all that can be sold in the domestic market, any price above $1.845 that could be obtained for the additional 10,000 units in the foreign market would be profitable. From Table 8-3 it will be seen that successive increases of production above 70,000 units are accompanied with still lower unit costs for the additional increments of production. If the production were increased to plant capacity, the unit cost of the entire 100,000 units would be $2.13, while the increment cost of each of the 10,000 above 90,000 would only be $1.775. Suppose it is estimated that 30,000 units could be sold annually on the foreign market for $2.05 each. This is less than the unit cost of $2.17 for producing *all* the 90,000 total production. Determine whether the company should try to sell the extra 30,000 units

Solution (based on increment costs)

Increment revenue for 30,000 units: 30,000 × $2.05		$61,500
Cost of producing 90,000 units	$195,250	
Cost of producing 60,000 units	− 140,400	
INCREMENT COST (for 30,000 units)		$54,850

Thus there would be an increment gain of $61,500 − $54,850 = $6,650, so the extra 30,000 units should be produced and sold.

Solution (based on total costs)

Income from sales:	
60,000 units at $2.27	$136,200
30,000 units at $2.05	61,500
Total income	$197,700
Cost of producing 90,000	195,250
TOTAL PROFIT	$ 2,450

This total profit of $2,450 should be compared to the total loss of $4,200 which would be incurred if only 60,000 units were produced and sold on the domestic market. The difference is $2,450 − (− $4,200) = $6,650, which is the same as the the incremental gain. Thus again we see that the 30,000 units should be sold on the foreign market even at the low price. ∎

The immediate advantages of such pricing and selling policies are apparent. However, in many cases the longer-range results may be somewhat complicated, as will be discussed shortly.

increment cost pricing in the public utility industry

The public utility industry has long been a leader in utilizing increment cost theory in setting the rates for its services. This is a quite natural development, inasmuch as one characteristic of the public utility industry is the very large amount of capital investment required, and this capital is used to provide long-lived facilities. Consequently, the fixed costs frequently are as much as 70% of the total unit cost. Under these conditions it is not possible to ignore unused capacity, and the utilities have made use of a number of very effective techniques, based on increment cost pricing, to achieve better utilization of their facilities. Some of these now will be examined, and their applicability in nonutility industries will be considered.

Figure 8-2 shows a typical daily load curve of a power company. Most public utilities not only have daily variations in demand, but also seasonal variations. Quite clearly, this utility must have installed capacity sufficient to meet the peak demand, which occurs at about 6 P.M. Since the provision of this capacity requires the investment of capital, the load curve may be thought of as representing the effectiveness of capital utilization. The situation is portrayed in Figure 8-3, where the shaded area represents unutilized capacity, and thus unutilized invested capital.

With a variable load as shown in Figure 8-2, a public utility naturally wishes to achieve three goals: (1) raise all portions of the load curve as close as possible to the level of the peak demand, (2) reduce the peak load, and (3) increase the total sales.

The total cost of providing a utility service is composed of three portions:

1. Costs proportional to plant capacity, determined by the *demands* of the customers.

Figure 8-2. Daily load curve of a power company on January 5.

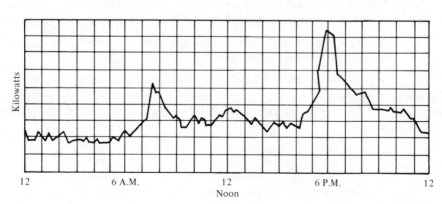

12   6 A.M.   12   6 P.M.   12
                        Noon

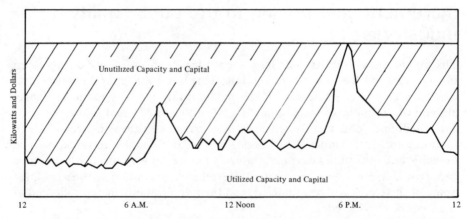

Figure 8-3. Relationship of a daily load curve to utilization of capacity and capital.

2. Costs proportional to service consumed by the customers.
3. Costs proportional to the number of customers.

On this basis, the cost of the service to a customer for a given period should theoretically be made up of three parts, as follows:

$$\text{cost of service} = Ax + By + C$$

where Ax = demand charge for demand of x during the period
By = energy charge for y units of service during the period
C = customer cost for the period

Such a three-part rate schedule is too complicated for convenient use and for customers to understand and so is rarely used. Furthermore, the principle of charging what the traffic will bear makes it inadvisable to adhere strictly to the theoretically correct procedure. This is simply recognition of the fact that some types of customers can afford to pay and will pay more than others for the service received; in other words, a given service is more valuable to some types of customers than to others. A two-part rate, commonly known as a *Hopkinson* type of rate, is sometimes used. An example of this type of rate is the following:

$1.50 per month per kW of measured maximum demand, plus
$0.015 per kWh of energy consumed

More common, however, is a *block* type of rate structure in which the rate a customer must pay for service is determined by the amount of the service he consumes. For example, the rate schedule of one company for domestic power is as follows:

First 40 kWh per month	$0.05 per kWh
(minimum monthly charge, $1.50)	
Next 200 kWh per month	0.02 per kWh
All above 240 kWh per month	0.01 per kWh

In order to set up such a rate schedule, the company had to determine the fixed and increment costs of producing and distributing power. Quite clearly, this rate schedule seeks to raise the main portion of the demand curve, but there is no direct incentive to prevent the peak demand also from increasing. Because it is a rate schedule for domestic power, the utility hopes the increased uses that residential customers might make of electricity, beyond a base amount, will not increase appreciably during the peak-demand hours.

Public utilities try to reduce their peak load demands, and thus reduce their capital investment and fixed costs, by two procedures. One is to offer considerably lower rates for service purchased during off-peak hours. For example, one power company built up a good off-peak demand by establishing a rate of $0.0075 per kilowatt-hour for power used for domestic water heating. Special electric heaters, with larger than normal storage tanks, were provided and equipped with devices that disconnected them between the hours of 4:30 P.M. and 9:30 P.M. Although the average rate received for round-the-clock energy by this utility was about $0.023 per kilowatt-hour, the $0.0075 rate for the off-peak energy was very profitable because of its low increment cost.

Another company uses the following clause to allow discounts for off-peak power:

> For current used for industrial power purposes ... the rate for peak hours shall be the same as for commercial purposes and for off-peak hours the rate shall be twenty-five per cent (25%) less on all loads of 21 horsepower and over. The peak hours shall be between 4:30 P.M. and 9:30 P.M.

A second procedure for controlling peak demands is the *Wright* type of rate structure, in which the amount of power a customer must purchase at a given rate, before being able to buy at a lower rate, is determined by his maximum demand. The following is an example of the rate schedule used by one company:

Service charge per month	$1.60
Energy charge (to be added to the service charge)	
First 1,000 kWh per month	0.027 per kWh
Next 2,000 kWh per month	0.023 per kWh
Next 5,000 kWh per month	0.019 per kWh
For all excess over 8,000 kWh per month:	
First 50 kWh per kW of maximum demand	0.017 per kWh
Next 150 kWh per kW of maximum demand, but	
not more than 85,000 kWh	0.012 per kWh
All excess	0.007 per kWh

This type of rate schedule encourages increased consumption of power but also encourages the customer to keep peak demands low.

Utility companies attempt to improve their diversity factors by obtaining customers having different demand characteristics. Different rates may be established to encourage the consumption of services by groups of customers that have desired demand characteristics but that would not use the services at normal rates. Obviously a thorough knowledge of its increment costs is essential if the utility is to establish rate schedules that will be effective in producing the desired demand and yet be profitable. This matter is of greater importance to public utilities than to other companies, inasmuch as a public utility nearly always must accept all customers who apply for service under an established schedule, and it is difficult for the utility to obtain permission to change a rate schedule once it is established.

Gas companies have an additional complicating factor in connection with the load factor of their pipelines. This is the effect of weather. Electric transmission lines are affected very little by changes of temperature, but the temperature has a great effect upon the amount of gas or oil that can be transmitted in pipelines. In some cases the transmission capacity may be maintained only at greatly increased pumping expense. In order to transmit oil or gas during cold weather, increased pumping costs must be met. Yet this is the time of year when the demands are heaviest. Thus these companies are faced with the problem of providing equipment that is adequate to meet their peak cold-weather demands. During the summer months this equipment will be operated at a very low capacity factor. In most localities it is very difficult to find customers who can use large quantities of gas or oil during only the summer months, no matter how low the price.

Although gas companies can provide for the storage of certain amounts of their product during the off-peak periods, the amount of storage that must be provided is determined by the load factor and diversity factor of the system. Therefore, these factors are of great importance to these companies.

the cost of service with block-demand rate schedules

Where electric power or other utility service is purchased under block-demand rate schedules, the effect of demand must be taken into account when we compute the cost of the power used not only by the proposed equipment, but also possibly that used by existing equipment.

EXAMPLE 8-5

A company that used an average of 20,000 kilowatt-hours of energy per month, purchased under the rate schedule shown on page 257 was going to purchase a spot welder that would consume 2,000 kilowatt-hours per month. The existing maximum demand was 55 kilowatts. The ABC machine, which cost $4,500, would increase the maximum demand by 15 kilowatts. The XYZ machine, costing $6,000, would increase the demand by only 5 kilowatts. The welder selected would have to be written off over a 5-year period, using a profit rate of 8%. Taxes and insurance on either machine would be 2% of first cost per year, and other costs were estimated to be the same for either welder. Determine the best alternative using the annual cost method.

Solution

It is evident that the difference in maximum demand for the two welders would not only affect the cost of the power actually used by the welders, but also the cost of the power used by other equipment. Therefore, the actual cost of power due to the use of the welders can be determined only by computing the total power bill before and after each welder is used. The existing monthly power bill was as follows:

Service charge	$ 1.60
First 1,000 kWh at $0.027	27.00
Next 2,000 kWh at $0.023	46.00
Next 5,000 kWh at $0.019	95.00
For service charge plus 8,000 kWh Subtotal	$169.60
Next 50 × 55 = 2,750 kWh at $0.017	46.75
Next 150 × 55 = 8,250 kWh at $0.012	99.00
Next 1,000 kWh at $0.007	7.00
For 12,000 kWh Subtotal	$152.75
TOTAL = $169.60 + $152.75	$322.35

If the ABC welder were used, the charges would be as follows:

Service charge plus 8,000 kWh	$169.60
Next 50 × 70 kWh = 3,500 kWh at $0.017	59.50
Next 150 × 70 kWh = 10,500 kWh at $0.012	126.00
TOTAL	$355.10

For the XYZ welder the charges would be as follows:

Service charge plus 8,000 kWh	$169.60
Next 50 × 60 kWh = 3,000 kWh at $0.017	51.00
Next 150 × 60 kWh = 9,000 kWh at $0.012	108.00
Next 2,000 kWh at $0.007	14.00
TOTAL	$342.60

It is thus apparent that the actual increased power cost due to the ABC welder would be $355.10 − $322.35 = $32.75 per month, and due to the XYZ welder would be $342.60 − $322.35 = $20.25 per month. The comparison between the two machines would then be as follows:

	ABC	XYZ
Annual costs:		
Depreciation + minimum profit		
	$4,500(A/P, 8\%, 5) = $1,127	$6,000(A/P, 8\%, 5) = $1,503
Power	$32.75 × 12 = 393	$20.25 × 12 = 243
Taxes and insurance		
	$4,500 × 2\% = 90	$6,000 × 2\% = 120
TOTAL	$1,610	$1,866

It is thus apparent that the ABC welder should be purchased. ∎

Since power and other utility services often can be purchased under more than one rate schedule, sometimes considerable economies can be effected by a careful analysis of the actual costs when changes are contemplated from existing loads. Likewise, when utility services are purchased under block schedules, with or without maximum demand clauses, a proper analysis is required to determine the true cost of the service due to an additional piece of equipment.

long-range problems associated with increment cost pricing

Although the immediate advantages to be derived from using increment cost pricing are apparent, we must keep certain possible long-range effects in mind. The increment cost analysis usually is correct only if the new product is, and will continue to be, a true by-product. When a new product is to be produced in this manner, it is assumed that the existing main products are the principal source of income and will therefore continue to pay the fixed costs of the company. It is expected that the new products are to be more or less temporary items, or are to be considered as side lines. Considered in this way, they are the tail of the dog, but in numerous cases the tail has wagged the dog.

A classic example of this occurred some years ago in a large sash-and-door mill in the northwest. The company was an old established manufacturer, and its various kinds of doors were well known throughout the trade. In an effort to use some of the pieces of scrap wood that resulted from the manufacturing processes, it was decided to produce a cheap door made by gluing these scrap pieces together to give the necessary thickness. It was thought that by selling these laminated doors at a price considerably below that of their other doors, the company might be able to sell enough to use up the waste pieces of wood. After a number of these cheap laminated doors had been sold and used, it was found they were superior to any of the ordinary doors the company produced, principally because the laminated structure prevented warping in damp climates. In a short time the main product of the plant was laminated doors that sold at a premium price because of their superiority. The company had reached the unacceptable condition where it was necessary to cut up large pieces of lumber in order to obtain small pieces that could then be glued together to make large pieces from which the doors could be made. Inasmuch as the by-product had become the main product, it was obvious that it would have to bear the true cost of the material being used and the proper share of the fixed costs of the business. It was necessary to revise the cost figures of the entire plant to fit the new conditions.

As was mentioned at the beginning of this chapter, in connection with the Christmas vacation trip of the four students, even very simple situations can have longer-range implications. In this instance, assume that the owner of the automobile agreed to accept $17.60 per person from the three riders, based on the costs that were purely incremental for the Christmas trip. What would happen if the three students, because of the low costs, returned and proposed another 800-mile trip the following weekend? And what if there were several more such trips on subsequent weekends? Quite clearly what started out to be a temporary change in operating conditions—from 15,000 miles per year to 15,800 miles—soon would become a standard operating condition of 18,000 or 20,000 miles per year. On this basis it would not be valid to compute increment cost per mile as $0.066. A more valid increment cost would be obtained by computing the total annual cost if the car were driven, say, 18,000 miles and, by subtracting the total cost for 15,000 miles of operation, determining the cost of 3,000 additional miles of operation. From this value an increment cost per mile for abnormal mileage could be obtained. In this instance the total cost for 15,000 miles of driving per year was

$$15,000 \times \$0.128 = \$1920$$

If the cost of 18,000 miles per year of service, which is due to increased depreciation, repairs, reduced gasoline mileage, and so on, turned out to be $2,190, it is evident that the cost of 3,000 miles of additional driving was $270. The corresponding increment cost per mile would be $0.090. Therefore, if the owner of the car suspected that several weekend trips might be made, he would be on much safer economic ground to quote an increment cost of $0.090 per mile for even the first trip.

There is no single, best way of taking possible long-range effects of increment cost pricing into account; the possible effects can be extremely varied. However, we never should adopt an increment cost pricing policy without giving adequate attention to the possible long-range effects and determining whether these effects can be controlled if they should arise.

sunk costs

A third type of cost that often must be considered in economy studies of going concerns is *sunk cost*. Sunk costs are different from other costs considered in economy studies in that they are costs of the past, rather than of the future. Virtually all economy studies deal with future costs, and this fact should immediately indicate that sunk costs have no place in them. Yet many misconceptions have existed, and still do exist, concerning sunk costs, and there is a necessity for recognizing them so that they may be handled properly.

Sunk costs may be defined in several ways, such as:

1. Any past expenditure (or commitment to expend).
2. The unrecovered balance (or book value) of an investment.
3. The book value minus selling price if sold.
4. Capital already invested that cannot be retrieved.

The principle of sunk costs may be illustrated by the following simple, extreme, and ludicrous example. John Jones, at the beginning of the winter season, purchased 1 quart of antifreeze for $1.00 and put it into the radiator of his car. From his past, and presently anticipated, driving habits he believed that 1 quart of antifreeze would provide the degree of protection needed. However, a month later he decided to make a trip to a colder locality and realized he would need 3 additional quarts of antifreeze to prevent the radiator from freezing during the trip. When he got to the service station, he found that the antifreeze was obtainable in quart cans costing $1.00 or in 1-gallon cans for $2.75. The service-station attendant suggested that he drain his radiator, purchase a gallon of antifreeze for $2.75, and refill the radiator with this gallon and the additional water as required. However, Mr. Jones insisted that he could not afford to throw away the quart of antifreeze that was already in the radiator and for which he had paid $1.00. So he bought 3 quart cans of antifreeze for $1.00 each.

It is apparent in this case that Mr. Jones spent $3.00 for the extra amount of antifreeze that he required, but if he had followed the service station attendant's suggestion he could have acquired approximately the same amount of antifreeze in the radiator of his car for only $2.75. The $1.00 that John Jones had previously spent for antifreeze was a sunk cost insofar as the immediate problem was concerned. What he had spent *in the past* had no bearing on the proper choice between two possible methods of acquiring 3 additional quarts of antifreeze: one by purchasing three 1-quart cans for $3.00, the other by purchasing a gallon for $2.75. John Jones' reluctance to forget about the $1.00 that he had spent previously, in order to make a correct immediate decision, is shared by many, including some who make economy studies and business decisions.

As another simple example of a sunk cost, suppose that Joe College finds a car that he likes on a Saturday and pays $40 "down payment," which will be applied toward the $1,300 purchase price but which must be forfeited if he decides not to take the car. Over the weekend, Joe finds another car, which he considers equally desirable, for a purchase price of $1,230. For purposes of deciding which car to purchase, the $40 is a sunk cost and thus would not enter into the decision. The decision then boils down to paying $1,300 − $40 = $1,260 for the first car versus $1,230 for the second car.

In many cases past expenditures are pertinent to a situation being studied, because they provide information for estimating future differences between alternatives. In many others they represent sunk costs that, like water gone over the dam, cannot be recovered. In such cases they have no place in an economy study and must be completely, though perhaps regretfully, forgotten.

sunk costs and depreciable assets

Sunk costs frequently arise in connection with depreciable† assets. Such a situation may be seen in the case of a small company that purchased an ABC dictating and transcribing machine for $350. At the time of purchase it was

---

† The meaning and computation of depreciation will be discussed in Chapter 9.

estimated that this machine would have a life of at least 5 years, and depreciation was charged on the accounting records on this basis, by use of the straight-line method. Upon being used, it was found that the machine contained certain design defects that could not be corrected and that it did not give satisfactory service. At the end of 3 years a new XYZ dictating machine, using magnetic tape and costing only $220, came on the market. The distributor who had sold the original machine wanted to sell the company a new model of the ABC machine for $375 and offered in his sales proposal to "... allow $100 for the old machine on the price of the new model. This will make it possible for you to recover $100 of the price originally paid for the old machine." The XYZ machine was judged to be at least as good as the new ABC machine, and the two were estimated to have equal lives. As far as could be determined, if the XYZ machine were purchased, nothing could be obtained for the old ABC machine.

Figure 8-4 shows what had happened to the value of the ABC machine. According to the accounting records, the book value† of the machine was $140. Actually, the machine had decreased in value at a faster rate than had been assumed when the depreciation account was established. The rate of depreciation and the value at the end of 3 years depended upon which new machine was to be purchased. If a new ABC machine were bought, there would be a sunk cost (defined in this case as book value minus disposal value) of $140 − $100 = $40. If the XYZ machine were purchased the sunk cost would be $140 − $0 = $140.

Figure 8-4. Effect of time and replacement on the value of the ABC dictating machine.

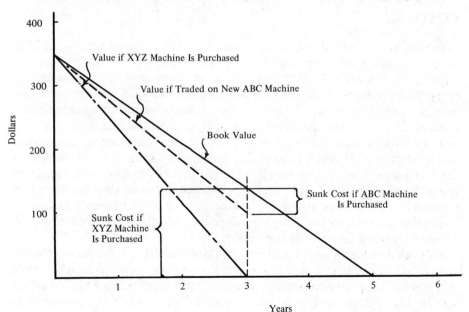

However, the most significant fact was that neither of these sunk cost values had any relationship to the decision that had to be made regarding the purchase of the new dictating machine. The only relevant factors were the costs of providing the *future* service. These were as follows:

ABC Machine		XYZ Machine	
List price	$375	List price	$220
Less trade-in value of old machine	100	Less trade-in value of old machine	0
NET COST	$275	NET COST	$220

This example illustrates the manner in which sunk costs may arise and that their magnitude may depend upon both previous assumptions and current decisions. At the same time, it shows that they have no relevance to decisions regarding future actions (except in after-tax analyses, which will be discussed later).

When depreciable assets are involved, sunk costs may rise through below-normal activity, an error in estimated salvage value, or an error in estimating the economic life. It should also be remembered that above-normal activity, higher-than-expected salvage value, and longer-than-anticipated life may result in an extra profit. Such a profit, like a sunk cost, has no relevance to decisions related to the future, except when income taxes are considered.

the pertinence of results purchased by sunk costs

Although sunk costs themselves are not relevant in economy studies, the benefits or results that have been obtained through them may be of considerable importance. For example, assume you have an automobile that is 4 years old. It has just developed differential trouble that will require a complete rear-end overhaul and considerable expense. You must decide whether it would be more economical to have the necessary repair work done or to trade the car in on a new one. Three months previously you paid $200 for a complete engine overhaul. Obviously, the $200 paid for the engine overhaul is a sunk cost with respect to the decision at hand. Yet the fact that the engine has been overhauled and will probably give very good and economical service for some time is a factor of considerable importance in deciding whether to keep the old car or to buy a new one. Thus the sunk cost of $200 is not relevant, but the possible effect of this expenditure upon future costs is pertinent.

This relationship of sunk costs to future disbursements is encountered quite frequently. For example, should more money be spent on a research project that has not yet produced positive results but still is promising? Should an oil well be dug 1,000 feet deeper or abandoned? In such cases it is apparent that what has been done must be considered in terms of future disbursements that may be made. To neglect what has taken place would be a gross error.

problems

8-1. Give as many reasons as you can for closely scrutinizing accounting data before using them in an engineering economy study.

8-2. Why might a company's purchasing department be a good source of estimates for equipment required 2 years from now for a modernized production line? Which department(s) could provide good estimates of labor costs and maintenance associated with operating the production line? What information might the accounting department provide that would bear upon the incremental costs of the modernized line?

8-3. Refer to a periodical in your library entitled *Engineering News-Record*. Develop an index that reflects cost changes for 600-MW fossil-fired boilers over the past 7 years. Assign the year of your choice an index value of 100 and plot relative changes in installed boiler cost on a graph. (Other valuable sources of information for preparing estimates include the Bureau of Labor Statistics and *Survey of Current Business*, published by the U.S. Department of Commerce.)

8-4. Contingency costs in construction projects represent an account to take care of unexpected costs that arise. If contingency costs amount to 10% of total project costs, what does this tell you about the accuracy of estimating procedures when (a) the total project is expected to cost $20,000, and (b) the total cost is expected to be $2,000,000? List some types of costs that would be placed in the contingency category.

8-5. Suppose that your brother-in-law has decided to start a company that produces synthetic lawns for lazy homeowners. He anticipates starting production in 18 months. In estimating future cash flows of the company, which of the following would be relatively easy versus relatively difficult to obtain? Also, suggest how each might be estimated with "reasonable" accuracy.
(a) Cost of land for a 10,000-square-foot building.
(b) Cost of the building (cinder block construction).
(c) Fixed capital investment.
(d) Initial working capital.
(e) First year's labor and material costs.
(f) First year's sales revenues.

8-6. In a new industrial park, telephone poles and lines must be installed. Altogether it has been estimated that 23 miles of telephone lines will be needed and each mile of line costs $14,000 (includes labor). In addition, a pole must be placed every 100 yards on the average to support the lines, and the cost of the pole and its installation is $210. What is the estimated cost of the entire job?

8-7. If an ammonia plant that produces 500,000 pounds per year cost $2,500,000 to construct 8 years ago, what would a 1,500,000-pound-per-year plant cost now? Suppose that the construction cost index has increased at an average rate of 10% per year for the past 8 years and that the cost-capacity factor to reflect economy of scale, x, is 0.6?

8-8. Your rich aunt is going to give you end-of-year gifts of $1,000 for each of the next 10 years. If inflation is expected to average 6% per year during the next 10 years, what is the total value of these gifts at the present time? The real interest rate is considered to be 4% per year.

8-9. In Problem 8-8, suppose that your aunt specified that the annual gifts of $1,000 are to be increased by 6% each year to keep pace with inflation. With a real interest rate of 4%, what now is the present value of the gifts?

8-10. Which of these situations would you prefer?

(a) You invest $2,500 in a certificate of deposit that earns an effective rate of 7% per year. You plan to leave the money alone for 5 years, and the inflation rate is expected to average 4% per year. Taxes are ignored.

(b) You spend $2,500 on an antique piece of furniture. In 5 years you believe the furniture can be sold for $3,600. Assume that the average inflation rate is 4% per year. Again, taxes are ignored.

8-11. Operating and maintenance costs for two alternatives have been estimated on different bases as follows.

Year in Which Cost Incurred	Alternative A—Costs Estimated in Actual (Inflated) Dollars	Alternative B—Costs Estimated in Real (Constant) Dollars
1	$120,000	$100,000
2	132,000	110,000
3	148,000	120,000
4	160,000	130,000

If the average inflation rate is expected to be 6% per year and money can earn a real rate of 9% per year, show which alternative has the least negative present worth at time 0.

8-12. It is known that $4,000 (in today's dollars) will be required to pay our firm's power bill for the month ending 1 year from now. If the annual inflation rate causes our power bill to increase by 4%, we now would need to place how much in a bank account earning 4% interest to pay that electric power bill 1 year hence?

8-13. A man desires to have $30,000 in a savings account when he retires in 20 years. This amount is to be equivalent to $30,000 in today's purchasing power. If the expected average inflation rate is 7% per year and the savings account earns 5% interest, what lump sum of money should the man deposit now in his savings account?

8-14. The Leatherex Company, several years ago, started manufacturing high-quality, handcrafted, leather sandals as a side line. While in earlier years the sales were higher, during the last 2 years the sales volume has leveled off at about 10,000 pairs per year, which it sells for $14.50 per pair. The product has brought a certain amount of prestige to the company, but the operation has not been satisfactory financially. A marketing consulting organization has rendered a report which states that domestic sales could be increased to 20,000 pairs per year if the selling price were reduced to $12 per pair. It also states that it believes that the company could sell 10,000 pairs in the export market at a price of $11 per pair.

The company has carefully investigated its costs and finds that those directly attributable to this product are as follows, for several levels of output:

Units	Labor	Material	Overhead
0 (temporary)	0	0	$35,000
10,000	$ 40,000	$ 43,000	37,000
20,000	78,000	83,000	39,000
30,000	113,000	122,000	40,000

What course of action would you recommend?

8-15. A company, which operates three plants, finds that its sales forecast for the following month shows that it can sell up to 1,500 units of its product at a price of $475 per unit. The monthly capacities and cost data for its plants are as follows:

	Plant A	Plant B	Plant C
Maximum capacity	1,000 units	500 units	750 units
Fixed costs	$175,000	$100,000	$205,000
Total cost per unit, at capacity	$475	$450	$500

Assuming that the variable cost at each plant varies linearly from zero to capacity, what production should the company schedule for the coming month, how should it be allocated between the plants, and what will be the resulting profit (or loss) for the company?

8-16. (a) What is the minimum sales volume and selling price that the company in Problem 8-15 has to obtain to justify operating all three plants?

(b) Would this method of operation be a good strategy? Explain why or why not.

8-17. A manufacturer has considerable excess capacity in his plant and is seeking ways to utilize it. He has been invited to submit a bid to become a subcontractor on a product which is not competitive with his but which, with the addition of $75,000 in new equipment, could readily be produced in his plant. The contract would be of 5 years duration at an annual output of 20,000 units.

In analyzing his probable costs, he estimates the direct labor at $1.00 per unit and new materials at $0.75 per unit. In addition, he finds that in each unit he can use 1 pound of scrap material from his present operations, which he now is selling for $0.30 per pound as scrap. He has been charging overhead at 150% of direct labor cost, but he believes that for this new operation the incremental overhead, above depreciation and maintenance and taxes and insurance on the new equipment, would not exceed 60% of the direct labor cost. He estimates that the maintenance costs on this equipment would not exceed $2,000 per year, and annual taxes and insurance would average 5% of first cost.

Although he can see no clear use for the new equipment beyond the 5 years of the proposed contract, he believes that it could be scrapped for at least $3,000 at that time. He estimates that the project will tie up $15,000 in working capital, and he wants to earn at least his usual 20% before-tax rate of profit on all capital utilized. What unit price should he bid?

8-18. A company has sufficient capacity to produce 2,000 units of a certain product per month. For several years it has produced and sold 1,700 units per month, at a selling price of $400 per unit. Under these conditions the total cost per unit is $350 and the fixed costs are $300,000 per month. A careful market survey indicates that if the selling price were reduced 10%, 2,000 units per month could be sold. Would you recommend lowering the selling price?

8-19. Sam Student has decided to buy a secondhand motorcycle. At a local dealer he finds one he likes for $500, but because he is not certain he should pay that much for such a luxury item, he gives the dealer $75 as a deposit to hold the motorcycle until the following Monday, with an agreement that the deposit will be applied toward the purchase price if he completes the purchase but will be forfeited if he does not complete the purchase. Over the weekend he finds the opportunity to purchase an identical motorcycle from a fellow student for $410. Which motorcycle should Sam purchase?

8-20. The Yusokem Company manufactures an extensive line of radios. In order to obtain better utilization of its facilities, it has decided to manufacture intercom communication units for home use. It finds that without adding to its facilities it could produce 900 units per month, but the operation will require the use of $50,000 as working capital to finance inventories. Direct labor costs are estimated at $4 per unit and materials at $8. The company has determined that its overhead costs are $500,000 per year plus 40% of direct labor cost. It presently has a total investment of $2,000,000 in plant and equipment.

The company ordinarily prices its products so as to return a profit of 10% on sales, which produces a before-tax profit of 30% on its capital investment. What price should it set on the intercom units?

8-21. The XYZ Company manufactures special tools and dies for the automobile industry. In connection with these operations, it has developed and patented an adjustable wrench that has received considerable acceptance from the few people who have seen and used it. Because it does not have a suitable marketing organization, it has contacted the operators of a large chain of retail stores, who have offered to buy 1,000 wrenches per month, on a 1-year contract, at a price of $3.50 each, or 1,500 per month at $3.00 each.

The XYZ Company finds that it can produce 1,000 of the wrenches per month without adding to its facilities, except for one piece of equipment costing $5,000. If it must produce more than 1,000 wrenches per month, the company would have to rent warehouse space at a cost of $250 per month. The company concludes that because of the uncertainty of renewal of the contract at the end of the year, it should write off any capital investment in the year. Labor and materials costs (per unit) would be

	1,000 per Month	1,500 per Month
Material	$0.90	$0.85
Labor	1.25	1.10

Annual overhead costs for the company are equal to $700,000 + 15% of direct material and labor costs. What would you advise the company to do?

8-22. A company that manufactured refrigerators decided to add a line of food freezers. It was able to produce up to 1,000 units per month by operating an extra shift and without adding any new factory space or equipment. The company did have to rent warehouse space at a cost of $1 per unit. The annual overhead costs of the company were $800,000 + 20% of the direct material and labor costs. Labor costs on each freezer unit were $60 and direct materials $30. In addition, compressors, costing $18 per unit, were purchased.

The company priced the freezers on an increment cost basis, with an inclusion of profit equal to 20% of the selling price. At the end of 1 year the company was selling 1,000 freezers and 2,000 refrigerators per month. At this time it was decided that the freezers should bear their share of the fixed overhead costs. What price adjustment has to be made in the selling price of the freezers to maintain the desired profit margin?

8-23. Breezy Wilson sells used cars, receiving a 4% commission on each sale. He has spent $35 entertaining a prospective customer who is interested in buying a car for $5,295. The customer now tells Breezy that she will buy the car only if a stereo tape player costing $125 is included in the deal at no extra charge. (a) Should Breezy accept the offer? (b) What is the maximum amount that Breezy could afford to spend on the stereo tape player?

8-24. An asset costing $20,000 was purchased 3 years ago. At that time it was estimated that it would have a 5-year life and a salvage value of $5,000. Consideration is being given to replacing it with a new asset costing $25,000, and a trade-in of $6,000 will be allowed on the old one. If straight-line depreciation has been used for accounting purposes, what is the sunk cost?

chapter 9

depreciation and valuation

Depreciation is a bothersome fact that must be dealt with in business and economy studies. It is the decrease in value of physical properties with the passage of time.† Although the fact that depreciation does occur is easily ascertained and recognized, the determination of its magnitude in advance, as must be done in economy studies, is not easy. In fact, the actual amount of depreciation can never be determined until the asset is retired from service. But because depreciation is a noncash cost that must be considered properly in economy studies, the analyst often encounters problems in dealing with it. At the same time, it is evident that depreciation, as contained in an economy study, will be an estimate, and it most likely will not be entirely accurate. However, the analyst possibly may find some solace in the fact that accountants and business managers face equally perplexing problems in dealing with depreciation.

That depreciation is not a new phenomenon may be seen in the following quotation from the writings of Vitruvius, probably written before A.D. 27‡:

> Therefore when arbitrators†† are taken for party-walls, they do not value them at the price at which they were made, but when from the accounts they find the tenders for them, they deduct as price of the passing of each year the 80th part, and so—in that from the remaining sum repayment is made for these walls—they pronounce the opinion that the walls cannot last more than 80 years. There is no deduction‡‡ made from the value of brick walls provided that they remain plumb; but they are always valued at as much as they were built for.

Despite the fact that the existence of depreciation has been recognized by some for centuries, it is interesting that accounting for it by business in the United States dates back only about 125 years.

---

† There are some exceptions—notably rare antiques, good works of art, some musical instruments and liquors, and, in most instances, land.

‡ Frank Granger (ed. and trans.), *Vitruvius on Architecture* (New York: G. P. Putnam's Sons, 1931–1934), Vol. 1, p. 117.

†† These arbitrators were used for building laws.

‡‡ This could only come in when the improved methods of brick building were established under the Empire.

270

Basically, from a business viewpoint, a physical asset has value because one expects to receive future monetary benefits through the possession and use of it. These benefits are in the form of future cash flows resulting from (1) the use of the asset to produce salable goods or services, or (2) the ultimate sale of the asset. It is because of these anticipated cash flows that the asset has commercial value. Depreciation, then, represents a decrease in value because the ability of the asset to produce these future cash flows decreases, as the result of one or more of several causes, with the passage of time.

definitions of value

Because depreciation is defined as decrease in *value*, it is necessary to give some consideration to the meaning of that term. Unfortunately, we discover that there are several meanings attached to it. Probably the best definition of value, in a commercial sense, is that it is the present worth of all the future profits that are to be received through ownership of a particular property. This undoubtedly excellent definition is, however, difficult to apply in actual practice, inasmuch as we can seldom determine profits far in advance. Thus several other measures of value are commonly used, some of which are approximations of the foregoing definition.

The most commonly encountered measure of value is *market value*. This is what will be paid by a willing buyer to a willing seller for a property† where each has equal advantage and is under no compulsion to buy or sell. The buyer is willing to pay the market price because he believes it approximates the present value of what he will receive through ownership *with some rate of interest or profit included*. In most matters relating to depreciation, it is market value that is used. For new properties the cost on the open market is used as the original value.

Next to market value, probably the most important kind of value is *use value*. This is what the property is worth to the owner as an operating unit. A property may be worth more to the person who possesses it and has it in operation than it would be to someone else who, if he purchased it, might have to spend additional funds to move it and get it into operation. Use value is, of course, very closely akin to the original definition of value that was stated previously. It is difficult to determine for the same reasons.

A third type of value is known as *fair value*. This usually is determined by a disinterested party in order to establish a price that is fair to both seller and buyer.

Book value is the worth of a property as shown on the accounting records of a company. It is ordinarily taken to mean the original cost of the property

† The term *property* is used in this chapter in a general sense. It includes buildings, machines, goods, and so on.

less the amounts that have been charged as depreciation expense. It thus represents the amount of capital that remains invested in the property and must be recovered in the future through the depreciation accounting process. It should be remembered, however, that because companies may use various depreciation accounting methods that produce different results, book value may have little or no relationship to the actual or market value of the property involved.

Salvage, or *resale*, *value* is the price that can be obtained from the sale of the property secondhand. Salvage value implies that the property has further utility. It is affected by several factors. The reason of the present owner for selling may influence the salvage value. If the owner is selling because there is very little commercial need for the property, this will affect the resale value; change of ownership will probably not increase the commercial utility of the article. Salvage value will also be affected by the present cost of reproducing the property; price levels may either increase or decrease the resale value. A third factor that may affect salvage value is the location of the property. This is particularly true in the case of structures that must be moved in order to be of further use. The physical condition of property will also have a great influence upon the resale price that can be obtained. A structure that has been well maintained and is in good condition will obviously be of greater value than one that has been neglected and would require considerable repair before it could be used.

Scrap value ordinarily is considered to be the amount that the property would bring if sold for junk. The utility of the article is assumed to be zero. Because for most materials, except the precious metals, the scrap price usually fluctuates considerably over a period of time, the fact that there is an existing scrap value does not assure that there will be in the future. Therefore, it is questionable practice to assume that a property will have more than a minimum scrap value at a future date. Unless it is certain that a stated scrap value will always exist, in most economy studies the future scrap value should be assumed to be zero.

It may be seen that the various definitions of value vary considerably. Although a person normally possesses property so that he may receive benefits from it, some of the benefits frequently are not in the form of money. This fact further complicates the setting of value in monetary terms in order to place an ordinary commercial value upon property.

value for rate setting

One exception, where value defined as a measure of future profits cannot be used, is in the determination of value for setting utility rates. These rates are usually set by a governmental agency so they will yield a fair profit upon the value of the property. The value assigned to the property for purposes of establishing rates is called the *rate base value*. If this value were measured by the present value of the future profits, a vicious circle would be established. The higher the rates that were set, the greater would be the profits. In turn, the greater the profits, the higher the rates would have to be in order to yield a reasonable return upon the value of the property. For rate setting some other method of determining value must be utilized. This problem will be discussed later in this chapter.

purposes of depreciation

Because property decreases in value, it is desirable to consider the effect that this depreciation has on engineering projects. Primarily, it is necessary to consider depreciation for two reasons:

1. To provide for the recovery of capital that has been invested in physical property.
2. To enable the cost of depreciation to be charged to the cost of producing products or services that result from the use of the property. *Depreciation cost is deductible in computing profits on which income taxes are paid.*

To understand these purposes, consider the following example.

EXAMPLE 9-1

Mr. Doe invested $3,000 in a machine for making a special type of concrete building tile. He found that with his own labor in operating the machine he could produce 500 tiles per day. Working 300 days per year he could make 150,000 tiles. He was able to sell the tiles for $50 per thousand. The necessary materials and power cost $20 per thousand tiles.

At the end of the first year he had sold 150,000 tiles and computed his total profit, at the rate of $30 per thousand, to be $4,500. This continued for 2 more years, at which time the machine was worn out and would not operate longer. To continue in business, he would have to purchase a new machine.

During the 3-year period, believing he was actually making a profit of $4,500 per year, he had spent the entire amount for his annual living expenses. He suddenly found that he no longer had his original $3,000 of capital, his machine was worn out, and he had no money with which to purchase a new one. What error had Mr. Doe made in his reasoning and accounting?

Solution

Analysis of the situation described in Example 9-1 reveals that Mr. Doe had not recognized that depreciation was occurring, and he had made no provision for recovering the capital invested in the tile machine. The machine, which was valued at $3,000 when purchased, had decreased in value until it was worthless. Through this depreciation, $3,000 of capital had been used in making tiles. Depreciation was just as much a cost of producing the tiles as was the cost of the material and power. However, depreciation differs from these other costs in that *it always is paid or committed in advance. Thus it is essential that depreciation be considered so that the capital that is used to prepay this cost may be recovered.* Failure to do this will always result ultimately in the depletion of capital.

Because capital must be maintained, it is necessary that the recovery be made by charging the depreciation that has taken place to the cost of producing whatever has been produced. Thus, in the case of the tile machine, production of 450,000 tiles "consumed" the machine. We might say that each thousand tiles produced decreased the value of the machine $3,000/450 = $6.67. Therefore, $6.67 should be charged as the cost of depreciation for making each thousand tiles. Adding this cost of depreciation to $20, the cost of materials and power, gives the true cost of producing 1,000 tiles. With the true cost known, the actual profit can then be determined. At the same time, with depreciation charged as a cost, a means for recovery of capital is provided. ∎

Thus depreciation accounting has a twofold purpose. First, it provides for the maintenance of capital. Second, it enables the proper amounts to be charged as the cost of depreciation in determining production costs, and ultimately in determining profits. It is this second purpose that is of primary importance to the engineer in making economy studies.

actual depreciation revealed by time

Depreciation differs from other costs in several respects. First, although its actual magnitude cannot be determined until the asset is retired from service, it always is paid or committed in advance. Thus, when we purchase an asset, we are prepaying all the future depreciation cost. Second, throughout the life of the asset we can only estimate what the annual or periodic depreciation cost is. Consequently, we must estimate the depreciation cost in economy studies. Obviously, it follows that such estimates will not be entirely accurate, but this should not be too disturbing inasmuch as the same is true of virtually all other cost items in an economy study.

A third difference is the fact that while much usually can be done to control the ordinary out-of-pocket costs, such as labor and material costs, relatively little can be done to control depreciation cost once an asset has been acquired, except, perhaps, through maintenance expenditures. Further, many of the factors that affect depreciation costs are external to the person or organization that owns the asset. If future conditions change and the demand for a product decreases, there may be a decline in the amount of material used, and probably a decrease in the profits. However, the depreciation cost, having been prepaid, may continue as before, and the result may be a loss of capital through failure to recover what has been prepaid. This possibility of resulting loss of capital is a primary reason why depreciation is more bothersome than are the other costs.

types of depreciation

Another bothersome feature of depreciation is the fact that the decrease in value has several causes, some of which are very difficult to predict or anticipate. Decreases in value with the passage of time may be classified as follows:

1. Normal depreciation: (a) physical, (b) functional.
2. Depreciation due to changes in price level.
3. Depletion.

Physical depreciation is due to the lessening of the physical ability of a property to produce results. Its common causes are wear and deterioration. These cause operation and maintenance costs to increase and output to decrease. As a result, the profits may decrease. Physical depreciation is mainly a function of time and use. It will be affected greatly by the maintenance policy of the owner. Some people contend that it is possible to maintain a property so that it remains "good as new." This subject is open to discussion, but it is doubtful if anything that is subject to depreciation can ever be as good as new, regardless of maintenance. A property might be improved so that it is more valuable than when it was new, but it is then not the same as it was originally. Improvement has been confused with maintenance.

Functional depreciation, often called *obsolescence*, is more difficult to determine than physical depreciation. It is the decrease in value that is due to the lessening in the demand for the function that the property was designed to render. This lessening may be brought about in many ways. Styles change, population centers shift, more efficient machines are produced, or markets are saturated. Increased demand may mean that an existing machine is no longer able to produce the required volume. Thus *inadequacy* is a cause of functional depreciation. The engineer does much to bring about these changed conditions that cause functional depreciation. The result of such changes is to lessen the need for the output of a particular machine or property. This directly affects the profits to be derived from its use, and thus its value.

Although physical depreciation may be reasonably anticipated and estimated, functional depreciation is much more elusive. It is caused by events that have not yet occurred. Who can foretell what is to happen? Yet functional depreciation is very real and its importance is increasing. Modern industry is characterized by rapid improvement and change. Indeed, in many businesses the greater portion of the total depreciation cost is due to functional factors. Although it is difficult to determine, it cannot be ignored.

Depreciation due to changes in price levels is almost impossible to predict and is seldom accounted for in economy studies. Yet this type of depreciation is very real and troublesome. When price levels rise during inflationary periods, even if all the capital invested at the time of original purchase has been recovered through proper depreciation procedure, this recovered capital will not be sufficient to provide an identical replacement. Although there has been a recovery of the invested capital, the capital has decreased in value. Thus it is the capital, not the property, that has depreciated. This is a primary reason why such depreciation is not considered in economy studies. Another reason is that inflation of annual depreciation write-offs is not permitted in determining profits for income tax purposes.

depreciation and the internal revenue service

The Internal Revenue Service (IRS) is charged with the duty of collecting the proper amount of Federal income tax from individuals and corporations as required by the existing tax laws. Most states also collect income taxes and have an agency comparable to the IRS. Because the IRS recognizes that depreciation is a proper expense that should be deducted from revenue in order to determine the profit upon which income tax must be based, the regulations of this agency have a considerable effect upon depreciation practice. Obviously, one of the most important matters is the life over which the total depreciation can be spread.

The IRS has guideline procedures under the names *Asset Depreciation Range (ADR) System* and *Class Life System*, which were instituted in 1971. These are described in *IRS Publication 534*, and represent a great convenience to the taxpayer by minimizing disputes with the IRS regarding depreciable lives and estimated salvage values of property. Sample lives are shown in Table 9-1.

It should be noted that the guidelines given by the IRS do not necessarily have to be followed, either in accounting practice or in making economy studies. However, if they are not used in accounting, the burden of proving that other lives used are correct rests with the taxpayer. It therefore follows that the suggested lives do have a profound effect upon the actual depreciation write-off lives that are used by companies.

requirements of a depreciation method

From the standpoint of management, the depreciation method should:

1. Provide for the recovery of invested capital as rapidly as is consistent with the economic facts involved; known and computed salvage values should agree, if possible.
2. Not be too complex.
3. Assure that the book value will not be greater than actual value at any time.
4. Be accepted by the IRS.

These requirements are somewhat contradictory and are not easily met. As a result, numerous methods for computing depreciation have been devised. Each is based upon some hypothesis regarding loss of an asset's value versus time and is an attempt to solve the complex depreciation problem in a reasonably simple and satisfactory manner. Because there are conflicting factors involved, and because future, unknown factors exist, it can be expected that perfection will not be achieved through the use of any depreciation formula.

For economy study purposes the requirements of a depreciation method are somewhat different. Obviously, it should provide for the recovery of capital and the proper assignment of depreciation cost *over the estimated life of the asset.*

table 9-1
range of useful life allowed for the depreciation of selected classes of assets under the ADR system of the Internal Revenue Service, *Publication 534*, revised October 1974

Description of Depreciable Assets	Asset Depreciation Range (years)		
	Lower Limit	Guideline Period	Upper Limit
Transportation			
Automobiles, taxis	2.5	3	3.5
Buses	7	9	11
General-purpose trucks:			
Light	3	4	5
Heavy	5	6	7
Air transport	5	6	7
Petroleum			
Exploration and drilling assets	11	14	17
Refining and marketing assets	13	16	19
Manufacturing			
Sugar and sugar products	14.5	18	21.5
Tobacoo and tobacco products	12	15	18
Knitwear and knit products	7	9	11
Lumber, wood products, and furniture	8	10	12
Paper and paperboard	13	16	19
Chemicals and allied products	9	11	13
Cement	16	20	24
Fabricated metal products	9.5	12	14.5
Electrical equipment	9.5	12	14.5
Aerospace products	6.5	8	9.5
Communication			
Telephone:			
Central-office buildings	36	45	54
Distribution poles, cables, etc.	28	35	42
Radio and television broadcasting	5	6	7
Electric utility			
Hydraulic plant	40	50	60
Nuclear plant	16	20	24
Services			
Office furniture and equipment	8	10	12
Computers and peripheral equipment	5	6	7
Data handling—typewriters, copiers, etc.	5	6	7
Recreation—bowling alleys, theaters, etc.	8	10	12

But, equally important, it should account properly for the flow of capital funds that are recovered, and which thereby reduce the amount of capital remaining invested in a project. These recovered funds thus are available to the firm for other use or investment. Finally, the method used must permit the proper evaluation of the potential of an investment being considered in an economy study.

common methods of depreciation accounting

To gain an understanding of the proper use and effect of depreciation formulas in economy studies, several of the common ones will be applied in Example 9-2.

EXAMPLE 9-2

A new asset is purchased for $120 and is estimated to have a life of 10 years and a scrap value of $20 at the end of that time. What will be the depreciation cost or charge for the sixth year, and the book value at the end of the sixth year? Assume an interest rate of 3%.

The Straight Line Method

The *straight line method* of computing depreciation assumes that the loss in value is directly proportional to the age of the asset. This straight line relationship gives rise to the name of the method. Given

N = depreciable life of the asset in years
P = original cost
d_n = annual cost of depreciation in the nth year ($1 \leq n \leq N$)
BV_n = book value at the end of n years
F = value at the end of the life of the asset, the scrap value (including gain or loss due to removal)
D_n = total depreciation up to age n years

then

$$d = \frac{P - F}{N} \tag{9-1}$$

$$D_n = \frac{n(P - F)}{N} \tag{9-2}$$

$$BV_n = P - \frac{n(P - F)}{N} \tag{9-3}$$

Solution by Straight Line Method

Applying these equations to Example 9-2, we obtain

$$d = \frac{\$120 - \$20}{10} = \$10 \text{ per year}$$

$$D_6 = \frac{6(\$120 - \$20)}{10} = \$60$$

$$BV_6 = \$120 - \frac{6(\$120 - \$20)}{10} = \$60 \quad \blacksquare$$

This method of computing depreciation is widely used. It is simple and gives a uniform annual charge. Its proponents hold that inasmuch as other costs, as well as the depreciable life, must be estimated, there is little reason for attempting to use a more complex formula.

Declining Balance Method

In the *declining balance method*, sometimes called the *constant percentage method* or the *Matheson formula*, it is assumed that the annual cost of depreciation is a fixed percentage of the book value at the beginning of the year. The ratio of the depreciation in any one year to the book value at the beginning of that year is constant throughout the life of the asset and is designated by k. Thus

Depreciation during the first year:

$$d_1 = P \times k \tag{9-4}$$

Depreciation for the nth year:

$$d_n = (P_{n-1})k \tag{9-5}$$

Salvage value at age N years:

$$F = P(1 - k)^N \tag{9-6}$$

Book value at the end of n years:

$$BV_n = P(1 - k)^n = P\left(\frac{F}{P}\right)^{n/N} \tag{9-7}$$

Rate of depreciation so that $BV_N = F$.

$$k = 1 - \sqrt[N]{\frac{F}{P}} = 1 - \sqrt[n]{\frac{BV_n}{P}} \tag{9-8}$$

The declining balance procedure is rather simple to apply. However, it has two weaknesses. The annual cost of depreciation is different each year and, from a calculation viewpoint, this is inconvenient. Also, with this formula an asset can never depreciate to zero value. This is not a serious difficulty, and in actual practice computation of the theorerical depreciation rate k (Equation 9-8) seldom is made. Instead, a reasonable rate is assumed or taken from the IRS guidelines.

The declining balance depreciation rate allowed by the IRS depends on the type of depreciable property and is stated in terms of the alternate straight line depreciation rate, which is $1.0/N$ (for 0 salvage value).

Type of Property	Allowable Declining Balance Rate, k
All new depreciable property except real estate	Double straight line, $2/N$
All used depreciable property and new real estate property	$1\frac{1}{2}$ straight line, $1.5/N$
Used rental residential property	$1\frac{1}{4}$ straight line, $1.25/N$

When double the straight line rate is used, the method is called *double declining balance depreciation.*

Solution by Declining Balance Method

Applying the declining balance relationships in Equations 9-7 and 9-8 to Example 9-2,

$$k = 1 - \sqrt[10]{\frac{\$20.00}{\$120.00}} = 0.1641$$

$$BV_5 = \$120.00\left(\frac{\$20.00}{\$120.00}\right)^{5/10} = \$48.97$$

$$BV_6 = \$120.00\left(\frac{\$20.00}{\$120.00}\right)^{6/10} = \$40.94$$

$$d_6 = \$48.97 \times 0.1641 = \$8.03$$

If double declining balance depreciation were applicable,

$$k = 2/10 = 0.2$$
$$BV_5 = \$120(1 - 0.2)^5 = \$39.32$$
$$BV_6 = \$120(1 - 0.2)^6 = \$31.46$$
$$d_6 = \$39.32 \times 0.2 = \$7.86 \quad \blacksquare$$

Proponents of this method assert that the results more nearly parallel the actual secondhand sales value than do those obtained by the straight line method. This undoubtedly is true in the case of such things as automobiles, where new models and style changes are large factors in the establishment of the salvage value. However, it is not true of many industrial and commercial structures and some equipment.

The Sum-of-the-Years'-Digits Method

In order to obtain the depreciation charge in any year of life by the *sum-of-the-years'-digits method* (commonly designated as SYD), the digits corresponding to the number of each year of life are listed in reverse order. The sum of these digits is then determined.† The depreciation factor for any year is the reverse digit for that year divided by the sum of the digits. For example, for a property having a life of 5 years, SYD depreciation factors are given in the following table.

Year	Number of the Year in Reverse Order (digits)	SYD Depreciation Factor
1	5	$\frac{5}{15}$
2	4	$\frac{4}{15}$
3	3	$\frac{3}{15}$
4	2	$\frac{2}{15}$
5	1	$\frac{1}{15}$
Sum of the digits = 15		

The depreciation for any year is the product of the SYD depreciation factor for that year and the depreciable value, $P - F$. The general expression for the annual cost of depreciation for any year n, when the total life is N, is

$$d_n = (P-F) \times \frac{2(N - n + 1)}{N(N+1)} \tag{9-9}$$

Solution by SYD Method

When these equations are applied to the data of Example 9-2, the results are:

$$\text{Sum of the years' digits} = 55$$
$$\text{Depreciation factor for the sixth year} = \tfrac{5}{55}$$
$$d_6 = (\$120.00 - \$20.00)\tfrac{5}{55} = \$9.09$$

Since $\tfrac{45}{55}$ of the depreciable value will have been written off by the end of the sixth year, the value at the end of that year will be

$$BV_6 = \$120.00 - (\$120.00 - \$20.00)\tfrac{45}{55} = \$38.18 \quad \blacksquare$$

In Figure 9-1 it will be noted that the SYD method, like the declining balance method, provides for very rapid depreciation during the early years of life. Further, this method enables properties to be depreciated to zero value and is easier to use than the declining balance method. This method also tends to reduce chances that the book value of an asset will exceed actual, or resale, value at any time. However, use of the SYD, or any other accelerated depreciation method, in effect, reduces the computed profits of a corporation during the early years of asset life and thus reduces income taxes in those early years. This will be discussed further in Chapter 10.

† The sum of the digits for a life N can be calculated by the relationship $N(N + 1)/2$.

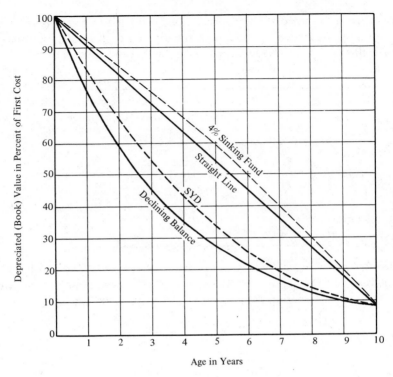

Figure 9-1. Comparison of depreciated (book) values obtained by various depreciation formulas.

The Sinking Fund Formula

The *sinking fund formula* assumes that a sinking fund is established in which funds will accumulate for replacement purposes. The total depreciation that has taken place up to any given time is assumed to be equal to the accumulated value of the sinking fund at that time.

With this formula, if the estimated life, scrap value, and interest rate on the sinking fund are known, a uniform yearly deposit can be computed. The cost of depreciation for any year is the sum of this deposit and accumulated interest for that year.

$$d = (P - F)(A/F, i\%, N) \tag{9-10}$$

$$d_n = d(F/P, i\%, n - 1) \tag{9-11}$$

$$D_n = (P - F)(A/F, i\%, N)(F/A, i\%, n) \tag{9-12}$$

$$BV_n = P - (P - F)(A/F, i\%, N)(F/A, i\%, n) \tag{9-13}$$

Solution by Sinking Fund Method

Applying these equations to the data of Example 9-2 produces the following results:

$$d = (\$120.00 - \$20.00)(A/F, 3\%, 10)$$
$$= (\$100.00)(0.0872) = \$8.72$$

$$d_6 = (\$8.72)(F/P, 3\%, 5) = (\$8.72)(1.1593) = \$10.11$$

$$BV_6 = \$120.00 - (\$120.00 - \$20.00)(A/F, 3\%, 10)(F/A, 3\%, 6)$$
$$= \$120.00 - \$100.00 \times 0.0872 \times 6.4684 = \$63.60 \ \blacksquare$$

Although the sinking fund method almost never is used for accounting purposes because of its low depreciation charges in the early years of asset life, it has importance relative to economy studies, especially when using the E.R.R.R. method.

The Service Output Method

Some companies attempt to compute the depreciation of equipment on the basis of its output. When equipment is purchased, an estimate is made of the amount of service it will render during its economic life. Depreciation for any period is then charged on the basis of the service that has been rendered during that period. The *service output method* has the advantages of making the unit cost of depreciation constant and giving low depreciation expense during periods of low production. That it is difficult to apply may be understood by realization that not only the depreciable life, but also the total amount of service that the equipment will render during this period, must be estimated.

For some types of equipment the service output method is recognized by the Internal Revenue Service. For example, a depreciation rate of 2 cents per mile may be considered reasonable for taxicabs.

The *machine-hour method* of depreciation is a variation of this procedure. Following is an example.

EXAMPLE 9-3

A machine costs $220,000 and is expected to last a total of 10,000 hours over a period of 10 years and then have a $20,000 salvage value. In the first year, it was utilized for 1,500 hours. If depreciation is based on hours of use, what is the depreciation charge for the first year and the book value at the end of the first year?

Solution

Depreciation/hour:

$$\frac{\$220,000 - \$20,000}{10,000 \text{ hours}} = \$20/\text{hour}$$

Depreciation for first year:

$$\$20/\text{hour} \times 1,500 \text{ hours} = \$30,000$$

Book value at end of first year:

$$\$220,000 - \$30,000 = \$190,000 \ \blacksquare$$

In before-tax economic analyses all methods of depreciation will yield the same results. That is, if the retirement of an asset takes place at the age predicted and at the salvage value predicted, the equivalent annual sum of depreciation and interest on the undepreciated balance for any method of depreciation can be shown to be equal to the capital recovery cost calculated by many methods, such as by Equation 5-1, 5-2, or 5-3. An example problem early in Chapter 5 demonstates this. However, in *after-tax* analyses, the method of depreciation does affect how much income taxes are paid when. Thus different methods of depreciation are not equivalent in after-tax economic analyses. This will be demonstrated in Chapter 10.

multiple asset depreciation accounting

To this point we have considered depreciation on individual assets. Very frequently, for accounting purposes, identical or even somewhat dissimilar assets may be considered as a group in regard to depreciation. This group procedure has some advantages.

Although it may be known that the *average* life of assets in a group will be N years, it is recognized that some items will last less than N years, and some will last longer than N years. If N years is assumed to be the depreciable life of an individual asset, and if it should be retired from service in less than N years at a salvage value less than the book value, there would be a loss of capital. On the other hand, if it should be used for more than N years and then sold for any amount, a gain of capital would be realized. By using group, or *multiple asset,* accounting, accounting for such losses or gains is avoided, since such a "dispersion effect" in the lives is recognized when the average life for the group is adopted.

It is fairly common practice, for tax purposes, to use item accounts for buildings, structures, and high-value equipment and to use group accounts for general equipment. In the following portions of this book we shall assume item depreciation, since this will provide some degree of simplicity and will highlight the effects of income taxes relative to gains and losses that may occur.

depletion

When natural resources are being consumed in producing products or services, the term *depletion* is used to indicate the decrease in value of the resource base that has occurred. The term is commonly used in connection with mining properties, oil and gas wells, timber lands, and so on. In any given parcel of mineral property, for example, there is a definite quantity of ore, oil, or gas available. As some of the mineral is mined and sold, the reserve decreases and the value of the property normally diminishes.

However, there is a difference in the manner in which the amounts recovered through depletion and depreciation must be handled. In the case of depreciation, the property involved usually may be replaced with similar property when it has. become fully depreciated. In the case of depletion of mineral or other natural resources, such replacement usually is not possible. Once the gold has been removed from a mine, or the oil from an oil well, it cannot be replaced. Thus, in a manufacturing or other business where depreciation occurs, the principle of maintenance of capital is practiced, and the amounts charged for depreciation expense are reinvested in new equipment so that the business may continue in operation indefinitely. On the other hand, in the case of a mining or other mineral industry, the amounts charged as depletion cannot be used to replace the sold natural resource, and the company, in effect, may sell itself out of business, bit by bit, as it carries out its normal operations. Such companies frequently pay out to the owners each year the amounts recovered as depletion. Thus the annual payment to the owners is made up of two parts—(1) the profit that has been earned, and (2) a portion of the owner's capital that is being returned, marked as depletion. In such cases, if the natural resource were eventually completely consumed, the company would be out of business, and the stockholder would hold stock that was theoretically worthless but would have received back all his or her invested capital.

In the actual operation of many natural resource businesses, the depletion funds may be used to acquire new properties, such as new mines and oil-producing properties, and thus give continuity to the enterprise.

Although the theoretical depletion for a year would be

$$\frac{\text{cost of property}}{\text{number of units in the property}} \times \text{units sold during year}$$

in actual practice the depletion is based upon a percentage of the year's income as permitted by the IRS. Depletion allowances on oil, gas, and mineral properties may be computed as a percentage of the gross income, provided that the amount charged for depletion does not exceed 50% of the taxable income before deduction of the depletion allowance. Typical percentage depletion allowances are as follows:

Oil and gas wells	22%
Antimony, bismuth, cadmium, cobalt, lead, manganese, nickel, tin, tungsten, vanadium, zinc, sulfur, uranium, asbestos, bauxite, graphite, mica	22%
Other metal mines not in the 22% group	15%
Coal, lignite, sodium chloride	10%
Brick and tile clay, peat, pumice, sand	5%

It is apparent that the total amount that can be charged for depletion over the life of a property under this procedure may be far more than the original cost.

High depletion allowances often are defended as being necessary for the encouragement of the discovery and development of mineral resources, on the basis that such ventures involve a high degree of risk and uncertainty. However, the high mortality rate of new businesses, and even some older ones, in the manufacturing and service industries is rather strong proof that they also involve considerable risk and uncertainty. There thus remains considerable evidence that much political pressure is involved in the special tax consideration given to depletion.

accounting for depreciation funds

The procedure by which accounting is made of the flow of depreciation funds and their subsequent reinvestment in the business often somewhat mystifies engineers who have not had any training in accounting. Quite probably, some of the terminology that is applied to some of the accounts involved has added to the confusion.

The basic procedure can be illustrated by the simplified balance sheets shown in Figure 9-2. These relate to a man, John Doe, who starts out with an investment

Figure 9-2. Balance sheets, showing the manner in which cash is converted into depreciable assets and recovered through depreciation accounting.

BALANCE SHEET (a)

Assets			Liabilities and Ownership		
Cash		$3,000	John Doe, Ownership		$3,000
	TOTAL	$3,000		TOTAL	$3,000

BALANCE SHEET (b)

Assets			Liabilities and Ownership		
Cash		$ 0	John Doe, Ownership		$3,000
Truck		3,000			
	TOTAL	$3,000		TOTAL	$3,000

BALANCE SHEET (c)

Assets			Liabilities and Ownership		
Cash and other assets		$1,000	John Doe, Ownership		$3,000
Truck	$3,000				
Less amount charged for depreciation to date	1,000				
		2,000			
	TOTAL	$3,000		TOTAL	$3,000

of $3,000 in cash; this condition is portrayed in balance sheet (a) of Figure 9-2. Mr. Doe then uses his $3,000 to purchase a truck. The relationship of his assets, liabilities, and ownership for this condition is shown in balance sheet (b). At the end of a year he computes the depreciation on his truck to be $1,000, assuming a 3-year life and zero salvage value and using straight line depreciation. He therefore sets aside, out of revenue, this $1,000 and reinvests it in the business, perhaps keeping some of it in the form of cash but using some of it for the purchase of other needed assets. This new financial state is portrayed in balance sheet (c), which at the same time shows the original value of the truck and also its depreciated value. Thus, in this general manner, the flow of depreciation money into the company, and the fact that it is retained and used, either in the form of cash or other assets, are recorded. Unfortunately, the item labeled "Less amount charged for depreciation to date" in Figure 9-2 is called by a variety of names, including *reserve for depreciation*, a term that is very often confusing.

We must always make certain that the scrap or resale value of a property is taken into account in computing depreciation *if there will be such value*. In many cases the probable scrap value is very difficult to determine, or there is considerable likelihood that there may be no scrap value. Unless there are very definite indications of scrap value, the usual practice in making economy studies is to assume that the scrap value is zero.

valuation

Frequently, an engineer is called upon to decide the value of engineering properties. An adequate discussion of the methods used to arrive at the correct value of any property would require at least a good-sized volume; it is obviously beyond the scope of this book. However, a few of the principles involved will be considered, inasmuch as they are so intimately connected with the subject of depreciation.

The reasons for determining the value of property vary. Similarly, a valuation that is correct for one need not be correct for another. The need for determining the value of property often occurs when a private buyer is purchasing the property from a private owner. Again, the value of property may need to be known to serve as a tax base. When a municipality wishes to purchase a privately owned utility plant, the value must be decided upon. In establishing utility rates, the regulating bodies must arrive at a fair value of the property that is used to render the service. If a company examines the book value of its property at periodic intervals to determine whether the established depreciation rates are adequate, it must have some method of estimating the correct value of the property.

Obviously, both physical and functional depreciation affect the value of a property. Additions to and deletions from the property also will affect its value. But another, and very troublesome, factor is the matter of price changes due to inflation and deflation.

Theoretically, the value of a property should be a measure of the present worth of the future net profits that can be derived through ownership. Such a determination would, of course, necessitate the ability to predict accurately the future profits, and some interest rate would have to be agreed upon and used. Thus such a procedure is difficult to apply. Consequently, several generalized methods are used, each being based upon a certain hypothesis and producing results that are dependent upon the conditions contained in the hypothesis. It thus is well to keep in mind the dictum of the United States Supreme Court that stated that there is no single way to determine value. The following several methods are in widespread use today.

historical cost less depreciation

The first method that might be used is sometimes called *historical cost less depreciation*. In arriving at value by this method, we consider the actual costs that have been incurred in obtaining the property being valued. This historical cost is then reduced by the amount of depreciation that appears to have occurred. This depreciation must be considered in terms of the ability of the present property to render service. The depreciated historical cost is then taken as the true value.

It is apparent that the historical-cost method neglects many factors. For example, no consideration is given to advances in technological methods that have occurred since the property was acquired. It is obvious that if an equivalent plant could be built, using modern methods and equipment, that would render the future service at much less unit cost, the existing plant could never be worth more than the more modern plant, regardless of how much might have been expended in acquiring the old plant.

The historical-cost method also gives no consideration to changes in price levels that have taken place since the time of original construction. Drastic changes in price have sometimes made it possible to build a new plant at much less cost than was required for the old plant. It is apparent that such a factor should not be neglected in determining value.

Another serious objection to the historical-cost method is that no consideration is given to the fact that the owners may have made unwise decisions and investments in the old property. Such a situation implies that unwise expenditure in acquiring the property adds to its secondhand value. This condition, of course, is absurd.

Perhaps the most serious difficulty is that some depreciation method must be used. Obviously, the results can be greatly affected by the choice of method.

reproduction-cost-new less depreciation

So that the weaknesses of the historical-cost method can be overcome, the fictitious equivalent plant or property is usually assumed to be obtained by building it by the most modern methods, according to the most efficient design. This theoretical plant is then depreciated until it would be capable of rendering only the same amount of future service as the existing plant. This method is often called *reproduction-cost-new less depreciation*. By its use, technological progress and changes in price level are given consideration.

The results of this method are of obvious advantage to the seller of industrial property when price levels have risen; consequently, it gained considerable favor during the period 1900–1930. The method, however, does not consider whether the existing property is of the correct size for actual future demand or whether it is the result of unwise investments policy. The question also remains as to whether or not the owners should be given full advantage of increases in price levels that may have occurred.

Again, this method requires the application of some depreciation formula.

determination of value by replacement theory

In certain cases where suitable cost and/or revenue data are available, it is possible to determine the economic value of properties by the application of replacement theory. This method is discussed in Chapter 11.

intangible values

In the determination of the value of industrial property or equipment, four intangible items are often encountered. The first of these is *goodwill*. This item arises out of the public or trade favor that an enterprise may have earned through the service it has rendered. This may be of very real value to the business. In the sale of a structure, goodwill is a legitimate item and is of value to the buyer. In establishing value for rate setting, however, little if any consideration is given to this item. In this case, goodwill must have come through the business done with the consumers. The courts do not allow goodwill to be included, since this would require the consumers to pay the utility a profit upon something they have bestowed upon the business.

The second intangible item is value of *franchises*. This is also granted to the utility by the consumers; therefore, it cannot be considered in rate cases. Where value is being determined for sale purposes, franchises have a real value to the buyer and may be considered.

The fact that an enterprise that is actually doing business is worth more than a similar one that possesses the same physical assets, but is not operating, is recognized in *going value*. Money must be spent to get the business into an operating condition. The courts recognize this fact and allow this item to be included in all valuation work. As a rule, from 8 to 15% of the value of the assets is allowed for going value.

The fourth intangible is *organization cost*. Some money must be spent in organizing almost any business and arranging for its financing and building. These are legitimate expenses and are recognized as such by the courts.

determination of property life from mortality data

As has been pointed out in this chapter, one of the most difficult problems in connection with estimating depreciation costs is the determination of the life that a property may be expected to have. Obviously, it would be desirable to have a better method for determining probable life than estimating or guessing. For certain properties and under certain conditions, well-established statistical techniques can be used. However, it must be pointed out that the number of cases to which these techniques can be applied is very limited. This limitation is due, primarily, to two conditions. First, it is necessary to wait until substantially all of a group of identical properties have been retired from service, or have in service a substantial number of identical properties of all possible ages. Second,

table 9-2
mortality data for electric lamps

Life (hours)	Lamps in Service (Number Surviving)	Remaining Life Expectancy	Probable Life
1,999.5	0	0	1,999.5
1,899.5	755	50.0	1,949.5
1,799.5	1,897	89.8	1,889.3
1,699.5	4,237	112.5	1,812.0
1,599.5	7,559	141.1	1,740.6
1,499.5	11,334	177.2	1,676.7
1,399.5	15,637	214.9	1,614.4
1,299.5	20,620	250.5	1,550.0
1,199.5	26,131	287.4	1,486.9
1,099.5	31,969	325.8	1,425.3
999.5	37,857	367.5	1,367.0
899.5	43,745	411.2	1,310.7
799.5	49,483	457.7	1,257.2
699.5	54,994	506.8	1,206.3
599.5	59,977	560.6	1,160.1
499.5	64,280	619.7	1,119.2
399.5	68,055	682.5	1,082.0
299.5	71,377	748.5	1,048.0
199.5	73,717	823.1	1,022.6
99.5	74,859	909.8	1,009.3
−0.5	75,614	1,000.2	999.7

where rapid technological progress is occurring, or functional depreciation is a primary, determining factor, these techniques will be of little help. Thus these procedures are of definite, but limited, assistance in solving the problem of economic life in economy studies.

Just as human beings are born, live, grow old, and die, physical properties are produced, put into service, render service, and are removed from use. The same procedures that are used to determine the probable mortality, average life, and life expectancy for human beings may also be used for determining and estimating the life that may be expected from physical property, *provided that certain conditions exist*. These basic conditions are as follows:

1. A sufficient number of basically identical units must be involved so that averages may be used.
2. The service conditions must be the same for future units as for those for which the mortality data were obtained.
3. The study period must be long enough to assure valid data.

Table 9-2, for example, gives mortality data for a certain type of electric lamp, and Figure 9-3 shows the survivor and probable life curves for these lamps.

Figure 9-3. Survivor and probable life curves for electric lamps.

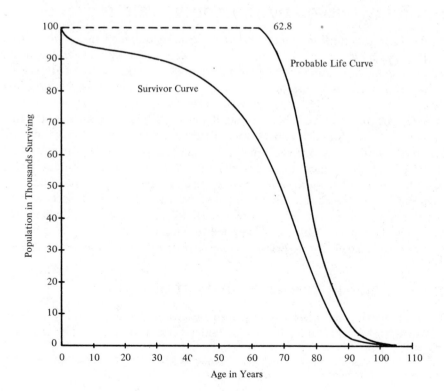

Note that the average life of the entire group is shown (approximately 1,000 hours). Also shown is that lamps which have attained an age of 700 hours have a probable life of 1,200 hours.

mortality curves by the individual-unit method

The simplest method for compiling life experience data of a given type of physical property is to record the age in years of each individual unit of property of that class as it goes out of service. This is known as the *individual-unit method* of mortality table compilation. When a large number of such individual lives has thus been recorded, the data can be summarized and presented in a table.

Although this method is simple, it requires that a substantial number of identical units be placed in service at the same time, and the results cannot be obtained until all the units have been removed from service, or at least a sufficient number to establish the shape of the survivor curve with reasonable accuracy. This frequently requires quite a period of time. In many cases, by the time the required information is available, the need for it has passed. Very commonly the future properties and service conditions will be different from those of the past so that data that are obtained over a long period will not be usable.

mortality curves by the annual-rate method

Under certain conditions mortality curves can be obtained in a shorter time by the *annual-rate method*. To do so, it is necessary to have in service, at the same time, a considerable number of substantially identical units of all possible ages—from zero up to the oldest on record. A study period is then selected, usually 3 to 5 years.† During each year of the study period a record is kept of all removals from service. These are tabulated according to their age at the time of removal. Thus, for each year, it can be determined what percentage of the total number of units in service was removed at age 1, 2, 3, ..., n. At the end of the study period the values are averaged; this gives the mortality data for an average year. These data show what percentage of all units attain, and are removed at, each possible age, and thus give the information necessary for plotting survivor curves. The desired mortality information is obtained in a relatively short period and can be applied to those units remaining in service or to new ones to be placed in service, *as long as the service conditions do not change.*

limitations in the use of mortality data

In the preceding discussion it has been pointed out that mortality data are obtained for specific units in service under certain specific conditions. It is

† A single year sometimes is used, but a longer period is desirable to assure that representative results are obtained.

apparent that they can be applied with validity only to the same type of units that will be subjected to the same service conditions. This places very severe limitations upon the use of such data for determining life for depreciation purposes. There are relatively few physical properties that, because of technological advances or changed service requirements, are not considerably changed with the passage of time. For example, a 1978 model automobile is vastly different from a 1950 model of the same make. Therefore, it cannot be expected that mortality data for 1950 automobiles can be applied to 1978 cars for the same service conditions. Likewise, service conditions frequently change. If a new engine lathe made in 1945 were subjected to typical 1978 operating conditions, its life would be much different from that predicted by mortality data obtained under 1945 conditions. Therefore, it is apparent why relatively few statistically valid mortality data are available. Furthermore, considerable caution should be exercised when available data are used, to make certain the units and conditions are the same as those for which the data were obtained.

It should be pointed out that a considerable amount of the mortality data published in handbooks and other sources is not based on actual statistical studies, only on someone's estimate or personal experience.

Although mortality data cannot be used for most types of property, they are very useful for a number of items, such as wooden telephone and power-line poles, railroad ties, telephone cable, and electric lamps, Where such data are available and apply, they certainly should be used.

problems

9-1. It has been said that depreciation is probably one of the least understood but most important aspects of engineering economic analyses. Explain why this is likely to be true.

9-2. The accounting need for depreciation arises for several reasons. Describe what these reasons are. How are they related to definitions of the *value* of an asset?

9-3. How is the cost of depreciation different from other production or service expenses such as labor, material, and overhead?

9-4. Why would a company want to depreciate a capital investment rapidly in its early years of productive life? Which depreciation methods allow for accelerated write-offs?

9-5. Explain the difference between depreciation and depletion. Does an uncertain future favor a company that depreciates its physical assets or one that depletes its nonrenewable resources? Explain your reasoning.

9-6. What is the difference between individual asset (item) accounting for depreciation purposes and multiple asset (group) accounting?

9-7. On accounting balance sheets, "reserves for depreciation" are often confusing. Because taxable income can be legally reduced each year by the total amount of depreciation that the IRS allows, depreciation reserves represent cash that a firm supposedly has on hand for use in replacing wornout or obsolescent assets. What actually happens to these "cash reserves" in practice, and how does a firm acquire new assets when they are needed?

9-8. An asset for drilling that was purchased by a petroleum refinery has a first cost of $60,000 and an estimated salvage value of $12,000. The guideline period for useful life is taken from *IRS Publication 534* (Table 9-1). Compute the depreciation amount in the third year and the book value at the end of the fifth year of life by each of these methods:
(a) The straight line method.
(b) The sum-of-the-years-digits method.
(c) The double declining balance method (with $k = 2/N$).
(d) The 4% sinking fund method.

9-9. In Problem 9-8, suppose that the company wanted to use a declining balance method of depreciation that results in a book value that exactly equals $12,000 at the end of the guideline period. (a) What value of k would be used and what would the depreciation write-off in the fourth year be? (b) What is the book value at the end of year 8?

9-10. By each of the methods indicated below, calculate the book value of a high-post binding machine at the end of 12 years if the item originally cost $1,800 and had an estimated salvage value of $400 at the end of its expected 20-year life:
(a) Straight line depreciation.
(b) 6% sinking fund depreciation.
(c) The double declining balance method ($k = 2/N$).
(d) Sum-of-the-years'-digits depreciation.

9-11. An asset costs $24,000 and is expected to be used 10 years and have a salvage value of $4,000 at that time. Find the depreciation charge for the third year and the book value at the end of the third year using the following depreciation methods: (a) straight line, (b) sum-of-the-years' digits, and (c) declining balance (where $k = 1 - \sqrt[N]{F/P}$).

9-12. A piece of equipment that cost $5,000 was found to have a trade-in value of $4,000 at the end of the first year, $3,200 at the end of the second year, $2,560 at the end of the third year, $2,048 at the end of the fourth year, and $1,638 at the end of the fifth year. (a) Determine the depreciation that occurred during each year. (b) Using a 10% rate, compute the interest on the remaining investment each year. (c) Determine the sum of depreciation and interest on the unrecovered investment for each year. (d) Compare the results from (c) with the uniform year-end amount that represents capital recovery with interest.

9-13. A company purchased a machine for $30,000 and sold it at the end of 5 years for $8,000. What uniform, year-end net revenue did it have to receive from the use of the machine to cover the cost of depreciation and a return of 10% on the capital invested in the machine?

9-14. An asset was purchased 6 years ago for $6,400. At that time its life and salvage value were estimated to be 10 years and $1,400, respectively. If the asset is sold *now* for $1,500, what is the difference between its market value of $1,500 and its present book value if depreciation has been by:
(a) The straight line method?
(b) The sum-of-the-years'-digits method?

9-15. A mining venture involved investment of $2,000,000 in the purchase and development of a mineral property. At the end of the following 5 years, the stockholders were paid $780,000, $708,000, $628,000, $552,000, and $476,000 as payment for depletion and profits. The property was then sold for $100,000, and this amount was also paid to the stockholders. The property had been depleted at a uniform rate throughout the period. What uniform rate of profit did the stockholders receive on the capital they had invested in the property at the end of each year?

9-16. A company purchased a machine for $15,000. It paid shipping costs of $1,000 and nonrecurring installation costs amounting to $1,200. At the end of 3 years it had no further use for the machine, so it spent $500 for having it dismantled and was able to sell it for $1,000. (a) What are the total investment and depreciation costs for this machine? (b) The company had depreciated the machine on a straight line basis, using an estimated life of 5 years and $1,000 salvage value. By what amount did the recovered depreciation fail to cover the actual depreciation?

9-17. The Ajax Plumbing Corporation uses the sum-of-the-years'-digits method of depreciation. It purchases an automatic soldering machine for $18,000. The Corporation plans to transfer the machine to a subsidiary after 6 years of service for $11,000. Determine the specified life, N, of the machine so that the book value after 6 years will be $11,000 if the salvage value after N years should be $4,000 according to IRS guidelines.

chapter 10

the effects of income taxes in economy studies

Before discussing income taxes in economy studies, we should make a clear distinction between income taxes and several other common taxes:

1. *Income taxes* are assessed as a function of net income or profits minus certain allowable deductions and exemptions. They are levied by the Federal, most State, and occasionally municipal, governments.
2. *Property taxes* are assessed as a function of the "value" of real estate, business, and personal property. Hence they are independent of the income or profit of an individual or business. They are levied by municipal, county, and/or State governments.
3. *Sales taxes* are assessed as a function of purchases of goods and/or services, and are thus independent of net income or profits. They are normally levied by State, municipal, and/or county governments.
4. *Excise taxes* are Federal taxes assessed as a function of the sale of certain goods or services often considered "nonnecessities," and hence are independent of the income or profit of an individual or business. Although they are usually charged to the manufacturer or original provider of the goods or service, the cost is passed on to the consumer.

In this chapter we will be concerned only with income taxes. Rarely are other types of taxes important in economy studies, with the exception of property taxes, which are normally annual disbursements.

Thus far in this book there has been no consideration of possible effects of income taxes in economy studies. That is, only *before-tax* studies have been made. This has been for two reasons. First, in the types of studies considered thus far income taxes often do not have any effect on the decisions that would be made; the accuracy of the study would not be improved by their inclusion. Second, by not complicating the studies with income tax effects, we could place primary emphasis on basic economy study principles. However, there are numerous economy study conditions wherein income taxes do affect the results and thus after-tax studies should be made.

296

An approximation of the before-tax rate of return requirement to include the effect of income taxes in studies using before-tax cash flows can be determined from the following relationship:

(before-tax R.R.)[1 − (effective income tax rate)] $\cong$ (after-tax R.R.)

Thus

$$\text{Before-tax R.R.} \cong \frac{\text{after-tax R.R.}}{1 - \text{(effective income tax rate)}} \qquad (10\text{-}1)$$

If the property is nondepreciable and there is no capital gain or loss or investment tax credit, the relationship above is exact, not an approximation. If salvage value is less than 100% of first cost and if life of the property is finite, the depreciation method selected for income tax purposes affects the timing of income tax payments and, therefore, error can be introduced by use of the relationship in Equation 10-1. Indeed, we will show through the results of Examples 10-4 through 10-8 that the earlier (sooner) in an asset's life that depreciation charges are made, the earlier income taxes are saved, and the higher the after-tax rate of return. In practice, it is usually desirable to make after-tax analyses for any income tax-paying enterprise unless one knows that income taxes would have no effect on the alternatives being considered.

After-tax analyses can be performed by exactly the same methods (P.W., I.R.R., etc.) as before-tax analyses. The only difference is that after-tax cash flows should be used in place of before-tax cash flows by adjustment for either tax disbursements or tax savings, and the calculation made using an after-tax minimum attractive rate of return.

The mystery behind the sometimes-complex computation of income taxes is reduced when one recognizes that income taxes paid are just another type of disbursement, while income taxes saved (due to business deductions or expenses or direct tax credits) are the same as any other kind of disbursement savings. The Federal (and most State) income tax regulations can be very complex, but the provisions that apply to most economic analyses of capital investments generally can be understood and applied without great difficulty. However, sometimes the help of a tax specialist is needed. This chapter is not intended to be a comprehensive treatment of Federal tax law. Rather, herein is described only some of the more important provisions of the tax law as of 1976, followed by illustrations of general procedures for computing after-tax cash flows and making after-tax economic analyses. In making actual economy studies, the analyst should, of course, use the existing or anticipated applicable rates and regulations.

basic income tax principles and calculation of effective rates

Income taxes are levied on both personal and corporate incomes. Although the regulations of most of the states with income taxes have the same basic features as in the Federal regulations, there is great variation in the tax rates. State income taxes are in most cases much less than the Federal taxes, and very often can be closely estimated as a constant percentage of Federal taxes. Therefore, no attempt will be made to discuss State income taxes. An understanding of the applicable Federal income tax regulations usually will enable the analyst to apply the proper procedures if State taxes must be considered.

As an example of an effective (combined Federal and State) incremental tax rate for a corporation, suppose that the Federal income tax incremental rate is 48% and the State income tax incremental rate is 10%. Further assume the common case in which income is computed the same way for both taxes except that State taxes are deductible from taxable income for Federal tax purposes but Federal taxes are not deductible from taxable income for State tax purposes. Thus the net effective tax rate is $48\% + 10\% - 0.48(10\%) = 53.2\%$.

It is the effective tax rate(s) on any increment of *taxable income* that is of importance herein. Taxable income is defined next.

taxable income of individuals

An individual's total earned income is essentially what is called *adjusted gross income* for Federal tax purposes. From this amount individuals may subtract for personal exemptions and allowable deductions to determine the *taxable income*. Personal exemptions are provided at the rate of $750 for the taxpayer and each dependent, with extra exemptions permissible for blind persons and persons 65 years of age or over. Allowable deductions are provided for such items as excessive medical costs, State and local taxes, interest on borrowed money, charitable contributions, and casualty losses, within some limits. A standard deduction may be taken by people who do not itemize their allowable deductions. As of 1976, the amount of the standard deduction was 16% of adjusted gross income, with a minimum deduction of $1,700 and a maximum of $2,400 for unmarried taxpayers.† Thus, for individual taxpayers,

$$
\begin{aligned}
\text{taxable income} = {} & \text{adjusted gross income} \\
& - \text{personal exemption deductions} \\
& - \text{other allowable deductions} \qquad \text{(10-2)}
\end{aligned}
$$

taxable income of business firms

The revenue of a firm is essentially what is called its *gross income*. From this amount firms may subtract all ordinary and necessary expenses to conduct the business (including very restricted contributions) except for capital expenditures. Business capital expenditures can be declared as an expense for each period

---

† For married couples filing a joint return the minimum was $2,100, with a maximum of $2,800.

(year) only to the extent of allowable depreciation or depletion charges on those capital expenditures. Thus, for business firms,

taxable income = gross income
            − all expenditures except capital expenditures
            − depreciation and depletion charges   (10-3)

The taxable income of a firm is often virtually the same as the *before-tax net profit*. After income taxes are subtracted from the before-tax net profit, the remainder is often called the *after-tax net profit*. There are two types of income for tax computation purposes—*ordinary income* (*and losses*) and *capital gains* (*and losses*). Each of these will be explained below.

ordinary income (and losses)

Ordinary income is the net profit that results from the regular business operations (such as the sale of products or services) performed by a corporation or individual. For Federal tax purposes, virtually all ordinary income is taxable income and is subject to a graduated rate scale with provision for higher rates with higher income. For corporations in 1976, the Federal rates for taxable income were

      20% for taxable income up to $25,000
      22% for taxable income between $25,000 and $50,000
      48% for taxable income in excess of $50,000

These are graphically illustrated in Figure 10-1.

For individuals, the Federal rates increase progressively from 14% to 70% of taxable income, as shown in Table 10-1, which happens to be for a category of unmarried taxpayers. Rates for married persons filing a joint return are somewhat lower for a given taxable income than the rates shown in Table 10-1.

Figure 10-1. The 1976 Federal income taxes (and rates) for corporations.

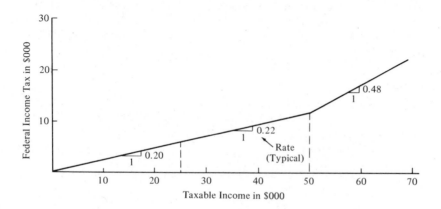

table 10-1

1976 federal income tax rates for persons categorized as "unmarried and not head of a household"

If Your Taxable Income Is:		The Tax Is:		
Over	But Not Over	This	Plus Following Percentage	Of the Taxable Income Over This
$ 0	$ 500	$ 0	+14%	$ 0
500	1,000	70	+15%	500
1,000	1,500	145	+16%	1,000
1,500	2,000	225	+17%	1,500
2,000	4,000	310	+19%	2,000
4,000	6,000	690	+21%	4,000
6,000	8,000	1,110	+24%	6,000
8,000	10,000	1,590	+25%	8,000
10,000	12,000	2,090	+27%	10,000
12,000	14,000	2,630	+29%	12,000
14,000	16,000	3,210	+31%	14,000
16,000	18,000	3,830	+34%	16,000
18,000	20,000	4,510	+36%	18,000
20,000	22,000	5,230	+38%	20,000
22,000	26,000	5,990	+40%	22,000
26,000	32,000	7,590	+45%	26,000
32,000	38,000	10,290	+50%	32,000
38,000	44,000	13,290	+55%	38,000
44,000	50,000	16,590	+60%	44,000
50,000	60,000	20,190	+62%	50,000
60,000	70,000	26,390	+64%	60,000
70,000	80,000	32,790	+66%	70,000
80,000	90,000	39,390	+68%	80,000
90,000	100,000	46,190	+69%	90,000
100,000		53,090	+70%	100,000

EXAMPLE 10-1

Jane Doaks files a Federal income return on an adjusted gross income of $44,750. She can claim only one personal exemption of $750 and have allowable personal deductions of $4,000. What is her taxable income and Federal income tax owed (i.e., tax liability)?

Solution

Taxable income:

$$\$44,750 - \$750 - \$4,000 = \$40,000$$

Income taxes:

On first $38,000	$13,290
On next ($40,000 − $38,000) = $2,000 at 55%	1,100
TOTAL	$14,390

EXAMPLE 10-2

A corporation now is expecting an annual taxable income of $45,000. It is considering investing an additional $100,000, which is expected to create an added annual net cash flow (receipts minus disbursements) of $35,000 and an added annual depreciation charge of $20,000. What is the corporation's Federal tax liability:
(a) Expected now?
(b) Expected if the new investment is made?

Solution

(a) Income taxes/year:

On first $25,000 at 20%	$5,000
On next $45,000 − $25,000 = $20,000 at 22%	4,400
TOTAL	$9,400

(b) Taxable income/year:

Before added investment	$45,000
+ added net cash flow	+35,000
− depreciation	−20,000
TOTAL	$60,000

Income taxes/year:

On first $25,000 at 20%	$ 5,000
On next $25,000 at 22%	5,500
On next $60,000 − $50,000 = $10,000 at 48%	4,800
TOTAL	$15,300

As an added note on Example 10-2, subtracting the results of (a) from (b) would indicate that the net investment is expected to increase the Federal tax liability by $15,300 − $9,400 = $5,900. The determination of change in tax liability can usually be determined most readily by an incremental approach. For example, this problem involved changing the taxable income from $45,000 to $60,000 as a result of the new investment. Thus the change in income taxes per year could be calculated as:

First $50,000 − $45,000 = $5,000 at 22%	$1,100
Next $60,000 − $50,000 = $10,000 at 48%	4,800
TOTAL	$5,900

capital gains (and losses)

When an asset other than those sold as part of normal business operations is disposed of for more (or less) than its book value, the resulting *capital gain* (or *capital loss*) often is taxed (or saves taxes) at a rate which is different from that for ordinary income. The amount of tax effect depends upon the amount of capital gain or loss, type of property, whether it is *long term* or *short term*, whether other capital gains or losses are involved, and the applicable rate or rates. In formula form,

$$\text{capital gain (loss)} = \text{net selling or disposal price} - \text{book value}$$

or

$$\text{capital gain (loss)} = \text{net selling or disposal price} - \text{original first cost} + \text{accumulated depreciation charges} \qquad (10\text{-}4)$$

short-term and long-term gains and losses

As of 1976, the distinction between long-term and short-term capital gains and losses was as follows: The holding period required for a gain or loss to qualify for "long-term" tax treatment is shown in the table at the top of p. 303.

table 10-2
tax treatment of capital gains and losses

	Short-Term	Long-Term
For Individuals:		
Capital gain	Taxed as ordinary income.	50% of the capital gain taxed as ordinary income. Alternatively, up to $50,000 of the capital gain can be taxed at 25%, and the balance of the gain taxed at a higher rate if that is to taxpayer's advantage.
Capital loss	Subtract capital losses from any capital gains; balance may be deducted from ordinary income, but not more than $1,000[a] per year.	Subtract capital losses from any capital gains; half the balance may be deducted from ordinary income, but not more than $1,000[a] per year.
	Any balance may be carried over to succeeding years.	
For Corporations:		
Capital gain	Taxed as ordinary income.	Taxed at ordinary income tax rate (20%, 22%, or 48%) or 30%, whichever is smaller.
Capital loss	Corporations may deduct capital losses only to the extent of capital gains. Any capital loss in the current year that exceeds capital gains is carried forward to subsequent years until it is completely absorbed.	

[a] Limitations given are for married taxpayers filing a joint return. Married persons filing separate returns and unmarried taxpayers are entitled to only one-half the limitations shown. As of 1976, the $1,000 limitation was programmed to be changed to $2,000 for property sold in 1977 and to $3,000 for property sold in 1978 and thereafter.

For Property Sold	Minimum Holding Period (months)
1976 or before	6
In 1977	9
1978 and after	12

Gains or losses on capital assets sold after holding for a period less than the applicable minimum (above) are subject to "short-term" tax treatment. The rules for tax treatment of capital gains and losses are quite involved. These are summarized in Table 10-2. Discussion and examples are given in Appendix 10-A.

investment tax credit

A special provision of the Federal tax law, which, when in force, may have direct bearing on economic analyses, is the *investment tax credit*. This allows businesses to subtract from their tax liability as much as 10%† of what they invest in "qualifying" property. By "qualifying" is meant depreciable equipment or tangible personal property (not including buildings) used in a business. This is subject to limitations according to the expected life of the qualifying assets, the amount of used property, and the total tax credit claimed. The limitation according to expected life per 1976 Federal tax law is summarized in the following table:

If the Expected Life Is:		The Maximum Possible Investment Credit (as a Percentage of Investment) Is
At Least	But Less Than	
7 years		10%
5 years	7 years	$\frac{2}{3}$ of 10%
3 years	5 years	$\frac{1}{3}$ of 10%
0	3 years	0%

A further limitation is that at most $100,000 of the cost of used property may be counted in any one year for purposes of the investment credit.

EXAMPLE 10-3

A firm invested $250,000 in property expected to last for 10 years, $150,000 in property expected to last 5 years, $450,000 in property expected to last 4 years,

---

† During the 1960s and early 1970s, the investment tax credit (with a maximum rate of 7% instead of 10% as shown above) was made law and then later repealed at least two times. The maximum rate was increased from 7% to 10% in the mid-1970s. As of 1977, the law provides that the maximum rate will revert to 7% (4% for public utilities) for property acquired after December 31, 1980.

and $200,000 in property expected to last 2 years. What is the maximum possible tax credit if all the property is new?

Solution

$250,000 × 10%	$25,000
$150,000 × $\frac{2}{3}$ × 10%	$10,000
$450,000 × $\frac{1}{3}$ × 10%	$15,000
TOTAL	$50,000 ∎

The allowable investment tax credit in any year cannot exceed the tax liability. If the tax liability is more than $25,000, the tax credit cannot exceed $25,000 plus 50% of the tax liability above $25,000. However, the investment tax credit can be carried back as many as 3 years and forward as many as 7 years.

EXAMPLE 10-4

Suppose that a firm in a given year has a tax liability of $65,000 before consideration of credits and a maximum possible investment credit of $50,000 (such as calculated in the last example). What is the firm's tax liability for that year?

Solution

Maximum allowable investment credit:

$$\$25,000 + 50\%(\$65,000 - \$25,000) = \$45,000$$

Tax liability for year:

$$\$65,000 - \$45,000 = \$20,000 \text{ ∎}$$

It should be noted that for the above example, the difference between the maximum possible tax credit and the allowable investment credit for the year is $5,000 (= $50,000 − $45,000) and can be carried backward or forward to reduce taxes paid or payable in other years.

As an additional note, even though the investment tax credit results in a cash savings and, therefore, in a reduction in the net investment, 1976 Federal tax law provides that the credit does not reduce the investment to be depreciated. Thus a firm can depreciate an item based on the original purchase price of P even though their net cash outlay or commitment for the item was as little as $0.9P$. (i.e., $P − $0.1P$).

EXAMPLE 10-5

Given a $20,000 investment which is expected to have a life of 10 years and a $500 salvage value, what is the net investment, assuming that the full investment tax credit can be used? What is the depreciable investment and the annual straight line depreciation charge?

Solution

Investment tax credit:

$$\$20,000 \times 10\% = \$2,000$$

Net (effective) investment:

$$\$20,000 - \$2,000 = \$18,000$$

Depreciable investment:

$$\$20,000 - \$500 = \$19,500$$

Straight line depreciation charge/year:

$$\$19,500/10 = \$1,950 \ \blacksquare$$

The above discussion provides only a minimal description of some main provisions of the Federal income tax law which are important to economic analyses. It is by no means complete, but is intended to provide a basis for illustrating after-tax (i.e., after income tax) economic analyses. In general, the analyst should either search out specific provisions of the Federal and/or State income tax law† affecting projects being studied or seek information from persons qualified in tax law.

The remainder of the chapter will illustrate various after-tax economy studies using a suggested tabular form for computing after-tax cash flows.

general procedure for making after-tax economic analyses

After-tax economic analyses can be performed by exactly the same methods as before-tax analyses. The only difference is that after-tax cash flows should be used in place of before-tax cash flows by inclusion of disbursements or savings due to income taxes and then making calculations using (or determining) an after-tax minimum attractive rate of return. While the tax rates and governing regulations may be complex for particular analysis conditions and both are subject to changes, once those rates and regulations can be translated into effect on after-tax cash flows, the remainder of the after-tax analysis is straightforward.

† Some applicable publications are

Tax Guide for Small Business, U.S. Internal Revenue Service Publication 334, published annually.

Your Federal Income Tax, U.S. Internal Revenue Service Publication 17, published annually.

J. K. Lasser, *Your Income Tax* (New York: Simon and Schuster, Publishers), published annually.

J. R. Canada and J. A. White, *Capital Investment Decision Analysis for Management and Engineering* (Englewood Cliffs, N.J.: Prentice-Hall, Inc., in press).

A suggested tabular format to facilitate the computation of after-tax cash flows is

Year	(A) Before-Tax Cash Flow	(B) Depreciation	(C)=(A)+(B) Taxable Income	(D)=−(C)×Rate Cash Flow for Income Taxes	(E)=(A)+(D) After-Tax Cash Flow

Column A contains the same information used in "before-tax" analyses—the cash revenues or savings minus the cash direct, indirect, and overhead expenses. Column B contains any depreciation amounts that can be declared as expenses for tax purposes. Column C is the income or amount subject to taxes. Column D is the income taxes paid or saved. Column E contains the "after-tax" cash flows to be used directly in after-tax economic analyses just the same as the cash flows in column A are used in before-tax economic analyses. The column headings indicate the arithmetic operations for computing columns C, D, and E for annual operating results. It is intended that the table be used with the conventions of "+" for cash inflow or savings and "−" for cash outflow or opportunity forgone. Slight deviations from this are the conventions of assigning a "−" to a depreciation charge even though it is not a cash expense and, in column C, using "+" for increases and "−" for decreases in taxable income. When all events affecting cash flow are the same for more than 1 year, a time-saving convention is to use one line to describe events for all of the years indicated in the left-hand column.

illustration of computations of after-tax cash flows for various common situations

The following problem series (Examples 10-6 through 10-10) illustrates the computation of after-tax cash flows as well as various common situations affected by taxes. All problems herein include the common assumption that any tax disbursement or savings occurs at the same time (year) as the income or expense which affects the taxes. For purposes of comparison of the effect of various situations, the after-tax internal rate of return will be shown for each example. One can observe from the results of example problems 10-6 through 10-10 that the faster (earlier) the depreciation write-off, the higher the after-tax rate of return.

EXAMPLE 10-6

Certain new machinery is estimated to cost $180,000 installed. It is expected to reduce net annual operating disbursements by $36,000 per year for 10 years and to have $30,000 salvage value at the end of the 10th year. (a) What is the before-tax I.R.R.? (b) Show the cash flow tabulation and calculate the after-tax I.R.R. if straight line depreciation is used and the ordinary income tax rate is 48%.

Solution

(a)

$$-\$180,000 + \$36,000(P/A, i'\%, 10) + \$30,000(P/F, i'\%, 10) = 0$$

at $i' = 15\%$: $\quad -180,000 + \$36,000(5.019) + \$30,000(0.2472) = +\$7,950$

at $i' = 20\%$: $\quad -180,000 + \$36,000(4.192) + \$30,000(0.1615) = -\$24,180$

$$i' = \text{I.R.R.} = 15\% + \left(\frac{7,950}{7,950 + 24,180}\right)(20\% - 15\%) = 16.2\%$$

(b)

$$\text{Annual depreciation} = \frac{\$180,000 - \$30,000}{10} = \$15,000$$

Year	(A) Before-Tax Cash Flow	(B) Depreciation	(C)=(A)+(B) Taxable Income	(D)=−(C)×48% Cash Flow for Income Taxes	(E)=(A)+(D) After-Tax Cash Flow
0	− $180,000				− $180,000
1–10	+ 36,000	− $15,000	+ $21,000	− $10,080	+ 25,920
10	+ 30,000				+ 30,000^a

^{a} When salvage value equals book value, there is no capital gain or loss, and hence no income tax.

$$-\$180,000 + \$25,920(P/A, i'\%, 10) + \$30,000(P/F, i'\%, 10) = 0$$
By trial-and-error, $i' = \text{I.R.R.} = 8.9\%$. ∎

EXAMPLE 10-7

This is the same problem as Example 10-6(b) except that sum-of-the-years'-digits depreciation is used over the 10-year life.

Solution

$\text{SYD} = 1 + 2 + 3 + \cdots + 10 = 55$

First-year depreciation $= \frac{10}{55}(\$180,000 - \$30,000) = \$27,300$

Decrease in depreciation each year thereafter $= \frac{1}{55}(\$180,000 - \$30,000) = \$2,730$

Year	(A) Before-Tax Cash Flow	(B) Depreciation	(C)=(A)+(B) Taxable Income	(D)=−(C)×48% Cash Flow for Income Taxes	(E)=(A)+(D) After-Tax Cash Flow
0	− $180,000				− $180,000
1	+ 36,000	− $27,300	+ $8,700	− $4,180	+ 31,820
Each year 2–10	+ 36,000	Decreases by $2,730 each year	Increases by $2,730 each year	$1,313 more negative each year	Decreases by $1,313 each year
10	+ 30,000				+ 30,000

$- \$180,000 + \$31,820(P/A, i'\%, 10) - \$1,313(P/G, i'\%, 10)$
$$+ \$30,000(P/F, i'\%, 10) = 0$$

By trial and error, $i' = 9.7\%$. ∎

EXAMPLE 10-8

This is the same problem as Example 10-6(b) regarding the cash flows expected to happen over the 10-year life. However, tax regulations will permit the equipment to be depreciated over a 4-year period using straight line depreciation and zero salvage value. Assume that any annual operating loss reduces taxable ordinary income and that any capital gain, being due to recovery of previously charged depreciation, is taxed at 48%.

Solution

$$\text{Annual depreciation} = \frac{\$180,000 - \$0}{4} = \$45,000$$

Year	(A) Before-Tax Cash Flow	(B) Depreciation	(C)=(A)+(B) Taxable Income	(D)=−(C)×48% Cash Flow for Income Taxes	(E)=(A)+(D) After-Tax Cash Flow
0	− $180,000				− $180,000
1–4	+ 36,000	− $45,000	− $ 9,000	+ $ 4,320	+ 40,320
5–10	+ 36,000	0	+ 36,000	− 17,250	+ 18,750
10	+ 30,000		30,000 [a]	− 14,400 [b]	+ 15,600

[a] Capital gain = selling price − book value = $30,000 − $0 = $30,000.
[b] Tax on capital gain = $30,000(48%) = $14,400.

$- \$180,000 + \$40,320(P/A, i'\%, 4) + \$18,750(P/A, i'\%, 6)(P/F, i'\%, 4)$
$$+ \$15,600(P/F, i'\%, 10) = 0$$

By trial and error, $i' = 10.6\%$. ∎

EXAMPLE 10-9

This is the same problem as Example 10-6(b) regarding expected cash flows, except that the equipment is expected to have no salvage value after 10 years. Even though the equipment is serviceable for only 10 years, assume that tax regulations are expected to require that the equipment be amortized over a 15-year period using straight line depreciation and zero salvage value. Assume that any annual loss reduces taxable ordinary income.

Solution

$$\text{Annual depreciation} = \frac{\$180,000}{15} = \$12,000$$

Year	(A) Before-Tax Cash Flow	(B) Depreciation	(C)=(A)+(B) Taxable Income	(D)=−(C)×48% Cash Flow for Income Taxes	(E)=(A)+(D) After-Tax Cash Flow
0	− $180,000				− $180,000
1–10	+ 36,000	− $12,000	+ $24,000	− $11,550	+ 24,450
1–15		− 12,000	− 12,000	+ 5,760	+ 5,760

$$-\$180{,}000 + \$24{,}450(P/A, i'\%, 10) + \$5{,}760(P/A, i'\%, 5)(P/F, i'\%, 10) = 0$$
By trial and error, $i' = 7.5\%$. ∎

EXAMPLE 10-10

This is the same problem as Example 10-6(b) regarding estimated cash flows over the 10-year life. However, the actual salvage value obtained at the end of the tenth year turns out to be $8,000 rather than the $30,000 estimated for depreciation purposes using the straight line method. Assume that any capital gain is taxable at 30% and that any capital loss reduces taxable ordinary income.

Solution

$$\text{Annual depreciation} = \frac{\$180{,}000 - \$30{,}000}{10} = \$15{,}000$$

Year	(A) Before-Tax Cash Flow	(B) Depreciation	(C)=(A)+(B) Taxable Income	(D)=−(C)×Rate Cash Flow for Income Taxes	(E)=(A)+(D) After-Tax Cash Flow
0	− $180,000				− $180,000
1–10	+ 36,000	− $15,000	+ $21,000	+ $10,080	+ 25,920
10	+ 8,000		− 22,000[a]	+ 10,560[b]	+ 18,560

[a] Capital loss = book value − selling price = $30,000 − $8,000 = $22,000.
[b] Taxes saved on capital loss = $22,000(48%) = $10,560.

$$-\$180{,}000 + \$25{,}920(P/A, i'\%, 10) + \$18{,}560(P/F, i'\%, 10) = 0$$
By trial and error, $i' = 8.5\%$. ∎

The after-tax I.R.R. results from the examples above are summarized as follows. Note the marked increase in the I.R.R. with increased rates of depreciation.

Example	Method of Depreciation (with Special Treatment)	After-Tax I.R.R.
10-6(b)	Straight line (10 years)	8.9%
10-7	Sum-of-the-years'-digits (10 years)	9.7%
10-8	Straight line (over 4 years)	10.6%
10-9	Straight line (over 15 years)	7.5%

illustration of after-tax economic comparisons using different methods

EXAMPLE 10-11

It is desired to compare the economics of two alternative material handling systems. The pertinent data are as follows:

	System X	System Y
First cost	$80,000	$200,000
Economic life	20 years	40 years
Annual before-tax cash disbursement	$18,000	$6,000
Salvage value at end of life	$20,000	$40,000

Use a table to compute the after-tax cash flows based on straight line depreciation and a 30% tax rate to cover both state and Federal income taxes. Assuming a 6% minimum attractive after-tax rate of return, show which alternative is best by the (a) internal rate of return (I.R.R.) method; (b) explicit reinvestment rate of return (E.R.R.R.) method; (c) annual cost (A.C.) method; (d) present worth–cost (P.W.-C.) method.

Before applying the various economy study methods, it should be recognized that the solutions shown below for each method employ the common simplifying assumptions used for comparisons of alternatives having different economic lives as discussed in Chapter 7. That is:

1. The period of needed service for which the alternatives are being compared is either indefinitely long or a length of time equal to a common multiple of the lives.

table 10-3
determination of after-tax cash flows for example 10-11

Year	(A) Before-Tax Cash Flow	(B) Depreciation	(C)=(A)+(B) Taxable Income	(D)=−(C)×30% Cash Flow for Income Taxes	(E)=(A)+(D) After-Tax Cash Flow
System X					
0	− $ 80,000				− $ 80,000
1–20	− 18,000	− $3,000 [a]	− $21,000	+ $6,300	− 11,700
20	+ 20,000				+ 20,000
System Y					
0	− 200,000				− 200,000
1–40	− 6,000	− 4,000 [b]	− 10,000	+ 3,000	− 3,000
40	+ 40,000				+ 40,000

[a] Annual depreciation for system X = ($80,000 − $20,000)/20 = $3,000.
[b] Annual depreciation for system Y = ($200,000 − $40,000)/40 = $4,000.

2. What is estimated to happen in the first life cycle will happen in all succeeding life cycles, if any.

Table 10-3 shows the calculation of after-tax cash flows to be used in the following solutions.

Solution by I.R.R. Method

The I.R.R. cannot be found for either alternative alone, but one can use the calculated I.R.R. to determine if the increment investment is justified. Since the economic lives differ, the easiest way is to find the $i'\%$ at which the annual costs of the two alternatives are equal (or the difference in the annual costs for the two alternatives is zero). Thus

$$\text{A.C.}_X = \$80,000(A/P, i'\%, 20) - \$20,000(A/F, i'\%, 20) + \$11,700$$
$$\text{A.C.}_Y = \$200,000(A/P, i'\%, 40) - \$40,000(A/F, i'\%, 40) + \$3,000$$

The calculated results are as follows:

	$i' = 5\%$	$i' = 10\%$
A.C.$_X$	$17,516	$20,850
A.C.$_Y$	14,330	21,370
A.C.$_X$ − A.C.$_Y$	$ 3,186	−$ 520

Thus

$$i'\% = \text{I.R.R.} = 5\% + \left(\frac{3,186}{3,186 + 520}\right)(10\% - 5\%) = 5\% + 4.3\% = 9.3\%$$

Since 9.3% > 6% M.A.R.R., the incremental investment is justified, and system Y is the indicated choice. ∎

Solution by E.R.R.R. Method

Reinvestment of the recovered depreciation capital is made at some assumed after-tax rate of return which is usually the after-tax M.A.R.R.

	System X	System Y
Annual expenses		
Disbursements	$11,700	$3,000
Depreciation: ($80,000 − $20,000)(A/F, 6%, 20)	1,630	
($200,000 − $40,000)(A/F, 6%, 40)		1,040
TOTAL ANNUAL EXPENSES	$13,330	$4,040

Δ Total annual expenses (savings) = $13,330 − $4,040 = $9,290
Δ Investment = $200,000 − $80,000 = $120,000

$$\text{E.R.R.R. on } \Delta \text{ investment} = \frac{\$9,920}{\$120,000} = 7.7\%$$

Since 7.7% > 6% M.A.R.R., system Y is again the indicated choice. ∎

Solution by A.C. Method

$$\text{A.C.}_X = \$80,000(A/P, 6\%, 20) - \$20,000(A/F, 6\%, 20) + \$11,700$$
$$= \$80,000(0.0872) - \$20,000(0.0272) + \$11,700 = \$18,130$$
$$\text{A.C.}_Y = \$200,000(A/P, 6\%, 40) - \$40,000(A/F, 6\%, 20) + \$3,000$$
$$= \$200,000(0.0665) - \$40,000(0.0065) + \$3,000 = \$16,040$$

Since A.C.$_X$ > A.C.$_Y$, system Y is again the indicated choice. [*Note:* Of course, these answers could have been obtained by adding the annual interest on investment ($P \times i$) to the total annual expenses in the "Solution by E.R.R.R. Method" shown above.] ∎

Solution by P.W.-C. Method

Using the lowest common multiple of lives = 40 years as the study period, we obtain the following data:

	System X	System Y
Annual disbursements:		
$11,700($P/A$, 6\%, 40)$	$176,041	
$3,000($P/A$, 6\%, 40)$		$ 45,139
Original investment	80,000	200,000
First replacement: ($80,000 − $20,000)($P/F$, 6\%, 20)$	18,708	
Less: salvage after 40 years		
−20,000(P/F, 6\%, 40)$	− 1,944	
−40,000(P/F, 6\%, 40)$		−3,889
TOTAL P.W.-C.	$272,805	$241,250

Since P.W.-C.$_X$ > P.W.-C.$_Y$, system Y is again the indicated choice. ∎

table 10-4

data and after-tax annual cash flow computations for after-tax comparison of the four molding presses in example 10-12 (original data given in example 7-1)

	Press			
	A	B	C	D
Investment	− $6,000	− $7,600	− $12,400	− $13,000
Economic life	5 years	5 years	5 years	5 years
Salvage value	0	0	0	0
Annual amounts, years 1–5:				
Before-tax cash flow	− $7,800	− $7,282	− $ 6,298	− $ 5,720
Depreciation:				
(investment − $0)/5 years	− 1,200	− 1,520	− 2,480	− 2,600
Total taxable income	− $9,000	− $8,802	− $ 8,778	− $ 8,320
Income taxes at 50%	+ $4,500	+ $4,401	+ $4,389	+$ 4,160
After-tax cash flow	− $3,300	− $2,881	− $1,909	−$ 1,560

EXAMPLE 10-12

It is desired to compare the four presses in Example 7-1 based on after-tax analyses using (a) the E.R.R.R. method, and (b) the A.C. method. Assume a tax rate of 50%, that straight line depreciation is used for tax purposes, and that the after-tax M.A.R.R. is 5%. The essential data, together with a tabular computation of the after-tax annual cash flows for each press (in a revised format), are shown in Table 10-4. This revised format is often convenient when there are multiple alternatives and the investment and salvage value amounts are the same both before and after taxes.

Solution by E.R.R.R. Method

	Press			
	A	B	C	D
Investment	$6,000	$7,600	$12,400	$13,000
After-tax annual expenses:				
Cash flow	3,300	2,881	1,909	1,560
Depreciation:				
investment × $(A/F, 5\%, 5)$	1,085	1,375	2,245	2,353
TOTAL ANNUAL EXPENSES	$4,385	$4,256	$4,154	$ 3,913
Increment considered	A → B		B → C	B → D
Δ Investment	$1,600		$4,800	$5,400
Δ Total annual expenses (savings)	129		102	343
E.R.R.R. on Δ investment	8.1%		2.1%	6.4%
Is increment justified (> 5%)?	Yes		No	Yes

Hence, press **D** is the choice, being the highest investment alternative for which each increment will earn at least the 5% after-tax M.A.R.R. ∎

Solution by A.C. Method

	Press			
	A	B	C	D
After-tax annual costs				
Cash flow	$3,300	$2,881	$1,909	$1,560
Depreciation + minimum required profit:				
investment × $(A/P, 5\%, 5)$	1,385	1,755	2,865	3,003
TOTAL ANNUAL COST	$4,685	$4,636	$4,774	$4,563

Hence press **D** is again shown to be the choice, this time based on the fact that it has the minimum after-tax annual cost. ∎

It should be noted that press D was the indicated choice also in the before-tax analyses shown in Example 7-1. This is to be expected since all alternatives

table 10-5
capital recovery provided by an oil well, with cost and percentage depletion allowances used in computation of income taxes

(1) Year	(2) Barrels of Oil Sold	(3) Gross Income (Cash Flow)	(4) Net Income (Cash Flow) Before Income Taxes	(5) 50% of Net Income	(6) Cost Depletion at $4.00 per Barrel	(7) Depletion Allowance at 22% of Gross Income	(8) [=(4) − either (5), (6), or (7)] Net Taxable Income	(9) [=−(8) × 70%] Income Tax	(10) [=(4) + (9)] Net Cash Flow After Taxes
1	70,000	$1,400,000	+ $800,000	$400,000	$280,000	$308,000*a*	$492,000	− $344,000	$456,000
2	60,000	1,200,000	+ 700,000	350,000	240,000	264,000*a*	436,000	− 305,000	395,000
3	45,000	900,000	+ 480,000	340,000	180,000	198,000*a*	282,000	− 198,000	282,000
4	20,000	400,000	+ 240,000	120,000	80,000	88,000*a*	152,000	− 106,000	134,000
5	5,000	85,000	+ 25,000	12,500*b*	20,000	18,700	12,500	− 8,750	16,250

a Percentage depletion shown deducted.
b Only amount shown can be deducted because of percentage depletion exceeding 50% of net income.

314

were subject to the same depreciation method and tax rate and there were no gains or losses on disposal.

the effect of depletion allowances

As was discussed in Chapter 9, income from investment in certain natural resources is subject to depletion allowances before income taxes are computed. Under certain conditions, notably where the taxpayer is in a relatively high tax bracket, depletion provisions in the tax law can provide considerable savings.

As an example, consider the case of a single individual who has a present net taxable income of $200,000. He spends $800,000 to drill and develop a new oil well that has an estimated reserve of 200,000 barrels. Oil is produced and sold in accordance with the schedule shown in column 2 of Table 10-5 to produce the gross cash flow shown in column 3. Column 4 shows the net cash flow after cash production costs have been deducted.

The depletion allowance that may be deducted in a given year may be based on a fixed percentage of the gross income (22% for oil), provided that the deduction does not exceed 50% of the net income before such deduction. Depletion, computed in this manner, is shown in column 7 of Table 10-5. Another method is to base depletion on the estimated investment cost of the product. In this case the estimated 200,000 barrels of oil in the well cost $800,000. Depletion may therefore, if desired, be charged at the rate of $4.00 per barrel, as shown in column 6 of the same table.

The net taxable income resulting from the most favorable application of these depletion allowances is shown in column 8. The advantage of depletion allowances stems from the fact that the total depletion which can be claimed often exceeds the depreciable capital investment. In this case, the summation of the deductions claimed (denoted a in column 7 and b in column 5) is $870,500, even though the total investment was only $800,000. Inasmuch as the taxpayer has a net taxable income of $200,000 even before returns from the oil well, use of Table 10-1 shows that the incremental tax rate is 70%, thus giving the income tax shown in column 9. Column 10 shows the net after-tax cash flow provided to the investor. By equating the sum of the present values of the cash flows shown to the $800,000 of investment, we find that the I.R.R. after taxes is roughly 25%. It is of interest to note that the before-tax I.R.R.—by use of the net cash flow information in column 4—is roughly 55%.

after-tax comparison of property fully depreciable by straight line method with property subject to depletion allowances

If the individual in the oil well example (above) were to invest his $800,000 in ordinary business property that is *fully depreciable* over a 5-year period and brings the same before-tax I.R.R. of 55%, the before-tax net annual cash flow would be $800,000(A/P, 55\%, 5) = \$800,000(0.625) = \$500,000$. If we assume straight line depreciation, the annual depreciation expense is $800,000/5 = \$160,000$. Hence the taxable income would be $500,000 - \$160,000 = \$340,000$, and the income taxes would be $340,000 \times 70\% = \$238,000$. Thus the after-tax net cash flow would be $500,000 - \$238,000 = \$262,000$. The after-tax I.R.R. can be found as the $i'\%$ at which

$$-\$800,000 + \$262,000(P/A, i'\%, 5) = 0$$

Solving by interpolation in interest tables, we obtain $i' = 19\%$. When this 19% is compared with the after-tax I.R.R. of roughly 25% for the oil well having the same investment, economic life, and before-tax I.R.R., the advantages of depletion allowances become apparent.

after-tax comparison of fully depreciable property with nondepreciable property

If the same individual as the one discussed above were to invest his $800,000 in a nondepreciable ordinary business venture that would produce the same before-tax I.R.R. of 55% (a net annual cash flow and taxable income of $800,000 \times 0.55 = \$440,000$), the annual income tax would be $440,000 \times 70\% = \$308,000$, an after-tax profit of $440,000 - \$308,000 = \$132,000$, or 16.5% I.R.R. on the $800,000 investment.

Comparison of the 19% after-tax I.R.R. for the fully depreciable property with the 16.5% after-tax I.R.R. for nondepreciable property having the same investment, economic life, and before-tax I.R.R. makes it evident that if one is comparing depreciable versus nondepreciable alternatives, income taxes should definitely be considered.

tax-exempt income

Interest received on bonds or other obligations of states, United States territories or the District of Columbia, or political subdivisions thereof (such as cities and counties) is exempt from Federal income tax. Obviously, where one alternative in an economy study involves income upon which no tax is paid and the income from another alternative is taxable, these facts must be given proper consideration. Ordinarily, the interest rate of such tax-exempt securities is from 4 to 6%. An individual who derives his entire income from such securities would thus pay no income tax. In some instances the net income so derived, despite the relatively low interest rates received, may be greater than the income that would be obtained from investments yielding larger interest rates but subject to income tax. For an individual with a moderately high income, tax-exempt investments may be quite attractive.

EXAMPLE 10-13

Consider the case of an unmarried individual who already has a net taxable income of $44,000 and is planning to invest $5,000 of capital. Opportunity A offers a prospective before-tax return of 7% per year on a nondepreciable investment. Opportunity B is to buy tax-exempt bonds that pay interest at the rate of 4%. From Table 10-1 it is seen that this individual will have to pay income taxes at the rate of 60% on the incremental taxable income. Her net income, after taxes, from opportunity A would be as follows:

Gross income = $5,000 × 0.07		$350
Additional income tax = $350 × −0.60		− 210
NET INCOME AFTER TAXES		$140

This net after-tax income is a return of only $140/$5,000 = 2.80%, so the after-tax 4% offered by opportunity B is higher.

In most business enterprise investment problems, however, the alternative of investing capital in tax-exempt securities is not of sufficient attractiveness to merit much consideration. Nevertheless, investment in tax-exempt securities is a possibility that should not be overlooked, particularly if low risk of loss is desired.

problems

10-1. (a) Last year, Cat Tractor, Inc., had a gross income of $1,753,000 and their expenses, including depreciation, totaled $848,000. How much Federal income taxes would be payable for the year?

(b) A corporation has estimated that its taxable income will be $57,000 next year. It has the opportunity to invest in a business venture that is expected to add $8,000 to next year's taxable income. How much in Federal taxes will be owed *with* and *without* the proposed venture?

10-2. An unmarried man presently earns $19,800 per year as his adjusted gross income. He is offered another job that pays an extra $3,500 per year. This man expects to take the standard deduction (amounting to 10% of his gross adjusted income) that is allowed in filing his annual tax return and will claim one exemption. By using the appropriate tax table in Chapter 10, answer the following questions:

(a) What is the man's effective income tax rate if he chooses to accept the new job?

(b) What is the extra amount of tax that he will pay with the new job?

(c) At his higher salary, what is this man's effective income tax rate and tax owed *if* he could claim a long-term capital loss of $2,000 next year?

10-3. Bob Brown had a taxable income for 1976 of $30,000, derived from his salary and interest from investments. For the same year, Joe Jones had a taxable income of $27,000 composed of $12,000 from salary and a long-term capital gain of $15,000.

(a) Which man had the greater after-tax income?

(b) What effective tax rate did Joe pay on the $15,000 capital gain?

10-4. Rocky Baritone, who is unmarried, had such a large income that he was in the 70% income tax bracket. His business manager convinced him to purchase a farm for $75,000, with the idea of improving it over a 4-year period and then selling it for a capital gain. Rocky bought the farm, and he immediately spent an additional $15,000 for structural improvements. During the next 4 years the annual operating costs of the farm exceeded the revenues by $15,000 per year. However, Rocky was able to deduct these annual losses on his annual income tax returns and also to claim $5,000 each year as depreciation expense. At the end of the fourth year he sold the farm for $150,000.

(a) What was the before-tax internal rate of return on the investment in the farm?

(b) What was the after-tax rate of return if his gain was taxed at a rate of 25%?

(c) What was the after-tax rate of return if he was charged 70% for "recovery of previously charged depreciation?" (See Appendix 10-A.)

10-5. A corporation built a warehouse at a cost of $100,000, estimating that it would have a depreciable life of 10 years and a salvage value of $50,000 at that time. It charged depreciation on a straight-line basis. At the end of 5 years it had no further need for the warehouse and was able to sell it for $120,000. Assuming that the corporation is in the 48% effective income tax bracket, what was the net after-tax result from the sale? Long-term capital gains are taxed at 30%. Refer to Appendix 10-A concerning "recovery of previously charged depreciation."

10-6. A corporation that has an incremental Federal income tax rate of 48% and a State income tax rate of 4% invests $100,000 in a piece of equipment that will reduce annual operating costs by $25,000 for a period of 10 years, but thereafter it will have no functional use and will have no salvage value. The equipment will be depreciated by straight-line depreciation and a 10% investment tax credit can be claimed at the time of purchase. What will be the before- and after-tax internal rate of return produced by the investment?

10-7. (a) What would be the after-tax rate of return in Problem 10-6 if the equipment were written off over a 5-year period even though the savings are obtained over the entire 10-year period?

(b) What would be the after-tax rate of return in Problem 10-6 if the annual savings of $25,000 were achieved during only the first 5 years, but the equipment was depreciated over the entire 10-year period? Assume there are other earnings against which the depreciation costs could be charged.

10-8. (a) Miss Amanda Plumrose bought some common stock 10 years ago for $5,000. She decides to sell the stock now for a long-term capital gain amounting to $3,000. Furthermore, the taxable income from her job is $8,000. At the end of the year, how much in taxes will she owe the Federal government?

(b) A certain taxpayer has a taxable income that is subject to an average tax rate of 38% and an incremental rate of 45%. He has the chance to either invest in tax-free school bonds yielding 3% or to lend money at interest to a company. If he lends to the firm, what interest rate is necessary to yield him the same after-tax return as the municipal bonds?

(c) An asset originally bought for $10,000 has already been depreciated for tax purposes for 3 years at $500/year. It is now sold for $9,000. Find the amount of the capital gain on this transaction.

(d) Above what figure should an individual's incremental tax rate be before he applies the "25% maximum" provision to his capital gains that are subject to income tax?

10-9. An individual who was, and is, in the 70% income tax bracket, purchased some nondepreciable property for $20,000. Two weeks later he had an opportunity to sell it for $23,000. He held the property for 4 years and then sold it for $35,200. In the meantime he paid annual property taxes of $600. His objective in keeping the property was to achieve a 10% return on his capital, after taxes. Did he achieve his objective?

10-10. A corporation buys a new asset for $10,000. The life is expected to be 5 years. A salvage value of $2,000 has been estimated. Assume that the corporation saves 40% in income taxes each year from depreciation (i.e., tax savings = 0.4 × depreciation for the year).

(a) Calculate the present worth of the tax savings if straight line depreciation is used.

(b) If double declining balance depreciation is used, calculate the present worth of the tax savings.

(c) If sum-of-the-years'-digits depreciation is used, calculate the present worth of the tax savings.

(d) Which depreciation method do you recommend?

10-11. An asset with an estimated life of 10 years and a salvage value of $5,000 is purchased by a firm for $60,000. The firm's effective tax rate is 50% and the after-tax M.A.R.R. is 10%. A 10% investment tax credit applies to this purchase. The gross income from the asset before depreciation and taxes will be $8,500/year. Calculate the present worth of the *taxes* for (a) the straight line method of depreciation, and (b) the sum-of-the-years'-digits method of depreciation.

10-12. A small, local TV repair shop is planning to invest in some new circuit testing equipment. The details of the proposed investment are

First cost = $5,000. Does not qualify for an investment tax credit.
Salvage value = 0
Extra revenue = $2,000/year
Extra expenses = $800/year
Expected life = 5 years
Effective income tax rate = 20%

(a) If sum-of-the-years'-digits depreciation is used, calculate the present worth of after-tax cash flows when the M.A.R.R. (after taxes) is 12%. Should the equipment be purchased?

(b) If double declining balance depreciation is used, calculate after-tax cash flows and recommend whether the new equipment should be purchased if the after-tax M.A.R.R. is 12%. Use the present worth method.

10-13. A single woman, having a taxable income of $30,000 per year, is considering reinvesting $15,000 of her capital, which now is in a savings and loan association drawing 7% effective annual interest. One possibility is to invest it in tax-exempt bonds that would pay $4\frac{1}{2}$% interest. Another possibility is to invest it in some property that probably could be sold at the end of 1 year for $16,000. Capital gains tax, at the minimum rate, would have to be paid on the gain. Which use for the capital will be most profitable for her?

10-14. A used machine that does not qualify for an investment tax credit can be purchased now for $27,000 and will have a salvage value of $6,000 in 6 years.

It will be depreciated using the sum-of-the-years'-digits method. Income will be $6,000 per year for each of the 6 years. Expenses will be $800 per year. Determine the prospective after-tax I.R.R. for the machine if the effective income tax rate is 50%. Is this a good investment opportunity?

10-15. A company must purchase a particular asset for $10,000. The life of the asset is expected to be 5 years, and it will have a salvage value of zero in computing its straight line depreciation charges. An investment tax credit of two-thirds of 10% can be applied to this asset. However, management of this company believes the asset will have an actual market value of $2,000 at the end of its 5-year life. The annual operating and maintenance costs are $2,000 the first year and increase by $200 per year thereafter. This asset will be purchased entirely with borrowed money at 10% effective annual interest. The principal of the loan and all compounded interest will be repaid in one lump sum at the end of year 5. With M.A.R.R. = 12% after taxes and an effective tax rate of 0.50, calculate the after-tax equivalent annual cost of this asset. Assume that the company is profitable in its other activities, and that long-term capital gains are taxed at 30%.

10-16. You can purchase your own machine for $12,000 to replace a rented machine. The rented machine costs $4,000 per year. The machine that you are considering would have a life of 8 years and a $5,000 salvage value at the end of the life. By how much could annual operating expenses of the purchased machine increase and still provide a return of 10% after taxes? Assume that you are in a 50% tax bracket and the revenues produced with either machine are identical. Assume that straight line depreciation will be utilized to recover the investment in the machine.

10-17. John Jones, who is a widower with no dependents, has operated his own business for 30 years and has built it to the point where it produces a net taxable income of $25,000 per year. He is considering retiring, and he also would like to realize a capital gain by selling the business, which has a book value of $175,000.

A large corporation has offered him $400,000 for his company. He has made up his mind that if he keeps the business he will retire in 10 years, at which time he believes he could sell it for at least twice its book value at that time, which he expects to be about the same as at present.

If he sells the business now, he will invest the proceeds in tax-exempt bonds, which will earn an average of 4%. Assuming that his net tax on any capital gains will be 20%, what would you advise him to do?

10-18. Your boss asked you to determine whether the company can justify the purchase of a special piece of earth-moving equipment. The data for the problem are as follows:

First cost of equipment:	$20,000 (equity financing)
Depreciation:	Straight line with $2,000 salvage value, 10-year life
Property taxes and insurance:	$500/year
Maintenance cost:	$400 the first year, increasing $40 per year thereafter
Operating savings (due to increased productivity):	$7,000 the first year, decreasing $300 per year thereafter

Even though government (IRS) guidelines suggest that the salvage value should be no less than 10% of the equipment's first cost (i.e., $2,000) the boss expects there to be *no* salvage value in reality. Assume the boss is correct, and make an after-tax study for the equipment when the effective income tax rate is 0.50, after-tax M.A.R.R. = 10%, and capital losses can be used to offset capital gains, which are taxed at 30%. Utilize a 10% investment tax credit in your analysis.

10-19. An automatic lathe was purchased 3 years ago for $20,000. It has been depreciated on the books using the declining balance method with a depreciation rate of 0.15. The machine can now be sold for $15,000. If it is kept for another 10 years, it is estimated that the sale price will be reduced to $2,500. The company has an effective income tax rate of 50%. The lathe is depreciated in an item account, and long-term capital gains are taxed at 30%. With an after-tax rate of return of 8%, compute the after-tax capital recovery cost of extending service another 10 years.

10-20. A construction company, operating as a corporation, has a contract to build an underground aqueduct, which it will complete in the current tax year. If conventional methods are used, it expects to make a profit of $120,000 on the job. However, it believes that a new and untried procedure can be used, and if this method is successful it should result in a net saving of $60,000 (i.e., $60,000 in additional profits). If the method is not successful, it would add $30,000 to the cost of doing the job. The chief engineer of the company believes the odds that the new method will be successful are at least 2:1. The corporation averages about $300,000 per year in before-tax profits. (a) Use the tax rates given in this chapter and make an after-tax analysis of the proposal, and give your recommendations. (b) What advantage, if any, is gained in this case by making the study on an after-tax basis rather than on a before-tax basis?

10-21. (a) A corporation uses equity funds to purchase a $60,000 asset with an estimated life of 4 years and expected salvage value of $3,000. If after 4 years the asset is sold for $20,000, what is the capital gain (or loss) when the double declining balance depreciation method is used? Assume that all accounting adjustments between actual and estimated salvage value occur at the time the asset is sold.

(b) If the effective tax rate of this firm is 40% of the taxable income, calculate the present worth of after-tax cash flow. Let M.A.R.R. = 12% and suppose that annual net cash flow before taxes is $20,000. Assume a 30% capital gains (or losses) tax rate and that the firm is profitable in its operations.

10-22. A temporary building can be built to house a certain operation required for a special contract. You can depreciate it on a straight line basis for 10 years with no salvage value. If the building costs $40,000 and eliminates $10,000 annual rent, how many years must the contract run for you to get a 10% after-tax return? Assume that the building is useless to you after the contract runs out; however, you cannot shorten the depreciation time. Therefore, if the contract runs out before the life of the building expires, the building just sits there idle and incurs no out-of-pocket expenses. Round your answer off to the closest year that will guarantee an after-tax return of 10%. The effective income tax rate is 50%.

10-23. A company is investigating the possible installation of electronic equipment that will automate most of its accounting, billing, and inventory control functions. It also will supply management with some additional services not now available. It is estimated that this equipment will replace four people whose salaries total $32,000 annually. In addition, there will be a saving of 12% of these salaries in fringe benefits.

Two types of equipment are being considered. The ABC equipment can be leased on a 5-year basis for $2,000 per month. This price includes full maintenance. The XYZ equipment would have to be purchased at a cost of $60,000. A maintenance contract can be obtained at a cost of $200 per month. Taxes and insurance on owned equipment is about 4% of first cost per year. If either type of equipment is installed, one employee, whose annual salary is now $8,000, will be upgraded with an accompanying salary increase of $1,200 per year, plus associated fringe-benefit costs.

Because of the rapid technological development in the field, the company feels that if the XYZ equipment is purchased, it should be fully written off within 5 years. Assume an income tax rate of 48% and make an after-tax analysis of both alternatives. The after-tax M.A.R.R. = 12%. Which, if any, equipment would you recommend?

10-24. Individual industries will use energy as efficiently as it is economical to do so, and there are several incentives to improve the efficiency of energy consumption. One incentive for the purchase of more energy-efficient equipment is to reduce the time allowed to write off the initial cost. Another "incentive" might be to raise the price of energy in the form of an energy tax.

To illustrate these two incentives, consider the selection of a new motor-driven centrifugal pump for a refinery. The pump is to operate 8,000 hours per year. Pump A costs $1,600, consumes 10 hp, and has an overall efficiency of 65% (it delivers 6.5 hp). The other available alternative, Pump B costs $1,000, consumes 13 hp, and has an overall efficiency of 50% (delivers 6.5 hp).

Compute the after-tax rate of return on extra investment in pump A, assuming an effective income tax rate of 50%, a tax life of 10 years [part (a) only], zero salvage values, and straight line depreciation for each of these situations:

(a) The cost of electricity is $0.04/kWh.
(b) A 6-year depreciation write-off period is allowed, the expected life of both pumps is still 10 years, and the cost of electricity is $0.04/kWh.

appendix 10-A: tax computations involving capital gains and losses

Table 10-2 summarized tax treatment of capital gains and losses. The following discussion provides further explanation and numerical examples.

For tax computation purposes, short-term gains and losses are added separately, and the total of one is subtracted from the total of the other to determine whether a net short-term gain or loss results. The long-term gains and losses are treated in the same manner to determine the net long-term gain or loss. The *total net gain or loss* is then determined by merging the net short-term capital gain or loss with the net long-term capital gain or loss. A numerical example for one case is given below.

EXAMPLE 10-A-1

During the year an unmarried individual had the following transactions:

Long-term capital gain	$5,400	
Long-term capital loss	(1,750)	
Net long-term capital gain		$3,650
Short-term capital gain	$2,600	
Short-term capital loss	(3,790)	
Net short-term capital loss		($1,190)
Excess of net long-term capital gain over net short-term capital loss		$2,460

Thus the $2,460 would be taxed as a long-term capital gain. What is the individual's incremental (added) Federal tax liability due to the above if his taxable income *before* consideration of the capital gains and losses is (a) $23,000? (b) $63,000?

Solution (using Tables 10-1 and 10-2)

(a) $(\$2,460)(\frac{1}{2})(0.40) = \$492$

(b) Regular tax computation:

$$(\$2,460)(\tfrac{1}{2})(0.64) = \$787.20$$

Alternative tax computation:

$$(\$2,460)(0.30) = \$738$$

Since the alternative tax is lower than the regular tax, the $738 increment would be paid. ∎

Corporations, as well as individuals, treat a net short-term capital gain the same as ordinary income for tax purposes. Further, any net long-term capital gain should be taxed at the full rate for ordinary income or, alternatively, at a flat rate of 30%, whichever is smaller.

A corporation may deduct capital losses only to the extent of its capital gains in any given year. However, any capital loss in the current year that exceeds its capital gains for that year may be carried over to subsequent years until it is completely absorbed. When carried over, the loss will retain its original character as long term or short term; thus a long-term capital loss carried over from a previous year will offset long-term gains of the current year before it offsets short-term gains of the current year.

EXAMPLE 10-A-2

A corporation had a taxable income of $120,000, which included the following:

Long-term capital gain		$65,000
Long-term capital loss		(18,000)
Net long-term capital gain,		$47,000
Short-term capital gain	$10,000	
Short-term capital loss	(27,000)	
Net short-term capital loss		($17,000)
Excess of net long-term capital gain over net short-term capital loss		$30,000

What is the total income tax to be paid?

Solution

Regular tax computation:

First $25,000 at 20%	$ 5,000
Next $25,000 at 22% (taxable income in excess of $25,000)	5,500
Remaining ($120,000 − $50,000) at 48%	33,600
TOTAL TAX	$ 44,100

Alternative tax computation:

Taxable income	$120,000
Less: excess of net long-term capital gain over net short-term capital loss (see above)	30,000
Ordinary income	$ 90,000
Tax on ordinary income:	
First $25,000 at 20%	5,000
Next $25,000 at 22% (taxable income in excess of $25,000)	5,500
Remaining ($90,000 − $50,000) at 48%	19,200
TOTAL	$ 29,700
Tax on net capital gain:	
30% of $30,000	9,000
TOTAL TAX	$ 38,700

Since the alternative tax of $38,700 is less than the regular tax of $44,100, the corporation would choose to pay the alternative tax. ▌

When depreciable property such as buildings and equipment used in a business is sold, it can result in gains that are taxed partly as capital gains and partly as ordinary income, according to how much depreciation has been charged. The determination of how such a gain or loss affects taxes can be complicated and can be explained best by reference to Figure 10-2 and by assuming that straight

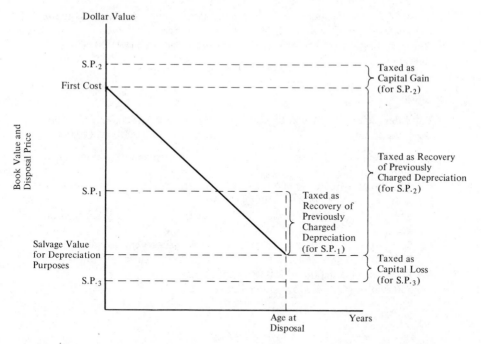

Figure 10-2. Tax categories for various possible gains and losses on disposal of depreciable property used in a business.

line depreciation has been used for accounting and tax purposes.† If we sell the asset for a price less than the original cost but greater than the book value (S.P._1 in Figure 10-2), the excess of selling price over book value represents a recovery of previously charged, and deducted, depreciation cost. Consequently, this excess is considered to be ordinary income and is taxed as such, since this amount had been deducted from ordinary income in prior years. If the selling price is greater than the original cost (S.P._2 in Figure 10-2), the difference between the original cost and the book value again is recovery of previously charged depreciation, and it is taxed as ordinary income, but the excess of the selling price above the original cost is taxed in a manner similar to a capital gain. On the other hand, if we sell the asset for a price that is less than the book value (S.P._3 in Figure 10-2), a loss results which is treated as a capital loss.

Example 10-A-3

A corporation has equipment that cost $200,000 5 years ago. At that time a straight line depreciation schedule was set up based on an estimated 15-year life

---

† Some complications exist for assets purchased prior to 1964 and for cases in which accelerated depreciation methods are used. When such conditions are encountered, the specific IRS rules should be consulted.

and $20,000 salvage value. The incremental tax rate on ordinary income is 48%
and on capital gains or losses is 30%. Determine the taxes to be paid (or saved)
on the gain (or loss) if the property is sold now for (a) $160,000; (b) $240,000;
(c) $90,000.

Solution

Total depreciation charged to now = $(\frac{5}{15})$($200,000 − $20,000) = $60,000
Book value now   = $200,000 − $60,000 = $140,000

(a) $160,000 − $140,000 = $20,000 gain due to recovery of previously charged
     depreciation.
     $20,000 × 48% = $9,600 taxes to be paid.
(b) $240,000 − $140,000 = $100,000 gain.
     $200,000 − $140,000 = $60,000 is recovery of previously charged depreci-
     ation.
     $240,000 − $200,000 = $40,000 is gain subject to capital gain rate.
     $60,000 × 0.48 + $40,000 × 0.30 = $40,800 taxes to be paid.
(c) $140,000 − $90,000 = $50,000 capital loss.
     $50,000 × 0.30 = $15,000 taxes saved. ∎

chapter 11

replacement studies

Business firms and individuals almost constantly must decide whether existing assets should be continued in service or whether available new assets will better and more economically meet current needs. These decisions must be made with increasing frequency as the dynamic pace of business quickens and technology produces more rapid changes. Thus the *replacement problem*, as it commonly is called, requires careful economy studies if we are to arrive at sound decisions.

Unfortunately, replacement problems sometimes are accompanied by unpleasant financial facts. Often it appears that earlier decisions, particularly as to the anticipated economic life of existing assets, were not as good as might be desired, especially when hindsight can be applied. Frequently, new and apparently superior assets have unexpectedly appeared on the scene; their arrival may not be timed optimally with respect to the financial position of the firm. Sometimes replacement situations arise because there have been completely unanticipated, and even unreasonable, changes in market conditions. Consequently, there is a tendency to look on the entire area of replacement as a bugaboo, whereas in fact it often represents economic opportunity.

Economic studies of replacement alternatives are performed in the same manner as economic studies of any two alternatives—the only difference is that one alternative is to keep an *existing (old) asset*, which can be descriptively called the *defender*; and there are one or more alternative *replacement (new)* assets, which can be called *challengers*.

Replacement decisions are critically important to a firm. A hasty decision to get rid of old equipment or faddishly to buy the most modern or elaborate available equipment can be a serious drain on capital available. On the other hand, a firm may go to the other extreme by delaying replacements until it becomes noncompetitive or there is no other way to continue production.

reasons for replacement

It should be recognized that not all existing assets replaced for a given service are physically scrapped or even disposed of by the owner. Frequently, new

327

assets are acquired to perform the services of existing assets, with the existing assets not retired but merely transferred to some other use—often an "inferior" use, such as standby service. Of course, many assets are sold by their existing owners and are used by one or more other owners before reaching the scrapheap.

For purposes of replacement studies, the following is a distinction between economic life and other types of lives for typical assets.

Economic life (also known as service life) is the period of time extending from date of installation to date of retirement (by demotion or disposal) from the primary intended service. The need for retirement is signaled by an economy study indicating when the equivalent cost of a new asset (challenger) is less than the equivalent cost of keeping the present asset (defender) for an additional period.

Ownership life is the period between date of acquisition and date of disposal by a specific owner. A given asset may have several service lives for a given owner. For example, a car may serve as the primary family car for several years and then serve only for local job commuting for several more years.

Physical life is the period between original acquisition and final disposal of an asset over its succession of owners. For example, the car above may have several owners over its existence, and only one of those owners was the one described above as using it first for a family car and then as a commuting car.

Depreciable life is the period over which depreciation charges are scheduled to be made.

The four major reasons for replacement are as follows:

1. *Physical impairment.* The existing asset is worn out, owing to normal use or accident, and no longer will render its intended function unless extensive repairs are made.
2. *Inadequacy.* The existing asset does not have sufficient capacity to fill the current and expected demands. Here, clearly, the requirements have changed from those that were anticipated at the time the asset was acquired. This condition does not necessarily imply physical impairment, but to meet the new demands the asset either must be supplemented or replaced.
3. *Obsolescence.* This may be of two types: (a) functional, or (b) economic. Both types result in loss of profits. In the case of functional obsolescence, there has been a decrease in the demand for the output of the asset; thus a loss in revenue follows. For example, the market may wish a higher quality product than it previously demanded. Economic obsolescence is the result of there being a new asset that will produce at lower cost than can be obtained with the old asset. Engineers are notorious contributors to this situation.
4. *Rental or lease possibilities.* This is a variation of obsolescence, except that the replacement asset does not necessarily have to be different, in any respect, from the existing asset. The possible economic advantage is due to advantageous financial factors that sometimes may accrue from leasing. These usually involve income tax considerations.

income taxes in replacement studies

The replacement of assets often results in capital gains or losses, or gains or losses from the sale of land or depreciable property, as was discussed in Chapter 10. Consequently, to obtain a complete economic analysis in such cases the studies must be made on an after-tax basis. In specific cases the situation may be complicated by whether the organization has had other such gains or losses within the tax year, inasmuch as such losses must be offset against any gains before the full tax-credit benefits apply. However, because it often is difficult to determine the existence of such gains or losses throughout an organization without undue loss of time, many companies follow the practice of considering each case separately and applying the full tax benefits or penalties to the gains or losses associated with each study, unless the amounts involved are very large. In many real situations, where an existing asset is replaced with a substantially identical new asset, there may be no tax-qualifying loss when the old asset is sold for less than its book value; the book loss must be written off over the future years or added to the depreciable value of the new asset. For simplicity, in all problems discussed in this chapter an ordinary income tax rate of 50% and a capital gains tax rate of 30% will be used.

If no gains or losses are involved, we generally may make replacement studies on either a before-tax or an after-tax basis with equal validity.

It is evident that the existence of a taxable gain or loss, in connection with replacement, can have a considerable effect on the results. A prospective gain from the disposal of assets will be reduced by 30% in most cases. Correspondingly the net after-tax effect of a loss from the disposal of asset is substantially diminished when 30% or 50% of the loss is recovered through tax credit. Consequently, one's normal propensity for disposal or retention of existing assets can be influenced considerably by these tax effects.

The correct consideration of the income tax aspects of retirement and replacement situations may well result in conclusions exactly opposite the conclusions one may reach by intuition. For example, intuition often favors disposal at an apparent profit. However, a taxable gain reduces the net after-tax cash realized from such a disposal. Further, although intuition often runs counter to disposal of an asset at an apparent loss, a loss that is deductible from taxable income increases the net after-tax cash realized from such a disposal.

a typical replacement situation

The following is a typical replacement situation that will be used to illustrate a number of factors that must be considered in replacement studies.

EXAMPLE 11-1

Five years ago a company purchased pump A, including a driving motor, for $1,700. At the time of purchase it was estimated that it would have a useful life of

10 years and a salvage value at the end of that time of 10% of the initial cost. Depreciation has been charged in the accounting records on this basis, by use of the straight line procedure. Considerable difficulty has been experienced with the pump; annual replacement of the impeller and bearings has been required at a cost of $175. Normal operating and maintenance costs have been $325. Annual taxes and insurance are 2% of the first cost. It appears that the pump will continue to operate for another 10 years—the foreseeable demand period—if the present maintenance and repair practice is continued. It is estimated that if it is continued in service its ultimate salvage value will be about $20. An alternative to continuing the existing pump in service is to sell it and to purchase a new and different type of pump, B, for $1,600. A cash salvage value of $75 would be obtained for the existing pump A. A 10-year write-off period would be assigned to the new pump and an estimated salvage value would be 10% of the cost. Operating and maintenance costs for the new pump are estimated at $300 per year. Annual taxes and insurance would total 2% of first cost.

The company is in a 50% tax bracket on ordinary income, and it has a M.A.R.R. of about 12% on its capital before taxes and 6% after taxes.†

factors that must be considered in replacement studies

The problem of the pump illustrates most of the factors that must be considered in replacement studies. These are as follows:

1. Recognition and acceptance of past "error."
2. The possible existence of a sunk cost.
3. Remaining life of the old asset (defender).
4. Economic life for the proposed replacement asset (challenger).
5. Method of handling unamortized values.
6. Possible capital gains or losses.

Once a proper viewpoint has been established with respect to these items, little difficulty is encountered in making replacement studies and in arriving at sound decisions.

recognition of past "errors" and sunk costs

In the case of the existing pump in Example 11-1 it is apparent that several events have not turned out to be in accord with conditions assumed at the time the decision was made to purchase it. First, the pump has not performed as expected; maintenance has been abnormally high. Second, it appears that the salvage value has decreased more, and at a faster rate, than was anticipated and provided for by the depreciation accounting. By straight line depreciation

† This problem is solved later in this chapter.

accounting, the book value at the end of 5 years is $935.† Yet it appears that the company can get only $75 for the old pump. If it should sell the pump, there would be an unamortized value of $935 − $75 = $860.

Such differences frequently have been designated as past "errors." Such a designation is unfortunate because in most cases, as in this one, the differences are not the result of errors but of honest inability to foresee future conditions. The distinction is important in establishing a proper state of mind that will accept an important set of facts:

> Any unamortized values arising from an existing asset under consideration for replacement are strictly the result of *past* decisions—the initial decision to invest in that asset and decisions as to the method and number of years to be used for depreciation purposes. Unamortized values are *sunk costs* and thus have no relevance to the replacement decisions that must be made (*except to the extent that they result in a tax saving*).

Some have found that acceptance of these facts may be made easier by posing a hypothetical question: "What will be the costs of my competitor whose similar property is completely amortized and who therefore has no past 'errors' to consider?" In other words, we must decide whether we wish to live in the past, with its errors and discrepancies, or to be in a sound competitive position in the future. A common reaction is: "I can't afford to take the loss in value of the existing asset that will result if the replacement is made." The fact is that the loss already has occurred, whether or not it could be afforded, and it exists whether or not the replacement is made.

It must be remembered, of course, that the existence of an unamortized value or sunk cost may mean that there will be a resulting income tax saving that should be considered. This will be illustrated in the after-tax solution to Example 11-1.

the investment value of existing assets for replacement studies

Recognition of the nonrelevance of book values leads to the proper value to be placed on existing assets for replacement study purposes. Clearly, the value of an existing asset at any time is determined by either (1) what amount of money will be tied up in it if it is kept rather than disposed of, or (2) the present worth of any income that can be obtained in the future through its profitable use. For replacement study purposes the second measure of value is not relevant; we do not yet know whether the asset should be retained—a requirement for it to earn future profits. Instead, we wish to determine whether it should be retained or replaced. Consequently, it is certain that the *minimum* present value upon which capital recovery costs (depreciation and profit) can be based is the amount

---

† Book value = original cost − Σ depreciation charges
    = $1,700 − (5)($1,700 − $170)/10 = $935

of capital that will be tied up in the asset if it is retained, but that could be recovered immediately if it were replaced. Thus the present realizable salvage value (modified by any income tax effects to be an after-tax salvage value for after-tax studies) is the correct investment amount to be assigned to an existing asset in replacement studies. A good way to reason that this is true is to make use of the *opportunity cost* or *opportunity forgone* principle. That is, if it should be decided to keep the existing asset, one is giving up the opportunity to obtain the net realizable salvage value at that time. Thus this represents the *opportunity cost* of continuing to keep the defender.

There is one addendum to the rationale stated above: If any new investment expenditure (such as for overhaul) is needed to upgrade the existing asset so that it will be competitive in level of service with the challenger, the extra amount should be added to the present realizable salvage value to determine the total investment in the existing asset for replacement study purposes.

Application of the foregoing principle does not rule out the possibility that the true value of an existing asset, *as an operating property*, may turn out to be greater than the investment value assigned to it for economy study purposes. If this happens, then one can say that, on the basis of the investment value used for replacement study purposes, the asset turned out to earn a greater rate of profit than anticipated.

methods of handling losses due to unamortized values

There can be no question that, although unamortized values are sunk costs and have no place in replacement studies, they do cause considerable concern on the part of business owners and managers. Therefore, an understanding of how they are handled may provide some peace of mind to those making economy studies. Several methods have been and are used for dealing with unamortized values. Some are sound; others are incorrect.

One incorrect method is, in effect, a pretense that no unamortized loss has occurred. This is accomplished by using book value for the old equipment in the replacement study. People who do this usually justify their action by saying that if the old equipment is kept, they can continue to charge depreciation on it until the entire amount of capital that has been invested is recovered. Actually, such wishful thinking could only be realized if one had no competition. If a competitive situation exists, one's competitor may obtain the more economical new equipment and be able to sell at lower prices and thus prevent any further recovery of capital, inasmuch as capital can be recovered only through the sale of goods or services produced by the old equipment. It must always be remembered that the value of any property is determined by the profits that can be obtained through its use. Thus attempting to recover or prevent unamortized values in this manner is entirely erroneous.

A second incorrect method of handling unamortized values involves adding them to the cost of the challenger and then computing future depreciation costs, if the replacement is made, with this total. Users of this procedure believe that they can thus require the challenger to repay the unamortized value "that the replacement has caused." This line of reasoning is fallacious on two counts. First, the unamortized value is a fact, regardless of whether the replacement is made or not. Second, if one has competition, we probably will not be able to make the challenger pay the sunk cost. Again, a competitor who did not have an old machine, and thus had no unamortized value to recover, would have much lower costs with a new machine and could thus sell at a lower price. One's selling price usually is dictated by competition and not by one's desire to recover a sunk cost.

The correct handling of unamortized values requires that they be recognized for what they are—losses of capital. One of the most common procedures is to charge them directly to the profit and loss account. This considers them a loss in the current operating period and thus deducts them from current earnings. This is a satisfactory procedure except for the fact that unusually large losses will have a serious effect upon the stated accounting profits of the current period.

A more satisfactory method is to provide a surplus account against which such losses may be charged. Thus a certain sum is set aside each accounting

table 11-1
summary of given information for typical replacement problem (example 11-1)

Existing pump A (defender)		
Investment when purchased 5 years ago		$1,700
Estimated useful life when originally purchased		10 years
Estimated salvage value at end of 10 years		$170
Annual disbursements:		
Replacement of impeller and bearings	$175	
Operating and maintenance	325	
Taxes and insurance $1,700 × 2\%$	34	
TOTAL ANNUAL DISBURSEMENTS		$534
Present realizable salvage value		$75
Now it is estimated that this pump could last 10 more years (5 years beyond originally estimated life) and experience same annual disbursements		
Salvage value at end of 10 more years		$20
Replacement pump B (challenger)		
Investment now		$1,600
Estimated useful life		10 years
Estimated salvage value at end of 10 years		$160
Annual disbursements:		
Operating and maintenance	$300	
Taxes and insurance $1,600 × 2\%$	32	
TOTAL ANNUAL DISBURSEMENTS		$332
Income tax rate $= 50\%$		
M.A.R.R. $= 12\%$ before taxes and 6\% after taxes		

period to build up and maintain this surplus. When any losses occur from unamortized values, they are charged against the surplus account and current profits are not affected seriously. The periodic contribution to the surplus account is in the nature of an insurance premium to prevent profits from being affected by losses from unamortized values.

solution of the typical replacement problem

If applied to Example 11-1 before the effect on income taxes is considered, the investment amount in the existing pump for purposes of deciding whether it should be kept is the cash salvage value of $75. Note that this investment amount specifically ignores such amounts as the $1,700 original purchase price and the $935 present book value.

With the use of the principles that have been discussed thus far in this chapter, a before-tax and an after-tax solution of the pump replacement problem presented in Example 11-1 will be made. The data for this problem are summarized in Table 11-1 (p. 333).

Solution of Example 11-1 by E.R.R.R. Method (Before Taxes)

	Keep Old Pump A	Replacement Pump B
Annual expenses		
Disbursements	$534	$332
Depreciation:		
($75 − $20)($A/F$, 12%, 10)	3	
($1,600 − $160)($A/F$, 12%, 10)		82
TOTAL ANNUAL EXPENSES	$537	$414

Δ Total annual expenses (savings) = $537 − $414 = $123
Δ Investment = $1,600 − $75 = $1,525
E.R.R.R. on Δ investment = $123/$1,525 = 8.1%

Since 8.1% < 12% M.A.R.R., the replacement pump apparently is not justified.

Solution of Example 11-1 by A.C. Method (Before Taxes)

	Keep Old Pump A	Replacement Pump B
Annual costs:		
Disbursements	$534	$332
Capital recovery		
($75 − 20)($A/P$, 12%, 10) + $20(0.12)	12.14	
($1,600 − 160)($A/P$, 12%, 10) + $160(0.12)		274.08
TOTAL	$546.14	$606.08

Since $546.14 < $606.08, the replacement pump apparently is not justified. We could also make the analysis using other methods (e.g., I.R.R.) and the indicated choice would be the same. ∎

It now must be acknowledged that a before-tax analysis often is not a valid basis for a replacement decision because of the effect of any significant capital gain or capital loss (unamortized value) on income taxes. An after-tax analysis that is a valid basis for such decisions follows.

Solution of Example 11-1 by E.R.R.R. Method (After Taxes)

Table 11-2 shows computations of the after-tax cash flows for the existing and replacement pumps of Example 11-1. The values given may readily be developed by a two-step procedure. The first step is to determine the tax effects that will result from disposal of the old pump, given a tax rate of 50%, and the after-tax investment that would remain in the old pump if it is kept.

STEP 1: One-time tax effects

Book value of old pump, $1,700 − ($1,700 − $170)5/10 = $935.
Capital loss if sold, $935 − $75 = $860.
One-time tax recovery, $860 × 50% = $430.
After-tax investment in old pump, if kept, $430 + $75 = $505.

The second step is the determination of the *annual* tax effects in using the old and replacement pumps, as would be shown on the accounting and tax records.

STEP 2: Annual tax effects

	Keep Old Pump A	Replacement Pump B (all for 10 years)
Disbursements	$534	$332
Depreciation:		
(1700 − 170)/10	153 (5 years only)	
(170 − 20)/5	30 (last 5 years)	
(1600 − 160)/10		144
Taxable income reduction	$687 (first 5 years)	$476
	$564 (last 5 years)	
Annual tax credit (50%)	$343.50 (first 5 years)	$238.00
	$282.00 (last 5 years)	
Net after-tax annual disbursement:		
First five years	$534 − 343.50 = $190.50	$332 − $238 = $94
Last five years	$534 − 282.00 = $252.00	

It will be noted that the effect of income taxes is to reduce the actual annual out-of-pocket disbursements in each case.

Table 11-2 shows the computations above to determine after-tax cash flows in tabular form corresponding to that recommended in Chapter 10. Note in Table

table 11-2
after-tax cash flow computations for replacement alternatives in example 11-1

Year	(A) Before-Tax Cash Flow	(B) Depreciation	(C) = (A) + (B) Taxable Income	(D) = −(C) × Rate Cash Flow for Income Taxes at 50%	(E) = (A) + (D) After-Tax Cash Flow
Keep old Pump A					
0	(−)$ 75	—	(+)−$860ᵃ	(−)+$430	(−)$505
1-5	− 534	− $153ᵇ	− 687	+ 343.50	− 190.50
6-10	− 534	− 30ᵇ	− 564	+ 282.00	− 252.00
10	+ 20	—	—	—	+ 20
Replacement Pump B					
0	− $1,600	—	—	—	− $1,600
1-10	− 332	− 144ᶜ	− 476	+ 238	+ 94
10	+ 160	—	—	—	+ 160

ᵃ Capital loss forgone = [$1,700 − (1,700 − 170)(5/10)] − $75 = $935 − $75 = $860.

ᵇ Depreciation for first 5 years is based on original depreciation schedule; for the last 5 years a straight line charge of ($170 − $20)/5 = $30 per year is made.

ᶜ Depreciation = ($1,600 − $160)/10 = $144.

11-2 that the amounts for the alternative "Keep old pump A" on the line for year 0 have reversed signs in parentheses. This is because, as a computational convenience, the amounts first were determined on the basis of selling or *replacing* the existing pump A. The reversed signs in parentheses indicate the opposite effect, which would result from *keeping* pump A.

Also shown in Table 11-2 are the after-tax investment in the replacement pump B and the after-tax salvage values for both alternatives. (*Note:* Because the salvage values are estimated to equal the respective book values at the end of the life for each, there is no capital gain or loss and hence no tax cost or savings at those times.)

After the one-time and annual tax effects have been determined, either by the two-step procedure shown above or through a tabulation of cash flows as in Table 11-2, the third step in an after-tax replacement study is the economy study, which basically is no different from any other study of alternatives except, of course, that an after-tax minimum attractive rate of return is used. The following is the after-tax E.R.R.R. analysis for Example 11-1.

	Keep Old Pump A	Replacement Pump B
Annual expenses		
Disbursements:		
$190.50 for first 5 years ⎫		
$252.00 for last 5 years ⎬	$216.86	$ 94.00
Annual equivalent at 6% ⎭		
Depreciation:		
($505 − 20)(A/F, 6%, 10)	36.80	
($1,600 − 160)(A/F, 6%, 10)		109.30
TOTAL ANNUAL EXPENSES	$253.66	$203.30

Δ Total annual expenses (savings) = $253.66 − $203.30 = $50.36
Δ Investment = $1,600 − $505 = $1,095
E.R.R.R. on Δ investment = $50.36/$1,095 = 4.6%

Since 4.6% < 6% M.A.R.R., the extra investment in the replacement pump is shown to be not justified. The following is an after-tax annual cost analysis for the same example.

	Keep Old Pump A	Replacement Pump B
Annual costs		
Disbursements		$ 94.00
$190.50 for first 5 years ⎫		
$252.00 for last 5 years ⎬	$216.86	
Annual equivalent at 6% ⎭		
Capital recovery:		
($505 − $20)(A/P, 6%, 10) + 20(0.06)	67.11	
($1,600 − $160)(A/P, 6%, 10) + 160(0.06)		205.30
TOTAL	$283.97	$299.30

Because $283.97 < $299.30, the extra investment in the replacement pump is again shown to be not justified. However, the economic advantage of keeping the old pump is marginal and other considerations (e.g., reliability of operation) may swing the decision in favor of purchasing a replacement pump B.

Although the answer described is consistent with the results of the before-tax analysis for the same problem, the consistency would not necessarily always be true, particularly for analyses in which one or more alternatives involve large capital gains or capital losses taxed at rates markedly different from the rates on ordinary income. ∎

We will now describe how to determine economic lives of depreciable assets and how economic analyses should be performed when the economic life of the defender differs from that for the challenger. Most of the examples illustrating these principles do not explicitly include income taxes. However, the principles used in determining economic lives apply on an after-tax basis whenever income taxes are applicable.

determining the most economic life of a new asset or challenger

In all economy studies up to this point, we have used assumed economic lives for all new alternatives under consideration, including challengers in replacement analyses. For any new alternative, if the various operating and maintenance costs are known along with its year-by-year salvage values, the most economical useful life may be computed. Example 11-2 illustrates a set of computations for determining the years of life at which the equivalent uniform annual cost (E.U.A.C.) is a minimum.

EXAMPLE 11-2

A new fork truck will require an investment of $20,000 and is expected to have year-end salvage values and annual out-of-pocket disbursements as shown in columns 2 and 5, respectively, of Table 11-3. If capital is worth 10%, how long should the asset be retained in service?

Solution

The solution to this problem is obtained by completing columns 3, 4, 6, and 7 of Table 11-3. In the solution the customary year-end occurrence of all events is assumed. The depreciation for any year is the difference between the beginning and year-end salvage values. The opportunity cost of capital is 10% of the capital remaining invested in the asset at the beginning of each year. The values in column 7 are the equivalent, uniform annual costs that would be incurred each year if the asset were retained in service for the number of years shown and replaced at the end of the year.

From the values shown in column 7 it is apparent that the asset will have minimum annual cost if it is kept in service only 3 years. ∎

table 11-3
determination of the most economic life of a new asset (challenger) (example 11-2)

| (1) | (2) | Cost of Service for Nth Year | | | | Equivalent Uniform Annual Cost if Kept N Years |
| | | (3) | (4) | (5) | (6) | (7) |
Year, N	Salvage Value at Year End	Depreciation for Year	Cost of Capital at 10%	Annual Disbursements	$= [(3) + (4) + (5)]$ Total Cost for Year	$= \sum [(6)_1(P/F, 10\%, 1) + \cdots + (6)_N(P/F, 10\%, N)](A/P, 10\%, N)$
0	$20,000	—	—	—	—	—
1	15,000	$5,000	$2,000	$ 2,000	$ 9,000	$9,000
2	11,250	3,750	1,500	3,000	8,250	8,637
3	8,500	2,750	1,130	4,620	8,500	8,600 ←minimum
4	6,500	2,000	850	8,000	10,850	9,082
5	4,750	1,750	650	12,000	14,400	9,965

The computation approach in the example above, as shown in Table 11-3, was to determine the total annual cost for each year (sometimes called *marginal* or *year-by-year cost*) and then to convert these into an E.U.A.C. The E.U.A.C. for any life can also be calculated using the more familiar capital recovery formulas. For example, for a life of 2 years, the E.U.A.C. can be calculated as

$(\$20,000 - \$11,250)(A/P, 10\%, 2) + \$11,250(10\%)$

$\qquad + [\$2,000(P/F, 10\%, 1) + \$3,000(P/F, 10\%, 2)](A/P, 10\%, 2) = \$8,637$

which checks with the corresponding row in column 7 of Table 11-3.

the remaining economic life of a defender

In replacement economic analyses, one problem is to select the remaining economic (or service) life that is most favorable to the defender. When a major outlay for defender alteration or overhaul is needed, the remaining defender life that will yield the least annual cost is likely to be the period that will elapse before the next major alteration or overhaul will be needed.

When there is no defender salvage value now or later (and no outlay for alteration or overhaul), and when defender operating disbursements are expected to increase annually, the remaining life that will yield the least annual cost will be 1 year (or possibly less).

When salvage values are expected to decline from year to year, it may be necessary to calculate the "apparent" remaining economic life. This is done in the same manner as in Example 11-2 for a new asset. The only difference is that the present realizable salvage value is considered to be the same as the initial investment for a new asset.

Regardless of how the "apparent" economic remaining life for the defender is determined, whenever a decision is made to keep the defender this does not mean that it should be kept only for this period of time. Indeed, the defender should be kept longer than the "apparent" economic life as long as its marginal cost (total cost for an additional year of service) is less than the minimum E.U.A.C. for the best challenger.

EXAMPLE 11-3

Suppose that it is desired to determine how much longer an old fork truck should remain in service before it should be replaced by the new fork truck (acting as a challenger) for which data were given in Example 11-2 and Table 11-3. The old truck (defender) is 2 years old, originally cost $13,000, and has a present realizable salvage value of $5,000. If kept, its salvage value and operating disbursements are expected to be as follows:

Year from Now	Salvage Value	Operating Disbursements
1	$4,000	$5,500
2	3,000	6,600
3	2,000	7,800
4	1,000	8,800

Determine the most economical period to keep the old truck before replacing it (if at all) with the present challenger.

Solution

Table 11-4 shows the calculation of total cost for each year (marginal cost) and E.U.A.C. for each year for the defender using the same format as that used in Table 11-3. Note that the minimum E.U.A.C. of $7,000 corresponds to keeping the old truck for 1 more year. However, the marginal cost for keeping the truck for the second year is $8,000, which is still less than the minimum E.U.A.C. for the challenger (i.e., $8,600, from Example 11-2). The marginal cost for keeping the defender the third year and beyond is greater than the $8,600 minimum E.U.A.C. for the challenger, so it would be most economical to keep the defender for 2 more years and then to replace it with the challenger. Figure 11-1 (p. 342) shows this. ∎

The example above assumes that there is only one new asset (challenger) alternative available. It exhibits the general relationship that if the old asset (defender) is retained beyond the point that its marginal (year-by-year) costs exceed the minimum E.U.A.C. for the challenger, the difference in costs continues to grow and replacement becomes more urgent, as illustrated to the right of the intersection in Figure 11-1.

Figure 11-2 (p. 343) illustrates the effect of improved new challengers in the future. If an improved challenger X becomes available before replacement with the new asset of Figure 11-1, then a new replacement study probably should take place to consider that improved challenger. If there is a possibility of a further-improved challenger Y as of, say, 4 years later, it may be better still to postpone

table 11-4
determination of the most economic life of an old asset (defender)

(1) Year from Now	(2) Depreciation for Year	(3) Cost of Capital at 10%	(4) Operating Disbursements	(5) Total Cost for Year (Marginal Cost)	(6) Equivalent Uniform Annual Cost
1	$1,000	$500	$5,500	$ 7,000	$7,000
2	1,000	400	6,600	8,000	7,475
3	1,000	300	7,800	9,100	7,966
4	1,000	200	8,800	10,000	8,406

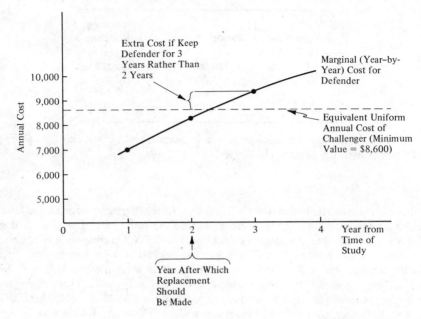

Figure 11-1.   Defender versus challenger fork trucks (based on Examples 11-2 and 11-3).

replacement until that challenger becomes available. Thus retention of the old asset beyond the breakeven point has a cost that may well grow with time, but this cost of waiting can, in some instances, be worthwhile if it permits purchase of an improved asset having economies that offset the cost of waiting. Of course, a decision to postpone a replacement may "buy time and information" also. Because technological change tends to be sudden and dramatic rather than uniform and gradual, new challengers with significantly improved features can arise sporadically and can change replacement plans substantially.

When replacement is not signaled by the economy study, more information may become available before the next "challenge" to the defender, and hence the next comparison should include that additional information. *Postponement* generally should mean a postponement of the decision "when to replace," not the decision to postpone replacement until a specified future date.

economic comparisons in which defender remaining life differs from challenger useful life

When the lives of the replacement alternatives are equal, as in Example 11-1, economic comparisons of alternatives can be made by any of the theoretically correct methods without complication. However, when the expected economic lives of the defender and challenger(s) differ, this should be taken into account and the analysis may be complicated.

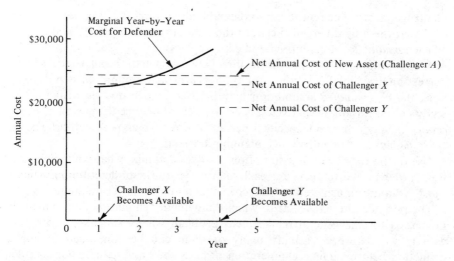

Figure 11-2. Old versus new asset costs with improved challengers becoming available in the future.

Chapter 7 described the two types of assumptions that are commonly used for economic comparisons of alternatives having different lives: (1) *repeatability*, and (2) *coterminated* assumptions. The repeatability assumption involves two main subassumptions as follows:

1. The period of needed service for which the alternatives are being compared is either indefinitely long or a length of time equal to a common multiple of the lives of the alternatives.
2. What is estimated to happen in the first life cycle will happen in all succeeding life cycles, if any, for each alternative.

For replacement analyses, the first subassumption may be acceptable, but normally the second subassumption is not reasonable for the defender. The defender is typically an older piece of equipment with a modest current realizable salvage value (selling price). An identical replacement, even if it could be found, probably would have an installed cost far in excess of the current salvage value of the defender.

Failure to meet the second subassumption can be circumvented if the period of needed service is assumed to be indefinitely long and if we recognize that the analysis is really to determine if *now* is the time to replace the defender. When the defender is replaced, it will be by the challenger—the best available replacement.

Example 11-3 involving the defender versus challenger fork trucks made use of the *repeatability* assumption. That is, it was assumed that the challenger would have a minimum E.U.A.C. regardless of when it replaces the defender. Figure 11-3 shows time diagrams of the cost consequences of keeping the defender for the optimal 2 remaining years versus adopting the challenger now, with the challenger costs to be repeated into the indefinite future. It can be seen in Figure 11-3 that the only difference between the alternatives is in years 1 and 2.

The *repeatability* assumption applied to replacement problems simplifies their economic comparison by permitting direct comparison of the alternatives by any correct analysis method. Whenever the *repeatability* assumption is not applicable, the *coterminated* assumption may be used, which involves using a finite study period for all alternatives. This study period may be the period of needed service or any arbitrarily specified period of time, which is sometimes descriptively called the "organization's planning horizon."

Use of the *coterminated* assumption involves detailing what, and when, cash flows are expected to occur for each alternative and then determining which is most economical, using any of the correct economic analysis methods.

The possible alternatives as to when the defender should be replaced can be numerous for the general replacement problem when using the *coterminated* assumption. However, usually many of them can be eliminated by logical deduction before detailed enumeration of cash flows and actual analysis calculations are performed.

Example 11-4

Suppose that we are faced with the same replacement problem as in Example 11-3 except that the period of needed service (or the firm's planning horizon) is (a) 3 years, and (b) 4 years.

Figure 11-3. Effect of *repeatability* assumption applied to alternatives in Example 11-3.

Keep Defender for Two Years:

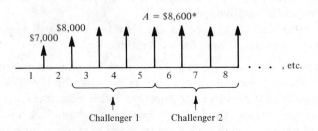

Replace with Challenger Now:

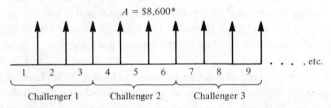

* The E.U.A.C. of $8,600 for Challenger is the result of marginal (year-by-year) costs of $9,000, $8,250, and $8,500 for years 1, 2, and 3 of each life cycle, respectively. From Table 11-3 the economic life of the Challenger was calculated to be 3 years.

Solution

(a) For a planning horizon of 3 years, one might intuitively think that either the defender should be kept for 3 years or it should be replaced immediately by the challenger to serve the next 3 years. From Table 11-4 the E.U.A.C. for the defender for 3 years is $7,966, and from Table 11-3 the E.U.A.C. for the challenger for 3 years is $8,600. Thus, following this reasoning, the defender would be kept for 3 years. However, this is not quite right. Focusing on the "Total Cost for Year" (column 5 of each table), one can see that the defender has the lowest cost in the first 2 years, but in the third year its cost is $9,100, whereas the cost of the first year of service for the challenger is only $9,000. Hence it would be economic to replace the defender after the second year.

This conclusion can be confirmed by enumerating all replacement possibilities and their respective costs and then computing the E.U.A.C. for each, as will be done for the 4-year planning horizon in part (b).

(b) For a planning horizon of 4 years, the alternatives and their respective costs for each year and the E.U.A.C. of each are given in Table 11-5.

table 11-5
determination of most economic life of defender for a planning horizon of 4 years

Keep Defender for:	Keep Challenger for:	Total (Marginal) Costs for Each Year				E.U.A.C. at 10% for 4 Years
		1	2	3	4	
0 years	4 years	$9,000	$8,250	$8,500	$10,850	$9,082
1	3	7,000	9,000	8,250	8,500	8,301
2	2	7,000	8,000	9,000	8,250 (min)→	8,005
3	1	7,000	8,000	9,100	9,000	8,190
4	0	7,000	8,000	9,100	10,000	8,406

Thus the most economic alternative is to keep the defender for 2 years and the challenger for 2 years. The decision to keep the defender for 2 years happens to be the same as when the repeatability assumption was used, but of course this would not be necessarily true in general. ∎

replacement versus augmentation

Conditions frequently arise wherein an existing asset is not of sufficient capacity to meet increased demands that have developed. In such cases there usually are two alternatives open: (1) to augment the existing asset, or (2) to replace it with a new asset of sufficient capacity to meet the new demands. A case of this kind occurs in the following example.

EXAMPLE 11-5

A steel highway bridge must be either reinforced or replaced. It cost $80,000 when built 28 years ago and has a present book value of $50,000. If the bridge is replaced now, the scrap value of the steel would exceed the demolition cost by $7,000. Reinforcement would cost $22,000 and would make the steel bridge adequate for another 10 years service, at which time the net salvage value would be $9,000.

A new prestressed concrete bridge would cost $203,000 and would meet foreseeable requirements for the next 40 years and have a $3,000 salvage value. It is estimated that the annual maintenance costs of the steel bridge would exceed those of the concrete bridge by $7,000. Assuming that money invested by the State Highway Department is worth 5%, which alternative should be chosen based on an annual cost analysis and using the *repeatability* assumption?

Solution

	Reinforce Existing Steel	Replacement Concrete
Depreciation:		
($7,000 + $22,000 − $9,000)(*A/F*, 5%, 10)	$ 1,590	
($203,000 − $3,000)(*A/F*, 5%, 40)		$ 1,660
Interest on capital:		
($7,000 + $22,000)(5%)	1,450	
$203,000(5%)		10,150
Extra annual maintenance costs	7,000	
TOTAL ANNUAL COSTS	$10,040	$11,810

Thus it is found that reinforcing the existing steel bridge is more economical. The same results would be obtained if some other standard method of economic analysis were used. Note that income taxes were not relevant for this study because it applied to a governmental agency. ∎

The example above will now be altered so that the *coterminated* assumption is applicable.

EXAMPLE 11-6

Suppose that in Example 11-5 the services of a bridge is needed only, say, for 40 years. Further, suppose that if the concrete bridge replaces the reinforced steel bridge after 10 years, it is expected to require an investment of $240,000 (rather than $203,000), and to have a salvage value of $15,000 (rather than $3,000) after 30 years life, which would be at the end of 40 years needed service. Show which alternative is best through before-tax analysis using (a) the P.W.-C. method, and (b) the A.W.-C. method.

Solution by P.W.-C. Method

This method is practicable because there is a 40-year study period for each alternative.

P.W.-C.

Reinforcing steel bridge, then replacing:
    Investment in reinforced steel: $7,000 + $22,000 $ 29,000
    Annual maintenance: $7,000$(P/A, 5\%, 10)$ 54,000
    Less: salvage for reinforced steel: $9,000$(P/F, 5\%, 10)$ −   5,520
    Investment in concrete: $240,000$(P/F, 5\%, 10)$ 147,500
    Less: salvage for concrete: − $15,000$(P/F, 5\%, 40)$ −   2,130

        TOTAL P.W.-C. $222,850

Replacing with concrete immediately:
    Investment $203,000
    Less: salvage: − $3,000$(P/F, 5\%, 40)$ −   425

        TOTAL P.W.-C. $202,675

Thus, with the changed conditions, it is found to be more economical to replace with the concrete bridge immediately. Note that this is a reversal of the results of Example 11-5, which was based on the repeatability assumption.

Solution by Annual Cost (A.C.) Method

The easiest means to determine the A.C., once given the P.W.-C., is to convert directly using the factor $(A/P, i\%, N)$.

Reinforcing steel, then replacing:
        A.C. = $222,850$(A/P, 5\%, 40)$ = $13,000

Replacing with concrete immediately:
        A.C. = $202,675$(A/P, 5\%, 40)$ = $11,810

It is not surprising that immediate replacement with a concrete bridge is also shown to be the best choice by the A.C. method. ▮

Example 11-7 provides another case involving augmentation versus replacement and includes the consideration of income taxes.

EXAMPLE 11-7

Five years ago a construction company purchased an air compressor for $2,000. Its life was estimated to be 10 years, with a salvage value of 10% of first cost. Annual out-of-pocket disbursements for operation, maintenance, taxes, and insurance are running at approximately $1,050. The demands for compressed air have increased to the point where either an additional small compressor, costing $1,000, will have to be added, or the old compressor will have to be replaced with a larger compressor, which will cost $4,000. It is estimated that the new, small compressor would have annual disbursements of $300, and that the annual disbursements for the larger, more modern compressor would be $500. The new small (augmenting) unit would have a 5-year life and a $550 salvage value. Even though depreciation charges for the large unit will be based on a 10-year life and an assumed salvage value of $400, it is expected that the actual salvage value of the large unit at the end of the 5-year study period (or period of needed service) would be $2,000. If the existing compressor is replaced, it will bring a present salvage value of $200. If the old compressor is used for another 5 years, it is believed that its salvage value at that time will not exceed $25.

The company uses straight line depreciation in its accounting, and it earns about 20% on its capital before income taxes and about 10% after taxes, being in a 50% income tax bracket. Should the existing compressor be augmented or replaced based on an after-tax annual cost analysis and a 5-year study period?

Solution

Table 11-6 shows after-tax cash flow computations for Example 11-7. The following is an explanation of some of the main results shown in Table 11-6.

If the existing compressor is replaced, there would be a book loss of $900, inasmuch as its current book value is $1,100 and only $200 would be received as salvage value. This would produce an income-tax savings of $450, assuming a 50% tax rate. Hence the net after-tax cash flow at year 0 if the old compressor is sold would be + $200 salvage value + $450 taxes saved = + $650. If the old compressor is kept, not sold, the opposite happens, which is reflected by the changed signs in parentheses. Thus, if the old compressor is kept, there is a net investment *opportunity cost* (reflected as a cash flow) of − $650. This, together with the − $1,000 for augmentation, results in a net after-tax cash flow for an investment of − $1,650.

If the existing compressor is retained, and the depreciation adjusted to write off the difference between the present book value of $1,100 and the final salvage of $25, there would be a tax-free recovery of $25 when it is disposed of after 5 years. Similarly, if the small (augmenting) unit is depreciated so that its book value at the end of 5 years equals its salvage value of $550, there is no capital gain or capital loss at that time, so the after-tax recovery (cash inflow) would be + $550.

The new compressor would have a book value of $2,200 at the end of 5 years. If it were sold at the estimated resale price of $2,000, there would be a tax loss of $200, providing a tax credit of $100. Thus, on an after-tax basis, there would be an incoming cash flow of $2,100 from the disposal of the compressor.

By using the results of Table 11-6, the following is the after-tax annual cost comparison.

	Augment Existing Compressor	New Replacement Compressor
Annual expenses		
Disbursements	$522	$ 70
Depreciation:		
($1,650 − $575)($A/F$, 10%, 5)	176	
($4,000 − $2,100)($A/F$, 10%, 5)		311
Minimum required profit:		
$1,650(10%)	165	
$4,000(10%)		400
TOTAL ANNUAL COST	$863	$781

Thus the new larger compressor is shown to be the more economic choice. ∎

table 11-6
after-tax cash flow computations for replacement versus augmentation alternatives in example 11-7

	Year	Before-Tax Cash Flow	Depreciation	Taxable Income	Cash Flow for Income Taxes at 50%	After-Tax Cash Flow
Augment Old Compressor:						
Old	0	(−)$ 200		(+)−$ 900[a]	(−)$450	(−)$ 650
Augmented	0	− 1,000				− 1,000
TOTAL		− 1,200				− 1,650
Old	1–5	− 1,050	−215[b]	− 1,265	+ 633	− 417
Augmented	1–5	− 300	− 90[c]	− 390	+ 195	− 105
TOTAL		− 1,350				− 522
Old	5	+ 25				+ 25
Augmented	5	+ 550				+ 550
TOTAL		+ 575				+ 575
New Larger Compressor:						
	0	− 4,000				− 4,000
	1–5	− 500	−360[d]	− 860	+ 430	− 70
	5	+ 2,000		− 200[e]	+ 100	+ 2,100

[a] Book loss = $2,000 − ($2,000 − $200)5/10 − $200 = $900.
[b] Depreciation = ($1,000 − $25)/5 = $215.
[c] Depreciation = ($1,000 − $550)/5 = $90.
[d] Depreciation = ($4,000 − $400)/10 = $360.
[e] Book loss = $4,000 − ($4,000 − $400)5/10 − $2,000 = $200.

retirement without replacement

In some cases it may be decided that although an existing asset is to be retired from its current use, it will not be replaced or removed from all service. Although it may not be able to compete economically in its existing usage, it may be desirable and even economical to retain it as a standby unit or for some different use. The cost to retain it under such conditions may be quite low, owing to its relatively low present realizable resale value and perhaps low costs for operation and maintenance. There often are income tax considerations that also bear on the true cost of retaining the asset. This type of situation is illustrated in the following example.

EXAMPLE 11-8

What would be the cost of retaining, as a standby unit for 5 years, the existing compressor discussed in Example 11-7? Assume that the before-tax annual cost for operation and maintenance as a standby would be $155.

Solution

First, it is evident that if the compressor is retained, the company will forego an after-tax recovery of $650, which would have resulted from its disposal, and will receive, at the end of 5 years, $25 from its ultimate disposal. Thus the annual cost of depreciation and minimum profit would be ($650 − $25)($A/P$, 10%, 5) + $25(10%) = $167.

The $155 for operation and maintenance plus the $215 depreciation charge result in a total annual reduction in taxable income of $370. At the 50% rate, taxes would be reduced by $185. Thus the net effect is to have an annual after-tax saving of $185 − $155 = $30.

Thus the total equivalent annual cost of keeping the old compressor on standby service for 5 years is $167 − $30 = $137. ∎

short-term replacement due to decreasing efficiency

There are some situations where the economic replacement period is quite short —usually less than a year—for which a simplified form of analysis is satisfactory for the purposes involved. For such cases interest is neglected and average values used, in place of equivalent costs. One example of this type is the case of special tooling that has a rather short economic life, because wear causes an increase in costs for rejects and repairs. The following is such an example.

EXAMPLE 11-9

Certain special tooling costs $500. The number of defects resulting with its use increases each month, requiring costly reworking. The resulting figures are shown in Table 11-7. How frequently should the tooling be replaced?

table 11-7

cost analysis for example 11-9—determination of the most economical replacement period for special tooling that decreases in efficiency

(1) Month Number	(2) Tooling Cost	(3) Cost of Rework Required During Month	(4) Total Cost for Month	(5) Total Cost to End of Month, Σ (4)	(6) Average Monthly Cost to End of Month [(5) $\div$ (1)]
1	$500	$ 25	$525	$ 525	$525
2	—	75	75	600	300
3	—	175	175	775 (min)→	258
4	—	300	300	1,075	269
5	—	450	450	1,525	305

Solution

It is noted that minimum cost will be obtained by replacing the tooling at the end of 3 months. Including interest and computing equivalent uniform values would not change the results appreciably. More important in such situations is the very practical matter of the length of time the asset will be used. The foregoing analysis assumes that the tooling will be needed indefinitely, or a length of time that is a multiple of 3 months. Quite clearly, if the tooling were only needed for 4 months, the theoretical 3-month replacement cycle would not be economical. ∎

replacement where present salvage value is zero or unknown

Sometimes replacement studies must be made where salvage or market values of existing assets are zero or very uncertain. Such cases basically are no different from those discussed previously if we remember that the purpose of the study is to determine whether the existing asset can compete, economically, under the most favorable conditions that realistically can be assigned to it. The following example, made on a before-tax basis, illustrates this principle.

EXAMPLE 11-10

In a certain process in an oil refinery, several types of crude stock are put through a particular piece of equipment. When a change is made from one stock to another, this portion of the plant must be shut down so that the equipment can be cleaned before the next stock is processed. This cleaning requires 20 hours, during which no production is obtained from this portion of the refinery. A manufacturer has developed a new piece of equipment that makes it possible to change from one stock to another in 2 hours.

The management estimates that it costs them $50 for each hour this portion of the refinery is not in operation. The existing equipment was installed 2 years previously at a cost of $20,000. Depreciation has been figured by the straight line method, based upon an expected life of 5 years. The new equipment would cost $32,000. Although it would undoubtedly last at least 5 years, the refinery managers have decided that, because of recent rapid changes in the industry, any new equipment at this time must be fully depreciated within 2 years.

During the past 2 years the system has been changed from one stock to another twice each month. This is expected to continue. Capital is worth 8% before taxes. Should the change in equipment be made at this time?

Solution

Obviously, the most favorable conditions for the old equipment would be those in which the depreciation charge assessed against it as a cost of operation would be zero. If it cannot compete with the new equipment under these conditions, it is apparent that it cannot compete when its operating costs are increased by the addition of depreciation expense. Thus the before-tax replacement study would be made as follows:

Old equipment:
    Depreciation $ 0
    Cost of making stock changes: $20 \times \$50 \times 24$ 24,000

        TOTAL COST PER YEAR $24,000
New equipment:
    Depreciation: $32,000(A/F, 8\%, 2)$ $15,385
    Cost of making stock changes: $2 \times \$50 \times 24$ 2,400

        TOTAL COST PER YEAR $17,785
    Annual savings = $24,000 - $17,785 = $6,215
    E.R.R.R. = $6,215/$32,000 = 19.4\%

Since 19.4% > 8%, it is clear that the old equipment should be replaced. ∎

If the study had shown that the return on the required investment under these conditions was not sufficient to justify making the investment, or if the annual operating costs of the old equipment were less than for the new equipment, this would have meant that *as an operating unit*, in competition with the new equipment, the value of the old equipment was greater than zero.

replacement where salvage value is greater than book value

Occasionally, a replacement study involves an existing asset that is found to have actual salvage value greater than the book value. An extreme form of this situation is the complete amortization of the value of the old asset, resulting in a zero book value. Under such conditions, it might be argued that no problem exists concerning the equipment, since the invested capital has been fully recovered and it can therefore be disposed of without loss. Actually, however, such is not the case. Obviously, if the existing asset is sold for an amount greater than the book value, there will be a gain from the sale of a depreciable asset, and this gain is subject to a gains tax. Also, it is possible that there may be a loss, either in connection with the replacement being considered or from some prior transaction during the year. Consequently, it is best that such replacement studies be made on an after-tax basis. The following example is of this type.

EXAMPLE 11-11

A small factory, making a single product, has utilized a casting as the main body of the article. This is the only casting in the product. It has been making its own castings, having a small electric furnace and other necessary foundry equipment. The furnace cost $4,000 when new 4 years ago. Straight line depreciation has been used to write off this equipment over a 10-year period. The other foundry equipment has been entirely written off the books. The owners now find that the casting can be eliminated by welding rolled shapes. The resulting saving in labor, material, maintenance, and power will be $0.50 per article. The annual output is 600 machines. The necessary welding equipment will cost $2,000 and will have an estimated life of 8 years. It is estimated that the existing equipment will have zero salvage value after 6 years, and the welding equipment will have zero salvage value after 8 years. The electric furnace can be sold now for $2,000 and the remainder of the foundry equipment can be sold now for $700. The company is in the 50% tax bracket and has a M.A.R.R. on its capital of about 20% before taxes and 10% after taxes. Determine if the change in methods should be made based on an after-tax annual cost comparison using the *repeatability* assumption.

Solution

Table 11-8 (p. 354) details the computation of after-tax cash flows. The following is an explanation of several of the most important inclusions. It is evident that a recovery (gain) of $700 in previously charged depreciation can be obtained from the sale of the miscellaneous foundry equipment, and a $400 long-term loss from the disposal of the electric furnace, since the book value is $2,400. Because the loss and the gain must be offset against each other, there would be a net gain of $300. Since this net gain is the result of the recovery of previously charged depreciation, it should be taxed at 50%, which is $150. Consequently, the net one-time after-tax recovery from the disposal of existing assets would be $2,700 − $150 = $2,550. Reversal of signs (shown in parentheses) indicates the effect if the existing equipment is kept rather than disposed.

Using straight line depreciation for accounting purposes, we find that the annual reduction in taxable income with the existing equipment would be $400 for depreciation and $300 for excess production costs, a total of $700 per year. If the new welding method is used there would be a depreciation charge of $250 per year. As for previous after-tax examples, these reductions in annual taxable income save taxes and can be combined with the respective before-tax cash flows to determine the after-tax cash flows.

table 11-8
after-tax cash flow computations in example 11-11

	Year	Before-Tax Cash Flow	Depre- ciation	Taxable Income	Cash Flow for Income Taxes	After-Tax Cash Flow
Existing	0	(−)$2,700		(−)$300	(+)−$150	(−)$2,550
equipment	1–6	− 300	−$400^b	− 700	+ 350	+ 50
	6	0				0
Welding	0	− $2,000				− $2,000
equipment	1–8		−$250^c	− $250	+ $125	+ 125
	8	0				0

^{a} For electric furnace: book value = $4,000 − 4($4,000/10) = $2,400; capital loss = $2,400 − $2,000 = $400. Net gain on disposal of electric furnace and remainder of foundry equipment = $700 − $400 = $300.
^{b} Depreciation: $4,000/10 = $400.
^{c} Depreciation $2,000/8 = $250.

The after-tax annual cost comparison utilizing the results of Table 11-8 is

	Existing Equipment	Welding Equipment
Depreciation + minimum profit		
($2,550 − $0)($A/P$, 10%, 6)	$585	
($2,000 − $0)($A/P$, 10%, 8)		$375
Annual operation	− 50^a	− 125^a
TOTAL ANNUAL COSTS	$535	$250

^{a} The annual operation results are cash savings and hence are subtracted from other costs.

Thus the new welding equipment is shown to be economically advantageous. ▮

miscellaneous considerations in replacement decisions

There are six factors, any or all of which may be important in replacement decisions. The *first* is the possibility that if replacement is deferred, assets available in the future may be improvements over those presently available. When technology is changing rapidly, we hesitate to acquire an asset that very soon may be obsolete economically. Consequently, in making replacement decisions, a knowledge of the state and trend of technological development in the area involved is important. Knowledge that important improvements or break-throughs are expected will tend to make us delay replacement, in the hope of being able to take advantage of superior assets.

The *second* factor relates to probable variations in the future salvage value of an asset. If an economy study shows that the most economical disposal date is a considerable number of years distant, moderate changes in the assumed disposal value are not likely to make any substantial change in the calculated date. On the other hand, if only a few years are involved, prospective changes in the salvage value can have a considerable effect on the year-to-year economy.

Where the trade-in or disposal value is changing rapidly, we must be certain that a sufficient number of years is considered in determining the economy of the old asset. For example, on the basis of keeping the old asset 1 more year, it might appear to be more economical to make the replacement now. If, however, a 3- or 4-year retention period is used, in order to gain the smaller amounts of depreciation during the additional years, the old asset might be more economical than a new one. Thus, in general, slow rates of decrease in salvage value, with increasing operation and/or maintenance costs, tend to favor early replacement, whereas rapid rates of decrease in salvage value favor later replacement.

A *third* factor is the probability that the enterprise will grow in the immediate future. If it is evident that such growth will require increased output unobtainable from the existing asset, it is obvious that replacement or augmentation will have to be made in the near future. The resulting problem is basically to determine whether replacement or augmentation is cheaper, and at what time either should occur. The only complicating factor is the possibility that the desired equipment may not be available when needed if replacement is deferred, or that the increased demand may develop earlier than anticipated. To avoid these possibilities, companies sometimes prefer to make the replacement somewhat earlier than necessary and then attempt to develop the market.

A *fourth* factor is somewhat counter to the previous one. Replacement frequently results in excess capacity, since new machines usually are more efficient than those they replace. If there is no actual use for such excess capacity, it has no value and thus should not be given any consideration. On the other hand, if the excess capacity can be used for some function not rendered by the old asset, this should be considered in the replacement study. This can easily be done by deducting the annual net profit *after income taxes* derived from the added service from the annual costs for the new machine.

The possibility that the purchase price of the replacement asset will change in the future is a *fifth* factor that must be considered. Although it is difficult to predict what price changes will occur in the future, the long history of inflation that has occurred causes many to believe that the trend is apt to continue. Such a condition almost always favors earlier replacement, which decreases the initial cost of the new asset. However, this practice should not be followed blindly, particularly where the need for the service is expected to continue for many years. Two additional facts must also be considered. First, the earlier the replacement, the sooner the new asset will have to be replaced—at higher first cost. Second, it is likely that technological improvements will occur, and by deferring replacement it is possible that a more efficient asset may be obtained. Therefore, before too much emphasis is given to probable price increases, other possible and probable future changes also should be considered.

Budgetary and personnel considerations are a *sixth* factor that frequently affects replacement decisions. Most companies do not have unlimited funds, and many projects are usually competing for the funds that they do have. In such cases replacement studies supply information on the basis of which decisions can be made regarding the timing of replacements as funds become available.

Similarly, in the case of governmental agencies, where financing is done through bond issues, replacements may have to be timed in accordance with bond market conditions. In some instances a governmental unit may have reached the limit of its legal indebtedness, and replacements must wait until further financing can be done.

More and more replacements are timed with due consideration to personnel factors. Replacement equipment frequently involves a temporary reduction in labor requirements, or a change in skills required. So that undesirable layoffs of working personnel or unnecessary tensions among employees will be avoided, replacements are timed to coincide with upswings in business activity; thus temporarily displaced workers may be retained for other activities. The usual turnover in working force ordinarily can be counted on to absorb a temporary excess of personnel within a short period of time.

determination of value by replacement theory

As was pointed out in Chapter 9, it frequently is necessary to determine (estimate or assess) the value of old assets, often a considerable number of years after their acquisition. During the years since the assets were acquired, price levels will have changed and technological progress have occurred. As a result, the valuation process is difficult.

A common practice is to use the reproduction-cost-new-less-depreciation procedure. This method contains a serious defect in that, even if the cost of reproducing the asset with proper inclusion of the results of price changes and technological progress can be determined accurately, arbitrary depreciation methods must then be applied. Since the usual depreciation methods are not expected to give accurate salvage values at intermediate intervals in the life of an asset, but are merely empirical accounting devices, the resulting value obtained will be accurate only by chance.

If proper cost data are available, a more accurate value of old assets can be obtained by applying replacement theory. As has been pointed out in this chapter, an asset has economic value only if it can be operated profitably in competition with the most economically efficient asset available. The economic value of an existing asset is the maximum amount on which depreciation and interest can be charged while permitting it to compete on an even basis with the most efficient new asset. Such a comparison may be illustrated as follows.

EXAMPLE 11-12

An old asset could last for 5 more years and have annual disbursements of $17,000. A potential new asset would require an investment of $50,000 and have annual disbursements of $14,000 over a 5-year life. If either alternative is expected to have zero salvage value at the end of 5 years and capital should earn 8% before taxes, what is the present value of the old asset at which it is equally as economical as the new asset?

Solution

Letting V = value of the old asset, we obtain:

	Old Asset	New Asset
Depreciation + minimum profit:		
$V(A/P, 8\%, 5)$	$0.2505 V$	
$50,000(A/P, 8\%, 5)$		$12,525
Annual disbursements	17,000	14,000
TOTAL ANNUAL COSTS	$17,000 + $0.2505 V$	$26,525

Equating the total annual costs and solving, we find that V = $38,000. This is a good measure of the present economic value of the old asset if the new asset is the best alternative available.

Because the alternatives have the same economic lives, it would also be easy to obtain the same answer by equating present worths of costs over the 5-year period. Thus

$$V + \$17,000(P/A, 8\%, 5) = \$50,000 + \$14,000(P/A, 8\%, 5)$$
$$V = \$38,000 \quad \blacksquare$$

importance of true costs in replacement studies

Because replacement of an asset obviously involves a change from existing conditions, it is of the utmost importance in replacement studies that the costs used be factual. Usually, the accounting cost data, particularly relating to the existing costs, are based on conditions that were assumed some time previously, rather than as they are determined to be at the time the replacement study is made. Also, as has been discussed previously, they may contain arbitrary allocations that are not relevant to the particular conditions at hand. Consequently, we should make certain that cost information is scrutinized carefully before it is used in replacement studies.

problems

11-1. One year ago a machine was purchased at a cost of $2,000 to be useful for 6 years. However, the machine has failed to perform properly and has cost $500 per year for repairs, adjustments, and shutdowns. A new machine is available to accomplish the functions desired and has an initial cost of $3,500. Its maintenance costs are expected to be $50 per year during its service life of 5 years. The approximate market value of the presently owned machine has been estimated to be roughly $1,200. If the operating costs (other than maintenance) for both machines are equal, show whether it is economical to purchase the new machine. Perform a before-tax study using an interest rate of 8%, and assume that terminal salvage values will be negligible.

11-2. A pipeline contractor is considering the purchase of certain pieces of earth-moving equipment to reduce his costs. Alternatives are as follows.

Plan A—Retain a backhoe already in use. The contractor still owes $10,000 on this machine but can sell it on the current open market for $15,000. This machine will last another 6 years, with maintenance, insurance, and labor costs totaling $28,000 per year.

Plan B—Purchase a trenching machine and a highlift to do the trenching and backfill work and sell the current backhoe on the open market. The new machines will cost $55,000, will last 6 years with a salvage value of $15,000, and will reduce annual maintenance, insurance, and labor costs to $20,000.

If money is worth 8% before taxes to the contractor, which plan shouldhe follow? Use the present worth procedure to evaluate both plans on a before-tax basis.

11-3. An engineer proposes that an automatic lathe be purchased to replace a presently owned engine lathe and submits the following data:

Present lathe:

First cost 6 years ago	$20,000
Labor costs	One operator at $6.00 per hour for a 40-hour week
Property taxes, insurance, maintenance	3% annually on first cost
Other operating costs	$4,000 annually
Salvage value	Now: $6,000; 10 years hence: $1,000

New lathe:

First cost	$30,000
Service life	10 years
Estimated salvage value in 10 years	$6,000
Labor costs	One operator at $8 per hour for 20 hours per week
Property taxes, insurance, maintenance	$4\frac{1}{4}$% annually on first cost
Other operating costs	$2,000 annually

The company uses a 14% before-tax rate of return on its capital. The plant operates 52 weeks during the year, and the study period is 10 years.

Would you make the replacement if income taxes are neglected? Make a comparison of annual costs assuming that each lathe has the same production capacity and quality.

11-4. Company M has offered a duplicating machine to the Department of Industrial Engineering for $945, with a trade-in allowance of $365 on a 10-year-old machine. Ten years from now the old machine will have a scrap value of $25. The new machine in 10 years would have an approximate value of $460.

Extra maintenance on the old machine, based on past records, is considered to be $20 per year.

(a) Make an annual cost (or present worth) comparison and decide whether the IE Department should buy the new machine at this time. Assume a before-tax M.A.R.R. of 6%.

(b) The sales representative for the company made the following comparison. What is wrong with it, if anything?

To keep old machine:

$365	present trade-in value of old machine
−25	approximate trade in value of old machine in 10 years
$340	loss in value over 10 years (cost of $34 per year to keep old machine)

To buy new machine:

$580	cost of new machine (including allowance for trade-in)
−460	approximate value of machine in 10 years
$120	cost over 10-year period ($12 per year to purchase the new machine)

11-5. A 3-year-old asset that was originally purchased for $4,500 is being considered for replacement. The new asset under consideration would cost $6,000. The engineering department has made the following estimates of the operating and maintenance costs of the two alternatives.

Year	Old Asset	New Asset
1	$2,000	$ 500
2	2,400	1,500
3	—	2,500
4	—	3,500
5	—	4,500

The dealer has agreed to place a $2,000 trade-in value on the old asset if the new one is purchased now. It is estimated that the salvage value for either of the assets will be zero at any time in the future. If the before-tax rate of return is 12%, make an annual cost analysis of this situation and recommend the best course of action that can be taken now. Use the repeatability assumption and give a short written explanation of your answer.

11-6. You have a machine that cost $30,000 2 years ago—it has a present market value of $5,000. Operating costs total $2,000 per year so long as the machine is in use. In 2 more years it will no longer be useful and you can sell it for $500 scrap value. You are considering replacing this machine with a new model incorporating the latest technology—this new model will cost you $20,000 now. It has an annual operating cost of $1,000 with a useful life of 8 years and negligible salvage value. If you delay the purchase of the new model, the cost will be $24,000 because of the installation difficulties that do not exist now. When your present machine is retired, you have no hopes of getting another one like it, since it was a very limited model. If money is worth 10% before taxes, what should you do?

11-7. (a) The replacement of a boring machine is being considered by the Reardorn Furniture Company. The new improved machine will cost $30,000 installed, will have an estimated economic life of 12 years, and a $2,000 salvage value. It is estimated that annual operating and maintenance costs will average $16,000 per year. The present machine has a book value of $6,000 and a present market value of $4,000. Data for the present machine for the next 3 years are as follows:

Year	Salvage Value at End of Year	Book Value at End of Year	Operating and Maintenance Costs During the Year
1	$3,000	$4,500	$20,000
2	2,500	3,000	25,000
3	2,000	1,500	30,000

Using a before-tax interest rate of 15%, make an *annual-cost comparison* to determine whether it is economical to make the replacement now.

(b) If the operating and maintenance costs for the present machine had been estimated to be $15,000, $18,000, and $23,000 in years 1, 2, and 3, respectively, what replacement strategy should be recommended?

11-8. Robert Roe has just purchased a 1-year-old used car, paying $3,000 for it. A friend has suggested that he should determine in advance how long he should keep the car so as to assure the greatest overall economy. Robert has decided that, because of style changes, he would not want to keep the car longer than 4 years, and he has estimated the out-of-pocket costs and trade-in values for years 1 through 4 as follows:

	Year 1	Year 2	Year 3	Year 4
Operating costs	$ 950	$1,050	$1,100	$1,300
Trade-in value at end of year	2,250	1,800	1,450	1,160

If Robert's capital is worth 7%, at the end of which year should he replace the car?

11-9. Determine the economic service life of a machine that has a first cost of $5,000 and estimated operating costs and year-end salvage values as follows. Assume that the interest rate is (a) 0%; and (b) 10%.

Year, N	Operating Cost for Year	Salvage Value
1	$ 800	$4,000
2	900	3,500
3	1,100	3,000
4	1,100	2,000
5	1,300	1,500
6	1,400	1,000

11-10. Find the economic service life for a tractor with an expected disbursement pattern of the following form.

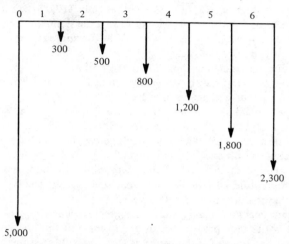

Assume that the salvage value is $5,000 - 250t^2$ at the end of any given year and suppose that the interest rate is 4%.

11-11. Consider a piece of equipment that initially cost $8,000 and has these estimated annual operating and maintenance costs and salvage values:

End of Year	Operating and Maintenance Costs for Year	Salvage Value
1	$3,000	$4,700
2	3,000	3,200
3	3,500	2,200
4	4,000	1,450
5	4,500	950
6	5,250	600
7	6,250	300
8	7,750	0

If the cost of money is 7%, determine the most economical time to replace this equipment.

11-12. Five years ago the QED Corporation installed a diesel-electric unit at a remote plant at a cost of $50,000 because no dependable electric power source was available from a public utility. It has computed depreciation on a straight line basis, using an economic life of 10 years and zero salvage value. Annual operation and maintenance costs are $16,000, and taxes and insurance charges total $2,000 per year.

Dependable power service is now available at an estimated annual cost of $30,000. The company wishes to know whether it would be more economical to dispose of the diesel-electric unit now, when it can be sold for $35,000, or to wait 5 years, when it would have to be replaced and have no salvage value. The company has a total income tax rate of 50%, a long-term capital gains tax of 30%, and tries to limit its capital investments to opportunities that will earn at least 15% after taxes. What would you recommend?

11-13. (a) Work Problem 11-12, but assume that only $10,000 can be realized from disposal of the diesel-electric unit, and the company must pay a tax of 30% on any capital gains. Assume that the company has other off-setting capital gains for the year.

(b) If your recommendation differs from that made for Problem 11-12, explain what caused the difference, since the same diesel-electric unit and annual disbursements are involved.

11-14. You have a machine that was purchased 2 years ago and was set up on a 5-year sum-of-the-years'-digits depreciation schedule with no salvage value. Original cost was $150,000. The machine can last for 10 years or more. A new machine is now available at a cost of only $100,000—it can be depreciated over 3 years with the sum-of-the-years'-digits method—no salvage value. Annual operating cost of the new machine is only $5,000 while operating cost of the present one is $20,000/year.

The new machine has a useful life greater than 10 years. You find that $40,000 is the best price you can get if you sell the present machine now. Your best projection for the future is that you will need the service provided by either of the two machines for the next 5 years. The salvage value of the old machine is estimated at $2,000 in 5 years, but the salvage value of the new machine is estimated at $5,000 in 5 years. If the after-tax rate of return required is 10%, should you sell the old machine and purchase the new one? You do not need both. Assume that the company is in a 50% tax bracket, and that long-term capital gains (losses) are taxed at 30%.

11-15. The company for which you are working must improve certain of its facilities to meet increasing sales. The existing facilities, which have been fully depreciated, can be used for another 5 years provided that some new materials-handling equipment, costing $25,000, is added. If this procedure is followed, neither the old nor the new equipment would have any salvage value at the end of 5 years, when it all would have to be replaced.

An alternative is to dismantle the existing facilities and build entirely new ones at a cost of $600,000. The new installation would have a depreciable life of 20 years, and it would result in a reduction of at least $109,000 per year in out-of-pocket disbursements, as compared with the first alternative. The company has a total income tax rate of 50% and has a 20% M.A.R.R. before taxes and a 10% M.A.R.R. after taxes. It uses straight line depreciation for accounting and tax purposes and finds that its book values agree quite well with actual salvage values. (a) Make a before-tax analysis and recommendation, using the E.R.R.R. procedure. (b) Make an after-tax analysis using the E.R.R.R. procedure. Would your recommendation be the same as in (a)?

11-16. (a) On the basis of the results obtained in Problem 11-15, give your opinion as to how the after-tax rate of return, determined by an I.R.R. analysis, would differ from that obtained by the E.R.R.R. method. (b) Make an after-tax analysis of Problem 11-15 using the I.R.R. method and compare the results with the results of Problem 11-15(b).

11-17. Four years ago the Attaboy Lawn Mower Company purchased a piece of equipment for their assembly line. Because of increasing maintenance costs for this equipment, a new piece of machinery is being considered. The cost characteristics of the defender (present equipment) and the challenger are shown below:

Defender	Challenger

Original cost = $9,000
Maintenance = $300 in year 1,
  increasing by a gradient of $200
  per year thereafter
Original estimated salvage value = 0
Original estimated life = 9 years
	Purchase cost = $13,000
	Maintenance = $150/year
	Salvage value = $3,000
	Estimated life = 5 years

Suppose that a $3,200 market value is available for the defender. Perform an after-tax analysis, using an after-tax M.A.R.R. of 10%, to determine which alternative to select. The effective tax rate is 50% on ordinary income, capital gains (losses) are taxed at 30% and straight line depreciation is applied to both assets.

11-18. The Quick Manufacturing Company, a large profitable corporation, is considering the replacement of a production machine tool. A new machine would cost $3,700, have a 4-year useful and depreciable life, and no salvage value. If purchased, the firm would be allowed a 3.33% investment tax credit. For tax purposes sum-of-the-years'-digits depreciation would be used. The existing machine tool was purchased 4 years ago at a cost of $4,000, and has been depreciated by straight line depreciation assuming an 8-year life and no salvage value. It could be sold now to a used equipment dealer for $1,000 or be kept in service for another 4 years, at which time it would have no salvage value. An analysis indicates the new machine would save about $900 per year in operating costs compared to the existing machine.

Compute the incremental after-tax rate of return (I.R.R.) for the new machine compared to the old machine.

11-19. A tool-and-die shop, operating as a corporation and having a total income tax rate of 50%, has two milling machines, which it purchased 2 years ago at a cost of $14,000 each. At that time it estimated their depreciable life to be 5 years, with a salvage value of $4,000 each, and it has charged straight line depreciation on that basis. Each machine is operated two shifts per day, and each requires an operator whose annual wage, including fringe benefits, amounts to $10,000.

The company is considering replacing the milling machines with a tape-type N/C machining center, which would have at least the capacity of the two milling machines. This machine would cost $120,000, have an estimated useful life of 5 years, with an estimated salvage value of at least $20,000 at that time. Straight line depreciation would be used. The machining center would require one operator on each shift, who would be paid the same rate as the operators of the milling machines. It is estimated that preparation of the programming tapes for this machine would cost $4,000 per year.

A market value of $10,000 each has been established for the milling machines. Taxes and insurance on any type of equipment amounts to 5% of first cost per year. In reviewing the income tax situation, the company finds that it will have unrelated long-term capital gains of $10,000 for the year. It wishes to obtain at least 10% on its capital, after taxes. Although its budget for new equipment is considerably less than needed, it is sufficient to finance this replacement if it is justified. What do you recommend on the basis of an after-tax annual cost analysis?

11-20. An engineering company purchased 10 small, desk-type electronic calculating machines 2 years ago for $3,500 each. It estimated that these would have a depreciable life of 5 years, and depreciation has been charged on an SYD basis with no salvage value. It has experienced some increasing maintenance costs with the machines, and annual maintenance costs are running about $300 per machine. It has been proposed that these calculators be replaced by installing small computer terminals, at an installed cost of $350 each, which would be hooked in with a central computer on a time-sharing basis. This service would cost $100 per month per terminal, and it is felt that somewhat more extensive service would be available without any appreciable delays or maintenance costs during the next 4 years, after which time it is believed some completely different type of computer service would be used.

The company would depreciate the computer terminals on an SYD basis with no salvage value. It finds that it could sell the calculators for $750 each. Annual taxes and insurance on any type of equipment total 4% of first cost. With the company being in the 48% tax bracket and demanding an after-tax return of at least 12% on its capital, what would you recommend on the basis of an after-tax annual cost analysis. Long-term capital gains (losses) are taxed at 30%.

11-21. Five years ago an airline installed a baggage conveyor system in a terminal, knowing that within a few years it would move to a new section of the terminal and that this equipment would then have to be moved. The original cost of the installation was $120,000, and through accelerated depreciation methods the company has been able to write off the entire cost. It now finds that it will cost $40,000 to move and reinstall the conveyor. This cost could be depreciated over the next 5 years, which the airline believes is a good estimate of the remaining economic life of the system if moved. It finds that it can purchase a somewhat more efficient conveyor system for an installed cost of $160,000, and this system would result in an estimated reduction in annual operating and maintenance costs of $7,000, during its estimated 10-year depreciable life. In both cases straight line depreciation is used with a salvage value of $0. A small airline, which will occupy the present space, has offered to buy the old conveyor for $135,000.

Annual taxes and insurance on the present equipment have been $1,500, but it is estimated that they would increase to $1,700 if the equipment is moved and reinstalled. For the new system it is estimated that these would be about $1,750 per year. All other costs would be about equal for the two alternatives. The company is in the 48% income tax bracket and pays 30% on any long-term capital gains. It wishes to obtain at least 10%, after taxes, on any invested capital. What would you recommend?

11-22. Ten years ago a corporation built a warehouse at a cost of $400,000 in an area that since has developed into a major retail location. At the time the warehouse was constructed it was estimated to have a depreciable life of 20 years, with no salvage value, and straight line depreciation has been used.

The corporation now finds it would be more convenient to have its warehouse in a less congested location, and can sell the old warehouse for $250,000. A new warehouse, in the desired new location, would cost $500,000, have a depreciable life of 20 years with no salvage value, and there would be an annual saving of $4,000 per year in the operation and maintenance costs.

Taxes and insurance on the old warehouse have been 5% of the first cost per year, while for the new warehouse they are estimated to be only 3% per year. The corporation has a 48% income tax rate, and a 30% long-term capital gains tax rate. Capital is worth not less than 12% after taxes. What would you recommend on the basis of an after-tax I.R.R. analysis?

11-23. A company is considering replacing a turret lathe with a single-spindle screw machine. The turret lathe was purchased 6 years previously at a cost of $8,000, and depreciation has been figured on a 10-year, straight line basis, using zero salvage value. It can now be sold for $1,500, and if retained would operate satisfactorily for 4 more years and have zero salvage value.

The screw machine is estimated to have a useful life of 10 years. Straight line depreciation would be used with $F = 0$. It would require only 50% attendance of an operator who receives $3.75 per hour. The machines would have equal capacities and would be operated 8 hours per day, 250 days per year. Maintenance on the turret lathe has been $300 per year; for the screw machine it is estimated to be $500 per year. Taxes and insurance on each machine would be 2% of the first cost annually. If capital is worth 10% to the company, after taxes, and the company has a 50% tax rate, what is the maximum price it can afford to pay for the screw machine? Capital gains (losses) are taxed at 30%, and an investment tax credit of 10% can be taken if the screw machine is purchased.

chapter 12

economy studies for public projects

Public projects are those authorized, financed, and operated by governmental agencies—Federal, State, or local. Such public works are numerous, may be of any size, and frequently are much larger than private ventures. Because they require the expenditure of capital, such projects have economy aspects, both with respect to acquisition and operation. However, because they are public projects, a number of important and special factors exist which ordinarily are not found in privately financed and operated businesses. To some extent these problems are based on personal philosophies and desires that are subject to change with time. Thus, to make and interpret economy studies of public projects, it is essential to have some understanding of these unique problems.

Some basic differences between privately owned and publicly owned projects may be noted by considering the following factors:

	Private	Public
1. Purpose	Provide goods or services at a profit Provide jobs Promote technology Improve living standards	Protect health Protect lives and property Provide services (at no profit)
2. Sources of capital	Private investors and lenders	Taxation Private lenders
3. Method of financing	Individual ownership Partnerships Corporations	Direct payment from taxes Loans without interest Loans at low interest Self-liquidating bonds Indirect subsidies Guarantee of private loans
4. Multiple purposes	Rarely	Common, such as electrical power, flood control irrigation, and recreation
5. Life of individual projects	Usually relatively short (5 to 20 years)	Usually relatively long (20 to 60 years)

366

	Private	Public
6. Relationship of suppliers of capital to project	Direct	Indirect, or none
7. Conflict of purposes	Usually none	Quite common (dam for flood control and power)
8. Conflict of interests	Usually none	Very common (between agencies)
9. Effect of politics	Little to moderate	Frequent factors; short-term tenure of decision makers. Pressure groups. Financial and residential restrictions, etc.
10. Measurement of efficiency	Rate of return on capital	Very difficult; no direct comparison with private projects

As a consequence of these differences, it often is not possible to make economy studies and investment decisions for public works projects in exactly the same manner as for privately owned projects. Different decision criteria must be used, and this creates problems for the public, who pays the bill, for those who must make the decisions, and for those who must manage public works projects.

During peacetime, public projects tend to be either of a durable nature, such as highways or flood control projects, or of a national-interest nature, such as research in new sources of energy or space exploration, which are of such magnitude and uncertain outcome that paramount public interest overrides the large degree of uncertainty and probable absence of profit. There are, of course, exceptions to the foregoing. It appears likely that the number and magnitude of public projects will increase in the future. Partly this is caused by the public's increasing demand for governmental services and partly by the fact that, to take advantage of many of the remaining undeveloped natural resources and to develop new and needed services, very large expenditures of capital will be required, or two or more states or political bodies will be concerned, or multiple purposes will be involved. For example, in the development of economical means for desalting seawater, it appears likely that the operation of large nuclear plants, which also will generate large quantities of electric energy, will be involved. It is not likely that such multipurpose plants can be constructed and operated profitably by private capital for many years.

In this chapter attention will be devoted mostly to projects that are financed by the Federal government, primarily because the basic principles are more readily set forth. Later these principles will be related to State and local projects, where the problems usually are somewhat simpler.

the relationship of engineers to public works

Engineers as a group have a greater interest in, and more responsibility for, public projects than most other people. Since a large proportion of public works involve engineering structures or equipment and their subsequent operation, many engineers are employed by governmental agencies to design, construct, operate, or manage various projects. These engineers have a direct responsibility to see that the projects are accomplished as economically as possible. Other engineers may be employed by private companies with which some governmental projects compete. To these engineers the economic basis of such governmental activities obviously is important.

All engineers, regardless of by whom they are employed, are taxpayers and thus have a very real stake in many governmental projects; they should take an active interest in the economy of such activities. But, as citizens, engineers also have a responsibility to the community in helping to interpret the economic facts regarding public works to those who are not as well trained to understand the problems and issues involved. Many of the issues and problems concerning public works are not black or white. As will be pointed out later, many involve highly technical matters as well as basic principles of engineering economy. It therefore is to be expected that those without the engineer's specialized knowledge may not understand all the factors and thus may arrive at misinformed conclusions, particularly if misled by an interested group or a politician with an axe to grind. As professional men, engineers have a special responsibility to aid in interpreting the problems and issues with respect to public projects for the general public.

self-liquidating projects

The term *self-liquidating project* is often applied to a governmental project that is expected to earn direct revenue sufficient to repay the cost in a specified period of time. Such projects usually provide utility services, such as water, electric power, sewage disposal, or irrigation water. In addition, toll bridges, tunnels, and highways are built and operated in this manner.

Self-liquidating projects are not expected to earn profits or pay income taxes. Neither do they pay property taxes, although in some cases they do make *in-lieu* payments in place of the property or franchise taxes that would have been paid to the cities, counties, or states involved had the project been built and operated by private ownership. For example, the U.S. government agreed to pay the States of Arizona and Nevada $300,000 each annually for 50 years in lieu of taxes that might have accrued if Hoover Dam had been privately constructed and operated. In most cases the in-lieu payments are considerably less than the actual property or franchise taxes would be. Furthermore, once the in-lieu payments are agreed upon, usually at the time the project is originated, they virtually never are changed thereafter. Such is not the situation in the case of property taxes.

Whether self-liquidating projects should or should not pay the same taxes as privately owned projects is not within the scope of this text. Such payments undoubtedly would make it much easier to compare the economy of similar publicly and privately operated activities. However, that public projects do not have to earn profits or pay certain taxes is a fact that must be given proper consideration in making economy studies of such projects and in making decisions regarding them.

multiple-purpose projects

Many governmental and private projects have more than one purpose or function. A governmental project, for example, may be intended for flood control, irrigation, and generation of electric power. A private project may utilize waste gases in a petroleum refinery to generate steam and electricity for refinery use and to cogenerate electricity for public sale. Such projects commonly are called *multiple-purpose projects*. By designing and building them to serve more than one purpose, greater overall economy can be achieved. This is very important in projects involving very large sums of capital and utilization of natural resources, such as rivers. It is not uncommon for a public project to have four or five purposes. This usually is desirable, but, at the same time, it creates economy and managerial problems because of overlapping utilization of facilities and, sometimes, conflict of interest between the several purposes and agencies involved. An understanding of the problems involved in such situations is essential to anyone who wishes to make economy studies of such projects or to understand the cost data and political issues arising from them.

A number of basic problems may arise in connection with multi-purpose public works. These may be illustrated by the simple example of a dam, shown in Figure 12-1, which is to be built in a semiarid portion of the United States, primarily to provide control against serious floods. It is at once apparent that,

Figure 12-1. Schematic representation of a multiple-purpose project involving flood control, irrigation, and power.

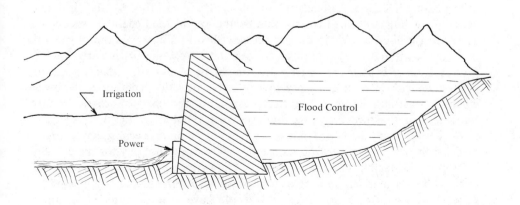

if the flow of the water impounded behind the dam could be regulated and diverted onto the adjoining land to provide irrigation water, the value of the land would be increased tremendously. This would result in an increase in the nation's resources, and it thus appears desirable to expand the project into one having two purposes—flood control and irrigation.

The existence of a dam with a high water level on one side and a much lower level on the other side, at least during a good portion of the year, immediately suggests that some of the nation's resources will be wasted unless some of the water is permitted to run through turbines to generate electric power that can be sold to customers in the adjacent territory. Thus the project is expanded to have a third purpose—the generating of power.

In the semiarid surroundings the creation of a large lake, such as would exist behind the dam, would provide valuable recreation facilities for hunting, fishing, boating, camping, and so on. This gives a fourth purpose to the project. All of these purposes have desirable economic and social worth, and what started as a single-purpose project has ended by having four purposes. Not to develop the project to fill all four functions probably would mean that valuable national resources would be wasted. On the other hand, however, the potential loss of land (e.g., for mining minerals and retaining ecological balance, etc.) must be factored into the analysis as disbenefits.

If the project is built for four purposes, the important fact that one dam will serve all of them will result in at least three basic problems. The *first* is the allocation of the cost of the dam to the four purposes that it will serve. Assume, for example, that the dam will cost $10,000,000. What portion of this amount should be assigned to flood control? What amounts to irrigation, power generation, and recreation? If the identical dam will serve all four purposes, deciding how much of its cost should be assigned to any one of the purposes obviously might present considerable difficulties. One extreme would be to decide that, since the dam was required for flood control, its entire cost should be assigned to this purpose. If this were done, the cost of the electric power and the irrigation water would obviously be far less than if a considerable part or all of the $10,000,000 had been assigned to these functions.

The cost allocation problem is complicated considerably by the different ways in which the allocated costs would be financed by the Federal government. The costs for flood control in the project described would be paid out of general tax monies without any attempt to recover them from the beneficiaries of the project. The costs assigned to irrigation might be recovered, possibly with no interest or with interest at a very low rate. Because the generation of power would be self-liquidating in nature, presumably the costs allocated to this function would be recovered with interest. Most, if not all, of the costs assigned to recreation would be financed directly from tax monies, because any charges made to the public for recreational use probably would not pay for the current upkeep of the facilities. With the costs of the various purposes being financed in the different manners indicated, it is obvious that the amounts allocated to flood control and recreational facilities would have a considerable effect upon the resulting unit costs and selling prices for power and irrigation water. It is only to be expected that those who would be direct beneficiaries of these two purposes would exercise their influence to have as large a proportion as possible of the dam cost allocated to flood control and recreation.

Another extreme would be to assign each purpose with an amount equal to the total cost of the dam. Such a procedure would have two effects. First, it would undoubtedly make some of the purposes so costly—for example, recreation and irrigation—that they could not be undertaken. Second, by using this procedure, the cost of the dam might be returned to the government several times. Although many taxpayers would appreciate such a profitable occurrence, it is hardly feasible.

The *second* basic problem in multiple-purpose projects is the matter of conflict of interest among the several purposes. This may be illustrated by the matter of the water level maintained behind the dam. For flood control it would be best to have the reservoir empty most of the year to provide maximum storage capacity during the season when floods might occur. Such a condition would be most unsatisfactory for power generation. For this purpose it would be desirable to maintain as high a level as possible behind the dam during most of the year. For recreational purposes a fairly constant level also would be desirable. The requirements for irrigation probably would be somewhere between those for flood control and for power generation. Furthermore, while the lake created might be ardently approved by boating and recreation enthusiasts, it might be opposed with equal vigor by ecology-minded groups. Thus some very definite conflicts of interest usually arise in connection with multiple-purpose works. As a result, compromise decisions must be made. These decisions have arbitrary effects on the economy of the various portions of a multiple-purpose project. This is a fact that never should be overlooked in evaluating the economy of such projects and the costs of the services produced.

A *third* problem connected with multiple-purpose public works is the matter of politics. Since the various purposes are likely to be desired or opposed by various segments of the public and by various interest groups that will be affected, it is inevitable that such projects frequently become political issues. This often has an effect upon cost allocations and thus on the overall economy of these projects.

A basic motive of every political candidate or officeholder is to get elected or reelected. Although candidates may honestly claim that their basic objective is to "serve the people," they cannot serve them unless they are in office. One primary way of obtaining votes is to demonstrate that the candidate is able to obtain direct benefits for the electorate. Such benefits as cheap irrigation, water, or electric power are easily recognized. As a result, there is a rather natural tendency for some politicians to exert influence in the allocation of costs of multiple-purpose public projects so that undue portions are assigned to such purposes as flood control and recreation; thus the services provided by the self-liquidating purposes, such as power generation, may be sold to the public at very low rates.

The net result of these three factors is that the cost allocations made in multiple-purpose public projects are arbitrary. As a consequence, production and selling costs of the services provided also are arbitrary. Because of this, they cannot be used as valid "yardsticks" with which similar private projects can be compared to determine the relative efficiencies of public and private ownership. A recognition of this fact will do much to eliminate meaningless claims and arguments and to encourage the development and use of sound procedures that will assure that only economical and needed multiple-purpose public works are authorized and that they are operated with maximum efficiency.

why economy studies of public works?

In view of the various problems that have been cited, engineers and other individuals sometimes raise the question as to whether economy studies of governmental projects should be attempted. It must be admitted that such economy studies frequently cannot be made in as complete and satisfactory a manner as in the case of studies of privately financed projects. However, decisions regarding the investment and use of capital in public projects must be made—by the public, by elected or appointed officials and, very importantly, by managers, who usually cannot use the same measures of operating efficiency that are available to managers of private ventures. The alternative to basing such decisions on the best possible economy studies is to base them on hunch and expediency. It is therefore essential that economy studies be made in the best possible manner, but with a sound understanding of the nature of such activities and of all the background, conditions, and limitations connected with them.

Even in areas where national security and national policy are paramount, such as in the defense-space industry, the use of proper economy studies can assure that the government and the taxpayer are getting maximum results from the money expended. In a recent survey it was found that companies engaged in defense and space work were making as much, or more, use of economy studies as those engaged solely in the production of civilian goods.

It must be recognized that economy studies of public works can be made from several viewpoints, each applicable for certain conditions. Many studies are made from the point of view of the governmental body involved. In comparing alternative structures or methods for accomplishing the same objective, such a point of view is satisfactory. The point of view of the citizens in a restricted area is used in other studies, such as local self-liquidating or service projects where those who must pay the costs are the direct beneficiaries of the project. In other cases the entire nation may be affected, so the economy study should reflect a very broad point of view. Such studies are apt to involve many intangibles and benefits that cannot be assigned to specific persons or groups. In each case as many factors as possible should be evaluated in monetary terms, because this usually will be most helpful in making decisions regarding the expenditure of public funds. However, it must also be recognized that non-monetary factors often exist and are frequently of great importance in final decisions.

the difficulties inherent in economy studies of public works

There are a number of difficulties inherent in public works which must be considered in making economy studies and economic decisions regarding them. Some of these are as follows:

1. There is no profit standard to be used as a measure of financial effectiveness. Most public works are intended to be nonprofit.
2. There is no easy-to-quantify monetary measure of many of the benefits provided by a public project.
3. There frequently is little or no direct connection between the project and the public, who are the owners.
4. Politics: Whenever public funds are used, there is apt to be political influence. This can have very serious effects on the economy of projects, from conception through operation, and knowledge of its actual or possible existence is an important factor to be considered.
5. Indefinite motivation of personnel: The usual profit motive as a stimulus to effective operation is absent. This may have a marked effect on the effectiveness of a project, from conception through operation. In some cases the major motivation may be toward inefficient operation and the creation of unnecessary services because an individual's advancement depends upon having more people under his supervision. This does not mean that all public works are inefficient or that many managers and employees of them are not trying to do, and are actually doing, an effective job. But the fact remains that the direct stimuli present in privately owned companies are lacking and that this lack may have a considerable effect.
6. Legal restrictions: Public works usually are much more circumscribed by legal restrictions than are private companies. Their ability to obtain capital usually is restricted. Frequently, their area of operations is restricted as, for example, where a municipally owned power company cannot sell power outside the city limits. There often are severe restrictions with respect to hiring, firing and paying employees.
7. In many cases decisions concerning public works, particularly their conception and authorization, are made by elected officials whose tenure of office is very uncertain. As a result, immediate costs and benefits may be stressed, to the detriment of long-range economy.

what interest rate should be used for public projects?

The interest rate should play the same formal role in the evaluation of public sector investment projects that it does in the private sector. The rationale for its use is somewhat different. In the private sector, it is used because it leads directly to the private sector goal of profit maximization or cost minimization. In the public sector, its basic function is similar, in that it should lead to a maximization of social benefits, provided that these have been appropriately measured. Choice of an interest rate will lead to determination of how available funds may be allocated best among competing projects, particularly considering the scale and capital intensity of those projects.

Three main considerations may bear upon what interest rate to use in governmental economy studies:

1. The interest rate on borrowed capital.
2. The opportunity cost of capital to the governmental agency.
3. The opportunity cost of capital to the taxpayers.

In general, it is appropriate to use the borrowing rate only for cases in which money is borrowed specifically for the project(s) under investigation and use of that money will not cause other worthy projects to be forgone.

The opportunity cost (interest rate) encompasses the annual rate of profit (or other social or personal benefit) to the constituency served by the government agency or the composite of taxpayers who will eventually pay for the government project. If projects are chosen so that the return rate on all accepted projects is higher than that on any of the rejected projects, the interest rate used in the economic analysis is that associated with the best opportunity foregone. If this is done for all projects and investment capital available within a government agency, the result is a *governmental opportunity cost*. If, on the other hand, one considers the best opportunities available to the taxpayers if the money were not obtained through taxes for use by the governmental agency, the result is a *taxpayers' opportunity cost*.

Theory suggests that in usual governmental economic analyses the interest rate should be the largest of the three listed above. Generally, the taxpayers opportunity cost is substantially the highest of the three. As an indicative example, a Federal government directive† in 1972 specified that an interest rate of 10% should be used in economy studies for a wide range of Federal projects. This 10%, it can be argued, is at least a rough approximation of the average return taxpayers could be obtaining from the use of that money. In any case, it is greater than the 5 to 9% that the Federal government has been paying for the use of borrowed money.

† Office of Management and Budget, Circular A-95, March 27, 1972. The 10% was an inflation-free interest rate, which means that the *combined* interest rate would be higher.

the benefit–cost (B/C) ratio

Basically, an economy study of a public project is no different from any other economy study. Any of the common methods could be used for such a study. From a practical viewpoint, however, because profit almost never is involved and most projects have multiple benefits, some of which cannot be measured precisely in terms of dollars, many economy studies of public projects are made by comparing equivalent costs or by determining the ratio of the equivalent benefits to the equivalent costs.

The *benefit–cost (B/C) ratio* can be defined as the ratio of the equivalent worth of benefits to the equivalent worth of costs. The equivalent worths are usually A.W.s or P.W.s, but they can also be F.W.s. In the case of governmental agencies, the benefits normally accrue to the user public and the net costs of supplying the benefits accrue to the government agency. The B/C ratio is also referred to as the *savings-investment ratio* (SIR) by some governmental agencies.

Two commonly used formulations of the B/C ratio (expressed in terms of equivalent *annual worths*) are as follows:

1. Conventional B/C ratio:

$$\text{B/C} = \frac{\text{A.W. (net benefits to user)}}{\text{A.W. (total net costs to supplier)}} = \frac{B}{\text{C.R.} + (O + M)} \qquad (12\text{-}1)$$

where A.W.($\cdot$) = annual worth of ($\cdot$)

B = annual worth of net benefits (gross benefits minus costs) to user

C.R. = capital recovery cost or the equivalent annual cost of the initial investment, including any salvage value

$O + M$ = uniform annual net operating and maintenance disbursements to supplier

2. Modified B/C ratio:

$$\text{B/C} = \frac{B - (O + M)}{\text{C.R.}} \qquad (12\text{-}2)$$

The numerator of the modified B/C ratio expresses the equivalent worth of the net benefits minus operating and maintenance costs; the denominator includes only the investment costs.

The conventional B/C ratio method appears to have been supplanted by the modified B/C ratio method among a number of user groups. Although both B/C methods give consistent answers regarding whether the ratio is > 1.0 or < 1.0, and both yield the same recommended choice when comparing mutually exclusive investment alternatives, they can yield different rankings (e.g., which is best, second best, etc.) for independent investment opportunities.

An individual investment *opportunity* is deemed a worthwhile investment if its B/C ratio ≥ 1.0. However, when comparing mutually exclusive investment *alternatives*, an incremental approach is required.

computation of B/C ratios for a single project

EXAMPLE 12-1

First cost	$20,000
Project life	5 years
Salvage value	$ 4,000
Annual benefits	$10,000
Annual O and M disbursements	$ 4,400
Interest rate	8%

Solution

The conventional B/C ratio and modified B/C ratio, based on equivalent annual worths, are computed as follows:

$$C.R. = (\$20,000 - \$4,000)(A/P, 8\%, 5) + \$4,000(0.08) = \$4,326$$

$$\text{conventional B/C ratio} = \frac{B}{C.R. + (O + M)} = \frac{\$10,000}{\$4,326 + \$4,400} = 1.146 > 1.0$$

$$\text{modified B/C ratio} = \frac{B - (O + M)}{C.R.} = \frac{\$10,000 - \$4,400}{\$4,326} = 1.294 > 1.0$$

Since B/C $\geq$ 1.0 (by either ratio), the individual investment opportunity is a worthwhile investment. ∎

Hereafter, unless otherwise noted, only the modified B/C ratio will be used and it will be denoted, simply, B/C ratio.

comparison of independent opportunities by B/C ratios

EXAMPLE 12-2

It is desired to compare the two investment projects in Example 7-5 by the B/C ratio method, with annual revenue to be considered as annual benefits. Assume that the two are independent opportunities rather than mutually exclusive alternatives. The minimum attractive rate of return is 10%, and a restatement of the problem appears below:

	A	B
Investment	$3,500	$5,000
Annual net benefits to user	1,900	2,500
Annual net O + M disbursements to supplier	645	1,383
Estimated life	4 years	8 years
Net salvage value	0	0

Solution

	A	B
Annual net benefits minus net O + M disbursements:		
$1,900 − $645	$1,255	
$2,500 − $1,383		$1,117
C.R. costs:		
$3,500($A/P$, 10%, 4)	1,104	
$5,000($A/P$, 10%, 8)		937
B/C ratio:		
$1,255/$1,104	1.14	
$1,117/$937		1.19

Thus investment opportunity B is the better of the two, but both are satisfactory because the B/C ratios are > 1.0. ▮

comparison of mutually exclusive alternative opportunities by B/C ratios

Mutually exclusive projects and alternative levels of investment in a given project also can be evaluated by using B/C ratios.

When different levels of investment and cost may be employed in carrying out a specific objective, usually with somewhat different levels of results, confusion and error sometimes exist in interpreting the corresponding benefit–cost ratios and in making the proper decision as to which alternative to adopt. The proper selection between the alternatives using the B/C ratio involves the same principles of incremental return as was discussed in Chapter 7. Specifically:

1. Each increment of cost should justify itself by a sufficient B/C ratio (generally ≥ 1.0) on that increment.
2. Compare higher cost alternative against a lower cost alternative only if that lower cost alternative is justified.
3. Choose the alternative requiring the highest cost for which funds are available and for which each increment of cost is justified (by a B/C ratio ≥ 1.0).

A typical problem and the correct method of analysis follow.

EXAMPLE 12-3

In dealing with a certain problem, seven alternatives are available, the first being to take no corrective action and thus to accept an annual loss of $500,000 per year. The annual benefits and costs and the B/C ratios are as shown in Table 12-1. Because all the B/C ratios exceed 1, it is not at once apparent which alternative should be selected. We might be tempted to select alternative C because it has the highest B/C ratio and its annual costs are much less than some of the other

table 12-1

annual benefits, costs, and benefit–cost ratios for
seven alternatives

Alternative	Annual Benefits ($000s)	Annual Costs ($000s)	B/C ($000s)
A	—	500	—
E	900	600	1.50
C	1,600	800	2.0
B	1,660	850	1.95
D	1,200	900	1.33
F	1,825	1,000	1.83
G	1,875	1,100	1.70

alternatives. On the other hand, if a project is justified when its B/C ratio exceeds 1, then we might be tempted to select alternative G because it produces the largest annual benefits. Actually, neither of these decisions would be correct. Which is the best alternative?

Solution

If the data are computed and arranged according to increasing annual cost as shown in Table 12-2, the proper decision becomes clear. In accordance with the principles set forth in Chapter 7, alternative A is the basic alternative because it involves the minimum expenditure of capital. The incremental benefit and incremental costs, *compared with the acceptable alternative having the next lower annual costs*, are computed. Then the *incremental* B/C ratios are computed, as shown in the next-to-last column of Table 12-2. It will be noted that, as in the case of computing incremental rates of return, whenever an incremental B/C ratio is less than 1.0, that alternative is eliminated from further consideration, and each successive alternative is compared with the last, previous alternative having an acceptable B/C ratio. Thus, in Table 12-2, alternatives D and G are not acceptable. Based on the incremental B/C ratios, and assuming capital is not limited, it is clear that

table 12-2

incremental benefit–cost ratios for seven alternatives

Increment Considered	Incremental Benefit ($000s)	Incremental Cost ($000s)	Incremental B/C Ratio	Justified?
A	—	—	—	—
A → E	900	100	9.0	Yes
E → C	700	200	3.5	Yes
C → B	60	50	1.2	Yes
B → D	−460	50	Negative	No
B → F	165	150	1.1	Yes
F → G	50	100	0.5	No

alternative F should be selected because it provides maximum benefits while having a B/C ratio greater than 1.0 on each increment of cost. Thus it is clear that neither alternative C nor alternative G would be a correct choice, as could be concluded if one were basing the study on B/C ratios for whole projects as in Table 12-1 rather than on incremental costs and benefits. ▮

Before a public project is undertaken, some governing body—the Congress in the case of Federal projects—must approve that it has social usefulness. Its economic justification then is based on the B/C ratio. Typically, a project may be designed and carried out at different levels. If only a portion of the project is undertaken, the resulting B/C ratio may be less than 1.0. If developed more extensively, the ratio may rise rapidly and be substantially greater than 1.0. Further expansion of the project may result in a decrease in the B/C ratio, with the project ultimately becoming uneconomic if overexpanded. Diminishing returns are just as applicable to public projects as to private projects.

cost allocation for economy studies of highways

Not only do economy studies of public works often involve problems of allocating costs of projects to multipurposes but also the problem of allocating the costs of a single project among various beneficiaries is frequently encountered. This problem can be illustrated by the case of highways, where several types of users are involved. In the first place, a decision must be reached as to the proportion of the total cost that should be borne by the direct users. For example, some believe that a part of the cost of all streets and highways should be paid out of general tax money. One proposal has been that the direct users should pay 80% of the cost of state highways, 32% of county and local highways, and 28% of city streets. Where such a policy is followed, it is obvious that the percentages are arbitrary. In recent years many states have tried to charge all of the cost of state highways to the users. Financing is by means of vehicle registration fees, gasoline taxes, and sometimes a load-mile tax on trucks of more than a certain weight.

Regardless of the percentage of total cost that is to be paid by the direct users, there remains the problem of determining the portion of the total cost that should be allocated to and paid by each type of user. For instance, if a highway is built and is used by passenger cars, medium trucks, and heavy trucks, how much of the total cost of construction and maintenance should be paid by each type of vehicle? This is a matter that has been and still is hotly debated. A concrete pavement 5 inches thick and with 7% grades might be entirely satisfactory for passenger cars. For medium trucks a $6\frac{1}{2}$-inch pavement with 5% grades might be satisfactory. However, if heavy trucks are to use the highway, a 9-inch pavement and maximum grades of 4% might be required. The costs of each type of highway would be considerably different.

Many studies have been made to determine the costs caused by each type of vehicle, but there is far from general agreement as to the correctness of the results, or as to the method by which these costs should be collected from the users. It is generally agreed that all types of vehicles should share in the costs of a basic highway which would be adequate for ordinary passenger cars, and that any type of heavier vehicles, primarily trucks, should pay the additional costs of constructing and maintaining the roads built to heavier standards required for their use. It is not difficult to determine the added construction costs, but the determination of the maintenance costs is very difficult. Probably the only way that this could be done with complete accuracy would be to construct parallel highways, one adequate for and used only by passenger cars, and the other adequate for and used only by trucks. For completely conclusive cost data, there would have to be several parallel highways, one for trucks of each weight group. Such a project would be unreasonably costly and physically difficult, and probably will never be carried out. As a consequence, the cost allocations must be based, to a large extent, on theoretical studies, and the arguments as to their correctness will undoubtedly continue.

Most studies have concluded, first, that the costs are a function of the vehicle-miles of usage, and, second, that the cost per vehicle-mile is the sum of a constant plus a variable that depends on the vehicle weight. For example, one state concluded that the annual cost of providing and maintaining highways for vehicles weighing up to 60,000 pounds was

$$c = 0.78903w + 2.70554$$

where c was the cost in mills ($0.001) per vehicle-mile, and w the registered gross weight of the vehicle in 1,000 pounds. This study also suggested that the minimum value of c should be 6 mills per vehicle-mile, this being the amount that should be assigned to passenger cars.

If such a formula is assumed to be valid, it is then possible to compute the cost that should be allocated to each class of vehicle. There remains, however, the problem of determining the manner of collecting the allocated costs. At this point politics and political pressure often enter the picture, not only in the case of highways but also for other types of public works. For highways two methods are in general use—gasoline taxes and ton-mile taxes on vehicles above a specified weight. Usually, each is in addition to a flat, or somewhat graduated, vehicle registration fee, which pays a portion of the basic fixed road cost. A brief discussion of these two methods will serve to point out some of the factors to be dealt with and some of the compromises that usually must be made to arrive at a cost allocation that is reasonably equitable, as well as acceptable and workable.

collection of highway costs by gasoline taxes

The method of determining the amount of gasoline tax required to pay for highway costs may be illustrated by the following example. For simplicity it is assumed that there are only three types of vehicles: passenger cars, medium trucks, and heavy trucks.

Type of Vehicle	Number	Total Miles per Year	Miles per Gallon
Passenger cars	1,000,000	10×10^9	15
Medium trucks	40,000	1.6×10^9	10
Heavy trucks	10,000	1×10^9	6

The thickness and incremental cost of the highways required for these vehicles are as follows:

Type of Vehicle	Pavement Thickness (in.)	Incremental Cost per Year
Passenger cars	5	$32,000,000
Medium trucks	$6\frac{1}{2}$	4,000,000
Heavy trucks	8	5,000,000

If we use the assumption that the highway costs should be allocated on the basis of the number of vehicles that cause and use each increment of pavement, the annual cost for each type of vehicle would be as follows:

Allocation of Increment Cost per Vehicle	Passenger Cars	Medium Trucks	Heavy Trucks
$32,000,000/1,050,000	$30.45	$ 30.45	$ 30.45
$4,000,000/50,000	—	80.00	80.00
$5,000,000/10,000	—	—	500.00
TOTAL	$30.45	$110.45	$610.45

The number of gallons of gasoline used per year by each vehicle of each type and the corresponding tax per gallon that would be required in each case would be as follows:

Type of Vehicle	Number of Gallons	Tax per Gallon
Passenger cars	667	$0.0457
Medium trucks	4,000	0.0276
Heavy trucks	16,667	0.0367

It is nominally considered not feasible to charge a different gasoline tax for each class of vehicle. A number of solutions could be devised. For example, setting the gasoline tax at $0.025 per gallon and the registration fees at $13.77 for passenger cars, $10.45 for medium trucks, and $193.45 for heavy trucks would theoretically allocate the costs. It may be seen that this solution probably would encounter some objections because the registration fee for medium trucks is lower than that for passenger cars. Thus some arbitrary compromise probably would have to be adopted. Furthermore, it usually is necessary to recognize more than three classes of vehicles in making the cost allocation.

state and local public works studies

Economy studies of state and local public works projects basically are no different from those for Federal projects. Some of the conditions tend to be more bothersome. Usually, such projects have fewer purposes and tend to be single-purpose. Financing by bond issues is much more common, and frequently the bonds are backed by the revenues from the project. Legal limitations tend to be more pronounced, such as requirements that employees of a city project must live within the city limits, local power companies cannot serve customers outside the city limits, and so on. Political pressures tend to be more acute and, consequently, decisions are frequently made on a short-term basis. If such special circumstances are taken into account, economy studies of these projects offer no special difficulties.

summary

From the discussion and examples presented in this chapter, it is apparent that many public works are essential and desirable. However, because of the methods of financing, the absence of the tax and profit requirements, and political and social factors, the same criteria frequently cannot be applied to such works as are used in evaluating privately financed projects. Neither should public projects be used as "yardsticks" with which to compare private projects. Nevertheless, whenever possible, public works should be justified on an economic basis to assure that the public obtains the maximum return from the tax money that is spent. Whether an engineer is working on such projects, is called upon as a consultant to them, or only fills the role of a taxpayer, he or she is bound by the professional ethics to do the utmost to see that these projects are carried out in the best possible manner within the limitations of the legislation enacted for their authorization.

problems

12-1. (a) What is a multiple-purpose project?
   (b) Why is it difficult to make a logical assignment of costs in a multiple-purpose project?
   (c) Why is it difficult to assess the efficiency of operation in most multiple-purpose projects?

12-2. Discuss some of the considerations that go into determining an interest rate to be used in evaluating alternatives in the public sector.

12-3. List some of the factors that would have a significant effect on a state's decision to build a new highway through rural and urban areas. Which factors could be quantified and which would most likely be regarded as "irreducible" or "intangible" in nature?

12-4. Briefly indicate how you might go about quantifying each of the following items that is a consequence of a public project:
(a) Benefits of newly created recreation areas.
(b) Costs of increased air pollution.
(c) Reduced safety hazards to the public at large.
(d) Increased long-term noise associated with a public project.

12-5. Consider the following mutually exclusive alternatives:

Alternative	Equivalent Annual Cost of Project	Expected Annual Flood Damage	Annual Benefits
No flood control	$ 0	$100,000	$ 0
Construct levees	30,000	80,000	112,000
Build small dam	100,000	5,000	110,000

Which alternative would be chosen according to these decision criteria:
(a) Maximum benefit?
(b) Minimum cost?
(c) Maximum benefits minus costs?
(d) Largest investment having an incremental B/C ratio larger than 1.0?
(e) Largest B/C ratio?

12-6. Because there have been an average of two fatal and seven nonfatal automobile accidents each year in connection with a drawbridge that must be raised each time a ship passes through, considerable public discussion has arisen regarding replacing this drawbridge with a high-level bridge that would eliminate these accidents and the delays that are caused by each opening. An engineering study shows that the drawbridge, which was built 10 years ago at a cost of $1,200,000, has a net salvage value of $80,000. A new, high-level bridge would cost $2,000,000, and it probably would have a salvage value of not over $100,000 at the end of 15 years, after which neither bridge would likely be used because of new highways that are being planned. The cost of removing either bridge at the end of 15 years would probably equal its scrap value at that time. Annual operation and maintenance costs for the drawbridge are $75,000, and for the high-level bridge are estimated at $25,000.

The bridge is used by 4,000 vehicles per day, and an average of 40 boats per day pass under it. It is estimated that on the average each opening delays 25 vehicles for 3 minutes each. The lost-time and fuel costs associated with each vehicle delay is estimated at $0.06 per minute. The cost of fatal accidents has arbitrarily been put at $50,000, and for each nonfatal accident at $3,500.
(a) If money cost to the state is 5%, make an annual cost comparison and determine whether the drawbridge should be replaced.
(b) Determine the B/C ratio for this problem.

12-7. A state Resources Development Department has proposed building a dam and hydroelectric project that will remedy a flood situation on a mountain river, generate power, provide water for irrigation and domestic use, and provide certain recreational facilities for boating and fishing. The construction costs would be

Dam	$40,000,000
Access roads	2,000,000
Power plant	4,000,000
Transmission lines	1,500,000
Fish ladders and elevators	800,000
Irrigation and water canals	3,000,000

It is proposed to finance the project by issuing 5%, tax-exempt, 40-year bonds.

It is estimated that the annual operation and maintenance costs will be $1,250,000 for the power generating and distributing facilities and $750,000 for all other portions of the project. In addition, the state will pay $400,000 annually to the county where the project is located in lieu of property taxes. Estimates of the annual benefits and revenues are as follows:

Flood control	$ 900,000
Sale of power	2,700,000
Sale of water	1,600,000
Recreation benefits	800,000
Income from sports concessions	200,000

(a) Determine the B/C ratio for the project, using the estimated values of the benefits as stated. (b) Would the elimination of the benefits that you consider to be intangible leave the project justified economically?

12-8. It is being proposed that a new toll bridge be built over an inlet of a bay, to be financed by the state highway department by means of a self-liquidating, 20-year, 5% bond issue. The bridge would reduce the travel distance into a city by 8 miles. A careful traffic survey indicates that the $40,000,000 bridge would be used by an average of 40,000 private cars and 3,000 trucks and commercial vehicles each day. A policy has been established that trucks and commercial vehicles should pay five times the charge for private vehicles on such state toll bridges. If the estimated annual out-of-pocket costs for operation and maintenance for the bridge are $1,000,000, what toll fees would have to be paid by private vehicles and by commercial vehicles in order for the bridge to be self-liquidating?

12-9. A certain state is contemplating a highway development involving the construction of 10,000,000 square yards of pavement. Vehicle registration in the state is as follows:

Passenger cars	2,000,000
Light trucks	200,000
Medium trucks	60,000
Heavy trucks	10,000

The characteristics and pavements necessary to carry the vehicles are as follows:

Class of Vehicle	Pavement Thickness (in.)	Cost per Square Yard
Passenger cars	5.5	$11.80
Light trucks	6.0	12.50
Medium trucks	6.5	13.20
Heavy trucks	7.0	14.10

Assuming that paving costs should be distributed on the basis of the number of vehicles in each class and the incremental costs of paving required for each class of vehicle, what should be the taxes per vehicle for each vehicle class?

12-10. Ten years ago the port of Secoma built a new pier containing a large amount of steel work, at a cost of $300,000, estimating that it would have a life of 50 years. The annual maintenance cost, much of it for painting and repair caused by the environment, has turned out to be unexpectedly high, averaging $27,000.

The port manager has proposed to the port commission that this pier be replaced immediately with a reinforced concrete pier at an initial cost of $600,000. He assures them that this pier will have a life of at least 50 years, with annual maintenance costs of not over $2,000. He presents the following figures as justification for the replacement, having determined that the net salvage value of the existing pier is $40,000:

Annual Cost of Present Pier		Annual Cost of Proposed Pier	
Depreciation: $300,000/50	$ 6,000	Depreciation: $600,000/50	$12,000
Maintenance cost	27,000	Maintenance cost	2,000
TOTAL	$33,000	TOTAL	$14,000

He has stated that since the port earns a net profit of over $3,000,000 per year, the project could be financed out of annual earnings and there would thus be no interest cost, and an annual saving of $19,000 would be obtained by making the replacement. (a) Comment on the port manager's analysis. (b) Make your own analysis and recommendation regarding the proposal.

12-11. An area on the Colozona river is subject to periodic flood damage, which occurs, on the average, every 2 years and results in $2,000,000 loss. It has been proposed that the river channel should be straightened and deepened, at a cost of $2,500,000, to reduce the probable damage to not over $1,600,000 for each occurrence during a period of 20 years before it would have to again be deepened. This procedure also would involve annual expenditures of $80,000 for minimal maintenance.

One legislator in the area has proposed that a better solution would be to construct a flood-control dam at a cost of $8,500,000, which would last indefinitely with annual maintenance costs of not over $50,000. He estimates that this project would reduce the probable annual flood damage to not over $450,000. In addition, this solution would provide a substantial amount of irrigation water that would produce an annual revenue of $175,000 and recreational facilities, which he estimates would be worth at least $45,000 per year to the adjacent populace.

A second legislator believes that the dam should be built and that the river channel also should be straightened and deepened, noting that the total cost of $11,000,000 would reduce the probable annual flood loss to not over $350,000, while providing the same irrigation and recreational benefits.

If the state's capital is worth 6%, determine the B/C ratios and the incremental B/C ratio. Recommend which alternative should be adopted.

12-12. Five mutually exclusive alternatives are available for developing a certain public project. The following tabulation shows the annual benefits and costs for each:

Alternative	Annual Benefits	Annual Costs
A	$1,800,000	$2,000,000
B	5,600,000	4,200,000
C	8,400,000	6,800,000
D	2,600,000	2,800,000
E	6,600,000	5,400,000

(a) Assume that the projects are of the type for which the benefits can be determined with considerable certainty and that the agency is willing to invest money as long as the B/C ratio is at least 1. Which alternative should be selected? (b) If the projects involved intangible benefits which required considerable judgment in assigning their values, would this affect your recommendation?

12-13. In developing a publicly owned, commercial, waterfront area, five possible plans are being considered. Their costs and estimated benefits are as follows:

	Present Worths (000 omitted)	
Plan	Costs	Benefits
A	$105,000	$111,000
B	90,000	81,000
C	123,000	139,000
D	135,000	150,000
E	99,000	114,000

Which plan should be adopted, if any, if the controlling board wishes to invest any amount required provided that the B/C ratio on the required investment is at least 1.0? Solve by use of incremental analysis.

12-14. In a certain city the fire insurance premium rate is $0.68 per $100 of the insured amount. There are approximately 10,000 dwellings of an average valuation of $25,000 in the city. It is estimated that the insured amount represents 80% of the value of the dwellings. The city commissioners have been advised that the fire insurance premium rate will be reduced to $0.55 per $100 if the following improvements are made to reduce fire losses.

Improvement	Cost	Life (years)
Increase capacity of trunk water lines from pumping station	$ 80,000	30
Increase pressure in pumps	12,000	20
Purchase two additional fire trucks and related equipment	100,000	20
Add two firemen	30,000/year	

The city is in a position to increase its bonded indebtedness and can sell bonds bearing 7% interest. Should the improvements above be made on the basis of prospective savings in insurance premiums paid by the homeowners? What is the benefit–cost ratio?

12-15. In a western community of 28,000 people, living in 7,000 homes, the water is quite "hard." About one-half of the homeowners have installed water "softeners" provided on a rental basis by a local company. These systems cost $80 for the initial installation and $4 per month for rental, chemicals, and service. It has been proposed that the municipally owned water company install a central system to treat all the water. Such a system would require an investment of $1,200,000 and have annual operating costs of $24,000 for chemicals, $40,000 for labor, $2,000 for increased pumping costs, and $8,000 for maintenance.

From the experience of other communities, it is estimated that the degree of softening of the water obtained will reduce the soap consumption by 5 pounds per person per year, and that soap costs an average of $1.80 per pound. The water-treating system would be financed by a 20-year, 5% issue of bonds. It has been proposed that the cost of the system be met by increasing each customer's water bill by $4 per month. Make an economy study and a recommendation.

chapter 13

economy studies for public utilities

Privately owned, regulated public utilities are an important part of the United States economy and constitute a unique form of business enterprise. Because of their unique characteristics, economy studies in such firms can be performed in a manner slightly different and often somewhat more easily than for the non-regulated sector of the economy. To understand these differences, and the reasons for them, it is helpful to have an understanding of the nature of privately owned, regulated utilities. It is the purpose of this chapter to point out the characteristic features of such utility firms, and then to show how these characteristics affect their economy studies.

Some utility firms are owned and operated by governmental bodies, usually by cities. However, some are state-owned or even Federally owned, such as the TVA. They are basically the same as any other governmentally owned and operated activity, and their economy studies are the same as those discussed in Chapter 12.

the nature of public utilities

Public utilities provide services, such as gas, electric power, water, telephone and radio communication; and transportation, including air, rail, bus, and pipeline. To provide these services, public property, such as streets, highways, or air space, must be utilized, or the utility must be given the right of *eminent domain* to acquire property where and when needed.

Because a utility must have unusually large amounts of capital invested in fixed plant and equipment, economy of operation for the company and lower rates to the public can be brought about only by high use factors for such assets. This means that unnecessary duplication of such facilities would not be economical or in the public interest. For example, two competing electric power companies in the same area would not provide the most economical service. If each had customers on the same street, power-line poles, transformers, distribution

388

lines, and many clerical functions would have to be duplicated. One of the companies could serve all the customers with very little increase in its investment in fixed assets. Similarly, many of the customer costs, such as those for meter reading, would be increased very little. The same general conditions apply to virtually all utilities.

In some types of utility services it is virtually impossible for satisfactory service to be obtained by the public except through a single utility, or at least a group of closely coordinated and noncompeting companies. For example, if two competing telephone companies served the same area and one company would not accept calls from the customers of the other, a completely unsatisfactory situation would result.

Recognizing the advantages to the public—the utility's customers—of avoiding wasteful duplication and competition, the public—again the customers—usually grants a utility an exclusive franchise for its particular service in a given geographical area. Thus most utilities have, in effect, monopolies with respect to the particular service that they render. However, it should be noted that many regulated utilities are not devoid of competition, in that there may be different, but competing, services. For example, a gas company may compete with an electric company in respect to domestic water heating and cooking.

Because the public—acting as cities, states, or the Federal government—grants the monopolistic position to the utility, it retains the right to regulate the utility. The public controls the rates the utility may charge, so that no undue profit will be made as a result of the privilege it has granted. It is obvious that without competition the price to be charged for the services cannot be established through the usual process of competition. Therefore, a public regulatory body is established by the people to exercise the desired control through various types of measures. As a consequence, there is the unique situation of customers determining, by direct regulation, what the supplier of a service can charge.

how utilities are regulated

Intrastate public utilities usually are regulated by a state agency, commonly known as a public utilities commission. The members of the commission may be appointed or elected, depending upon the laws of the individual state. Utilities that operate interstate, such as railroads, bus lines, air lines, telephone companies, and pipeline companies, are regulated by Federal agencies. These include primarily the Interstate Commerce Commission, the Federal Communications Commission, and the Federal Power Commission.

Governmental regulation was established originally to prevent public utility companies from discriminating between customers as to service provided and prices charged. The functions of regulatory agencies have been expanded so that at present they are concerned with the setting of rates and the establishment and maintenance of standards of service.

Basic to the setting of rates for public utility services is the concept that the utility companies must be able to earn profits and to pay dividends sufficient to assure the obtaining of the capital necessary to provide the assets and working capital that are required for rendering the service. If an adequate rate of profit is not obtained, the necessary capital will not be forthcoming from investors, and, as a result, the public will not be able to have the desired utility service. On the other hand, the utility operates at the permission of the public, through necessary franchises or other grants of permission, and thus is granted a monopolistic position by its customers. The public, therefore, has a right to expect that it should pay no more for the services than will permit an efficiently operated utility to earn the minimum required rate of profit that will assure the continuity of service at the level of quality desired. Thus the regulatory commissions have a delicate task of sensing the desires of the public as to the level of service desired and its willingness to pay for it, the price that must be paid for capital in the current money market, and the efficiency of the plans and operations of the utility.

Utility commission control is exercised primarily through the setting or approving of rates that may be charged for utility services. These are set so as to permit the utility to earn the required rate of profit on the capital utilized in rendering the service. This amount of capital is sometimes called the *rate-base value*. The rates approved or set by the commission are subject to appeal to the courts through usual legal procedures, but relatively few cases are taken to the courts.

To carry on their functions properly, the regulatory commissions usually prescribe certain accounting procedures that must be followed by the utilities under their jurisdiction. This assures that the same accounting procedures are followed from year to year and enables some comparisons to be made among utilities. As might be expected, the task of such a regulatory body is no bed of roses. Ordinarily, the public is not in favor of an increase in service rates, even though the regulatory commission knows that they are necessary and in the long-range interest of the public. Similarly, a utility is seldom pleased when a commission orders a decrease in rates, although it may be evident that its profits are above the rate required to obtain necessary capital.

Although the matter of public utility regulation is complex and sometimes seems to be quite ponderous, it has worked rather well. Some of the best-managed companies in this country can be found among public utilities in states where there are strong, but fair, regulatory bodies. Services of excellent quality are consistently provided by these companies at low cost. The greatest difficulties are experienced where regulatory bodies are weak, are not farsighted or aware of current developments, or tend to impose managerial restrictions rather than merely regulation.

characteristics of public utilities

Because of the inherent nature of the services they render, their monopolistic position, and the regulation to which they are subject, public utilities have a number of economic characteristics differentiating them from other businesses which must be taken into account in making economy studies. The major ones are discussed in the following paragraphs.

1. The investment per worker and the ratio of fixed costs to variable costs are very high. Correspondingly, the investment per dollar of gross revenue for typical electric utilities is about three times that of steel companies and about five times that of general manufacturing companies. This results in high fixed costs. It is not uncommon for the fixed costs to be 70% of the total unit cost. This means that careful attention must be given to investment problems and to assuring an adequate flow of capital for expansion.

2. Utilities must render whatever service is demanded by customers, within established rate schedules. Subject to regulatory safeguards, a utility must expand to meet the growth of the community it serves.

3. A utility is *required* to keep abreast of technical developments in its field which would permit reduction in the cost to the customer for its services or would improve the quality of the service if such improvement is demanded by the customers. A utility should be prepared to improve the quality of the service, even though not demanded immediately by the customers, in order to maintain public goodwill and to protect its monopolistic position.

4. The rates charged for a utility's services are based on total costs, including a fair return, after income taxes, on the rate-base value of its property.

5. The earnings of a public utility are virtually limited by the rate base. As a result, profit on sales is of very little significance. If sales income is increased as the result of increased operating costs—for example, by increased and more effective advertising—it may not produce any long-term profit increase. Profits for the current year might be realized, but if the increase were to result in a return that was judged by the regulatory commission to be greater than necessary, a rate reduction would be ordered. Thus the benefits of the improved operation in terms of financial gain to the company would be eliminated. The same situation exists with respect to increased earnings due to improved efficiency of operation.

This situation might appear to, and possibly does, reduce the incentive of a public utility to improve the efficiency of its operations. However, it is a condition that a utility accepts in return for the preferred position granted to it by the public. The public, in turn, has the right to expect the best possible efforts on the part of a utility, and a utility must take a long-range view and recognize that it can progress only through the goodwill of the public. Income increases for a given year are retained by the utility, and lowered rates almost always assure an expansion of service demands and greater *total* profits in the future. This does not, however, alter the fact that the rate of earnings per dollar of investment probably will not change to any great extent.

6. Utilities have much greater stability of income than other companies. The upper limit of earnings, after income taxes, usually is not permitted to exceed about 7 to 8% on equity capital. It should be noted that, although there is a maximum limit put on earnings, there is no guarantee of any such profits, and there is no assurance against loss. However, if the utility can show that it is

operating efficiently, it usually can obtain permission to increase its rates when needed to produce a fair profit, and thus be able to attract needed capital. In effect, the cost of capital to a well-managed utility sets the lower limit of the rate of profit it is permitted to earn, and the upper limit usually is only a little higher. Consequently, for the well-managed utility company there is a stability of earnings that does not exist for the nonutility company, and this has considerable effect in lowering the rate the utility must pay for both borrowed and equity capital.

7. Because of the stable nature of their business and earnings, utilities commonly employ a higher percentage of borrowed capital than do nonutility companies. Whereas most nonutilities seldom use more than about 30% of debt capital, many utilities use 50 to 60% of borrowed capital.

8. The assets of a utility, on the average, involve longer write-off periods than those of nonutilities. This is caused by the physical nature of the assets and by the fact that the monopolistic situation results in less functional depreciation.

9. Utilities must rely on a larger proportion of new capital for expansion than do other companies. This follows from the fact that earnings are so regulated that they are sufficient to pay only for the cost of capital. Therefore, if the earnings are just high enough to meet the payment of dividends demanded by investors, it follows that very few profits can be retained as surplus to provide for expansion. For example, one utility having equity capital of over $1,375,000,000 has only $104,000,000 in earned surplus, accumulated over more than 40 years.

10. Public utilities are much less limited as to the availability of capital than are nonutility companies. This is because of their greater stability of revenues and earnings and the fact that the regulatory agencies recognize that they must be permitted to earn a return that will assure an adequate flow of capital.

some general concepts regarding public utility economy studies

Because of the characteristics of regulated public utilities which have just been discussed, there are some general features that usually are contained in economy studies of such companies. These are as follows:

1. Whereas economy studies of regular companies usually reflect the viewpoint of the owners, those of regulated public utilities more nearly reflect the interests of the customer. Capital investment and current expenditures should result in lowest rates for the customer while achieving or maintaining a desired quality of service and yielding an adequate return on invested capital. Otherwise, proposed projects are not justified.

2. Public utility economy studies usually involve alternative ways of, or alternative programs for, *doing* something. Because a utility is obligated to provide the service demanded by its customers, studies are seldom made of the

economy of doing versus not doing. Instead, it is more often a matter of how to do it most economically.

3. Administrative and general supervision costs frequently are not included. Because these expenses will be about the same for each alternative, they may be omitted.

4. Revenue data seldom are included. Here, again, the revenue ordinarily will be the same, regardless of the method of providing the service.

5. The cost of money, depreciation, income taxes, and property taxes are usually expressed in terms of the capital invested, except in replacement studies. In replacement studies, present salvage value is used for determining cost of interest on capital, depreciation, and income taxes for existing assets. Property taxes are based on actual assessed value, if known, otherwise on original cost.

In other respects, economy studies of public utilities are essentially the same as those for other industries.

computation of income taxes as percent of investment for utility economy studies

Figure 13-1 shows the relationship of operating revenues and costs for a public utility, with emphasis on taxable income (i.e., profits before taxes) and income taxes as a cost.

It is common in the United States that utilities are allowed to earn up to a certain return (i.e., "fair" return) on their rate base, which is generally increased by the amount of any new investment. If the taxable income is a constant proportion of investment and if the effective income tax rate (i.e., income taxes/

FIGURE 13-1. Relationships of operating revenues and costs for a public utility.

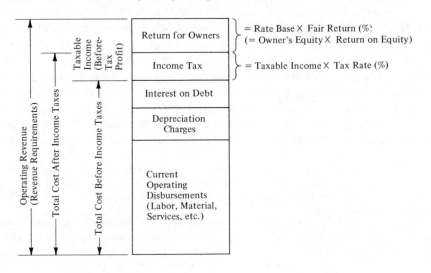

taxable income) is constant, then income taxes comprise a constant proportion of investment. That is,

$$\frac{\text{income taxes/year}}{\text{investment}} = \frac{\text{taxable income/year}}{\text{investment}} \times \frac{\text{income taxes/year}}{\text{taxable income/year}} \quad \text{(13-1)}$$

This results in a great computational convenience for making most after-tax utility economy studies. That is, rather than having to go through the tabular method for determining after-tax cash flows for all alternatives as explained in Chapter 10, one can merely add in the annual cost for income taxes as a percentage of the investment for each alternative.

The following is a typical formula for determining the annual cost of income taxes as a proportion of investment for a particular set of circumstances:

$$T = \left(\frac{t}{1-t}\right)\left[i + (A/F, i\%, N)(1 - c) - \frac{1}{N}(1 - c)\right]\left(1 - \frac{Bb}{i}\right) \quad \text{(13-2)}$$

where T = annual charge rate for income taxes
  (= income taxes/investment)
t = effective income tax rate
  (= income taxes/taxable income)
i = M.A.R.R. on invested capital
  (normally = weighted average cost of capital for utilities)
N = life of project
c = salvage value proportion
  (= salvage value/investment)
B = proportion of debt capital
  (= debt capital/total capital)
b = average interest rate on debt capital
  (= debt interest/year/debt capital)

Equation 13-2 is applicable to situations in which straight line depreciation is used for the determination of taxable income and for the determination of the utility's accounting (or book) profits wherein a fair return is allowed on a rate base equal to the depreciated book value. The term in brackets gives the "levelized after-tax return rate/year" on investment that is adjusted for the timing effects of depreciation (= minimum required profit + sinking fund depreciation − straight line depreciation),† all expressed as a proportion of the investment; while the $(1 - Bb/i)$ term adjusts for the proportion of the return that goes to the owners. Similar equations can be developed for other circumstances,‡ such as when an accelerated depreciation method is used for com-

---

† Since the amount of the depreciation cost computed by straight line procedure would be tax deductible each year, the difference between that amount and the amount shown in the economy study, computed by the sinking fund procedure, should be deducted before computing the income tax.

‡ See, especially, P. H. Jeynes, *Profitability and Economic Choice* (Ames, Iowa: Iowa State University Press, 1968), pp. 181–223, for formal derivations using almost the same notation as in Equation 13-2.

putation of taxable income, even though accounting depreciation is based on the straight line method.

Let us now do some manipulating and defining for later use. Transposing terms in Equation 13-2, we find that

$$T = \left(\frac{t}{1-t}\right)\left(1 - \frac{Bb}{i}\right)\left[i + (A/F, i\%, N)(1 - c) - \frac{1}{N}(1 - c)\right] \quad (13\text{-}3)$$

Letting

$$\phi = \left(\frac{t}{1-t}\right)\left(1 - \frac{Bb}{i}\right) \quad (13\text{-}4)$$

which we shall call the *adjusted income tax factor*, we obtain

$$T = \phi \text{ [after-tax return rate/year on the investment]}$$

$$= \phi\left(\frac{\text{after-tax return/year}}{\text{investment}}\right) \quad (13\text{-}5)$$

The income taxes/year can be computed merely as

$$\text{income taxes/year} = T \text{ (investment)} \quad (13\text{-}6)$$

Substituting Equation 13-5 into Equation 13-6 results in

$$\text{income taxes/year} = (\phi)\left(\frac{\text{after-tax return/year}}{\text{investment}}\right)\text{(investment)} \quad (13\text{-}7)$$

$$= (\phi)(\text{after-tax return/year}) \quad (13\text{-}8)$$

where

$$\text{after-tax return/year} = \text{minimum required profit/year}$$
$$+ \text{ sinking fund depreciation/year}$$
$$- \text{ straight line depreciation/year}$$

EXAMPLE 13-1

What is the annual income tax that will result from a nondepreciable investment of $1,000 where the debt ratio (i.e., debt to total capital) is 0.3, the interest rate on debt capital is 5%, the after-tax rate of return on equity capital is 9.29%, and the income tax rate is 48%?

Solution

The return on total capital may be found by completing a table as follows:

	Ratio	Capital	Return Rate (%)	Return Amount (= capital × return rate)
Debt	0.3	$ 300	5.0	$15
Equity	0.7	700	9.29	65
TOTAL	1.0	$1,000	8.0	$80

Using Equations 13-4 and 13-8,

$$\phi = \left(\frac{0.48}{1 - 0.48}\right)\left(1 - \frac{0.3 \times 0.05}{0.08}\right) = 0.75$$

income tax $= 0.75 \times \$80 = \$60$ ∎

an example of alternative new installations

EXAMPLE 13-2

A public utility must extend electric power service to a new shopping center. A decision must be made as to whether a pole-line or an underground system should be used. The pole-line system would cost only $4,860 to install, but because of numerous changes that are anticipated in the development and use of the shopping center, it is estimated that annual maintenance costs would be $2,300. An underground system would cost $31,500 to install, but the annual maintenance costs would not exceed $550. Annual property taxes are $1\frac{1}{2}\%$ of first cost. The company operates with 33% borrowed capital, on which it pays an interest rate of 4%. Capital should earn about 7% after taxes. The utility is in the 48% income tax bracket. A 20-year study period is to be used.

Solution

The adjusted income tax factor for the utility, using Equation 13-4, is

$$\phi = \left(\frac{0.48}{1 - 0.48}\right)\left(1 - \frac{0.33 \times 0.04}{0.07}\right) = 0.749$$

An after-tax cost comparison between the alternatives is as follows:

	Pole-Line System	Underground System
Depreciation:		
$4,860(A/F, 7\%, 20)$	$ 118	
$31,500(A/F, 7\%, 20)$		$ 769
Minimum required profit:		
$4,860 (7\%)$	340	
$31,500(7\%)$		2,205
Maintenance	2,300	550
Property taxes:		
$1\frac{1}{2}\% \times \$4,860$	73	
$1\frac{1}{2}\% \times \$31,500$		473
Income taxes:		
ϕ (Minimum required profit + sinking fund depreciation − straight line depreciation)		
$0.749[(\$340 + \$118) - \$4,860/20]$	161	
$0.749[(\$2,205 + \$769) - \$31,500/20]$		1,048
	$2,992	$5,045

There would be a very substantial saving in favor of the pole-line system. Of course, this is a case in which intangibles such as appearance and resistance to storm damage might swing the decision to the underground system. ∎

an example of a utility replacement problem

EXAMPLE 13-3

Because of earth slippage, a flume operated by a privately owned company requires annual maintenance costing $10,000. It is estimated that this flume will not be needed beyond 5 years because of changes that are contemplated in the system. A question has arisen as to whether it would be advisable to replace the flume immediately with a new steel pipe costing $40,000, which would require no maintenance expense during the next 5 years, or to continue the high maintenance costs of the old flume. The old flume has been fully depreciated and has no scrap value but is assessed for property tax purposes at $10,000. The new pipeline would have to be written off in 5 years and probably would have no salvage or scrap value because of the high cost of removal. Property taxes are 1% of new or assessed value. The company uses 30% borrowed capital for which it pays an interest rate of 4%. After-tax cost of total capital is 6%. This company has a 50% total income tax rate. It is desired to make after-tax annual cost analyses using the adjusted income tax factor.

Solution

	Retain Old Flume	Replace Now	
Depreciation	$ 0	$40,000(A/F, 6%, 5)	$ 7,100
Maintenance	10,000		0
Property taxes	100		400
Minimum required profit	0	$40,000(0.06)	2,400
Income tax[a]	0		1,245[a]
TOTAL ANNUAL COST	$10,100		$11,145

[a] Using Equations 13-4 and 13-8, we obtain

$$\phi = \left(\frac{0.5}{1 - 0.5}\right)\left(1 - \frac{0.3 \times 0.4}{0.7}\right) = 0.83$$

income tax = $0.83[\$40,000(0.06) + \$40,000(A/F, 6\%, 5) - \$40,000/5] = \$1,245$

These figures show a modest annual advantage ($1,045) in keeping the old flume.

This problem raises a matter that sometimes causes confusion. Why is no income tax included for the old flume? Although it is true that both alternatives will provide the same service, which it is expected will be profitable and produce taxable income, this economy study is concerned only with the difference in taxable income and after-tax profits that might result from replacing the flume. ▮

Results comparable to the above can be obtained by using the tabular format illustrated in Chapters 10 and 11 to obtain after-tax cash flows and then determining the equivalent annual cost for each alternative.

calculation of fixed charge rates

Fixed charge rates are a computational convenience often used in utility economy studies. These rates normally include two or more annual cost components that can be expressed as a constant percent of the investment. In Equations 13-3 through 13-6, T was a fixed charge rate to cover income taxes. A rate to cover the annual cost of depreciation plus return on investment is merely the capital recovery cost expressed as a percent of the investment. Such annual costs as ad valorem (property) taxes and sometimes maintenance can be legitimately estimated as a constant percent of investment. Further, the effect of any investment tax credit can be converted into an annual percent that *reduces* the annual charge rate, since it is a credit against costs.

EXAMPLE 13-4

It is desired to determine a total fixed charge rate of evaluating alternative steam-generating units having an economic life of 28 years and zero salvage value. This fixed charge rate should include the following cost components:

1. Depreciation plus minimum return (interest).
2. Income taxes.
3. Ad valorem taxes.
4. Routine maintenance.
5. Investment tax credit.

The combined effective rate for Federal and State income taxes is 51.1% of the taxable income.† Annual ad valorem taxes are 1% of investment and annual routine maintenance is 5.6% of investment. The investment tax credit is assumed to be a one-time 10% of the investment. The firm uses its weighted average cost of capital of 12% as its M.A.R.R., and its adjusted income tax factor, $\phi = 0.59$.

Solution

Depreciation plus minimum return:

$(A/F, 12\%, 28)(1 - c) + i = 0.0052 + 0.12$ 0.125

Income taxes:

From Equation 13-5,

$$T = (0.59)[0.12 + (A/F, 12\%, 28)(1 - 0) - \tfrac{1}{28}(1 - 0)]$$
$$= 0.59(.09)$$ 0.053

Ad valorem taxes (given)	0.010
Routine maintenance (given)	0.056
Investment tax credit: $-0.10(A/P, 12\%, 28)$	-0.013
	0.231

Thus the fixed charge components add to an annual rate of 0.231, or 23.1%. This means that for each dollar invested in this type of plant, additional revenue of 23.1 cents is needed each year over the 28-year life just to cover the costs included in the fixed charge. ∎

---

† Based on the assumption that state tax (i.e., 6% rate) is deductible from Federal tax (i.e., 48% rate), the combined rate is $6\% + 48\% - 0.48(6\%) = 51.1\%$.

using fixed charge rates in economy studies

Fixed charge rates can be used to make after-tax comparisons of alternatives by an equivalent worth method quite easily if the conditions for which the fixed charge rates are derived are applicable. For cases in which alternatives having different lives are compared using fixed charge rates, the "repeatability" assumptions as discussed in Chapter 7 must be applicable.

EXAMPLE 13-5

Suppose that it is desired to compare the economics of steam versus hydroelectric generating units on a per-kilowatt basis. We shall use a fixed charge rate of 23.1% for the steam unit as calculated in the example above. Similar calculations for hydro units having a life of 56 years result in a fixed charge rate of 21.0%. These and other data for the two alternatives, based on comparable kilowatt-hour generation, are summarized in the following table:

	Steam	Hydro
Investment/kW	$350	$800
Life	28 years	56 years
Other operating costs/year/kW not included in fixed charges	$115	$20
Annual fixed charge rate	23.1%	21%
Minimum return (interest)	12%	12%

Solution by A.W. Method

Fixed charges/kW		
$350 × 23.1%	$ 80.85	
$800 × 21.0%		$168.00
Other operating costs/kW	115.00	$ 20.00
TOTAL ANNUAL COSTS/kW	$195.85	$188.00 ∎

↑
better

Solution by P.W.-C. Method

	Steam	Hydro
Fixed charges/kW		
$350 × 23.1% × $(P/A, 12\%, 56)$	$ 672.50	
$800 × 21% × $(P/A, 12\%, 56)$		$1,397.40
Other operating costs/kW:		
$115(P/A, 12\%, 56)$	956.60	
$20(P/A, 12\%, 56)$		166.40
TOTAL PRESENT WORTH OF COSTS/kW	$1,629.10	$1,563.80

↑
better ∎

an example of immediate versus deferred investment

Many public utility economy studies involve immediate versus deferred investment to meet future demands, since utilities must always be prepared to meet the demands for service placed on them. The following is an example of this type.

EXAMPLE 13-6

A water company must decide whether to install a new pumping plant now and abandon a gravity-feed system, which has been fully depreciated, or wait 5 years, at which time it would have to be installed because of deterioration of the piping in the gravity system. Annual costs for operation, maintenance, and taxes for the gravity system are $45,000. The pumping plant will cost $375,000 to install, and it is estimated that it would have a salvage value of 5% of its initial cost at the time of removal from service 20 years hence, when a new and larger system will be installed. Annual operation, maintenance costs, and property taxes for the proposed plant would be $30,000. Capital costs the company 7% after taxes, and its income tax factor, as defined in Equation 13-4, is 0.85. The gravity-feed system has no salvage value now or later.

If the pumping plant is installed now, it would have an economic life of 20 years. If installed 5 years hence, its economic life would be only 15 years, but its salvage value would still be 5% of its initial cost. Using the present worth–cost method, determine which alternative is better.

Solution

Install new pumping plant now:

Present cost of pumping plant	$375,000	
Less P.W. of salvage: $375,000 × 0.05($P/F$, 7%, 20)	$-4,850$	
P.W. first cost		$370,150
Operation, maintenance, and taxes: $30,000($P/A$, 7%, 20)		312,000
Income taxes:†		
0.85[$370,150 − ($375,000 × 0.95)(1/20)(P/A, 7%, 20)]		158,300
TOTAL P.W.-COST		$840,450

---

† The first cost of an asset (investment) is the present worth of all future sinking fund depreciation plus profit on the investment. In annual cost, after-tax studies by this method there is included:

> Sinking fund depreciation (S.F.depr.)
> Minimum required profit (M.R.P.)
> Income tax = (M.R.P. + S.F. depr. − st. line depr.)

For P.W. cost studies there must be included:

$$\text{P.W.}\begin{pmatrix} \text{S.F. depr.} \\ + \\ \text{M.R.P.} \end{pmatrix} = \text{first cost} = \text{P.W. first cost}$$

$$\begin{aligned} \text{P.W.(income tax)} &= (\text{M.R.P.} + \text{S.F. depr.} - \text{st. line depr.})(P/A, i\%, N) \\ &= [\text{first cost} - (\text{st. line depr.})(P/A, i\%, N)] \end{aligned}$$

Defer installation of new pumping plant for 5 years:

P.W. pumping plant, deferred 5 years: $375,000$(P/F, 7\%, 5)$	$267,000
Less: P.W. of salvage: $375,000 \times 0.05$(P/F, 7\%, 20)$	$-4,850$
P.W. first cost	$262,150
Operation, maintenance, and taxes:	
Old plant: $45,000$(P/A, 7\%, 5)$	185,000
New plant: $30,000$(P/A, 7\%, 15)(P/F, 7\%, 5)$	195,000
Income taxes:	
P.W. first 5 years; (no investment)	0
P.W. last 15 years:†	
$0.85[$368,200 - ($375,000 \times 0.95)(1/15)(P/A, 7\%, 15)]$	
$\times (P/F, 7\%, 5)$	92,300
TOTAL P.W.-COST	$734,450

Thus it appears to be more economical to defer building the pumping plant for 5 years. ∎

the economic basis of utility rates

In establishing the rates for utility services, some arbitrary compromises frequently must be made. Theoretically, the costs of providing such services can be grouped into three classifications: *capacity costs*, *commodity* or *energy costs*, and *customer costs*. These are applicable to virtually any type of utility. Capacity costs are those resulting from the provision of the assets required to produce the commodity or energy sold, and to bring it to a point from which it can be diverted to the individual customers. For a power company, for example, this would include the costs for generators, dams, power plants, substations, transformers, a part of the labor at power plants and substations, transmission lines, and usually most of the distribution lines up to the point of step-down transformers to customers' voltage. For a gas company, these costs would include major pipelines, pumping stations, storage tanks, and a considerable portion of the major gas mains to customer areas. Capacity costs constitute a very large part of the total costs of providing most utility services, frequently at least one-half.

Commodity or energy costs include fuel, most of the labor at power plants and substations, the price paid for gas at the source, operating costs of pumping plants, and most maintenance costs. In a telephone company these costs are primarily those of labor.

Customer costs include investment charges on meters and customer service lines or pipes, a portion of the investment costs of distribution lines or pipe mains in the immediate customer area, and the cost of reading meters, billing, collecting, and the maintenance of business offices.

---

† P.W. of net investment at beginning of fifth year is $375,000 - [$375,000 \times 0.05 \times (P/F, 7\%, 15)] = $368,200$.

Of these three categories, the capacity costs tend to be a function of the maximum capacity that must be provided, the commodity or energy costs tend to be a function of the amount of commodity or energy produced and sold, and the customer costs tend to be proportional to the number of customers served (although there usually is a certain amount of relatively fixed customer cost). Therefore, in setting its rate schedules, a utility theoretically should charge each class of customers for its share of each of these three categories of costs. An attempt is made to do this in actual practice, but the theoretical cost allocation usually must be modified in order to construct a simpler and more workable rate structure and to take into account the practical reality that some types of customers will pay more for a given service than others. However, the actual rates are based upon an analysis of the actual costs.

The following tabulation shows the various annual costs of a gas company distributed according to the three classes.

	Capacity Cost	Commodity Cost	Customer Cost
Investment costs (cost of money, property taxes, income taxes, depreciation)	$23,500,000	—	$ 8,500,000
Production costs	1,500,000	$24,600,000	—
Transmission and distribution	3,500,000	750,000	8,130,000
Other costs	500,000	150,000	1,150,000
TOTAL	$29,000,000	$25,500,000	$17,780,000
Basis for distribution	Per MCF per day	Per MCF	Per customer
Base units for distribution	1,000,000 MCF per day	123,000,000 MCF	550,000 customers
Unit cost	$29.00 per year per MCF per day	$0.2075 per MCF	$32.30 per customer

table 13-1
allocation of costs for industrial versus general gas rates

	Industrial	General	Total
Estimated capacity requirements (MCF per day)	350,000	650,000	1,100,000
Annual consumption (MCF)	48,000,000	75,000,000	123,000,000
Number of customers	50,000	500,000	550,000
Capacity cost: $29,000,000(350/1, 100)	$10,150,000	$18,850,000	$29,000,000
Commodity cost: $25,500,000(48/123)	9,950,000	15,550,000	$25,500,000
Customer cost: $17,780,000(50/550)	1,615,000	16,165,000	$17,780,000
TOTAL COST	$21,715,000	$50,565,000	
Cost per MCF: $21,715,000/48,000,000	$0.462	$0.674	

It next is necessary to allocate the costs according to the two classes of customers. For this utility the customers were divided into two groups—*general* and *industrial*. The general classification included all residential customers, and the industrial customers used gas for any purpose except cooking of meals. The allocation of costs to these two groups was made as shown in Table 13-1 (with calculations included only for industrial customers).

Theoretically, a three-part rate would be required to assess properly each customer with his share of the capacity, commodity, and customer costs. As mentioned previously, this is seldom done because of the complexity of rate schedules that would result. Different types of utilities customarily use different types of schedules. Gas companies usually use quite simple block-type schedules. This company, for instance, utilized the following rate schedules:

	Per Customer per Month
For general gas service:	
First 200 ft^3 or less	$0.91
Next 2,300 ft^3 per 100 ft^3	0.06
Next 17,500 ft^3 per 100 ft^3	0.0575
Next 80,000 ft^3 per 100 ft^3	0.055
Over 100,000 ft^3 per 100 ft^3	0.053
For industrial gas service:	
First 100 MCF, per MCF	$0.51
Next 900 MCF, per MCF	0.499
Next 2,000 MCF, per MCF	0.488
Over 3,000 MCF, per MCF	0.476
Minimum billing charge	
800,000 ft^3 or less,	$40.00
Over 800,000 ft^3	$80.00

These schedules encourage industrial use and, due to the volume involved, yield good profitability for the utility. Furthermore, the only direct attempt to recover the fixed customer costs is through the minimum charge of $0.91 for the first 200 cubic feet or less in the general service schedule and the $40.00 or $80.00 minimum billing in the industrial schedule. This practice is generally followed by gas companies.

Electric companies often use a fixed service charge, as shown earlier in the rate schedule in Example 8-5. Water companies frequently employ rate schedules determined by the size of the installed meter connection. Either the constant monthly meter charge or the initial rate block in a schedule for a larger meter connection is higher than the charge when a smaller meter is used. This helps to meet the higher capacity costs caused by customers with potentially high demands.

In setting up rate schedules for utility services, two objectives are kept in mind. First, the costs for certain portions of services to customers are essentially incremental in nature; this is why most of the early economy studies relative to fixed and increment costs were made in public utility companies. Second, an attempt must be made, by proper cost allocation, to have the amounts paid for service by each class of customers cover the total costs caused by that class, so that one group does not carry a portion of the costs of another. This usually is required by the regulatory commissions. In some cases there is room for doubt as to whether the accounting and cost allocation procedures required by a commission permit this objective to be achieved.†

Cost allocation studies are, of course, very helpful in determining increment costs for purposes of setting up the rates to be charged for the various blocks of service and for special categories of service, such as off-peak power.

problems

13-1. (a) Describe the types of regulation to which public utilities, but not private industries, are subject. Why is regulation necessary?
   (b) How would economy studies differ in a government-owned utility as opposed to a privately owned utility?

13-2. (a) What advantages to the public result from utility companies?
   (b) What disadvantages might there be to public utilities?
   (c) How is regulation of public utilities that operate solely within an individual state achieved as opposed to those that provide services to many states (e.g., telephone companies and gas pipelines)?

13-3. Briefly summarize the basic characteristics that distinguish public utilities from nonregulated industries such as steel, automobile, and chemical manufacturing.

13-4. Why are most public utilities heavily financed with borrowed capital? What characteristics of this industry make it possible to attract large amounts of borrowed capital, and what advantages (disadvantages) are associated with the use of borrowed money?

13-5. Explain why it may be in the best interest of the consuming public for a regulatory agency to permit a public utility to charge sufficiently high rates to allow it to earn an adequate return on its capital.

13-6. (a) In a certain state a member of the Public Utilities Commission said, "I will oppose all rate increases. I am interested only in the rates the customers have to pay today." Comment on the results that could follow if all members of the Commission rigidly followed this concept.
   (b) Comment on this statement: "No company that provides an exclusive and required service, such as electric power, should be permitted to make a profit."

13-7. Is there justification for a privately owned, regulated public utility being permitted to include in its rates the cost of advertising that encourages the public to increase utilization of its service?

13-8. In your own words, give an economic interpretation of Equation 13-2. What

† The so-called "life-line" rates, adopted under pressure by some utilities, are examples. These provide very low rates to senior citizens and some others below a certain income level for a small, basic amount of service.

does the adjusted income tax factor, ϕ, accomplish in economy studies for public utilities?

13-9. A telephone company can provide certain facilities having a 20-year life and zero salvage value by either of two alternatives. Alternative A requires a first cost of $70,000 and $3,000 per year for maintenance. Alternative B will have a first cost of $48,000 and will require $6,000 annually for maintenance. Property taxes and insurance would be 4% of first cost per year for either alternative. The after-tax cost of capital is 8%, with 25% being borrowed at a 5% interest rate. The income tax rate is 48%. Which alternative will provide the lower after-tax cost? Use the adjusted income tax factor in determining taxes for each alternative.

13-10. A gas company must decide to build a new meter-repair and testing facility now or wait 3 years before doing so. It estimates that until the new facility is built its annual costs for these functions will be $90,000 greater than when the new facility is completed. The new facility would cost $900,000, would not be needed beyond 20 years from the present time, and would have an ultimate salvage value of $200,000 at that time. The company uses 30% borrowed capital, paying 7% interest for it, and the regulatory body permits it to earn 8.43% on its equity capital. Assuming that the company has a 48% income tax rate, make an after-tax, present worth–cost analysis, and recommendation.

13-11. The KATV Company is trying to decide how best to install a cable for a cable television system. If the entire installation is made at one time, it will cost $140,000. An alternative is to make the installation in three stages, the first being made immediately at a cost of $56,000, one segment 3 years later at the same cost, and the third segment at the end of 8 years, also costing $56,000. The best estimate is that this system will likely be used no longer than 30 years from the present time, and that neither alternative would have any net salvage value at that time. Annual property taxes and insurance are 4% of the installed cost. Only equity capital would be involved, which costs the company 7%. If the company has a 48% income tax rate, which alternative should be employed?

13-12. An electric utility company has an opportunity to build a small hydroelectric generating plant, of 20,000-kilowatt capacity, on a mountain stream where the flow is quite seasonal. As a consequence, the annual output of energy would be only 40,000,000 kilowatt-hours. The initial cost would be $2,000,000, and it is estimated that the annual operation and maintenance costs would be $32,000 during its estimated 30-year economic life. It is believed that the property would have a salvage value of $200,000 at the end of the 30-year period.

An alternative is to build a geothermal generating plant, which would have the same annual capacity, at a cost of $1,600,000. Because it would have to pay the owners of the property for the geothermal steam, the estimated annual cost for the steam and operation and maintenance is $120,000. A 30-year contract can be obtained on the steam supply, and it is believed that this period is realistic for the economic life of the plant but that the salvage value at that time would be little more than zero.

Property taxes and insurance on either plant would be 2% of first cost per year. The company employs 40% borrowed capital, for which it pays 7%

interest. It earns 8% after taxes on total capital, and it has a 48% income tax rate. Which development should be undertaken?

13-13. A telephone company is considering two plans to provide service required by present demand and the forecasted growth of demand for the coming 18 years.

Alternative A requires an immediate investment of $700,000 in property that has an estimated life of 18 years with 10% terminal salvage value. Annual disbursements for operation and maintenance will be $40,000. Annual property taxes will be 2% of first cost.

Alternative B requires an immediate investment of $400,000 in property that has an estimated life of 18 years with 20% terminal salvage value. Annual disbursements for its operation and maintenance during the first 8 years will be $42,000. After 8 years, an additional investment of $450,000 will be required in property having an estimated life of 10 years with 50% terminal salvage value. After this additional property is installed, annual disbursements (for years 9 to 18) for operation and maintenance of the combined properties will be $72,000. Annual property taxes will be 2% of the first cost of property in service at any time.

The regulatory commission is allowing a 6% "fair return" on depreciated book value to cover the cost of money to the utility. Assume that this rate of return will continue throughout the 18 years. The utility company's effective tax rate is 50%. Straight line depreciation is to be used for both rate regulation and income tax purposes. Half of the utility's financing is by debt with interest at $5\frac{1}{2}$%. Determine which plan minimizes equivalent annual revenue requirements after property and income taxes have been taken into account.

13-14. Capacity, commodity, and customer costs for an electric utility are as follows:

	Capacity Cost	Commodity Cost	Customer Cost
Investment charges	$ 800,000	—	$220,000
Production costs	150,000	$650,000	—
Transmission and distribution costs	75,000	16,000	140,000
Other costs	65,000	140,000	130,000
TOTAL	$1,090,000	$806,000	$490,000
Basis for distribution	Per kW	Per kWh	Per customer
Base units	25,000 kW	70,000,000 kWh	45,000 customers

The utility has 800 industrial customers, whose maximum demand is 7,000 kilowatts and who use 22,000,000 kilowatt-hours of power annually; 4,200 commercial customers, whose peak demand is 8,000 kilowatts and who use 14,000,000 kilowatt-hours annually; and 40,000 residential customers, whose peak demand is 10,000 kilowatts and who use 34,000,000 kilowatt-hours annually. Make a suitable allocation of the costs and determine an appropriate cost per kilowatt-hour for each type of customer.

chapter 14
capital budgeting and an overview of economy studies

Proper financing, budgeting, and management of capital represents the basic top management function of a firm and is crucial to its welfare. Capital budgeting may be defined as the series of decisions by individuals and firms concerning how much and where resources will be obtained and expended to meet future objectives. The scope of capital budgeting includes:

1. How the capital is acquired and from what sources (i.e., financing).
2. How individual capital projects (and combinations of projects) are identified and evaluated.
3. How standards for project acceptability are set.
4. How final project selections are made.
5. How postinvestment reviews are conducted.

This book has concentrated on correct methods of economic evaluation and selection among alternatives. This chapter will introduce other topics within the scope of capital budgeting and present an overview and summary of economy studies.

capital financing

Although the economy study analyst seldom engages in obtaining the capital for projects, the method by which the capital is to be obtained, and whether it is equity or borrowed capital, may be of great importance to him, inasmuch as the costs of obtaining capital and the restrictions that may be imposed upon its use can be quite different. Many well-engineered projects have failed because of improper or too costly financing.

It is possible to make economy studies from two distinctly different viewpoints relative to the capital used. One considers the *total* capital used without regard for source; this method, in effect, evaluates the *project* rather than the interests of any group of capital suppliers. The other looks at a proposed venture from the viewpoint of the suppliers of the equity capital; it thus is concerned with interests of the present owners of a business. Therefore, the engineer who is to make economy studies and economic recommendations should have an understanding of the various methods by which equity and debt capital are obtained and the consequences of the financing method used. The illustrations and problems in this book normally evaluate the *project* because in most engineering economic analyses the choice between alternatives can be made independently of sources of funds to be used. However, the sources of capital may be important in investment analyses in which different sources may be used for different alternatives (e.g., leasing vs. borrowing vs. equity), as well as when it is desired to know the return or cost of alternatives to the owners of a firm who have equity capital at stake.

basic difference between equity and borrowed capital

In previous chapters various differences in the uses of equity and borrowed capital have been mentioned. These can be summarized as follows:

1. Equity capital is supplied and used by its owners in the expectation that a *profit* will be earned. There is no assurance that a profit will, in fact, be gained or that the invested capital will be recovered. Likewise, there are no limitations placed on the use of the funds except those imposed by the owners themselves. There is no cost for the use of such capital, in the ordinary sense of a tax-deductible cost.

2. When borrowed funds are used, a fixed rate of interest must be paid to the suppliers of the capital, and the debt must be repaid at a specified time. The suppliers of debt capital do not share in the profits resulting from the use of the capital; the interest which they receive, of course, comes out of the firm's revenues. In many instances the borrower pledges some type of security to assure that the money will be repaid. Quite frequently, the terms of the loan may place some restrictions on the uses to which the funds may be put, and in some cases restriction may be put on further borrowing. Interest paid for the use of borrowed funds is a tax-deductible expense for the firm.

the corporation

The corporation is a form of organization that was originated to avoid as many as possible of the disadvantages of the individual and partnership forms of ownership of business enterprises. A corporation is a fictitious being, recognized by law, that can engage in almost any type of business transaction in which a real person could occupy himself or herself. It operates under a charter that is granted by a state, most states requiring three or more persons to sign the charter application and be interested stockholders. It enjoys certain privileges, important among which is perpetual life without regard to any change in the person of its owners, the stockholders. In payment for these privileges and the enjoyment of legal entity, the corporation is subject to certain restrictions. It is limited in its field of action by the provisions of its charter, which define the activities in which it may engage. In order to enter new fields of enterprise, it must apply for a revision of its charter or obtain a new one. Special taxes are also assessed against it.

The capital of a corporation is acquired through the sale of stock. The purchasers of the stock are part owners, usually called stockholders, of the corporation and its assets. In this manner the ownership may be spread throughout the entire world, and as a result enormous sums of capital can feasibly be accumulated. With few exceptions, the stockholders of a corporation, although they are the owners and are entitled to share in the profits, are not liable for the debts of the corporation. They are thus never compelled to suffer any loss beyond the value of their stock. Because the life of a corporation is continuous or indefinite, long-term investments can be made and the future faced with some degree of certainty. This makes debt capital (particularly long-term) easier to obtain and generally at a lower interest cost for corporations than for individual and partnership types of business organizations.

The widespread ownership that is possible in corporations usually results in the responsibility for operation being delegated by the owners to a group of hired managers. Frequently, the management may own very little or no stock in the corporation. At the same time, individual stockholders may exercise little or no significant influence in the running of the corporation and may be interested only in the annual dividends that they receive. As a result, the management sometimes tends to make decisions on the basis of what is best for themselves rather than what is best for the stockholders.

One factor that does not favor the corporation form of business organization is that, except in limited circumstances, the profits of a corporation are subject to double taxation of income.† That is, after the corporation income tax is paid, any remaining profits that are distributed as dividends to stockholders are again taxed as income to those stockholders. Thus, if the corporation's income tax rate is 50%, and all remaining profits are distributed to stockholders paying an average of 30% income taxes, $1.00 earned by the corporation becomes $1.00(1 - 0.50) = $0.50 distributed to the stockholder, which becomes $0.50(1 - 0.30) = $0.35 net after taxes. Hence the actual total tax rate is ($1.00 - $0.35)/$1.00 = 0.65 = 65%.

Stock certificates are issued as evidence of stock ownership. The value of stock commonly is measured in three ways. Market value is the price it will bring if sold on the market. Book value is determined by dividing the net equity (net worth) of the corporation by the number of shares outstanding, assuming that

---

† Profits of certain small corporations with no more than 10 stockholders can be taxed as only individual income to those stockholders, thereby avoiding the double taxation.

all the stock is of one type. A third measure is the price–earnings ratio, being the price divided by the annual net income (earnings).

There are a number of types of stock, but two are of primary importance. These are *common stock*, which represents ordinary ownership without special guarantees of return on investment; and *preferred stock*, which has certain privileges and restrictions not available for common stock.

financing with debt capital

There are many situations where the use of borrowed capital is preferable to the use of equity capital. Expansion through the use of equity capital requires either the existing owners to supply more capital or the sale of additional stock to others, which results in decreasing the percentage ownership of the existing stockholders. If the additional capital is needed for a fairly definite period of time and there is considerable assurance that the existing or future cash flow can readily pay the costs and provide for the repayment of borrowed capital, it may be to the advantage of the existing owners to obtain the needed capital by borrowing. It is common for 15 to 30% of the total capital of corporations to be from debt sources.

If additional debt capital is needed only for a short period of time, usually less than 5 years and more frequently less than 2 years, it may be borrowed from a bank or other lending agency by the signing of a short-term note. Such a note is merely a promise to repay the amount borrowed, with interest, at a fixed future date or dates. The lending agency may require something of tangible value as security for the loan, or at least it will make certain the financial position of the borrowing organization is such that there is minimal risk involved.

If capital is obtained through short-term borrowing, the corporation is faced with the necessity either of not needing the capital for long or of refinancing the loan every 2 years or so. Obviously, this prevents long-range planning and investment in projects that have long lives and that may ultimately be quite profitable but may not produce much cash flow during the first few years. Under such conditions there is considerable uncertainty as to whether the money for repayment of a short-term loan will be available when required, or whether refinancing will be available at reasonable cost if this is needed. Because long commitments of capital are required in most projects, corporations usually resort to bond issues for obtaining long-term debt capital.

financing with bonds

A *bond* is essentially a long-term note given to the lender by the borrower, stipulating the terms of repayment and other conditions. In return for the money loaned, the corporation promises to repay the loan and interest upon it at a specified rate. In addition, the corporation may give a deed to certain of its assets that becomes effective if it defaults in the payment of interest or principal as promised. Through these provisions the bondholder has a more stable and secure investment than does the holder of common or preferred stock. Because the bond merely represents corporate indebtedness, the bondholder has no voice in the affairs of the business, at least for as long as his interest is paid, and of course he is not entitled to any share of the profits.

Bonds usually are issued in units of from $100 to $1,000 each, which is known as the *face value* or *par value* of the bond. This is to be repaid the lender at the end of a specified period of time. When the face value has been repaid, the bond is said to have been *retired* or *redeemed*. The interest rate quoted on the bond is called the *bond rate*, and the periodic interest payment due is computed as the face value times the bond interest rate per period.

A description of what happens during the normal life cycle of a bond can be illustrated by the diagram and three-step explanation of Figure 14-1.

FIGURE 14-1.   Life cycle of financing a bond.

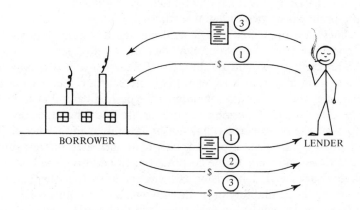

Description of Step	Supplementary Comments
(1) Bond sold by borrower to lender. Lender gets bond certificate.	Bonds are issued in even denominations (face values) like $250, $1,000, $10,000, etc., but amount paid by lender is determined by market supply and demand. The transaction is usually done through a broker.
(2) Periodic bond interest payments are made to lender.	The amount of each bond interest payment is computed as face value times bond interest rate.
(3) Borrower "redeems" bond by paying principal and getting bond certificate back.	Usually done at end of the stated bond life and the amount paid back is usually the face value.

Interest may be paid in either of two ways. The name of the owner of *registered bonds* is recorded on the record books of the corporation, and interest payments are sent to the owner as they become due without any action on his part. *Coupon bonds* have a coupon attached to the bond for each interest payment that will come due during the life of the bond. When an interest payment is due, the holder clips the corresponding coupon from the bond and can convert it into cash, usually at any bank. A registered bond thus requires no action on the part of the holder, but it is not as easily transferred to new ownership as is the coupon bond.

bond retirement

Because stock represents ownership, it is unnecessary for a corporation to make absolute provision for payments (dividends) to the stockholders. If profits remain after the operating expenses are paid, part or all of these can be divided among the stockholders. Bonds, on the other hand, represent debt, and the interest upon them is a cost of doing business. In addition to this periodic cost, the corporation must look forward to the day when the bonds become due and the principal must be repaid to the bondholders. Provision may be made for repaying the principal by two different methods.

If the business has prospered and general market conditions are good when the bonds come due, the corporation may be able to sell a new issue of bonds and use the proceeds to pay off the holders of the old issue. If conditions are right, the new issue may bear a lower rate of interest than the original bonds. If this is the case, the corporation maintains the desired capital at a decreased cost. On the other hand, if business conditions are bad when the time for re-financing arrives, the bond market may not be favorable, and it may be impossible to sell a new bond issue, or possible only at an increased interest rate. This could be a serious handicap to the corporation. In addition, the bondholders wish to have assurance that provision is being made so that there will be no doubt concerning the availability of funds with which their bonds will be retired.

When it is desired to repay long-term loans and thus reduce corporate indebtedness, a systematic program frequently is adopted for repayment of a bond issue when it becomes due. Some such provision, planned in advance, gives assurance to the bond holders and makes the bonds more attractive to the investing public; it may also allow the bonds to be issued at a lower rate of interest.

In many cases the corporation periodically sets aside definite sums which, with the interest they earn, will accumulate to the amount needed to retire the bonds at the time they are due. Because it is convenient to have these periodic deposits equal in amount, the retirement procedure becomes a sinking fund. This is one of the most common uses of a sinking fund. By its use the bondholders know that adequate provision is being made to safeguard their investment. The corporation knows in advance what the annual cost for bond retirement will be.

If a bond issue of $100,000 in 10-year bonds, in $1,000 units, paying 6% interest in semiannual payments, must be retired by the use of a sinking fund that earns 4% compounded semiannually, the semiannual cost for retirement will be as follows:

$$A = F(A/F, i, N)$$
$$F = \$100,000$$
$$i = \tfrac{4}{2} = 2\% \text{ per period}$$
$$N = 2 \times 10 = 20 \text{ periods}$$

Thus

$$A = \$100,000(0.04116) = \$4,116$$

In addition, the semiannual interest on the bonds must be paid. This would be calculated as follows:

$$I = \$100,000 \times \frac{0.06}{2} = \$3,000$$

$$\text{total semiannual cost} = \$4,116 + \$3,000 = \$7,116$$

$$\text{annual cost} = \$7,116 \times 2 = \$14,232$$

The total cost for interest and retirement of the entire bond issue over 20 periods (10 years) will be

$$\$7,116 \times 20 = \$142,320$$

callable bonds

In the problem just considered, no bond was retired until the end of 10 years. The money in the sinking fund earned interest at the rate of 4% compounded semiannually. At the same time, the corporation was paying 6% compounded semiannually on the money it had borrowed. It would obviously be advantageous for the company to be able to buy back (redeem) its outstanding bonds with the money in the sinking fund and obtain an interest saving of 2%. Because bonds usually have a higher rate of interest than can be obtained on sinking fund deposits in a bank or in short-term investments, it is an advantage for the corporation to be able to retire bonds whenever excess funds are available. Through such retirement of bonds the corporation is, in effect, investing its sinking funds in its own bonds. This is sometimes referred to as *serial retirement* of bonds.

Obviously, although the retirement of bonds before they are due usually is advantageous for the corporation, it usually is not desired by the bondholder. He does not want to relinquish a good bond that pays a satisfactory yield. Consequently, corporations frequently issue *callable bonds*. The terms of such bonds specifically permit them to be recalled prior to maturity. In some instances they are *callable by lot*; another procedure is to state that bonds having certain serial numbers will be retired at specified dates.

Now consider what saving would be effected if the bond issue cited previously had been callable. Since the interest rate on the sinking fund would be the same as the bond rate, the total semiannual cost would be

$$A + I = F[(A/F, i\%, N) + i] = F(A/P, i\%, N)$$

To retire the issue in 10 years:

$$F = \$100,000$$
$$i = \tfrac{6}{2} = 3\% \text{ per period}$$
$$N = 10 \times 2 = 20 \text{ periods}$$
$$A + I = \$100,000 \times 0.0672 = \$6,720$$

The interest that must be paid the first period is $3,000. The amount left for the bond retirement would be

$$\$6,720 - \$3,000 = \$3,720$$

Because the bonds are in units of $1,000, any retirement must be in a multiple of this amount. The multiple of $1,000 nearest to $3,720 is four. Thus at the

table 14-1
amortization schedule for a $100,000 bond issue

Interest Period	Principal	Interest at 3% per Period	Number of Bonds Retired	Principal Repaid	Total Payment for Period
1	$100,000	$3,000	4	$ 4,000	$ 7,000
2	96,000	2,880	4	4,000	6,880
3	92,000	2,760	4	4,000	6,760
4	88,000	2,640	4	4,000	6,640
5	84,000	2,520	4	4,000	6,520
6	80,000	2,400	4	4,000	6,400
7	76,000	2,280	4	4,000	6,280
8	72,000	2,160	5	5,000	7,160
9	67,000	2,010	5	5,000	7,010
10	62,000	1,860	5	5,000	6,860
11	57,000	1,710	5	5,000	6,710
12	52,000	1,560	5	5,000	6,560
13	47,000	1,410	5	5,000	6,410
14	42,000	1,260	5	5,000	6,260
15	37,000	1,110	6	6,000	7,110
16	31,000	930	6	6,000	6,930
17	25,000	750	6	6,000	6,750
18	19,000	570	6	6,000	6,570
19	13,000	390	6	6,000	6,390
20	7,000	210	7	7,000	7,210
			TOTAL	$100,000	$134,410

end of the first 6 months the bond interest would be paid, and, in addition, four $1,000 bonds would be retired. If a similar procedure is carried out at the end of each interest period, the results will be as shown in Table 14-1. From the $6,720 set aside each period, the interest on the outstanding bonds is first paid. With the money remaining, the corporation retires as many bonds as possible— using the number of bonds that have a redemption value nearest to the amount remaining. In some cases this will require slightly more than the normally provided amount and at other times less will be required. The difference will never be greater than one-half the price of the bond. Such a process of bond retirement is known as *amortization*,† and a tabulation like that shown in Table 14-1 is called an *amortization schedule*.

From the amortization schedule it is seen that the total cost of interest and bond retirement by this method is $134,410. By comparing this with $142,320, which was the total cost of the $100,000 issue of noncallable bonds, we can see that the different method of retirement resulted in a saving of $7,910 over the 10-year period. This is almost 8% of the face value of the bond issue. The potential advantage of callable bonds to the corporation is thus quite apparent.

In some bond issues provision is made in the amortization schedule that no bonds will have to be retired during the first few years. This recognizes the fact that bonds usually are sold to finance long-term expansion. Such programs ordinarily involve plants or other facilities that take some time to construct and place in operation and thus will not produce profits immediately. Bond retirement is thus deferred until the facilities have an opportunity to become productive.

In some cases the issuer of noncallable bonds can obtain the advantages of callable bonds by buying back his own bonds on the open market. He, in effect, invests his sinking funds in his own bonds, thus receiving the same interest rate upon the sinking funds as the bond rate. Of course, this relationship would be exactly true only when he is able to purchase his bonds on the open market at par. Frequently, he is not able to do this.

bond value

A bond provides an excellent example of commercial value being the present worth of the future net cash flows that are expected to be received through ownership of a property, as discussed in Chapter 9. Thus the value of a bond, at any time, is the present worth of the future payments that will be received. For a bond, let

Z = face, or par, value
C = redemption or disposal price (usually equal to Z)

---

† The term *amortization* may be applied to any method of extinguishing a debt, principal and interest, by a series of payments, usually at equal intervals.

r = bond rate per interest period
N = number of periods before redemption
i = investment rate per period
V_N = value of the bond N periods prior to redemption

The owner of a bond will receive two types of payments. The first consists of the series of periodic interest payments he will receive until the bond is retired. There will be N such payments, each amounting to Zr. These constitute an annuity of N payments. In addition, when the bond is retired or sold, he will receive a single payment in amount equal to C. Inasmuch as the present value of the bond is the present value of these two types of payments at the investment or yield rate,

$$V_N = C(P/F, i\%, N) + Zr(P/A, i\%, N) \tag{14-1}$$

EXAMPLE 14-1

Find the cost (present value) of a 10-year 6% bond, interest payable semianually, that is redeemable at par, if bought to yield 5%; par value is $1,000.

$$N = 10 \times 2 = 20 \text{ periods}$$
$$r = \tfrac{6}{2} = 3\% \text{ per period}$$
$$i = \tfrac{5}{2} = 2.5\% \text{ per period}$$
$$C = Z = \$1,000$$

Solution

Using Equation 14-1, we obtain

$$V_N = \$1,000(P/F, 2.5\%, 20) + \$1,000(0.03)(P/A, 2.5\%, 20)$$
$$= \$1,000(0.6103) + \$1,000(0.03)(15.5892)$$
$$= \$610.27 + \$467.67 = \$1,077.94 \quad \blacksquare$$

depreciation funds as a source of capital

As was explained in Chapter 9, the funds that are set aside out of revenue as the cost of depreciation are a part of the net cash flow and usually are retained and used in a business. These funds are available for reinvestment and must be utilized to the best advantage. Thus they are an important internal source of capital for financing new projects.

Because one of the purposes of depreciation accounting is to provide for replacement of a property when required, we might conclude that depreciation funds provide only for such replacement and never for new equipment of a different type. This, however, is only partially true. In a great many instances when a particular property or piece of equipment has become of no further economic value, the function for which it was originally purchased no longer exists. Under such conditions, we do not wish to replace it with a similar piece of equipment. Instead, other needs have developed, and different equipment or property is needed. The accumulated depreciation funds may thus be used to meet the new needs.

In other instances a piece of equipment may continue to be used after its original value has been recovered through normal depreciation procedures. Here again the accumulated funds are available for other use until the original equipment must be replaced. If depreciation procedures are used that provide for the recovery of a large portion of the initial cost during the first few years of life, such as the declining balance or the SYD methods, there usually will be excess funds available before the equipment must be replaced. This is one of the reasons why companies like to use these methods of accounting for depreciation.

In effect, the depreciation funds provide a revolving investment fund that may be used to the best possible advantage. The funds are thus an important source of capital for financing new ventures within an existing enterprise. They are particularly important because, not being subject to income taxes, they are available in their entirety. Obviously, the depreciation funds must be managed so that required capital is available for replacing essential equipment when the time for replacement arrives.

financing through retained profits

Another important source of internal capital for expansion of existing enterprises is retained profits that are reinvested in the business instead of being paid to the owners. Although this method of financing is used by most companies, there are three factors that tend to limit its use.

Probably the greatest deterrent is the fact that the owners (the stockholders in the case of a corporation) usually expect and demand that they receive some profits from their investment. Therefore, it usually is necessary for a large portion of the profits to be paid to the owners in the form of dividends. This is essential to assure the continued availability of equity (ownership) capital when it is needed. However, it usually is possible, and a good practice, to obtain part of the needed capital for expansion by retaining a portion of the profits. Often from 10 to 50% is retained. Although such a retention of profits reduces the immediate amount of the dividends per share of stock, it increases the book value of the stock and should also result in greater future dividends and/or market resale value for the stock.

Many investors prefer to have some of the profits retained and reinvested so as to help in increasing the value of their stock. The preferential treatment given to *capital gains*†—profits derived through the sale of stock for more than its cost—by the income tax laws is no small factor in this preference.

The second, also serious, limitation on the use of retained profits is the fact that income taxes must be paid upon them and deducted from them before they

† Income tax provisions for capital gains are discussed in Chapter 10.

may be used. With Federal taxes taking as much as 14 to 70% of the individual's income and 48% of the income of most corporations, this severely limits the amount available after dividends and taxes have been paid.

A third, and less important, deterrent is the fact that as profits are retained and used, the total annual profits and the profits per share of stock should increase. This gives the impression that a company is able to pay higher wage rates. Such an implication frequently is used by unions in wage negotiations without acknowledgment that the larger profits are due to the investment of increased amounts of capital by each shareholder through retained profits. This disadvantage may be quite easily avoided by issuing stock dividends in lieu of the retained profits, thereby maintaining a fairly constant rate of profit per share.

It should be noted here that regulated public utilities usually cannot retain a large percentage of their profits, since they are permitted to charge rates only sufficiently high to provide profits at a rate that will just assure a flow of required capital into the business.

leasing versus purchasing of assets

In recent years there has been a considerable increase in the practice of leasing rather than purchasing many assets. Most frequently, automobiles, trucks, buildings, or various types of equipment are involved. Such alternatives may arise either in connection with a consideration of the purchase of completely new assets or in connection with the possible replacement of existing assets. In most cases it is essential to consider the income tax implications so that we can compare accurately the alternatives of purchasing versus leasing or retention versus leasing.

The reasons companies decide to lease rather than purchase are not as simple and clear as might be expected. We quite naturally would think that a decision in favor of leasing would be based on the fact that it was more economical, yet this very often is not the basis for the decision. One study involving companies that had leased fleets of automobiles and trucks revealed that in the majority of cases, no detailed comparison of the costs of leasing versus purchasing had been made. The major reasons given by these companies were, in order of frequency, as follows:

1. Leasing freed needed working capital, or enabled needed equipment to be acquired without the company going into debt.
2. Leasing was thought to have some income tax advantages.
3. Leasing reduced maintenance and administrative problems; for example, it eliminated the necessity for a company to be a fleet operator, leaving this to the specialist.
4. Leasing was cheaper.

Inasmuch as leasing is only one of several ways of obtaining working capital, a decision to lease should consider the cost of obtaining capital by all possible methods. In the study cited, it was found that very few of the companies had made such a comparison. It is true that borrowing capital establishes some fixed obligations, but most leases cannot be canceled, or can be only with costly penalties. A lease is essentially a 100% mortgage on property.

A number of studies have shown that there is no real income tax advantage in leasing. This is particularly true, since the declining balance and SYD methods have been permitted for depreciation. Assuming a given purchase price, the lessor can charge no more for depreciation than can the owner of assets. If assets are leased, the annual lease payments are deducted in computing income taxes; if the assets are purchased, annual depreciation is deducted. Most companies have now come to realize that leasing does not offer tax advantages, but the myth still persists in some quarters.

There may or may not be savings in maintenance costs through leasing. This will depend on the actual circumstances, which should be carefully evaluated in each case. There is no doubt that leasing usually does simplify maintenance problems, and this may be an important factor. Also, many indirect costs, which frequently are difficult to determine, will usually be associated with ownership.

In many instances leasing can be shown to be cheaper than owning, but the actual comparative costs and all other factors should be considered before a decision to lease is made.

The following example illustrates a method of handling a lease versus purchase study on an after-tax basis.

EXAMPLE 14-2

A company is considering building a small office building at a cost of $100,000 on land that would cost $20,000. It is estimated that the land could be sold at its cost at the end of 50 years, and that the building, although fully depreciated on a straight line basis over the 50 years, probably would have a salvage value of $20,000. Annual disbursements for maintenance, taxes, insurance, and so on, are estimated at $5,250. An alternative is to lease a building for $16,500 per year for a 50-year period. The company ordinarily obtains about 20% on its capital before taxes and, being in a 50% tax bracket, about 10% after taxes. What should it do?

Solution (Using I.R.R. Method)

The following table for computing after-tax cash flows will facilitate understanding:

Year	(A) Before-Tax Cash Flow	(B) Depreci- ation	(C)=(A)+(B) Taxable Income	D=−(C)×50% Cash Flow for Income Taxes	(E)=(A)+(D) After-Tax Cash Flow
Purchase Building:					
0	− $120,000				− $120,000
1–50	− 5,250	− $2,000	− $ 7,250	+ $3,625	− 1,625
50	40,000		+ 20,000[a]	− 6,000	+ 34,000
Lease Building:					
1–50	− $ 16,500		− $16,500	+ $8,250	− $ 8,250

[a] Capital gain on building.

The I.R.R. on investment required to purchase rather than to lease the building is the $i'\%$ at which

$$-\$120,000 - [\$1,625 - (-\$8,250)](P/A, i'\%, 50) + \$34,000(P/F, i'\%, 50) = 0$$

The $i'\%$ can be found to be approximately 5.1%. Since 5.1% < 10%, this means that it would be better to lease the building. Of course, any of the other theoretically correct methods would result in the same recommendation. ∎

the average cost of capital and the effect of income taxes

Most businesses do not operate entirely on either equity capital or borrowed capital. Instead, in most instances good practice calls for most of the capital to be obtained from equity sources, a smaller portion being borrowed. To obtain the desired amount of equity capital, a sufficient rate of dividends must be maintained to make the investment attractive to investors in view of the risks involved. Because of the absence of security, the rate that must be paid for equity capital is virtually always greater than must be paid to obtain borrowed capital.

One factor that is usually important to decisions regarding debt versus equity capital is the probable effect of income taxes on the cost of that capital to the firm. Any interest paid for the use of money borrowed by a firm (or individual) is deductible from income or profits reported for Federal and state income tax purposes. Hence taxes are saved when borrowed funds are used to finance a project.

For example, suppose that $10,000 is borrowed at an annual interest rate of 8%, and the effective (combining Federal and state) income tax rate is 60%. The annual interest is

$$I = P \times i = \$10,000 \times 8\% = \$800$$

This $800 reduces the net profits or income on which taxes must be paid and hence results in *saving* income taxes in the amount of $800 × 60% = $480. Therefore, the net real after-tax cost of the interest is $800 − $480 = $320, and the after-tax rate of interest cost = $320/$10,000 = 3.2%. In other words, if there are other profits so that $480 of the interest cost can be charged against the taxable income on the profits, the $10,000 of debt capital need produce only $320 of added income to satisfy the demands of the suppliers of the capital. Without other taxable profits, the maximum the $10,000 would have to earn, in order to satisfy the interest demands, would be $800.

On the other hand, dividends paid by a firm for the use of equity capital are not deductible from the income or profits of the firm, on which income taxes must be paid. Hence, if a firm plans to pay 7% dividends in order to satisfy the stockholders who supply $10,000 of equity capital, the investment would

table 14-2

example of average cost of capital for various percentages of equity and borrowed capital, both before and after income taxes

	Borrowed Capital					
	0%	20%	40%	60%	80%	100%
(1) Total capital	$10,000	$10,000	$10,000	$10,000	$10,000	$10,000
(2) Borrowed capital [(1) × % borrowed]	0	2,000	4,000	6,000	8,000	10,000
(3) Before-tax interest cost [(2) × 8%]	0	160	320	480	640	800
(4) After-tax interest cost [(3) × (1 − 0.60)]	0	64	128	192	256	320
(5) Equity capital [(1) − (2)]	10,000	8,000	6,000	4,000	2,000	0
(6) Dividends cost [(5) × 7%]	700	560	420	280	140	0
(7) Before-tax earnings required to pay dividends [(6)/(1 − 0.60)]	1,750	1,400	1,050	700	350	0
(8) Total cost before taxes [(3) + (7)]	1,750	1,560	1,370	1,180	990	800
(9) Total cost after taxes [(4) + (6)]	700	624	548	472	396	320
(10) Average rate before taxes [(8)/(1)]	17.5%	15.6%	13.7%	11.8%	9.9%	8.0%
(11) Average rate after taxes [(9)/(1)]	7.00%	6.24%	5.48%	4.72%	3.96%	3.20%

421

have to earn \$1,750 [= \$10,000 × 7% ÷ (1.0 − 0.6)] before taxes. At a 60% income tax rate, \$1,700 × 0.60 = \$1,050 would be paid in income taxes, leaving \$700 to satisfy the stockholders. Consequently, in this case the before-tax rate for the capital is 17.5%, and the after-tax rate is 7%.

It is apparent that if the effects of income taxes are ignored, borrowed capital at 8% costs more than equity capital at 7%. However, the much more meaningful comparison, with income taxes being considered, shows that the borrowed capital at 3.2% (under most conditions) is much less costly for financing a project than equity capital at 7%.

When both types of capital are used for the general financing of an enterprise, the real cost of the capital is not that paid for equity capital or for borrowed capital. Instead, it is some intermediate rate, depending upon the proportions of each type of capital used. For example, suppose that for the case above involving \$10,000 capital that might be borrowed for 8% (with an income tax rate of 60%) or obtained from stockholders for 7%, one is considering various mixes of debt and equity capital. Table 14-2 (see p. 421) shows the average cost of several combinations of debt and equity capital both before and after considering the effect of income taxes.

In many economy studies it is clear that only borrowed capital or only owned capital will be employed. In such cases the corresponding cost of capital is sometimes used in the calculations. In most economy studies, however, all money is assumed to come from a common investment pool, and considerations of financing are taken up separately from the project economy study.

after-tax evaluations based on equity investment

To perform after-tax economic analyses on equity (owners') capital rather than total capital required for a project, one merely needs to convert all cash flows into *after-tax equity cash flow*. This can be done by adding a column for "Loan and Interest" in the standard after-tax cash flow table presented in Chapter 10, as will be illustrated in the following example:

EXAMPLE 14-3

This example is the same as Example 10-6(b) except that information is given on borrowed capital and it is desired to determine the after-tax I.R.R. on equity capital. It is restated with additional information as follows. Certain new machinery is estimated to cost \$180,000 installed. It is expected to reduce net annual operating disbursements by \$36,000 per year for 10 years and to have \$30,000 salvage value at the end of the 10th year. \$50,000 of the investment is to be borrowed at 5% annual interest, with the principal due at the end of the 10th year. Show a tabulation to determine the after-tax equity cash flow and calculate the after-tax I.R.R. on equity capital.

table 14-3
computations to determine the after-tax equity cash flow for example 14-3

Year	(A) Before-Tax Cash Flow	(B) Depreciation	(C) Loan and Interest	(D) = (A) + (B) + Interest portion of (C) Taxable Income	E = −(D) × 48% Cash Flow for Income Taxes	(F) = (A) + (C) + (E) After-Tax Equity Cash Flow
0	− $180,000		+ $50,000			− $130,000
1–10	+ 36,000	− $15,000	− 2,500	+ $18,500	− $8,880	+ 24,620
10	+ 30,000		− 50,000			− 20,000

Solution

Table 14-3 shows the recommended tabular format with the column headings indicating how the calculations are made. From the results of the right-hand column, the after-tax I.R.R. on equity investment can be calculated from solving the following:

$$-130,000 + \$24,620(P/A, i'\%, 10) - \$20,000(P/F, i'\%, 10) = 0$$

at $i' = 10\%$: $-\$130,000 + \$24,620(6.1446) - \$20,000(0.3855) = +\$13,570$

at $i' = 15\%$: $-\$130,000 + \$24,620(5.0187) - \$20,000(0.2149) = -\$10,789$

$$i' = \text{I.R.R.} = 10\% + \left(\frac{\$13,570}{\$13,570 + \$10,789}\right)(15\% - 10\%) = 12.8\%$$

Thus the after-tax I.R.R. on the equity investment, 12.8%, is substantially greater than the after-tax I.R.R. on the total investment for the project, which was shown to be 8.9% in Example 10-6(b). ∎

capital budgeting—selection among independent capital projects

Most companies constantly are presented with a number of opportunities in which they can invest capital. In most cases the amount of capital available is limited, or additional amounts can be obtained only at increasing incremental cost. They thus have a problem of budgeting, or allocating, the available capital to the various possible uses.

All methods that can be used for determining to which possible projects available funds should be allocated seem to require the exercise of judgment. The situation, and possible methods of solution, can be illustrated with a simple example.

EXAMPLE 14-4

Assume that a firm has five investment opportunities available, which require the indicated amounts of capital and which have economic lives and prospective rates of return as shown in Table 14-4. Further, assume that the five ventures are mutually independent of each other—investment in one does not prevent investment in any other, and none is dependent upon another being undertaken.

table 14-4
example prospective projects for a firm

Project	Investment	Life (years)	Rate of Return (%)
A	$40,000	5	7
B	15,000	5	10
C	20,000	10	8
D	25,000	15	6
E	10,000	4	5

table 14-5

example prospective projects of table 14-4 ordered
according to rate return

Project	Investment	Life (years)	Rate of Return (%)
B	$15,000	5	10
C	20,000	10	8
A	40,000	5	7
D	25,000	15	6

Now let it be assumed that the company has unlimited funds available, or at least sufficient funds to finance all these projects, and that capital funds cost the company 6%. For these conditions the company probably would decide to undertake all projects that offered a return of at least 6%, and thus projects A, B, C, and D would be financed. However, such a conclusion would assume that the risks associated with each project are reasonable in the light of the prospective profit rate or are no greater than those encountered in the normal projects of the company.

Unfortunately, in most cases the amount of capital is limited, either by absolute amount or increasing cost. If the total of capital funds available is $60,000, the decision becomes more difficult. Here it would be helpful to list the projects in order of decreasing profitability in Table 14-5 (omitting the undesirable project E). Here it is clear that a complication exists. We quite naturally would wish to undertake those ventures that have the greatest profit potential. However, if projects B and C are undertaken, there will not be sufficient capital for financing project A, which offers the next greatest rate of return. Projects B, C, and D could be undertaken and would provide an annual return of $4,600 (= $15,000 × 10% + $20,000 × 8% + $25,000 × 6%). If project A were undertaken, together with either B or C, the total annual return would be less.† A further complicating factor is the fact that project D involves a longer life than the others. It thus is apparent that we might not always decide to adopt the alternative that offers the greatest profit.

table 14-6

example prospective projects of table 14-5 ordered according to
overall desirability

Project	Investment	Life (years)	Rate of Return (%)	Risk Rating
C	$20,000	10	8	Lower
A	40,000	5	7	Average
B	15,000	5	10	Higher
D	25,000	15	6	Average

† This, of course, assumes that the leftover capital could not be used profitably.

The problem of allocating limited capital becomes even more complex when the risks associated with the various available projects are not the same. Assume that the risks associated with project B are considered to be higher than the average of projects undertaken, and that those associated with project C are lower than average. The company thus might rank the projects according to their desirability, as in Table 14-6 (see p. 425). Under these conditions the company might decide to finance projects C and A, thus avoiding one project with a higher-than-average risk and another having the lowest prospective return and longest life of the group.

selection among independent capital projects using risk categories

Another means of budgeting capital while explicitly considering risk is to place projects into two or more risk categories and to determine in advance what approximate proportion of the available capital will be invested in each category. Then one can provisionally pick which projects earn the highest returns within the capital available for each risk category. After this is done, if one judges that a certain provisionally rejected project with a certain risk and rate of return is more desirable than some other project with a relatively higher risk and higher rate of return (or relatively lower risk and lower rate of return), then the trade-off can be made as long as the total capital required and the resulting amount of capital invested in each risk category is reasonably in balance considering the objectives of the firm and the risks and returns involved.

EXAMPLE 14-5

Table 14-7 illustrates projects F through R categorized according to high, medium, or low risk. (The consideration of these projects is completely separate from the previous consideration of projects A through E in Example 14-4.)

table 14-7
example prospective projects with risk categories

High Risk			Medium Risk			Low Risk		
Project	Invest- ment	Rate of Return (%)	Project	Invest- ment	Rate of Return (%)	Project	Invest- ment	Rate of Return (%)
F	$2,000	30	K	$3,500	25	N	$1,500	21
G	2,000	28	L	1,500	24	O	5,000	19
H	1,000	23	R	1,000	22	P	3,500	17
J	1,500	16				Q	6,500	14
M	2,500	15						

Suppose that the firm has $15,000 capital and management desires to invest approximately ⅓, or $5,000, in each risk category as long as the return seems commensurate with the risk. Thus the firm would provisionally accept projects F, G, and H in the high-risk category and projects K and L in the medium-risk category. In the low-risk category the firm could provisionally accept either projects N and P or project O. Suppose it is judged that low-risk project N earning 21% on an investment of $1,500 is preferred to the provisionally accepted medium-risk project L earning 24% on the same investment amount. Hence project O is the apparent remaining provisional choice in the low-risk category. Other trade-offs between risk categories can now be considered for any combination of projects, keeping in mind the $15,000 total capital constraint. Suppose that after consideration of many combinations, the only remaining trade-off that is judged wise is to accept medium-risk project R, earning 22%, rather than high-risk project H, earning 23%.

Summarizing the results of these judgmental decisions, we find that the final acceptances are projects F and G, requiring $4,000 in the high-risk category; projects K and R, requiring $4,500 in the medium-risk category; and projects N and O, requiring $6,500 in the low-risk category. Thus the final allocation of capital among the risk categories is a bit more conservative than the initially planned one-third split.

setting minimum attractive return objectives for project acceptability

It is very common that a firm has limitations on the total capital available for investment in a given period, and that it needs to establish standard(s) for the minimum attractive rate of return to use as a criterion for choice in project economic analyses for that period. An oversimplified explanation of how this may be accomplished is as follows.

In general, the firm should estimate the project opportunities, including investment requirements and prospective rates of return for each, expected to be available for the coming period. Then the available capital should be tentatively allocated to the most favorable projects. The lowest prospective rate of return within the capital available then becomes the minimum acceptable rate of return for analyses of any projects during that period. This minimum attractive return should be at least as high as, and is usually much higher than, the actual cost of capital computed as in Table 14-2.

EXAMPLE 14-6

Suppose that the firm having tentative prospective projects A through E and $60,000 available capital as presented in Example 14-4 and Table 14-4 desires to set its minimum attractive rate of return for the period. The explanation accompanying Table 14-4 resulted in the indication that projects B, C, and D earning 10%, 8%, and 6% rates of return, respectively, would be selected. In this case the minimum acceptable rate of return for investment projects in that period would be

6%. This means that even though prospective projects other than A through E might become available, the firm would not be willing to invest in any projects earning less than 6%.

setting minimum attractive return objectives for various risk categories

A firm will often attempt to set various minimum acceptable rates of return according to level of risk, as in the following example.

EXAMPLE 14-7

Suppose that the firm with prospective projects F through R as presented in Example 14-5 (Table 14-7) and $15,000 available capital desires to set minimum attractive rates of return for each of the three risk categories. The accompanying explanation resulted in the indication that the following projects and rates of return would be accepted. Thus the minimum attractive rates of return for the various risk categories would be: high, 28%; medium, 22%; and low, 19%.

High Risk		Medium Risk		Low Risk	
Project	Rate of Return (%)	Project	Rate of Return (%)	Project	Rate of Return (%)
F	30	K	24	N	21
G	28	R	22	O	19

One other obvious problem that exists in such capital allocation decisions is the matter of *long-term* versus *short-term* investment. If available funds are invested in projects that have long lives, this may mean that some even more profitable, but now unforeseen, projects may have to be passed up because of the previous commitment of funds. On the other hand, if good, and immediately available, projects are passed over, in the hope that some better projects will arise in the future, we may be disappointed when no such projects appear. It is thus clear that the allocation of capital funds is not a matter that can be handled in a routine manner. Judgment and proper consideration of company goals and competition must always be exercised.

an overview of economy studies

The remainder of this chapter is devoted to the following considerations that are vital to the successful application of economy studies:

1. Preliminary planning and screening.
2. Use of standard and acceptable economy study methods.

3. Estimating and reliability of data.
4. Postaudits of actual versus estimated performance.
5. Objectivity of the analyst.

preliminary planning and screening

The most effective guarantee of the quality and usefulness of an economy study is to make sure that adequate preliminary planning and screening of alternatives are performed.

Preliminary planning includes study of the objectives to be satisfied, review of the facts pertinent to the problem, and identification of all logical plans or alternatives that will meet the objectives. Often this will result in a very large number of alternatives and subalternatives. These can be screened by preparing preliminary estimates and making rough economic comparisons, and dropping obviously inferior plans at once from further evaluation. The analyst then can make detailed economy studies of those alternatives that appear most promising.

Economy studies of alternatives, no matter how well performed, cannot result in the selection of the best plan if that potentially best plan is not identified and considered in the first place. There is a reasonable limit to the number of alternative plans that can be subjected to detailed economy studies. Hence it is vitally important that adequate preliminary planning which includes all viable alternatives be accomplished, and it makes good sense that these alternatives be screened before detailed economy studies are made.

use of standard and acceptable economy study methods

This book has concentrated on the use of theoretically correct and acceptable methods for economy studies. Many firms utilize standard formats and even standard forms to facilitate uniformity of analysis and ease of review by decision makers. As an example, the following is an illustrative segment of the excellently done *Instruction Manual-DCF Plan* by Giddings & Lewis, Inc.†

The full manual contains a very complete set of instructions on what to consider and how to complete the forms line by line as needed. The use of the manual facilitates the determination of what is called in this book the internal rate of return (I.R.R.) for a given project. The method of adjusting for the effect of depreciation on after-tax cash flows is different from that previously illustrated in this book, but the results are the same. The illustrated procedure should not be interpreted as necessarily the best available, and it certainly should not be emulated by any other firm without careful consideration of whether modifications are needed to meet the needs of the firm.

---

† Excerpts from *Instruction Manual-DCF Plan* dated May 1, 1969, reproduced by permission of Giddings Lewis, Inc., Fond du Lac, Wisconsin.

EXAMPLE (Using Forms in Figures 14-2, 14-3, and 14-4)

The illustrative example, for which all details will not be given, is an economic analysis of a machine requiring a $270,500 investment to be depreciated over 15 years, an additional investment of $15,000 to be declared as an expense in the year of expenditure, and an additional $2,000 for working capital, making a total investment of $287,500.

Once the capital expenditures have been estimated as above, the next step in the example is to determine the year-by-year estimates of before-tax cash flows such as is shown for the first year in Figure 14-2 and called "Pre-Tax Income from Operations." The net result on line 31 shows a net income or cash flow of $77,900. This $77,900 is broken down into $19,475 for 0–3 months and $58,425 for 3–12 months (details not shown) and the latter amounts entered in the second column of Figure 14-3. Similar analyses are done using separate forms such as in Figure 14-2 for each of the years 2, 3, 4, and 5, and the results entered in the second column of Figure 14-3. Because the same thing is assumed to happen for each of the years 6 through 10, the totals for that period can be lumped together as the $440,000 entry near the bottom of the second column of Figure 14-3.

The third column of Figure 14-3 converts pre-tax incomes (cash flows) into after-tax incomes using 47%, which is the residual after an assumed combined Federal and state tax rate of 53%.

The present worths can then be calculated using the 10%, 15%, 25%, and 40% discounts or present worth factors provided on the form in Figure 14-3. The 10-year totals are shown at the bottom for each trial percentage.

The present worths of income from operations over a 10-year period at 0% (which is the actual income or cash flow), 10%, 15%, 25%, and 40% are then entered on line 5 of the form in Figure 14-4. Other cash inflows are then entered in the "Trial No. 1—0% Interest Rate" column of Figure 14-4. A $7,950 tax credit is entered in line 7 for the expensed investment items. The $15,650 on line 8 is the after-tax proceeds from the sale of former assets (sold 8 months after time 0). Line 10 shows a depreciation tax credit of $143,365. Expected after-tax proceeds from the hypothetical liquidation of the proposed assets at the close of the comparison are shown at $79,975 on line 13. The total of all cash inflows at 0% interest is $654,181.

The investment requirements stated at the beginning of this example are entered in lines 1, 2, 3, and 4 of Figure 14-4.

The forms in Figures 14-3 and 14-4 utilize single-sum present worth factors which happen to be based on continuous compounding. Line 10 of Figure 14-4 contains specially derived present worth factors which adjust for the difference between the economic comparison period of 10 years and the years over which depreciation is to be charged, which is 15 years. Since all investment outlays occur within 3 months of time 0, the present worth factors for these costs (on lines 1, 2, and 3) are 1.0 at all interest rates.

The next step in Figure 14-4 is to multiply all cash flows by the present worth factors for all interest rates and to sum the present worths of cash outflows (line 4, called "A") and of cash inflows (line 14, called "B") for each interest rate. As explained in the lower part of Figure 14-4, the DCF rate of return (I.R.R.) is the rate at which the present worth of inflows (A) equals the present worth of outflows (B). A convenient interpolation chart is furnished for determining the final answer as the rate at which the ratio $A/B = 1.0$, which is 17% for the case illustrated.

FIGURE 14-2. Example of a form for determining revenues and expenses (pretax income) for a given year.

GIDDINGS & LEWIS Discounted Cash Flow Capital Investment Evaluation Procedure	Page No.
⑬ LONG FORM METHOD	5—L —

| SUMMARY SHEET : Pre-Tax INCOME from OPERATIONS | Period _____ |

INCREASE	DECREASE	Item No.	ANALYST _____ DATE _____
			Effect of Project on REVENUE
		1.	PRODUCT QUALITY (Potential loss or gain of profit through change is product quality.)
12,250		2.	PRODUCTION CAPACITY (Potential loss or gain as result of change in capacity.)
		3.	FLOOR DEMONSTRATION Hrs./Mo. (Potential profit from having G&L machine available for Customer Demonstration.)
12,250 X	Y	4.	TOTAL
			Effect of Project on DIRECT COSTS
	30,000	5.	Direct Labor Hrs. x (D.L. Rate + Fringes)
		6.	Direct Material
	7,500	7.	Sub-Contracting
		8.	Other
Y	37,500 X	9.	TOTAL
			Effect of Project on INDIRECT COSTS
	6,500	10.	Indirect Labor (Incl. Fringe) *(HANDLING)*
	1,500	11.	Overtime (Premium, etc.)
		12.	Indirect Materials & Supplies
	14,000	13.	Tooling *(FIXTURES)*
		14.	Tool Repairs
	2,000	15.	Scrap & Rework
	4,250	16.	Inspection
		17.	Maintenance
	2,400	18.	Down Time
		19.	Power & Utilities
	2,200	20.	Floor Space & Occupancy
4,700		21.	Property Taxes & Insurance
		22.	Inventory
		23.	Safety
		24.	Flexability
		25.	Working Conditions
		26.	Other
4,700 Y	32,850 X	27.	TOTAL
			COMBINED EFFECT
	12,250	28.	Net Increase in REVENUE (4x-4y)
	37,500	29.	Net Decrease in DIRECT Costs (9x-9y)
	28,150	30.	Net Decrease in INDIRECT Costs (27x-27y)
	77,900	31.	NET INCOME FROM OPERATIONS (28+29+30)

LJM Nov. 1968

FIGURE 14-3. Example of a form for calculating the present worth of income from operations.

Page No. 4—L	GIDDINGS & LEWIS Discounted Cash Flow Capital Investment Evaluation Procedure	
	LONG FORM METHOD: <u>Present Worth</u> of INCOME from OPERATIONS during Comparison Period.	

COMPARISON PERIOD ___10___ Yrs ZERO POINT (Date) ___6-1-70___

PERIOD AFTER ZERO POINT	INCOME FROM OPERATIONS Pages 5L X ___% PRE - TAX	AFTER TAX	0%	10% FACTOR	10% PRESENT WORTH	15% FACTOR	15% PRESENT WORTH	25% FACTOR	25% PRESENT WORTH	40% FACTOR	40% PRESENT WORTH
0 - 3 Mo.	19,475	9,153	9,153	1.000	9,153	1.000	9,153	1.000	9,153	1.000	9,153
3 - 12 Mo.	58,425	27,460		.952	26,142	.929	25,510	.885	24,302	.824	22,627
2nd Yr.	81,500	38,305		.861	32,481	.799	30,606	.689	26,392	.553	21,182
3rd Yr.	86,420	40,617		.779	31,641	.688	27,944	.537	21,811	.370	15,028
4th Yr.	88,050	41,384		.705	22,126	.592	18,579	.418	13,119	.248	7,783
5th Yr.	92,600	43,522		.638	27,767	.510	22,196	.326	14,188	.166	7,225
6th Yr.				.577		.439		.254		.112	
7th Yr.				.522		.378		.197		.075	
8th Yr.				.473		.325		.154		.050	
9th Yr.				.428		.280		.119		.034	
10th Yr.				.387		.241		.093		.023	
11th Yr.				.350		.207		.073		.015	
12th Yr.				.317		.178		.057		.010	
13th Yr.				.287		.154		.044		.007	
14th Yr.				.259		.132		.034		.005	
15th Yr.				.235		.114		.027		.003	
16th Yr.				.212		.098		.021		.002	
17th Yr.				.192		.084		.016		.001	
18th Yr.				.174		.073		.013		.001	
19th Yr.				.157		.062		.010		.001	
20th Yr.				.142		.054		.008		.001	
Sub-Totals											
5-Year Periods											
0 - 5 Yrs.				.787		.704		.571		.432	
5 - 10 Yrs.	440,000	206,800		.477	98,644	.332	68,658	.164	33,915	.0585	12,098
10 - 15 Yrs				.290		.157		.047		.0079	
15 - 20 Yrs				.176		.074		.013		.0011	
20 - 25 Yrs				.107		.0350		.0038		.0001	
25 - 30 Yrs				.0646		.0165		.0011		--	
30 - 35 Yrs				.0392		.0078		.0003		--	
35 - 40 Yrs				.0238		.0037		.0001		--	
TOTALS			407,241		248,454		202,646		142,280		95,096

*By Definition Here, "Income from Operations" excludes all tax credits, credit for salvage value of old asset, and terminal credits.

LJM Nov. 1968

ANALYST_____ DATE_____

FIGURE 14-4. Example of a form for summarizing the present worth of cash outflows and cash inflows to calculate the after-tax rate of return.

Ⓖ	GIDDINGS & LEWIS Discounted Cash Flow Capital Investment Evaluation Procedure	Page No.
	LONG FORM METHOD	3—L

CALCULATION OF TRUE RATE OF RETURN (AFTER TAX)

	COMPARISON PERIOD _10_ Yrs. ZERO DATE _6-1-70_	TRIAL NO. 1 0% INTEREST RATE ACTUAL AMOUNT		TRIAL NO. 2 10% INTEREST RATE FACTOR / PRESENT WORTH		TRIAL NO. 3 15% INTEREST RATE FACTOR / PRESENT WORTH		TRIAL NO. 4 25% INTEREST RATE FACTOR / PRESENT WORTH		TRIAL NO. 5 40% INTEREST RATE FACTOR / PRESENT WORTH	
	A CASH OUTFLOW										
1	Capitalized Items (1d)	270,500	1.00	270,500	1.00	270,500	1.00	270,500	1.00	270,500	
2	Expensed Items (2d)	15,000	1.00	15,000	1.00	15,000	1.00	15,000	1.00	15,000	
3	Working Funds (4c)	2,000	1.00	2,000	1.00	2,000	1.00	2,000	1.00	2,000	
4	TOTAL A (5a)	287,500		287,500		287,500		287,500		287,500	
	B CASH INFLOW										
5	Income from Operations (From Page 4-L)-------	407,241		248,454		202,646		142,880		95,096	
6	Investment Credit (6a)										
7	Expense Tax Credit (6b)	7,950	1.00	7,950	1.00	7,950	1.00	7,950	1.00	7,950	
8	Disposal-"Old" Asset (6c)	15,650	.952	14,899	.929	14,539	.885	13,850	.824	12,896	
9	Other (6d)										
10	Depreciation Tax Credit: 15 Yr. (1a) x 53 % Tax =	143,365	.604	86,592	.488	69,962	.340	48,744	.224	32,114	
11	___ Yr. (1b) x ___ % Tax =										
12	___ Yr. (1c) x ___ % Tax =										
13	Terminal Credits (9d)	79.975	.387	30,450	.241	19,274	.043	7,438	.023	1,839	
14	TOTAL B	654,181		388,845		314,571		220,862		149,889	
15	RATIO A/B	0.44 0%		0.74 10%		0.91 15%		1.30 25%		1.92 40%	

Note: The true DCF rate of return (profitability index) is the interest rate at which the sum of the present values of the receipts (B) is equal to the sum of the present values of all expenditures (A).

1. Plot A/B ratios at interest rates calculated.

2. Draw curve between points.

3. True DCF rate of return is where curve crosses heavy line 1.0.

INTERPOLATION CHART

% RETURN vs RATIO A/B

16	TRUE RATE OF RETURN (Profitability Index)	17 %		DCF Calculation by _____

LJM Nov. 1968

estimating and reliability of data

The most critical and usually most difficult aspect of economy studies is the estimating of the variables or factors—such as revenue, operating costs, and life —which are vital for those studies. The term *estimate*, when applied to economic analyses, can have a multiplicity of meanings. At one extreme it can be used to indicate a carefully considered computation of some quantity for which the exact magnitude cannot be determined. At the other extreme, it can be used to denote what are actually just offhand approximations that are little better than outright guesses.

The first part of Chapter 8 discusses sources of estimates and ways estimates are accomplished. Although engineers often need to obtain estimates and data from others who are more knowledgeable or are in the best position to make particular estimates, they should consider themselves responsible for the validity of the data they use. They should evaluate the basic data they obtain for reasonableness, investigate any indication of unreliability, and confer with those most knowledgeable for further clarification if needed.

In making estimates for economy studies, one must be wary of averages, standard data, and other regular published sources. One is generally evaluating specific plans, not average plans, and hence needs specific estimates rather than average estimates. For example, average building maintenance cost may be $0.50 per square foot per year, but for a particular case it may be as low as $0.20 or as high as $1.40, depending upon specific conditions. Average cost data can, however, be quite useful in verifying the reasonableness of specific cost data used for an economy study.

In evaluating the adequacy of an estimate or the extent to which effort ought to be devoted to generate the estimate, one should judge the importance of the estimate to the study in which it is used and the importance of the study results to the firm. Chapter 6 discussed the use of sensitivity concepts for judging the relative importance of one or more estimated variables on study results. The importance to the firm of a particular study depends upon many ramifications of the study—the effect upon objectives; the effect upon the organization, facilities, and finances; and the risks assumed.

Obviously, the more sensitive an economy study result is to a given estimate and the greater the importance of the economy study to the firm, the greater is the effort or attention justified for that estimate. This general principle must be moderated by the realities of just how much valuable information is potentially obtainable from additional estimating effort or attention, what resources are available for making estimates, and the degree of urgency for completion of the estimate studies.

postinvestment reviews of actual versus estimated performance

After projects have been undertaken or completed, the firm should conduct one or more postcompletion reviews or audits. These audits, sometimes called investment performance reports, can serve four purposes:

1. To verify any resulting savings or profit.
2. To reveal reasons for any project failure.
3. To check on the soundness of various managers' and engineers' proposals.
4. To aid in economy studies of future capital expenditure proposals.

The first two reasons are vitally important and are often sufficient justification for the postcompletion audit. The latter two reasons bear on the quality of economy studies, and hence should be examined more closely.

Managers and engineers who know they will be held to account for the results of their proposals tend to make every effort to ensure their reasonableness and accuracy. However, this accountability should not be overemphasized, for it will cause the managers and engineers to become overcautious and to avoid proposing projects that are really needed rather than risk exposure to censure.

The last-stated purpose of postaudits is to provide information for avoiding past pitfalls and for making better estimates of similar future projects. Such audits reveal the tendency of various managers or engineers to be overly cautious or pessimistic in their estimates and thus provide a basis for adjustments to bring estimates for future projects closer into line with reality.

objectivity of the economy study analyst

The objectivity of the economy study analyst (and any person furnishing him estimates) is a matter that is hardly challengeable in principle but sometimes compromised in practice.

It is too easy for one to consciously or unconsciously develop a bias favoring a particular project or alternative. One may then be convinced that it is in the firm's best interest to demonstrate the superiority of that particular alternative, thereby causing the person to slant his or her estimates in favor of that alternative. Such actions violate professional standards and destroy the integrity of economy studies.

The only condition in which economy studies can fulfill their intended function is for the analyst and others furnishing estimates to be wholly detached and let the results of the study be viewed objectively. The purpose of the study is accomplished only if it provides the best chance for selection of the plan or alternative that actually will turn out to be the most advantageous—not if it results in the "proving in" of some other favorite or preconceived best alternative.

problems

14-1. Explain how different methods of financing a project can affect its economic attractiveness.

14-2. (a) What is equity capital, and explain how it is different from debt capital.

     (b) Why do bondholders, on the average in the long-term, receive a lower return than do holders of common stock in the same corporation?

     (c) Why would a fast-growing company *not* desire to use large amounts of debt capital in its operation?

14-3. (a) List at least four characteristics of a corporation.

(b) What is the difference between a corporation's common stock and its preferred stock?

14-4. (a) Describe how *callable* bonds are repaid by a corporation.

(b) What advantage would there be to a company in issuing bonds that are callable?

14-5. (a) A corporation sold an issue of 20-year bonds, having a face value of $5,000,000, for $4,750,000. The bonds bear interest at 8%, payable semiannually. The company wishes to establish a sinking fund for retiring the bond issue and will make semiannual deposits that will earn 6%, compounded semiannually. Compute the annual cost for interest and redemption of these bonds.

(b) What amount could the corporation save, over the life of the bond issue, if the bonds had been callable?

14-6. (a) A company has issued 10-year bonds, with a face value of $1,000,000, in $1,000 units. Interest at 8% is paid quarterly. If an investor desires to earn 10% nominal interest on $10,000 worth of these bonds, what would the selling price have to be?

(b) If the company plans to redeem these bonds in total at the end of 10 years and establishes a sinking fund that earns 6%, compounded semiannually, for this purpose, what is the *annual* cost of interest and redemption?

14-7. What are five possible sources of capital for financing new projects?

14-8. A 20-year bond with a face value of $5,000 is offered for sale at $3,800. The rate of interest on the bond is 5%, paid semiannually. This bond is now 7 years old (i.e., the owner has received 14 semiannual interest payments). If the bond were purchased for $3,800, what effective rate of interest would be realized on this investment opportunity?

14-9. You bought a $1,000 bond that paid interest at the rate of 7%, payable semiannually, and held it for 10 years. You then sold it at a price that resulted in a yield of 8% nominal on your capital. What was the selling price?

14-10. The voters of a state approved the issue and sale of $250,000,000 in 6% bonds to finance higher education facilities. The bonds bore serial numbers such that they matured so as to make the sum of interest on the outstanding bonds and redemption of further bonds a constant annual amount over a 20-year period. How many dollars of interest and how many dollar's worth of bonds would be retired at the end of (a) the first year; (b) the 15th year?

14-11. During the first 5 years of its life, a small corporation, which has a capitalization of $2,000,000 represented by 2,000 shares of common stock, has paid no dividends, in order to finance its expansion out of retained profits. It now needs $1,000,000 in additional capital to finance and stock two new warehouses. Discuss briefly the advantages and disadvantages of obtaining the required capital from (a) selling additional common stock, (b) borrowing on a 5-year bank loan at 10% interest, (c) issuing 8%, 10-year bonds which would contain a provision that the corporation could not incur further indebtedness until the bond issue was retired.

14-12. An existing piece of equipment has been performing poorly and needs replacing. More modern equipment can be *purchased* for cash using retained earnings (equity funds) or it can be *leased* from a reputable firm. If purchased, the equipment will cost $20,000 and have a depreciable life of 5 years with no salvage value.

Because of improved operating characteristics of the equipment, raw materials savings of $5,000 per year are expected to result relative to continued use of the present equipment. However, labor costs for the new equipment will most likely increase by $2,000 per year and maintenance will go up by $1,000 per year. Straight line depreciation is used by the firm.

To lease the new equipment requires a refundable deposit of $1,500 and the yearly leasing fee is $6,000. Annual materials savings and extra labor costs will be the same when purchasing or leasing the equipment, but the company will provide maintenance for their equipment as part of the leasing fee.

The desired after-tax rate of return (I.R.R.) is 10% and the effective income tax rate is 50%. If purchased, it is believed that the equipment can be sold at the end of 5 years for $1,500 even through $0 was used in calculating depreciation. An investment tax credit of 10% can be used to offset income taxes at the time of the purchase (year 0). Determine whether the company should buy or lease the new equipment, assuming that it has been decided to replace the present equipment.

14-13. Determine the more economical means of acquiring a business machine if you may either (1) purchase the machine for $5,000 with a probable resale value of $1,000 at the end of 5 years, or (2) rent the machine at an annual rate of $900/year for 5 years with an initial deposit of $500 refundable upon returning the machine in good condition. If you own the machine, you will depreciate it for tax purposes at an annual rate of $800. All leasing rental charges are deductible for income tax purposes. As owner or lessee you will pay all expenses associated with the operation of the machine. (a) Compare these alternatives by use of the annual cost method. The after-tax minimum attractive rate of return is 10% and the effective income tax rate is 40%. (b) How high could the annual rental be such that the leasing option is still the more desirable alternative?

14-14. Determine the before-tax and after-tax average cost of capital for a firm that has this capital structure:

Amount	Source of Capital	Rate of Return (%)	Dollar Amount per Year
$3 million	Short-term bank loans	10	$0.30 million
$7 million	Mortgage bonds	7	$0.49 million
$4 million	Preferred stock	8	$0.32 million
$11 million	Common stock and retained earnings	13	$1.43 million

Assume that the firm's effective tax rate is 48% and that a 13% rate of return to purchasers of common stock represents a satisfactory opportunity cost of equity capital.

14-15. Suppose that a machine costing $11,000 can be financed entirely by borrowed funds or by equity capital. With borrowed funds, the loan is to be repaid at the rate of $2,000 at the end of each year for the first 4 years and $3,000 at the end of the fifth year. Interest charges are computed at 10% of the unpaid, beginning-of-year balance of the loan. Depreciation is calculated on a straight line basis, the depreciable life is 5 years, and the estimated salvage value is $1,000. The expected before-tax cash flow attributable to the machine *before* deducting interest charges and operating costs is $10,000, and the effective income tax rate is 50%. Operating costs will amount to $3,000 per year. (a) Determine the after-tax cash flows of both financing plans, and compute the present worth of each at a minimum attractive rate of return of 10%. (b) When borrowed funds are used to finance a project, what problems can arise when the I.R.R. method is utilized to compare alternative financing plans?

14-16. Rework Example 14-3 when the $50,000 worth of borrowed funds, at 5% interest, is to be repaid in equal end-of-year amounts such that no principal remains after the tenth payment is made. Be careful to note that interest and repaid principal will vary in amount from year to year in this situation.

14-17. A company has $300,000 for investment in new projects during the coming year. Projects currently being considered are as follows:

Project	Capital Required	Life (years)	Estimated Annual Rate of Profit (%)	Risk
A	$ 60,000	5	10	Low
B	100,000	3	20	Average
C	150,000	5	8	Average
D	110,000	8	15	High

The company follows a general policy of not committing capital for a longer period than 10 years, and it prefers 5 years or less. Uncommitted capital would remain temporarily in a bank, where it would earn at least 6%. Which projects do you recommend?

14-18. A company has $20,000 to invest in "would like to have" projects and has taken the position that at least half of these funds should be spent on medium- and low-risk projects. Based on the investments shown below, categorized by risk, what subset of these independent capital investment opportunities should be recommended?

Project	Investment	Prospective Annual Return After Taxes (%)
High Risk:		
M	$ 4,000	32
N	2,000	28
O	6,000	22
P	5,500	17
Medium Risk:		
S	8,000	22
T	7,500	20
U	10,000	16
V	5,000	12
W	6,200	10
Low Risk:		
C	12,000	16
D	8,000	14
E	10,000	13
F	5,000	8

14-19. (a) List some of the factors that a high-technology electronics firm might consider in establishing a minimum attractive rate of return for its investments.

(b) What other objectives, in addition to maximizing the future worth (or present worth) of the firm, would a high-technology company consider in its capital investment decision making?

14-20. Why are postinvestment reviews of actual versus estimated economic performance of projects important to the engineering economist?

Part III

Other Useful Methods for Minimizing Resource Requirements

chapter 15

minimum-cost formulas

This and the following three chapters deal with various types of engineering economy problems in which study periods tend to be rather short (1 year or less, usually). As a result, the compounding of interest is not one of the more important concerns that underlies these problems. Earlier, in Chapter 3, such problems were treated as present-economy studies. The subject of this chapter is the description and illustration of certain mathematical models that have been used in practice to determine minimum costs of alternatives having cost functions expressed in one (or more) variable. The topic of value analysis is described briefly in Chapter 16. Linear programming formulations of special types of resource allocation problems are presented in Chapter 17. In Chapter 18 present economy studies associated with critical path methods are illustrated.

minimum-cost economy

In the day-to-day operations of business organizations, many economy studies are made for the purpose of enabling costs to be minimized, where the total cost for an item or operation is the sum of a fixed cost, a variable cost that increases *directly* with respect to some design variable such as the number of units produced, and a second variable cost that is *inversely* proportional to the same design variable. Thus the total cost C is

$$C = ax + \frac{b}{x} + k \qquad (15\text{-}1)$$

where a, b, and k are constants and x is the design variable that can be controlled. A number of real situations commonly experienced in engineering practice can be dealt with by relating them to the theoretical relationship expressed in Equation 15-1. It is the purpose of this chapter to examine several typical problems of this type, including determination of economic order quantities, the most economical size of electrical conductors, and the most economical replacement interval for assets subject to sudden failures.

443

The condition for minimum cost can readily be determined by differentiating Equation 15-1 (the cost function) with respect to the design variable x, equating to zero, and solving for the value of x'. Thus

$$\frac{dC}{dx} = a - \frac{b}{x^2} = 0 \tag{15-2}$$

$$x' = \sqrt{\frac{b}{a}} \tag{15-3}$$

This optimum value of the design variable, x', is the point at which directly varying costs are equal to the inversely varying costs. This may be shown by noting that the increasing variable cost at x' is

$$ax' = a\sqrt{\frac{b}{a}} = \sqrt{ab}$$

and the decreasing variable cost is

$$\frac{b}{x'} = \frac{b}{\sqrt{\frac{b}{a}}} = \sqrt{ab}$$

Figure 15-1 depicts the cost relationships for the case just considered, where the costs are directly and inversely proportional to the decision variable.

economic purchase order size

A common situation involving the minimum-cost concept is the purchasing of batches of goods for use or sale. Ordinarily, the goods or supplies are purchased periodically throughout the year in lots, and each lot is received at one time and put into storage. Thereafter, the supply is used as needed, sometimes at a fairly uniform rate, until the supply is exhausted or diminished to a predetermined minimum inventory quantity, at which time a new order is placed. This situation is depicted in Figure 15-2. In such a situation the cost of originating, placing, and paying for an order is essentially constant, regardless of the quantity ordered. Consequently, the order cost *per unit* varies inversely with the size of the order. On the other hand, a number of costs increase as the size of the order increases: for example, the amount of storage space and its costs, the interest on the average inventory in stock, insurance, and taxes all increase directly with the order quantity. These directly varying costs are commonly known as *carrying* or *holding costs*. The cost per piece of the material purchased is, in many cases and for a considerable range of quantities, independent of order size. (Variable price schedules are discussed later.) Clearly, under these conditions there will be some

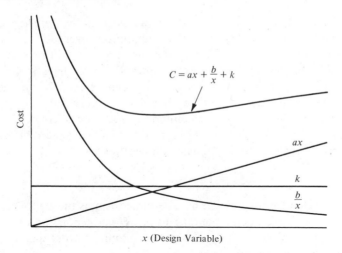

FIGURE 15-1. General relationship of the fixed cost (*k*), directly variable cost (*ax*), inversely variable cost (*b/x*), and total cost (*C*) in Equation 15-1.

order size that will be the most economical, and this *economic purchase order size* can be determined by the application of Equation 15-3. Thus, if

Q = most economic order quantity (lot size)
U = number of units used per year (at a constant rate)
S = cost of placing an order (setup cost)
C = commodity cost per unit
I = inventory carrying cost per year as a proportion of inventory value

FIGURE 15-2. Variation in inventory, with and without a safety stock, when lot quantities can be obtained instantaneously and use is at a constant rate.

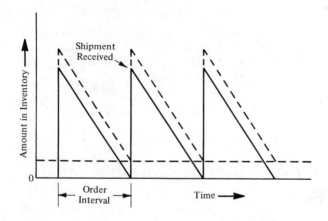

the total cost for 1 year's supply will be the sum of the carrying cost, the order cost, and the commodity cost. If no minimum inventory is maintained, the average inventory will be $Q/2$, and the carrying cost will be $QIC/2$. This, of course, assumes that exclusive storage space does not have to be reserved for each inventory item; the space for one item can be used for another as the stock diminishes. Similarly, the number of orders per year will be U/Q, and the annual order cost will be SU/Q. The commodity cost will, of course, be CU and thus will be unaffected by order size. The total cost (T.C.) of acquiring and storing 1 year's worth of the commodity thus can be written as

$$\text{T.C.} = \left(\frac{IC}{2}\right)Q + \frac{SU}{Q} + CU$$

It will be noted that this expression has the same form as the right-hand side of Equation 15-1, with $IC/2$ corresponding to a, SU corresponding to b, and CU corresponding to k. Consequently, the most economical order quantity may be obtained from Equation 15-3 as

$$Q = \sqrt{\frac{\text{inversely varying cost coefficient}}{\text{directly varying cost coefficient}}}$$

or

$$Q = \sqrt{\frac{2SU}{IC}} \qquad\qquad (15\text{-}4)$$

EXAMPLE 15-1

Determine the most economic purchase order size for the conditions applicable to Equation 15-4 when annual usage is 8,000 units, the unit commodity cost is \$2, the cost of placing an order is \$5, and the annual inventory carrying cost is 30% of the average inventory.

Solution

$$Q = \sqrt{\frac{2 \times \$5 \times 8,000}{0.3 \times \$2}} = 365 \text{ units}$$

To provide the required 8,000 units in 1 year, almost 22 lots of 365 each would have to be ordered. If exactly 8,000 units were required, this could be obtained in 21 lots of 365 each and 1 lot of 335 units. The total cost per year is

$$\frac{0.3(\$2)(365)}{2} + \frac{\$5(8,000)}{365} + \$2(8,000)$$

$$\text{T.C.} = \$16,219/\text{year} \quad\blacksquare$$

economic order size for variable price schedules

Price schedules that vary with lot size also may have a pronounced effect upon the economic size for lots that are purchased. For example, assume that the item mentioned in Example 15-1 can be purchased under the following price schedule:

Lot Size	Unit Price
1–200	$2.10
201–500	2.00
501 and over	1.85

When lot sizes of 1–200 are considered, 40 or more purchase orders per year are required. The minimum T.C. in this case is

$$\frac{0.3(\$2.10)(200)}{2} + \frac{\$5(8,000)}{200} + \$2.10(8,000) = \$17,063/\text{year}$$

For lot sizes of 201–500, orders placed per year range from 16 to 39, and T.C., which is almost constant in this range, is in the neighborhood of $16,250/year. Finally, when orders for more than 500 units are placed, 16 or less full lots are received each year and illustrative T.C.'s for the $1.85 unit price schedule appear below:

Number of Lots	Lot Size	Total Cost
14	571	$15,029
10	800	15,072
5	1,600	15,269

Figure 15-3 shows the cost of obtaining the year's requirement of 8,000 units under the indicated price schedule. As can be seen from this graph, minimum cost would be obtained by purchasing in lots of 501 pieces, requiring 16 lots.

Because there is never any assurance that the theoretical economical order quantity in any price range will fall within the lot size range for the particular price, or that this lot size will give the minimum total cost, problems of this type must usually be solved by a multiple-step procedure. (Of course, if curves such as are shown in Figure 15-3 are drawn, the correct solution is at once apparent.) The economical order quantity for one or more of the prices in the price–quantity schedule is computed until a result is obtained that falls within the corresponding range of the schedule. The annual cost resulting from the use of the corresponding number of lots is then computed. Next the annual cost that would result from using the smallest lot size at the next lower price is determined. If this cost is less, the theoretical lot size for this price range should be computed and the annual cost at this theoretical lot size ascertained, in order to determine whether the total cost curve is increasing or decreasing. If the cost increases as the lot size increases, the entire procedure—determining the cost for the most economical lot size and comparing it with the cost for the minimum lot size of the next lower price—must be repeated until the lowest annual cost is obtained.

FIGURE 15-3. Cost of obtaining a year's requirement of 8,000 units under a lot size pricing schedule.

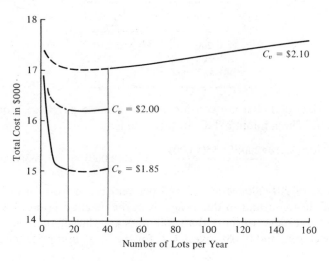

$C_v = \$2.10$

$C_v = \$2.00$

$C_v = \$1.85$

Number of Lots per Year

Total Cost in $000

Key: ――― Price schedule does not apply

economic production lot size

In many plants the parts that are used are produced internally. Typically, they are produced at a rate that is higher than the use rate, yet an entire production lot cannot be produced instantaneously. The resulting inventory situation is as depicted in Figure 15-4.

This condition is similar to that for determining economic purchase order quantities, with two exceptions. First, each time a lot is produced there are certain setup costs incurred. These correspond to the order costs S, where goods

FIGURE 15-4. Inventory quantity when production is in lots at a uniform rate considerably greater than the uniform use rate and no minimum (safety) stock is maintained.

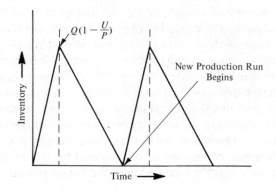

$Q(1 - \dfrac{U}{P})$

New Production Run Begins

Inventory

Time

are purchased. Second, inasmuch as the entire lot is not obtained instantaneously and some items are being used before the entire lot is produced, a correction must be made to account for this condition. If $P =$ the production rate per year, with production at a constant rate, the average inventory will be

$$\frac{Q}{2}\left(1 - \frac{U}{P}\right) \tag{15-5}$$

Again using the concept expressed in Equation 15-3, and assuming nonexclusive storage space, we find that the most economic lot size to produce is

$$Q = \sqrt{\frac{2SU}{I(1 - U/P)C}} \tag{15-6}$$

$$Q = \sqrt{\frac{2S}{I[(1/U) - (1/P)]C}} \tag{15-7}$$

the effect of risk and uncertainty on lot size

When a lot of goods is purchased or manufactured there may be uncertainty as to whether the entire quantity will in fact be used. Such uncertainty is due to several factors. Future demand for the item may be terminated. Design changes in a product may make parts for the old design obsolete. Consequently, it frequently is desirable to take such uncertainties into account in economic lot size formulas.

A little thought leads us to realize that the net effect of the risks just mentioned will be to make us want to have fewer parts on hand if there is a danger of their becoming obsolete. Numerous formulas have been derived that take such risks into account. All of them have the usual lot size form, but with some increase occurring in the denominator so as to decrease the economic lot size. A typical formula of this type is

$$Q = \sqrt{\frac{2SU}{(I + \alpha)C}} \tag{15-8}$$

where α is an annual obsolescence cost factor expressed as a fraction of the unit commodity cost.

the importance of the carrying cost rate in lot size formulas

A great amount of discussion has centered around the interest rate that should be used in economic lot size determinations. There are two aspects of this matter that should be considered. First, we should make certain that the holding or carrying cost does, in fact, cover the costs involved. Many studies have indicated that the charges used frequently are too low. The second aspect relates to the effect that interest rate can have on the economic lot size. Example 15-2 provides an illustration of the effect that a change in interest rate can have.

EXAMPLE 15-2

Consider again Example 15-1 where a carrying cost rate of 30% resulted in an economic lot size of 365 units. What would be the economic lot size if the carrying cost rate were reduced to 20%?

Solution

$$Q = \sqrt{\frac{2 \times \$5 \times 8,000}{0.2 \times \$2}} = 448 \text{ units}$$

Thus a decrease in the interest charge will increase the economic lot size, as might be expected, since the inventory holding costs are reduced. In common terms used for sensitivity studies, a one-third decrease in interest rate increases the economic lot size by $(448 - 365)/365 = 23\%$. ∎

minimum-cost situations without linear cost relationships

In problems thus far considered in this chapter, the cost relationships have been assumed to be either directly or inversely proportional to the number of items purchased or manufactured. Obviously, there are many situations where costs increase with some design variable and other costs decrease, but not in direct or inverse proportion. Such a condition exists in many "conductor" problems, where electric current or some fluid is forced through a conductor. Energy must be used to overcome the friction between the fluid, or current, and the conductor. If the size of the conductor is increased, with accompanying investment cost, the frictional loss, and the corresponding cost of overcoming it, will be reduced. Thus one cost increases, as conductor size is increased, and another decreases, and the total cost is the sum of the two. Since the two costs usually are not directly or inversely proportional to conductor size, Equation 15-3 cannot be applied. However, the costs can readily be computed and tabulated and the total cost and most economical conductor size determined.

Table 15-1 shows the cost relationships for a particular copper conductor that is to transmit 50 amperes for 4,500 hours per year.† It is shown that size 0 conductor would be most economical for this case. Figure 15-5 shows the cost curves for this electrical conductor problem, and it will be noted that the minimum-cost point does not occur at the point where the two component costs are equal.

---

† This problem illustrates the well-known *Kelvin's law*, which was derived for a somewhat idealized transmission-line situation and then generalized.

table 15-1

annual cost of lost power and investment charges for various sizes of copper wire (per 1,000 feet)

		000	00	0	1	2
A	Wire size					
B	Resistance (ohms at 20°C)	0.0618	0.0779	0.0983	0.124	0.156
C	Weight (lb)	508	403	320	253	201
D	KWh at 50 amperes and 4,500 hours per year	695	876	1,107	1,397	1,757
E	Cost of lost energy at $0.015 per kWh	$10.43	$13.14	$16.60	$20.95	$26.36
F	Investment at $0.35 per pound	$177.80	$141.05	$112.00	$88.55	$70.35
G	Investment costs (depreciation, taxes, interest, and insurance) at 16% of line F	$28.45	$22.57	$17.92	$14.17	$11.26
H	Total annual cost (line E + line G)	$38.88	$35.71	$34.52	$35.12	$37.61

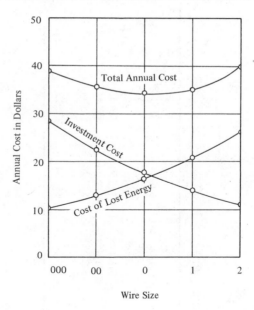

FIGURE 15-5. Breakeven chart for determining the most economical size of copper conductor. (Data from Table 15-1.)

production to meet a variable demand

Quite often a business must manufacture to meet a variable demand, which may be very low during certain parts of the year and very high at other times. Such a situation requires a decision to be made as to how the demand will be met. A single production unit may be able to meet the annual demand by operating continuously throughout the year, with the output being stored during the off-peak periods so as to meet the demands of the peak periods. Such a practice reduces the investment cost in production equipment, but the inventory and storage costs are increased. An alternative procedure would be to utilize more production units during a shorter period of time, thus increasing investment and possibly operation costs, but reducing inventory and storage costs. Clearly, this common type of situation presents a minimum-cost possibility by balancing increasing investment costs against decreasing inventory costs.

EXAMPLE 15-3

A company produces containers for the fruit-packing industry. The demand occurs during 3 months of the year, as shown in the first two columns of Table 15-2. It is customary to shut down the plant during October for vacations and overhaul of equipment, so that production occurs during only 11 months of the year. A machine can produce 250,000 units per month. The fixed costs, including depreciation, taxes, insurance, profit on capital, building charges, and so on, are

table 15-2

monthly demand and month-end inventory of containers (in 000s) when produced by one, two, or three machines (example 15-3)

Month	Demand	One Machine		Two Machines		Three Machines	
		Production	Inventory	Production	Inventory	Production	Inventory
Jan.	0	250	1,000	0	0	0	0
Feb.	0	250	1,250	0	0	0	0
Mar.	0	250	1,500	250[a]	250	0	0
Apr.	0	250	1,750	500	750	0	0
May	0	250	2,000	500	1,250	500[b]	500
June	700	250	1,550	500	1,050	750	550
July	1,100	250	700	500	450	750	200
Aug.	950	250	0	500	0	750	0
Sept.	0	0	0	0	0	0	0
Oct.	0	250	250	0	0	0	0
Nov.	0	250	500	0	0	0	0
Dec.	0	250	750	0	0	0	0
AVERAGE FOR YEAR			937.5		312.5		104.2

[a] One machine.
[b] Two machines.

$3,000 per year per machine, regardless of the amount of use. In addition, there are variable costs for each machine amounting to $700 per month. The containers have a value of $30 per thousand, and the annual inventory and storage costs amount to 20% of the value of the product.

The month-end inventories, resulting from using one, two, and three machines, are shown in Table 15-2, with the inventory flow depicted in Figure 15-6. It is desired to determine the total annual cost for each number of machines.

Solution

For one machine:

Fixed costs	$ 3,000
Variable costs: $700 × 11	7,700
Inventory costs: $30 × 937.5 × 0.20	5,625
TOTAL	$16,325

In the same manner, the annual cost using two machines would be $15,575, and using three machines would be $17,325. Thus it is clear that the use of two machines would give minimum annual cost. ∎

FIGURE 15-6. Monthly inventories resulting from using 1, 2, or 3 production units to meet a seasonal demand. (Data from Table 15-2.)

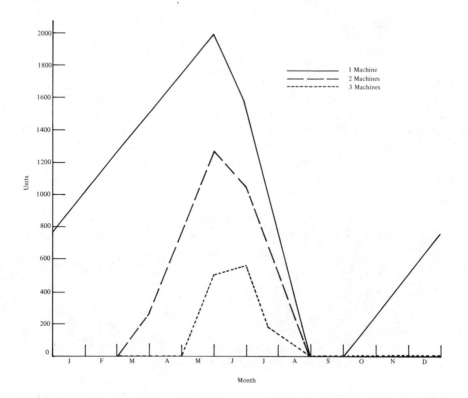

replacement of assets that fail suddenly

Certain types of assets fail suddenly while in service and do not exhibit significant deterioration in their useful capabilities until they fail. Examples of such assets are transistors, light bulbs, and jet engines. However, this type of asset is usually subject to an increasing rate of failure as its cumulative usage or age increases.

There are two general classes of costs involved in deciding whether to replace (or maintain) individual items upon failure or to replace (maintain) an entire group of these items at the same time: (1) costs that vary directly with time: for example, annual costs of maintenance for individually serviced items; and (2) costs that vary indirectly with time, such as annual costs of periodic group maintenance on all items. The aim in this type of replacement analysis is to determine answers for the following questions.

1. Should a group of assets subject to sudden failure be replaced in entirety, or should they be individually replaced upon failure?
2. If group replacement is the best policy, what is the most economical group replacement interval?

The general solution procedure for obtaining answers to these questions involves the comparison of cost of group replacement (maintenance) at different replacement intervals with the cost of individually replacing items as they fail in service. Often group replacement at a stated interval will prove to be the more economical of these two policies.

To illustrate how one might solve this type of replacement problem, an example problem is worked that demonstrates the general solution procedure. The time value of money is often omitted as an important consideration in these problems because optimum replacement intervals are frequently less than a year. They are, therefore, problems of the present-economy type.

EXAMPLE 15-4

A small air cargo service has 20 airplanes of the same make and each airplane has two engines on it. The research department of the manufacturer of these planes has collected past data on engine breakdowns based on 250 hours of flying time each month:

Months after maintenance	1	2	3	4	5
Probability of engine breakdown	0.2	0.1	0.2	0.3	0.4

The cost of remedial maintenance after a failure is $1,000 per engine, and if both engines of all planes were maintained as a group, it would cost $250 per engine. Assume that the maintenance schedule has no adverse affect on meeting air shipment schedules or revenues generated.

Let us assume that breakdowns occurring during a month are tallied at the beginning of the following month. Thus maintenance that occurs at the beginning of, say, the third month will be 1 month old at the beginning of the fourth month. During the first t time intervals, all failures are replaced as they occur. At the end of the tth month, all units are replaced regardless of their ages. The problem is to find that value of t which will minimize total cost per month. Because it is assumed that the entire replacement interval in question is of short duration such that the timing of money can be neglected, the total cost from time of group installation through the end of t months can be given by

$$K(t) = NC_1 + C_2 \cdot \sum_{x=1}^{t} f(x) \qquad (15\text{-}9)$$

where $K(t)$ = total cost for t months
$\quad C_1$ = unit cost of replacement in a group
$\quad C_2$ = unit cost of individual replacement after failure
$\quad f(x)$ = number of failures in the xth month
$\quad N$ = number of units in the group

Hence the objective is to find the value of t that minimizes $K(t)/t$. The relationship of the various costs is shown in Figure 15-7.

Table 15-3 illustrates how the total cost of maintaining aircraft engines during each month is calculated. Here $P_1, P_2, \ldots, P_5$ represents the probability of engine failure within $1, 2, \ldots, 5$ months after the previous maintenance was performed. For the data given above, and by using Table 15-3, an average cost of maintenance per month can be calculated as shown in Table 15-4. Here it is apparent that the minimum value of $K(t)/t$ occurs in month 3. Thus the most economic policy is to perform group maintenance on aircraft engines every 3 months and individually repair engines as they fail within this interval of time.

FIGURE 15-7. Replacement costs for Example 15-4.

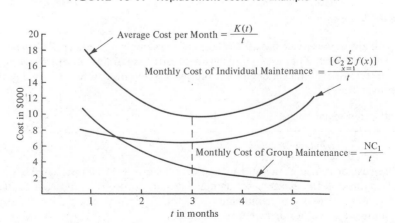

table 15-3
computational setup for determining total replacement costs for aircraft engines in example 15-4

Month, t	$f(x)$	$K(t)$
1	$f(1) = NP_1$	$K(1) = NC_1 + C_2 f(1)$ or $K(1) = NC_1 + C_2 NP_1 = N(C_1 + C_2 P_1)$
2	$f(2) = NP_2 + f(1)P(1)$	$K(2) = NC_1 + C_2 f(1) + C_2 f(2)$ $= NC_1 + C_2[f(1) + f(2)]$ or $K(2) = NC_1 + C_2 NP_1 + C_2 NP_2 + C_2 f(1)P_1$ $= NC_1 + C_2[N(P_1 + P_2) + f(1)P_1]$
3	$f(3) = NP_3 + f(2)P_1 + f(1)P_2$	$K(3) = NC_1 + C_2[f(1) + f(2) + f(3)]$ or $K(3) = NC_1 + C_2[N(P_1 + P_2 + P_3)$ $+ f(1)(P_1 + P_2) + f(2)P_1]$
4	$f(4) = NP_4 + f(3)P_1 + f(2)P_2 + f(1)P_3$	$K(4) = NC_1 + C_2[f(1) + f(2) + f(3) + f(4)]$ or $K(4) = NC_1 + C_2[N(P_1 + P_2 + P_3 + P_4) + f(1)(P_1 + P_2 + P_3)$ $+ f(2)(P_1 + P_2) + f(3)P_1]$
5	$f(5) = NP_5 + f(4)P_1 + f(3)P_2 + f(2)P_3 + f(1)P_4$	$K(5) = NC_1 + C_2[f(1) + f(2) + f(3) + f(4) + f(5)]$ or $K(5) = NC_1 + C_2[N(P_1 + P_2 + P_3 + P_4 + P_5)$ $+ f(1)(P_1 + P_2 + P_3 + P_4) + f(2)(P_1 + P_2 + P_3)$ $+ f(3)(P_1 + P_2) + f(4)P_1]$

table 15-4
calculations of replacement costs for aircraft engines in example 15-4

Month	Average Number of Engines to Be Maintained, $f(x)$	Cost of Individual Maintenance, $C_2 f(x)$	Cumulative Cost of Individual Maintenance, $C_2 \sum f(x)$	Average Cost of Individual Maintenance per Month, $[C_2 \sum f(x)]/t$	Average Cost of Group Maintenance per Month, $NC_1 t$	Average Total Cost of Maintenance per Month, $K(t)/t$
1	$40(0.2) = 8$	\$ 8,000	\$ 8,000	\$ 8,000	\$10,000	\$18,000
2	$40(0.1) + 8(0.2) = 5.6$	5,600	13,600	6,800	5,000	11,800
3	$40(0.1) + 5.6(0.2) + 8(0.1) = 5.92$	5,920	19,520	6,506	3,333	9,839 ← minimum
4	$40(0.2) + 5.92(0.2) + 5.6(0.1) + 8(0.1) = 10.544$	10,544	30,064	7,516	2,500	10,016
5	$40(0.4) + 10.544(0.2) + 5.92(0.1) + 5.6(0.1) + 8(0.2) = 20.861$	20,861	50,925	10,185	2,000	12,185

limitations in the use of formulas

Caution should be observed in the use of minimum-cost formulas as a method of making economy studies. Most economy study formulas are complex because of the necessity of making provision for all the possible cost factors that may be encountered. Some are so complex that electronic computers must be used to obtain a solution in a reasonable time.

Proponents of formulas usually justify them on the basis of the following:

1. They save time.
2. They can be used by persons who do not understand the principles on which they are based and who otherwise could not solve economy study problems.
3. They provide some assurance that pertinent factors will not be omitted from consideration.

Although all these claims *can* be true, they are not always so, and the following precautions should be kept in mind.

1. Formulas do not always save time.
2. The use of formulas by those who do not completely understand them can often be dangerous. (For example, how can they know whether a factor should be included or omitted?) Analyses or decisions seldom should be made by those who do not fully understand their implications.
3. There is no assurance that a given formula will make provision for all the factors that exist in a particular situation.
4. Formulas sometimes reveal cost relationships only at the breakeven point. There are many cases where the effect of not operating at the theoretically correct point is of prime importance in making a decision. Graphical solutions frequently are of great help in such problems.
5. Formulas give no consideration to intangible factors. Where there are important nonmonetary factors, formulas should be used with great caution.

Thus formulas are of value, but they should be used by those who thoroughly understand their implications and underlying assumptions.

problems

15-1. Use the economic lot size formula to determine the most economical size of a purchase order under these conditions: order preparation cost is $350, the cost of each item is $0.20, the annual demand for the item is 200,000 pieces, and the annual carrying charges are 25% of average inventory.

15-2. When the economic lot size formula is used, what relationship exists between the annual setup (or order) cost and the annual carrying (inventory) cost at the minimum-cost point?

15-3. The annual requirements for a certain type of plastic wrapping material are 80,000 pounds on a continuing basis. It is estimated that the average price of the material during the next few years will be about $0.42 per pound. At present it costs $30.00 to initiate and receive an order, and the percentage factor to cover interest, taxes, insurance, and storage is 0.20.

An improved system and equipment for initiating and receiving orders would cost $1,700, and it would reduce the "order" cost by $5.00 per order and reduce the percentage carrying charge by $\frac{1}{4}$. This investment would have to be recovered with interest in four years. If capital is worth 12% before taxes, and neglecting income tax effects, what do you recommend? No minimum inventory is maintained.

15-4. For a given set of conditions, how does each of the following affect the economic lot size?

(a) Reducing the setup, or order, cost by one-half.

(b) Tripling the cost of carrying inventory.

(c) The possibility of pilferage and/or obsolescence.

15-5. The Attaboy Lawn Mower Company can purchase component parts for one of its high-volume mowers according to the following price schedule:

Quantity per Order	Price per Unit
1–2,000	$1.20
2,001–5,000	0.95
Over 5,000	0.85

This company expects to use 20,000 of these parts per year on a continuing basis. The incremental cost of processing an order is $50 and the annual carrying cost per unit is 15% of the purchase price. If the company maintains no appreciable minimum inventory reserve, what is the most economical order quantity?

15-6. A company manufactures the deep-drawn metal cases for one of its products and uses them at a nearly constant rate of 8,000 per year. It can produce them at a rate of 40,000 per year. Because of the necessity for changing, aligning, and "proving" the dies each time a setup is made, the cost for making a setup and tearing it down at the conclusion of a run is $200. The variable costs, including material, die wear, and labor, are $4 per unit, and the annual "holding" cost is 25% of the unit variable cost. Because of the high production rate, the company maintains no appreciable inventory. Determine the most economical lot size for the production of the cases.

15-7. If the company in Problem 15-6 will invest $1,200 to modify the drawing dies used in making the cases, it can reduce the setup cost by $100. Assuming that (a) the dies have a 5-year life, (b) annual taxes and insurance amount to 3% of the first cost on all equipment and tooling, (c) the dies are used on three different presses, and (d) the company requires a before-tax return of at least 15% on such investments, would the modification be justified? Neglect possible income tax effects.

15-8. A small manufacturing firm can produce 400 ceramic fixtures each day, and the firm operates 250 days per year. These fixtures are required in the assembly of a product that the company sells at a uniform rate of 40,000 per year. If the setup cost for initiating production of these fixtures is $500, variable cost per piece is $40, and carrying charges are 30% of the average inventory, what is the most economical production quantity?

15-9. In Problem 15-8, suppose that obsolescence is expected to be high because of possible style changes. To account for this factor, it has been estimated that the annual obsolescence cost is roughly 20% of variable cost per fixture. How does this affect the most economic lot size?

15-10. You are required to make a manufacture versus purchase decision for your company involving a part which until now has always been manufactured. Ten thousand of the parts are used each year, and they have been produced in most economical lot sizes based on a setup of $80, all other increment costs of $5 per unit, and a percentage charge of 20% to cover the cost of carrying and storing inventory. No minimum inventory has been maintained, since the production and use rates are the same.

A supplier has offered to make the parts under a subcontract at a price of $6 per unit. It is estimated that the cost for placing and receiving an order would be $20, and it is planned to order them in economical lot size quantities. What should be done?

15-11. Assume that the prices and resistances of various sizes of electrical cable are the following:

Size	Resistance (ohms per 1000 feet)	Price (per 1000 feet)
1	0.1240	$ 750
0	0.0983	972
00	0.0779	1,086
000	0.0618	1,200

The cost of electrical energy to an industrial user of this cable is 2.5 cents per kWh. Annual capital recovery cost is based on a 25-year life with no salvage value, and the minimum attractive rate of return is 10%. What is the most economical size of cable to transmit 200 amperes for 1,800 hours each year?

15-12. Suppose that for a group of 10,000 electronic parts subject to sudden failure, the net cost of group replacement is $0.40 per unit, while the unit cost of individual replacement is $2.00. Further, the expected number of failures each period is shown in the following table. All failures that occur during each period are replaced only at the end of that time period.

total failures (replacements)
in each period t for 10,000
electronic parts

Period, t	Replacements Current $f(t)$	Cumulative, $\Sigma f(t)$
1	100	100
2	400	500
3	1,100	1,600
4	1,200	2,800
5	2,500	5,300
6	2,300	7,600
7	2,600	10,400
8	2,500	12,900

Determine whether group replacement is economical and, if so, what the optimum replacement interval is.

chapter 16

value engineering

Since the early days of mechanized industry, there has been a tendency for devices and products to be designed and made more complex and elaborate than was necessary. Inventors and designers ordinarily are intent primarily on achieving some functional objective, with insufficient attention being given to keeping the device simple, economical, or easy to use and service. The engineer's job has traditionally been involved with balancing *cost* and *quality* of a product—getting the most for the money. As Arthur Wellington observed in 1887, "The engineer is one who can do with a dollar what any bungler can do with two."

From the beginning of this century industrial engineers have given much successful attention to improving and simplifying methods, processes, and systems, but not much attention was given to rational simplification of designs—frequently because the product already was designed and in production. In recent years a procedure has developed for analyzing products and devices, from a design viewpoint, to determine and improve their economic value. This procedure has come to be known as *value analysis* or *value engineering*. It is widely used by many companies and agencies, some of which have reported savings of over $1,000,000 per year resulting from its application.

Briefly, value engineering is a method for examining the value of a product or service in relation to its cost with the aim of providing the required function(s) at the lowest overall cost. Ideally, value engineering seeks to provide the necessary function(s) during the design phase of a product's creation at the lowest cost, without lowering quality. Hence engineering economy studies, usually at the present-economy level, are an inherent and essential part of value engineering. Consequently, the purpose of this chapter is to introduce the reader to the basic concepts of value engineering and to illustrate the relationship between value engineering and engineering economy.

The origin of value engineering goes back to the late 1940s when the General Electric Company initiated a large-scale program to identify how material substitutions in many of their products would affect the functions performed by the products in addition to their costs and associated market value. It was discovered that many of the substitutions resulted in an improved product at a lower cost. Furthermore, lower costs were also achieved by closely examining rigidly enforced design specifications and standards to determine whether they were overly conservative. The person credited with the success of General Electric's program was L. D. Miles, who has since become known as the "father of value analysis." Later in 1954 the Bureau of Ships patterned its value engineering program after that developed at General Electric. Substantial savings in the cost of building ships for the U.S. Navy resulted and prompted the Department of Defense to require its prime contractors to initiate value engineering procedures. Subsequently, many nondefense industries were able to effectively utilize the principles of value engineering to improve the quality of their products while holding costs constant or even reducing them.

types of value

In Chapter 2 it was pointed out that goods have value because of their utility, and that consumers pay the purchase price for goods because they satisfy a need. However, consumers often do not analyze a product to determine exactly what needs it satisfies or what portion of the purchase price is paid for each need that it satisfies. For example, in early 1971 a large department store sold fur-trimmed "hot pants" for $24. What utility did they possess? One could list several, functions that they clearly fulfilled—warmth (limited), provide a certain degree of modesty, attract attention to the wearer. But what portion of the $24 purchase price would the buyer attribute to each function; would each purchaser make the same allocation; and could the same functions be achieved at a lesser cost?

In value engineering work, two types of value are recognized. *Use value* pertains to the properties and qualities that accomplish a use, work, or service. This type of value exists because without these properties and qualities useful work or service functions could not be achieved. *Esteem value*, on the other hand, pertains to the properties or qualities that make people want to possess the product or service. In somewhat oversimplified terms, we might say that use values cause a product to perform, and esteem values cause it to sell.

As examples of these types of values, consider a man's suit. The primary function of such a suit is to cover the body and to provide warmth and protection against the elements. A secondary function is to provide pockets in which objects may be carried. Fulfilling these functional requirements causes a suit to have use value, and a very plain, ill-fitting garment, made from cheap, coarse, colorless cloth could adequately provide these functions. However, it is doubtful that such a garment would sell very well in the United States. Most men prefer that their suits be made of better cloth, fit reasonably well, have good tailoring details and style, hold a press, and be attractive in color. These are items that make the garment desired; they are esteem values. Thus esteem values can be valuable attributes of goods or services, but only in fact if the customer will pay for them.

The objective of value engineering is to determine the most economical way of providing required use values, consistent with proper functional, safety, reliability, and quality standards and, at minimum cost, that degree of esteem value which the customer demands and for which he will pay.

methodology of value engineering

Value engineering is sometimes referred to as "just plain old-fashioned, every-day cost reduction with a new name." This sentiment indicates a lack of under-standing of both value engineering and the full meaning of cost reduction. The best product at minimum cost to the manufacturer can be delivered through the *simultaneous* use of conventional cost reduction and value engineering.

Before discussing the essential features of value engineering, consider some of the elements of a successful cost reduction program:

1. Systems and procedures for planning and scheduling work.
2. Organizational planning and analysis.
3. Methods improvement and work simplification.
4. Establishment of labor and materials standards.
5. Optimization of manufacturing processes.
6. Control of raw materials, in-process inventories, and finished-goods inventories.
7. Preventive maintenance programs.

These activities are primarily concerned with reduction of costs per se.

In contrast to conventional cost reduction programs, value engineering focuses attention on the inherent worth of the end product with a view toward better satisfying the user's functional requirements at the lowest overall cost to the firm. Thus value engineering is heavily oriented toward assuring the customer that his essential technical requirements are achieved at the minimum cost. Because of its close attention to function, quality, and worth of the product delivered, value engineering generally is not regarded as synonymous with cost reduction.

The value engineering methodology is based on these six fundamental questions:

1. What is it (e.g., the product or design being evaluated)?
2. What does it do (i.e., what functions are provided)?
3. How much does it cost?
4. Are the functions necessary?
5. How else could the functions be accomplished?
6. What would these alternatives cost and are any of them less expensive than the present design?

A certain amount of creativity and "free thinking" are required to explore each of these questions fully. Brainstorming sessions have successfully been used to

challenge acceptance of the status quo and to overcome roadblock excuses for
not delving into unfamiliar and untried design possibilities.

At this point some "idea stimulators" such as the following are often helpful
in initiating the generation of information during a value engineering brainstorming session.

1. Why does it (the present or proposed product) have this shape?
2. How much of this design is the result of custom? Opinion? Tradition?
3. What else would do the job (what *is* the job)?
4. Suppose this were left out?
5. Describe what the product is *not*.
6. Can it be made safer and/or easier to use?
7. What can be done to give extra value to the customer?
8. Is there a less costly part that will perform the same function?
9. Are there newly developed materials that could be used?
10. Are all machined surfaces necessary?
11. Are tolerances closer than they need to be?
12. Is this the best manufacturing process?
13. Why are we making (buying) it?

These questions may seem simplistic, but when seriously considered they
provide essential information for the three-step procedure that is often in value
engineering work. These three steps are

1. *Identify the functions required.*
2. *Determine value of the functions by comparison.*
3. *Develop feasible value alternatives.*

Value engineering starts with an assumption that a device or service is wanted.
With this assumption accepted, the *first step* is to identify the functions, primary
and secondary, that the device or service is to render. The primary function
nearly always can be expressed in two words, such as *open valve*, *grind garbage*,
support flywheel, *cover mechanism*, *exclude water*, or *provide light*. If a device
does not satisfy the primary functional need, no matter how good it may be
otherwise, it will not be produced. However, there frequently are secondary
functions. The primary function of a household electric refrigerator is to
preserve food; secondary functions are to make ice cubes, condition butter,
provide storage space, and so on. The primary function of a room air conditioner
is to provide cooling; secondary functions are to filter air and to dehumidify it.
Sometimes secondary functions may be extremely important. For example, an
electrical-appliance cord has the primary function of conducting electricity, but
two secondary functions, which are of absolute importance, are that it be flexible
and that it provide adequate electrical insulation. However, it is clear, in these
examples, that ability to completely satisfy the secondary functions would not
justify production of the devices if the primary need were not fulfilled.

When we identify and describe the primary and secondary functions of
complex devices, it frequently is helpful to divide the components into functional

groups, such as mechanical parts, electrical parts, enclosures, and so on. This procedure often will make clearer the primary and secondary functions of the parts. It also will help to expose useless components.

In the *second step*, determining value by comparison, we postulate that the real, or basic, value cannot exceed the cost of accomplishing the function by the least expensive method that also will meet the safety and reliability requirements. This least expensive method may be either one that exists or that can be developed. Obviously, these value comparisons require present economy studies, involving selection between alternative materials, designs, or processes. For example, portions of a shipping crate were held together by a bolt, a washer, and a nut. These cost $0.018 per set. It was determined that the parts could be held together satisfactorily by an ordinary nail that was clinched on the end. Such a nail cost $0.002. Obviously, the labor cost for using the bolt and nut differed from that for using the nail, and a hole also had to be drilled. The total-cost comparison was as follows:

	Bolt and Nut	Nail
Material	$0.018	$0.002
Labor	0.030	0.020
TOTAL	$0.048	$0.022

Because the nail appeared to be the lowest-cost method available, the maximum value that could be assigned to the fastening function was $0.022. This amount would be used as a target in evaluating all alternative methods that might be developed and considered for fastening the crate members. Any cost above the $0.022 target figure would have to be justified on some other basis than functional value.

Often a device may have more than one primary function and, consequently, more than one target alternative may be required for satisfying these functions. The spacer stud shown in Figure 16-1 is such an example. This stud had two primary functions—to hold and to space. It was made on a screw machine from hexagonal stock. The hexagonal portion of the part served the secondary func-

FIGURE 16-1.   Original design and functions of a spacer stud.

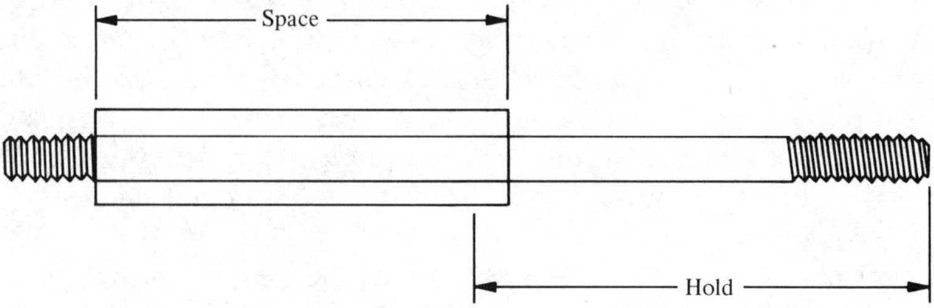

tion of providing a convenient means of holding it with a wrench while tightening the nuts that went on the ends. The cost of this spacer stud was $0.08. By value comparison it was found that the holding function could be achieved by a steel screw, costing $0.005. Therefore, the holding function of the spacer stud could not be worth more than $0.005. Similarly, it was found that the spacing function could be provided either by a cut-off length of tubing or by a rolled spacer, each of which could be obtained for $0.0025. Thus the total functional value provided by the spacer stud at a cost of $0.08 could be obtained by other means for $0.0075. Hence the target value was $0.0075.

The spacer stud example also illustrates the *third step* in the value-analysis process—causing value alternatives to be developed and selecting the most feasible alternative. While it is desirable to develop a number of alternatives, in most cases some will not be practicable, and most of those that are realistic possibilities will not have equal costs. Further, it is at this stage that esteem values become relevant. An alternative that might be completely satisfactory from a functional viewpoint might be lacking in sales appeal. One must decide what changes and additions are necessary to provide the required esteem value. But one also must be certain that the required esteem value is obtained at a minimum cost, and that no more is added than the customer is willing to pay for. In both decisions present economy studies are necessary.

In the case of the spacer stud, it was quite evident that if a satisfactory new cost, which would be only about one-tenth that of the old cost, was to be achieved, a screw-machine product could not be used and an entirely different type of design would have to be developed. Also, the use of a long screw and a piece of tubing was not an ideal solution, since two pieces were involved and the absence of a hexagonal section and threads on both ends would necessitate undesirable changes in the methods used in assembling the device in which the spacer stud was used. Thus a new and feasible design had to be developed, which would have a cost as close as possible to the target value.

In this instance, attention first was directed to obtaining, at minimum cost, the main body of the stud. An eight-penny nail was almost the same size in diameter and length, and it could be obtained for $0.001. It also had an upset head on one end. If the head could be moved from the end a sufficient distance to leave room for a thread, and if it could be made hexagonal in shape, and if threads then were rolled on each end, the primary function of holding and the secondary function of providing a wrench grip would be satisfied. The spacing function could be satisfied by upsetting a round head at the proper place along the shaft. Investigation proved that such a device, shown in Figure 16-2, could

FIGURE 16-2.   Modified design for spacer stud resulting from value analysis.

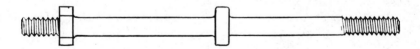

readily be produced, at a cost of $0.008. Thus a feasible alternative, close in cost to the target value, was achieved by the procedure of clearly defining the functions, evaluating the comparative value of these functions, and then developing a feasible alternative, based on means that would provide the required functions at or near target-value costs.

In this case, as frequently occurs in producers' goods, esteem value was of no consequence. Yet frequently, designs of such components needlessly incorporate details which can only be classified as adding esteem value, but for which the customer will not pay, if he has a choice. It is in this area that value analysis has made possible very large savings.

Lawrence D. Miles has applied colorful terms, "blast, create, and refine," to describe this three-step approach. The *blast* step involves determination of the comparative cost of something that will perform the major part of the functions and that will provide the base cost of the materials in simple form. This frequently requires radical departure from the existing design. The *create* step is the addition of items or features to the base so that the total functional requirements will be satisfied. The third step, *refine*, requires getting practical and realistic and making modifications that will take into account esteem values, reliabilities, ease of use, and so on. In effect, what is required is good, sound, economical engineering design, which should *always* be put into practice. Unfortunately, there are numerous examples to testify that such frequently is not the case.

In the development of feasible alternatives, it is extremely important that the person making the analysis be thoroughly acquainted with the application for which the device is intended. For example, in some applications a rivet might be a completely satisfactory and more economical substitute for a bolt and nut. On the other hand, if ease of disassembly were a requirement, such a substitution would not be feasible. For certain applications a low-cost Plexiglas cover might be a suitable substitute for a more costly molded-glass cover; in applications where high temperatures are involved, such a substitution would not be proper. Thus a thorough knowledge of the application not only aids in properly defining functions, but also helps to assure that the various alternatives will be satisfactory. However, we must make certain that familiarity with the application does not result in unwillingness to consider alternatives that may differ radically from the existing one. This attitude is one of the greatest deterrents to effective value analysis.

important cost concepts in value engineering

There are certain cost concepts that are of great importance in value engineering work. These basically are closely related to fixed and increment costs, which were considered in detail in Chapter 8.

Accurate and relevant cost records are a necessity in value engineering work. Likewise, the analyst needs to know the real reasons for design features and decisions. These are not always easily determined. A design decision that appears to have originated in the engineering department may actually have had its origin in the sales or purchasing department. Without a knowledge of the real reasons behind design and process decisions, it is difficult to determine the true costs or savings that may be associated with value alternatives.

Ordinary accounting costs seldom are of much help in such decisions. What is needed are *decision costs*. In *make versus buy* decisions we normally must consider labor costs, material costs, the variable overhead, and possibly some portion of the fixed overhead. Whether or not any fixed overhead should be included in the cost of production may vary in different cases. Frequently, this decision depends on the degree of plant utilization. If the facilities presently are not being used to capacity, some idleness or "stretch-out" probably exists. As a consequence, additional work can be absorbed with little or no actual increase in fixed overhead. Similarly, a decision to eliminate a product or process might result in "stretch-out" with no actual decrease in fixed overhead. On the other hand, if activity is at or near capacity any increase in output will add overhead.

In determining the cost of making a product one way versus the cost of making it another way, such as in value engineering analyses, there usually is little effect on fixed overhead.

One frequent and misleading cost situation that may occur is that where partially idle skilled workers may be used for operations that do not require their degree of skill—so as not to have high-cost workers idle. Such a situation results in excessively high, and unrealistic, costs, that may go undetected until a proper value analysis is made. A similar cost situation can result where a product is run in odd-sized lots when certain machines or workers are idle. The resulting frequent setup costs may be much too high. Thus in value engineering work and in engineering economy studies, it is most important that the true decision costs be obtained.

value analysis in purchasing

Although in this chapter most of the discussion has been in connection with its obvious application to the design of products and components, value engineering has an important role relative to purchasing. Obviously, it is much easier and cheaper to make design changes in products while they are on the drafting board or when writing purchase specifications than after they are in existence. Consequently, those who write purchase specifications, or who approve designs before orders are placed, are in a position to analyze designs from a value engineering viewpoint and to insist on changes so as to eliminate costly features which do not contribute to basic use value and required esteem value. By demonstrating that such features can be eliminated, the customer is in a position to force quoted prices to be reduced. The need to seek new suppliers and designs then becomes apparent.

Value engineering practitioners have been outstandingly successful in getting both designers and purchasing personnel to accept and apply value analysis and to make the necessary present-economy studies. This undoubtedly is because they have done a good job of selling and have shown the dollars-and-cents results that can be achieved. Value engineering concepts have even permeated certain governmental agencies to the point where they not only are practicing them, but they also are requiring their contractors to make value analyses of their products and operations. Certainly, every economy study analyst should applaud, aid, and abet these practices, recognizing the key role that engineering economy plays in these procedures.

problems

16-1. Explain the relationship between value engineering and engineering economy.

16-2. Why may the criterion of minimum production cost not be a proper basis for evaluating a design?

16-3. Why have companies and government agencies been eager to have value analysis applied at the design stage?

16-4. What are the basic differences between use value and esteem value?

16-5. (a) Explain why esteem values may knowingly be incorporated into products.
(b) What basic economic principle should be applied to esteem values in a given product?

16-6. Analyze a front bumper on a typical American automobile in terms of use and esteem values. Do the same for a stereo phonograph enclosed in a cabinet.

16-7. Name the primary use function and possible secondary use functions of the following: (a) shoes, (b) a sidewalk, (c) a desk clock, (d) a man's necktie.

16-8. Select some simple article that is on your desk or in your room. (a) Identify its primary use and other functions. (b) Evaluate the worth of these functions. (c) Determine the least expensive alternative for providing the primary use function. (d) Would this least-cost alternative be satisfactory to you? If not, why?

16-9. (a) What percentage of the selling price of a butane-fuel cigarette lighter would you assign to use and esteem values? (b) What absolute amount would you assign as the use value of such a lighter?

chapter 17

economy studies based on linear programming

Many industrial problems involve the allocation of limited resources for the purpose of obtaining the best possible results from their use. "Best possible results" usually means that the aim is to maximize profits or to minimize costs. It also implies that several alternatives exist from which to choose to accomplish a specific goal. Thus the problem is to allocate fixed and known amounts of resources in satisfying a given goal such that we maximize profits (or minimize costs) for feasible alternatives under consideration. This is really a general statement of what previous chapters have been dealing with.

For certain types of resource allocation problems, a technique known as *linear programming* (LP) can be used to great advantage. The purpose of this chapter, therefore, is to introduce the reader to linear programming formulations of selected economy problems and to indicate how these problems can be solved with graphical methods and/or enumerative methods. References are given for solution methods applicable to more complicated problems.

As we shall soon see, linear programming adequately represents a wide variety of real-world problems and can be quickly encoded for solution with digital computers. The *simplex method* of linear programming, developed in 1947 by George Dantzig, made the solution of large problems computationally tractable. Today linear programming is routinely used by many industries, including agriculture, steel, chemicals, airlines, petroleum, and utilities.

Several conditions must be met before linear programming can become a reliable tool. First, we are concerned with specifying nonnegative values of a set of variables that optimize a linear function expressed in terms of these variables. Second, the optimization of this function must also satisfy one or more linear constraints that mathematically take into account the availability of resources. Linearity implies, for example, that profit (or cost) per unit of output remains constant regardless of production level. Similarly, total resources consumed are assumed to be a linear function of the production level.

471

a simple, illustrative production problem

To illustrate how a simple resource allocation problem can be formulated as a linear programming situation and solved graphically, consider the manufacture of item a and item b by a small machine shop. Each unit produced requires a certain amount of machining time (i.e., standard time per operation) in each of three departments, as follows:

Item	Time (min) in Department:		
	1	2	3
a	40	24	20
b	30	32	24

Each department works a standard day consisting of 480 minutes, so it is clear that with no overtime there is a limit to the availability of machining time in each department. Because we cannot produce negative amounts of item a and item b, nor can we utilize a negative amount of time in their manufacture, none of the factors present in this problem can be negative. Suppose further that the profit per unit of item a and item b is \$5 and \$8, respectively.

In our simple problem, the aim is to maximize profit per day. This can be expressed mathematically by the following equation:

$$\text{Maximize } P = 5a + 8b$$

where a is the number of units of item a manufactured per day and b the number of units of item b produced each day. From a quick inspection it should be obvious that the function above is linear. This equation, known as the objective function, is expressed in terms of the *decision variables* (i.e., quantities that we can control so as to maximize profits). The *constraints* in this problem concern available machining time in each department and are also linear in terms of our two decision variables as seen below.

$40a + 30b \leq 480$ (constraint on available time in department 1)
$24a + 32b \leq 480$ (constraint on available time in department 2)
$20a + 24b \leq 480$ (constraint on available time in department 3)

If it is assumed that only two products are being manufactured and that all machining time in departments 1, 2, and 3 is available solely for this purpose, we can formulate this problem as a linear programming (LP) problem, since the special characteristics involving linearity of the objective function and constraints are present here. Thus the problem can be written:

$$\text{Maximize } P = 5a + 8b$$
$$\text{subject to} \quad 40a + 30b \leq 480$$
$$24a + 32b \leq 480$$
$$20a + 24b \leq 480$$
$$a \geq 0 \quad b \geq 0$$

Our task now is to determine values of a and b that maximize profits and at the same time satisfy the linear constraints.

Numerous alternatives are available for the solution of this problem. Consider, for example, what would happen if we decided to produce only item a *or* item b.

Depart-ment	Item a Only	Item b Only
1	$40a \leq 480$, or $a \leq 12.0$	$30b \leq 480$, or $b \leq 16.0$
2	$24a \leq 480$, or $a \leq 20.0$	$32b \leq 480$, or $b \leq 15.0$
3	$20a \leq 480$, or $a \leq 24.0$ No more than 12 units can be produced without violating the constraint on time of department 1. $P = 12$ units ($5/unit) = $60	$24b \leq 480$, or $b \leq 20.0$ No more than 15 units can be produced without violating the constraint on time of department 2. $P = 15$ units ($8/unit) = $120

Other alternatives involving various combinations of item a and item b could also be proposed and evaluated. However, in larger problems enumerating all possible combinations of the decision variables could be quite time-consuming.

Systematic solution procedures are available for solving large LP problems with the aid of a digital computer, and one of the best known is the simplex method. The interested reader is encouraged to learn more about this very useful method.†

Returning to our manufacturing problem, a solution can be discovered by first drawing a graph with units of item a along the ordinate and units of item b along the abscissa. If the constraint equations are then plotted on this graph, we would have Figure 17-1. The shaded area, called the *feasible region*, defines a convex polygon that contains the optimal solution to this problem. The feasible region is convex because a straight line connecting any two points in it will lie entirely within the region.

As one can easily verify, the feasible region permits these constraints to be satisfied:

$$a \geq 0 \quad b \geq 0$$
$$40a + 30b \leq 480$$
$$24a + 32b \leq 480$$

† The following references are suggested:

Bazaraa, M. S., and J. J. Jarvis, *Linear Programming and Network Flows* (New York: John Wiley & Sons, Inc., 1977).

Claycombe, W. W., and W. G. Sullivan, *Foundations of Mathematical Programming* (Reston, Va.: Reston Publishing Company, Inc., 1975).

Murty, K. G., *Linear and Combinatorial Programming* (New York: John Wiley & Sons, Inc., 1976).

Simmonnard, M., *Linear Programming* (Englewood Cliffs, N.J.: Prentice-Hall, Inc., 1966).

Strum, J. E., *Introduction to Linear Programming* (San Francisco: Holden-Day, Inc., 1972).

Zoints, S., *Linear and Integer Programming* (Englewood Cliffs, N.J., Prentice-Hall, Inc., 1974).

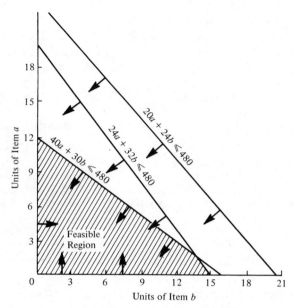

FIGURE 17-1. Feasible region.

Thus any point satisfying all the constraints and nonnegativity conditions is a feasible solution to our problem. The last constraint (available time in department 3) does not lie in the shaded region because it is not a binding constraint. This means there is no way of using all available time in department 3 to produce items a and b without violating constraints on time in departments 1 and 2. If it were possible to schedule overtime operation in these two departments, the constraint on department 3 could become binding (or "active"), but this possibility is not being considered in the present problem. Thus we do not regard the time available in department 3 as a limited resource, and only two constraints are necessary in this problem.

Now that the feasible region for a solution has been defined, we must attempt to maximize our objective function by specifying the optimal number of units

table 17-1
combinations of a and b
giving $120 profit

a	b	Profit (P)
10	8.75	$120
4.80	12	120
8	10	120
16	5	120

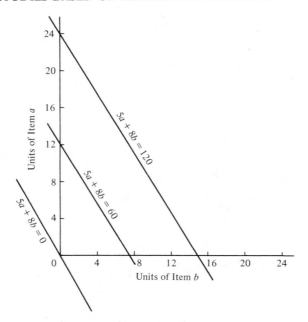

FIGURE 17-2. Objective function.

of items a and b to manufacture. This is done by superimposing the objective function on Figure 17-1 and moving it as far as possible from the origin without leaving the feasible region. (We move away from the origin, since we are trying to maximize profits.) This can be understood readily by referring to Figures 17-2 and 17-3.

In Figure 17-2 the objective function is plotted. When $5a + 8b = P$, where P is any constant value, there is a direct relationship between a and b that can be used to plot the objective function. For example, suppose that $P = \$120$. When $a = 0$ and $b = 15$, the profit is $120. But when $a = 24$ and $b = 0$, profit will also be $120. Table 17-1 illustrates a few of the other combinations of a and b resulting in a profit of $120. To plot the objective function, $120 = 5a + 8b$, we could solve for $a = 24 - \frac{8}{5}b$, which is a straight line with a slope of $-\frac{8}{5}$ and an a-intercept of 24. This line is plotted in Figure 17-2. Also shown are other members of a family of objective functions with slopes of $-\frac{8}{5}$ and a-intercepts of $P/5$ appearing as a series of parallel lines.

Some values of P, however, will cause the decision variables to violate one or more constraints. That is, all or a part of the line formed by $5a + 8b = P$ may not lie in the feasible region defined by our constraint set. The combination of a and b that we want to determine is the one allowing the maximum value of P to occur in the equation $5a + 8b = P$ while this same equation lies in the feasible region at one or more points.

An example of too large a value of P is given in Figure 17-3, where $P = 200$. This graph also illustrates a situation in which the objective function lies in

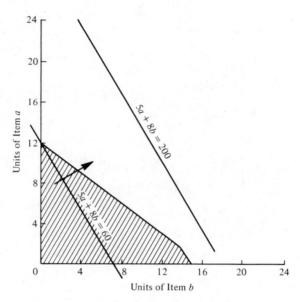

FIGURE 17-3.   Objective function and feasible region.

FIGURE 17-4.   Solution of the manufacturing problem.

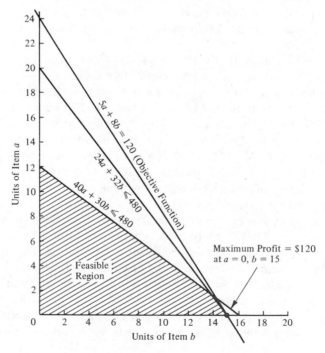

the feasible region but is not at its maximum possible value. In this case, the objective function, $5a + 8b = 60$, must move farther away from the origin to be maximized.

Finally, Figure 17-4 shows the maximum value of the objective function to be $5(0) + 8(15) = 120$ for $a = 0$ and $b = 15$. Note in Figure 17-4 that the objective function touches the feasible region at exactly one point. It is also clear that the feasible region depends only on the constraint equations and is unaffected by changes in the objective function. Moreover, the optimal solution is determined by the slope of the objective function, given a certain set of constraints. For example, suppose the objective function were $P = 9a + 3b$. The optimal solution would then be $a = 12$, $b = 0$, with a profit of \$108.

The optimal solution to our illustrative production problem could also have been determined by making inequalities into equalities and then solving a set of simultaneous linear equations. In this case there are only two equations to deal with simultaneously, since we have two decision variables. But when several constraints are imposed on the problem, it is often difficult to determine *which two equations* to solve to find the maximum (or minimum) value of the objective function. This raises another interesting possibility for solving linear programming problems, in addition to the graphical method illustrated above.

If we determine the points of intersection for every pair of constraint equations (equalities), it would be possible to find the point resulting in a maximum value of $P = 5a + 8b$. Suppose for the moment that we do not know that constraint 3 is nonbinding in the problem. All pairs of constraint equations and their points of intersection are shown below. There are $\binom{5}{2} = 5!/3!\, 2! = 10$ points of intersection, since we have five constraints and two decision variables. Also, $\binom{5}{2}$ is a combinatorial term that can be written in terms of factorials:

$$\binom{5}{2} = \frac{5!}{2!\,(5 - 2)!}$$

(5! is read "5 factorial" and is equal to the product $5 \times 4 \times 3 \times 2 \times 1 = 120$; the other factorials are computed in similar fashion.)

Pair 1:*	$40a + 30b = 480$	$a = \frac{12}{7}$	$P = 118\frac{2}{7}$
	$24a + 32b = 480$	$b = \frac{96}{7}$	
Pair 2:	$40a + 30b = 480$	$a = -8$	$P = $ undefined
	$20a + 24b = 480$	$b = \frac{80}{3}$	
Pair 3:	$24a + 32b = 480$	$a = 60$	$P = $ undefined
	$20a + 24b = 480$	$b = -30$	
Pair 4:	$40a + 30b = 480$	$a = 0$	$P = 128$
	$a = 0$	$b = 16$	
Pair 5:*	$40a + 30b = 480$	$a = 12$	$P = 60$
	$b = 0$	$b = 0$	

Pair 6:*	$24a + 32b = 480$	$a = 0$	$P = 120$
	$a = 0$	$b = 15$	
Pair 7:	$24a + 32b = 480$	$a = 20$	$P = 100$
	$b = 0$	$b = 0$	
Pair 8:	$20a + 24b = 480$	$a = 0$	$P = 160$
	$a = 0$	$b = 20$	
Pair 9:	$20a + 24b = 480$	$a = 24$	$P = 120$
	$b = 0$	$b = 0$	
Pair 10:*	$a = 0$		$P = 0$
	$b = 0$		

The solution to each pair of equations must now be inserted into the original set of constraints to ensure that values of a and b are feasible (i.e., values of a and b do not violate the constraints). Recall that our constraints are

$40a + 30b \leq 480$	(constraint 1)
$24a + 32b \leq 480$	(constraint 2)
$20a + 24b \leq 480$	(constraint 3)
$a \geq 0$	(constraint 4)
$b \geq 0$	(constraint 5)

The solution to pair 1 'satisfies the three constraints (3, 4, and 5 above) not used to determine the point of intersection at $a = \frac{12}{7}$ and $b = \frac{96}{7}$. That is,

$$20(\tfrac{12}{7}) + 24(\tfrac{96}{7}) < 480$$
$$\tfrac{12}{7} > 0$$
$$\tfrac{96}{7} > 0$$

Solutions to pairs 2 and 3 are not permissible, since a and b must be nonnegative. After evaluating other points of intersection in the same manner, it is apparent that solutions to pairs 1, 5, 6, and 10 satisfy all five constraints. Asterisks have been placed by each of these solutions.

These four points are termed *basic feasible solutions* to our linear programming problem and lie at the vertices of the feasible region formed by the constraints. The optimal solution will be located at one of these vertices (a single point). It is also true that all points lying on or within the feasible region are feasible solutions to the problem. If the objective function is coincident with (parallel with) one of the binding constraints there are an infinite number of solutions resulting in a profit of P.

The enumeration of intersection points above illustrates that *the optimal solution is the basic feasible solution that maximizes* (or minimizes) *the objective function.* In our problem it is seen that the solution to pair 6 yields the maximum profit of $120 at $a = 0$ and $b = 15$. Therefore, we would recommend that 15 units of item b be produced each day.

In making this recommendation to management, we could carry the analysis one step further and calculate idle time in each department:

Department 1: $480 - 40(0) - 30(15) = 30$ minutes/day
Department 2: $480 - 24(0) - 32(15) = 0$ minutes/day
Department 3: $480 - 20(0) - 24(15) = 120$ minutes/day

We may now want to suggest that management consider the production of item
c (a new product line), which would require machine work mainly in departments
1 and 3.

To complete this illustrative problem, suppose that management decides to
add a new product line (item c) that requires 20 minutes of machining time in
department 1 and 32 minutes of machining time in department 3. The constraints
and corresponding feasible region for the problem would now be

(1)	$40a + 30b + 20c \le 480$	Department 1 time
(2)	$24a + 32b \le 480$	Department 2 time
(3)	$20a + 24b + 32c \le 480$	Department 3 time
(4)	$a \ge 0$	⎧Nonnegativity
(5)	$b \ge 0$	⎨constraints on
(6)	$c \ge 0$	⎩each product line

The three-dimensional graph in Figure 17-5 shows the feasible region to be a
polyhedron.

FIGURE 17-5. Solution space for three decision variables.

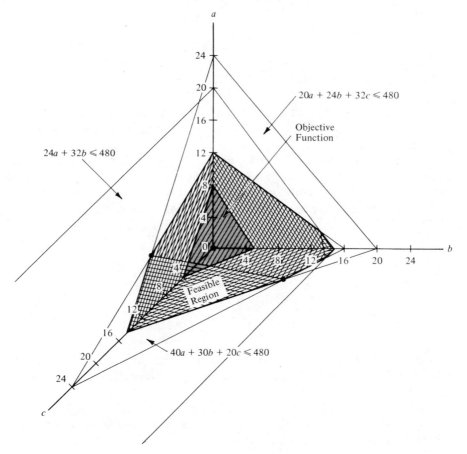

If the profit per unit of item c were \$8, the objective function is $P = 5a + 8b + 8c$. This is the shaded plane in the graph whose location depends on the value of P and whose slope is determined by the coefficients of the objective function. Because we are attempting to maximize P, the plane would move away from the origin until the optimal solution is reached.

It is difficult to determine from Figure 17-5 the point resulting in maximum profits. Apparently, the optimum solution lies in the b–c plane and is equal to 12 units of item b and 6 units of item c. The profit would be \$144, which exceeds our maximum profit of \$120 when only items a and b are manufactured.

The curious reader can verify that a maximum profit of \$144 per day results in this problem when $b = 12$ and $c = 6$. Instead of 10 pairs of equations that were presented earlier, a total of $\binom{6}{3} = 20$ sets of constraints would have to be evaluated. It is obvious that for larger problems, such enumerative search procedures can consume prohibitive amounts of time. Fortunately, the simplex method can greatly simplify the solution of large LP problems and is readily available for use on digital computers.

graphical solution of a cost problem

A problem is presented here that illustrates how an economy study involving cost minimization can be formulated as a linear programming problem and solved graphically.

Suppose that the Ajax Furniture Company buys its lumber from two sources, companies G and H, and classifies the lumber according to three different grades, A, B, and C. The following matrix shows the expected proportion of lumber in each grade from each source. Board-feet requirements are also shown.

Grade	Source G	H	Board-Feet Required
A	0.15	0.60	3,000
B	0.25	0.30	2,500
C	0.60	0.10	2,000

Company G charges \$1 per board-foot and H charges \$1.50 per board-foot. How much should be purchased from each supplier to satisfy requirements and minimize total cost?

We must first formulate our objective function and constraint equations. Let decision variable x_1 be the number of board-feet purchased from G, and x_2 be the board-feet from H. The objective is to:

Minimize $\$1.00x_1 + \$1.50x_2$
subject to $0.15x_1 + 0.60x_2 \geq 3,000$
$0.25x_1 + 0.30x_2 \geq 2,500$
$0.60x_1 + 0.10x_2 \geq 2,000$
$x_1 \geq 0 \qquad x_2 \geq 0$

The constraints and objective function are illustrated in Figure 17-6. Because we are minimizing costs, the optimal solution is the last point in the feasible region as the objective function moves toward the origin. This occurs at $x_1 = 5,716$ and $x_2 = 3,571$, with a minimum cost of \$11,072.

the transportation problem

This section deals with a special type of linear programming problem, the *transportation problem*, in which there are m origins, each capable of producing and shipping a known amount of goods; and there are n destinations which require a specified shipment of goods. The penalty of shipping goods from an

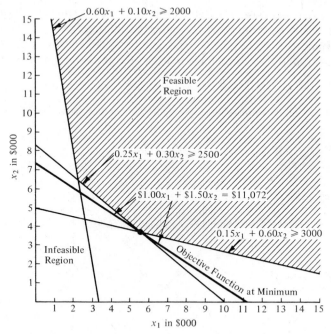

Minimum at:

Cost = \$11,072
x_1 = \$5,716
x_2 = \$3,571

FIGURE 17-6. Solution of the lumber problem.

origin to a destination is given on a per-unit basis (e.g., cost per ton-mile of goods shipped or miles over which a standard load is transported). The object is to determine the production shipping pattern that minimizes the total cost of transportation.

This problem can be simply illustrated as shown in Figure 17-7. As depicted in Figure 17-7, at plants A and B there are 6 and 8 truckloads, respectively, of cartons that are needed at warehouses C and D. Warehouse C requires 10 loads and warehouse D requires 4 loads of the cartons. The distances between the two plants and the two warehouses are as shown in Figure 17-7. How can the cartons at the plants be allocated between the two warehouses so as to minimize the loadmiles hauled? Table 17-2 shows how we might solve this problem by listing possible ways by which the required distribution might be achieved, and then computing the total mileage for each possibility. Using this approach, we can readily determine that by shipping 2 loads from plant A to warehouse C and 4 loads to warehouse D and by supplying the remaining requirements at warehouse C from plant B, a minimum mileage of 1,340 would be obtained.

This problem can also be represented in another manner, as shown in Figure 17-8. The availabilities and requirements are shown, and the transport distances are given in the small boxes.

Figure 17-9 shows how this transportation problem can be represented and solved graphically by computing the mileage resulting from the extreme conditions and connecting these values by straight lines, recognizing that the relationships are linear, and then adding the mileages to obtain the totals for those cases

FIGURE 17-7. Linear programming problem; transporting available loads to required destinations.

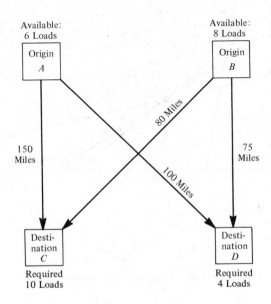

table 17-2
possible combinations of deliveries from plants A and B to warehouses
C and D with resulting load-miles

From:	A				B				
To:	C		D		C		D		
	(×150) Load-		(×100) Load-		(×80) Load-		(×75) Load-		Total Load-
Loads	Miles	Loads	Miles	Loads	Miles	Loads	Miles	Miles
6	900	0	—	4	320	4	300	1,520
5	750	1	100	5	400	3	225	1,475
4	600	2	200	6	480	2	150	1,430
3	450	3	300	7	560	1	75	1,385
2	300	4	400	8	640	0	— Min→	1,340

where the requirement of a total of 10 loads from the two plants to warehouse
C is met.

It is apparent that this problem is very simple, because there are only five
possible combinations of deliveries by which the required allocation of loads
could be achieved. Therefore, obtaining a solution by the method used in Table
17-2 was not unduly laborious. However, it is equally clear that for problems
more complex—and most real problems are considerably more complex—

FIGURE 17-8. Matrix representation of the illustrative problem shown in Figure 17-7.

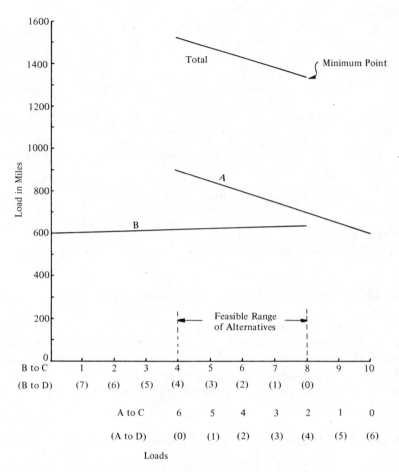

FIGURE 17-9. Graphical solution of Example 17-1.

determining all the possible combinations and computing the necessary values manually would be prohibitively costly in time and money.

A mathematical statement of the situation represented in Figure 17-8 can be written from the information given. From the availabilities at each plant, expressed in the rows of the matrix,

$$X_{AC} + X_{AD} = 6$$
$$X_{BC} + X_{BD} = 8$$

where the X values indicate the allocations—X_{AC} is the number of loads assigned from plant A to warehouse C, and so on. Similarly,

$$X_{AC} + X_{BC} = 10$$
$$X_{AD} + X_{BD} = 4$$

These equations express the *constraints* imposed by the availabilities at the origins and the demands of the destinations. They represent the allocation model. To complete the mathematical statement of the problem, we can write an equation that expresses the *objective* of the solution—the minimization of the total transport distance. This objective equation is

$$\text{Minimize } 150X_{AC} + 100X_{AD} + 80X_{BC} + 75X_{BD}$$

Special computational procedures have been developed for solving large transportation problems. Their discussion is beyond the scope of this text, but references given in the footnote on page 473 include complete coverage of such procedures.

the assignment problem

The *assignment problem* is a special case of the transportation problem. It can be viewed as a transportation problem with a demand of one at each destination and a supply of one at each origin (source), where all shipments must be either zero or one. A shipment of 1 indicates that a source is assigned to a destination. The solution involves assigning n sources to n destinations to minimize the total assignment cost. The assignment matrix must be a square ($n \times n$) matrix.

Many realistic problems can be solved with the assignment algorithm. If, for example, there are n machines to be assigned to n locations such that materials handling can be accomplished at minimum cost to the company, the machines may be viewed as sources and the locations as destinations. The costs for three machines and three locations are given by a matrix such as that of Table 17-3. It can be seen that the cost of having machine U in location 2 is 22 units. In the solution there must be exactly one assignment in each column, and one assignment in each row—there must be exactly one machine for each location.

table 17-3
assignment cost matrix

Machine	Location 1	Location 2	Location 3
U	43	22	28
V	40	38	37
W	24	25	36

The linear programming formulation of the assignment problem is straight-forward. By the use of Table 17-3, we find that the objective function is

$$\text{Minimize } Z = 43x_{U1} + 22x_{U2} + 28x_{U3} + 40x_{V1} + 38x_{V2}$$
$$+ 37x_{V3} + 24x_{W1} + 25x_{W2} + 36x_{W3}$$

where $x_{ij} = 1$ if machine $i\,(i = U, V, W)$ is assigned to location $j\,(j = 1, 2, 3)$
$\quad\quad x_{ij} = 0$ if machine i is not assigned to location j

To avoid assigning two locations to one machine or two machines to a single location, a set of constraint equations could be written:

$$\left.\begin{array}{l} x_{U1} + x_{U2} + x_{U3} = 1 \\ x_{V1} + x_{V2} + x_{V3} = 1 \\ x_{W1} + x_{W2} + x_{W3} = 1 \end{array}\right\} \begin{array}{l}\text{requires that every machine}\\\text{be assigned to exactly one location}\end{array}$$

$$\left.\begin{array}{l} x_{U1} + x_{V1} + x_{W1} = 1 \\ x_{U2} + x_{V2} + x_{W2} = 1 \\ x_{U3} + x_{V3} + x_{W3} = 1 \end{array}\right\} \begin{array}{l}\text{requires that each location be}\\\text{serviced by one machine only}\end{array}$$

In general, the assignment problem can be summarized as follows.

$$\text{Minimize } Z = \sum_{i}^{n} \sum_{j}^{n} c_{ij}x_{ij} \quad (c_{ij} = \text{materials-handling cost of assigning machine } i$$
$$\text{to location } j \text{ for a given production schedule})$$

subject to
$$\sum_{i=1}^{n} x_{ij} = 1$$

$$\sum_{j=1}^{n} x_{ij} = 1$$

$$x_{ij} = 1 \quad \text{or} \quad 0$$

For the simple problem above, assignments can be made by trial-and-error in an effort to discover the minimum cost. Unfortunately, for large problems, this approach would be impractical. Even when $n = 3$ (a small problem), there are six possible assignments for our machine–location problem, as follows:

Trial Number	Machine–Location Assignment	Cost
1	U–1, V–2, W–3	$117
2	U–1, V–3, W–2	105
3	U–2, V–1, W–3	98
4* (optimum)	U–2, V–3, W–1	83*
5	U–3, V–1, W–2	93
6	U–3, V–2, W–1	90

Thus the optimal assignment would be expressed as in Table 17-4 in matrix form.

In general, there are $n! = (n)(n-1)(n-2)\cdots(3)(2)(1)$ possible assignments to be considered. Obviously, when $n = 6$, it would be difficult to evaluate all 720 possibilities. For large assignment problems, a special algorithm has been developed that considerably reduces solution time relative to trial-and-error enumeration methods. Discussion of this algorithm appears in most of the references listed in the footnote on page 473.

table 17-4
optimal assignment

Machine	Location 1	Location 2	Location 3
U	43	22	28
	1		
V	40	38	37
		1	
W	24	25	36
	1		

$Z^* = 83$

capital budgeting problems

Linear programming is a useful technique for solving certain types of multiperiod *capital budgeting problems* when a firm is not able to implement all projects that increase its present worth. For example, constraints may exist on how much investment capital can be committed during each fiscal year, or interdependencies among projects may affect the extent to which projects can be successfully carried out during a specified planning horizon.

The final section of Chapter 7 dealt with the enumeration of mutually exclusive combinations of alternatives from sets of investment projects that can, themselves, be mutually exclusive, independent, and/or contingent. Suppose that the goal of a firm is to maximize its net present worth from the adoption of a capital budget that includes at least two mutually exclusive combinations. When the number of possible combinations becomes fairly large, manual methods that were described in Chapter 7 for determining the optimal investment plan tend to become quite complicated and time consuming. In such a situation, it is often worthwhile to consider linear programming as a solution procedure. The remainder of this section describes how simple capital budgeting problems can be formulated as LP problems. We hope that the reader will obtain some feeling of how more involved problems might also be modeled.

The objective function of the capital budgeting problem can be written as follows.

$$\text{Maximize net P.W.} = \sum_{j=1}^{n} B_j X_j$$

where B_j = net present worth of investment opportunity j during the planning period being considered

X_j = fraction of opportunity j that is implemented during the planning period (*Note:* In most problems of interest, X_j will be either 0 or 1); the X_j's are the decision variables

n = number of mutually exclusive combinations of alternatives under consideration

In computing the net P.W. of each mutually exclusive combination, a minimum attractive rate of return must be specified.

In view of the remaining notation defined as follows,

c_{tj} = cash outlay (e.g., initial investment or annual operating budget) required for opportunity j in time period t

C_t = maximum cash outlay that is permissible in time period t

there are typically two types of constraints present in capital budgeting problems:

1. Limitations on cash outlays for period t of the planning horizon

$$\sum_{j=1}^{n} c_{tj}X_j \leq C_t$$

2. Interrelationships among investment opportunities. The following are examples:

(a) If projects p, q, and r are mutually exclusive, then

$$X_p + X_q + X_r \leq 1$$

(b) If project r can be undertaken only if project s has already been selected, then

$$X_r \leq X_s \quad \text{or} \quad X_r - X_s \leq 0$$

(c) If projects u and v are mutually exclusive and project r is dependent (contingent) upon the acceptance of u or v, then

$$X_u + X_v \leq 1$$
$$X_r \leq X_u + X_v$$

For a very simple illustration, consider the capital budgeting problem presented earlier in Chapter 7 as Example 7-11. The linear programming formulation of that particular problem is the following.

Maximize

$13.4X_{B1} + 8.0X_{B2} - 1.3X_{C1} + 0.9X_{C2} + 9.0X_D$

subject to

$50X_{B1} + 30X_{B2} + 14X_{C1} + 15X_{C2} + 10X_D \leq 48$	(constraint on investment funds)
$X_{B1} + X_{B2} \leq 1$	($B1$, $B2$ mutually exclusive)
$X_{C1} + X_{C2} \leq X_{B2}$	($C1$ or $C2$ contingent on $B2$)
$X_D \leq X_{C1}$	(D contingent on $C1$)
$X_j = 0$ or 1	(no fractional projects allowed)

A problem such as the above could be solved quite readily by using the simplex method of linear programming if the last constraint ($X_j = 0$ or 1) were not present. With that constraint included, the problem is classified as a linear *integer* programming problem. There are many computer programs available for solving large linear integer programming problems.

As a second example, consider a three-period capital budgeting problem. Estimates of cash flows are as follows:

Investment Opportunity	Net Cash Flow ($000s), End of Year[a]:				Net P.W. ($000s) at 12%[b]
	0	1	2	3	
A1		150	150	150	
	− 225	(60)	(70)	(70)	+135.3
A2 mutually		200	180	160	
exclusive	− 290	(180)	(80)	(80)	+146.0
A3		210	200	200	
	− 370	(290)	(170)	(170)	+119.3
B1		100	400	500	
	− 600	(100)	(200)	(300)	+164.1
independent					
B2		500	600	600	
	− 1200	(250)	(400)	(400)	+151.9
C1		70	70	70	
mutually	− 160	(80)	(50)	(50)	+8.1
C2 exclusive and		90	80	60	
dependent on	− 200	(65)	(65)	(65)	− 13.1
acceptance of					
C3 A1 or A2		90	95	100	
	− 225	(100)	(60)	(70)	+2.3

[a] Estimates in parentheses are annual operating expenses (which have already been subtracted in determination of net cash flows).
[b] For example, net P.W. for $A1 = -225{,}000 + \$150{,}000(P/A, 12\%, 3) = +\$135{,}300.$

The M.A.R.R. is 12% and the ceiling on investment funds available is $1,200,000. In addition, there is a constraint on operating funds for support of the alternative selected, and it is $400,000 in year 1. From these constraints on funds outlays and the interrelationships among opportunities indicated above, we shall formulate this situation in terms of a linear integer programming problem.

First, the net present worth of each investment opportunity at 12% is calculated. The objective function then becomes

$$\text{Maximize net P.W.} = 135.3X_{A1} + 146.0X_{A2} + 119.3X_{A3} + 164.1X_{B1}$$
$$+ 151.9X_{B2} + 8.1X_{C1} - 13.1X_{C2} + 2.3X_{C3}$$

The budget constraints are the following.

investment funds constraint

$$225X_{A1} + 290X_{A2} + 370X_{A3} + 600X_{B1} + 1{,}200X_{B2}$$
$$+ 160X_{C1} + 200X_{C2} + 225X_{C3} \leq 1{,}200$$

first year's operating cost constraint

$$60X_{A1} + 180X_{A2} + 290X_{A3} + 100X_{B1} + 250X_{B2}$$
$$+ 80X_{C1} + 65X_{C2} + 100X_{C3} \leq 400$$

Interrelationships among the investment opportunities give rise to these constraints on the problem:

$$X_{A1} + X_{A2} + X_{A3} \leq 1 \qquad \text{A1, A2, A3 are mutually exclusive}$$

$$\left.\begin{matrix} X_{B1} \leq 1 \\ X_{B2} \leq 1 \end{matrix}\right\} B1, B2 \text{ are independent}$$

$$X_{C1} + X_{C2} + X_{C3} \leq X_{A1} + X_{A2} \qquad \begin{matrix} \text{accounts for dependence of} \\ \text{C1, C2, C3 on A1 } or \text{ A2} \end{matrix}$$

Finally, if all decision variables are required to be either 0 (not in optimal solution) or 1 (included in optimal solution), the last constraint on the problem would be written:

$$X_j = 0, 1 \qquad \text{for } j = A1, A2, A3, B1, B2, C1, C2, C3$$

As one can see, a fairly simple problem such as this one would require an inordinate amount of time to solve by listing and evaluating all mutually exclusive combinations as suggested in Chapter 7. Consequently, it is recommended that a suitable computer program be utilized to obtain solutions for all but the most simple capital budgeting problems.

problems

17-1. Determine the feasible region for the following LP problem:

$$\text{Minimize } C = 4x_1 + 6x_2$$
$$\text{subject to} \qquad 2x_1 + 5x_2 \geq 10$$
$$3x_1 + 2x_2 \geq 6$$
$$x_1 \geq 0 \qquad x_2 \geq 0$$

(a) Using graphical means, determine the optimal solution.
(b) If the objective function changes to $C = x_1 + 8x_2$, will there be a different optimal solution?

17-2. A small company binds books and has two bindings available. Binding A is a high-quality product that results in a profit of $1.80 per book. Binding B is a lower-quality product with a profit margin of $1.50 per book. If only the lower-quality binding were available, the company could bind 500 books each day. When binding A is requested, it requires 150% more time than does binding B. However, because of material shortages, only 350 books each day can be produced regardless of the type of binding. The high-quality binding requires a special glueing operation that has a maximum output of 250 books per day. Formulate this production problem as a linear programming problem to maximize profit and solve it graphically.

17-3. A person wishes to select a diet consisting of bread, butter, and/or milk, which has a minimum cost, but yields an adequate amount of vitamins A and B. The minimum vitamin requirements are 11 units of A and 10 units of B. The diet should not contain more than 13 units of A because more may be harmful. Furthermore, he likes milk and requires that the diet include at least 3 units of milk but is indifferent to the amount of butter and bread. A dietician has measured the vitamin contents and found, per unit of product, 1 unit of A and 3 of B in bread, 4 units of A and 1 of B in butter, 2 units of A and 2 of B in milk. The market price of 1 unit of bread, butter, and milk is 2, 9, and 1, cent, respectively. Formulate this problem as a linear programming problem. Be sure to specify the decision variables and the objective function. Solve this problem graphically.

17-4. A furniture manufacturer wants to determine how many tables, chairs, desks, and/or bookcases to make to optimize utilization of his available resources. These products use two different types of lumber, and he has on hand 1,500 board-feet of the first type and 1,000 board-feet of the second type. He has 800 man-hours of labor available for the total job. His sales forecast plus his back-orders require him to make at least 40 tables, 130 chairs, 30 desks and no more than 10 bookcases. Each table, chair, desk, and bookcase requires 5, 1, 9, and 12 board-feet, respectively, of the first type of lumber; and 2, 3, 4, and 1 board-feet, respectively, of the second type. A table requires 3 man-hours to make, a chair requires 2 man-hours, a desk 5 man-hours, and a bookcase 10 man-hours. The manufacturer makes a total profit of $12 on a table, $5 on a chair, $15 on a desk, and $10 on a bookcase. Formulate this situation as a linear programming problems such that profits are maximized.

17-5. A company supplies aggregate from three quarries, A, B, and C, to five premix concrete plants. The following matrix gives the dollar transportation costs from these quarries to the plants, and the capacities and requirements per day. Set this up as a transportation problem and attempt to enumerate several feasible solutions. The least-cost solution is $90. How close did you get to this minimum cost?

| | Plant | | | | | Capacity |
Quarry	1	2	3	4	5	(units)
A	4	2	3	2	6	9
B	5	4	5	2	1	10
C	6	5	4	7	3	13
Requirements:	4	5	7	8	8	

17-6. A company has three factories, A, B, and C, from which it supplies a product to four company-operated retail stores, D, E, F, and G. The monthly capacities at the factories are 120, 140, and 130 units, respectively, and it has found that the four stores can sell 80, 90, 110, and 160 units per month, respectively, if they are available. The unit profits that result from the supply and sale are as follows:

| | To | | | |
From	D	E	F	G
A	$42	$48	$38	$36
B	40	49	52	39
C	38	36	45	44

Formulate this situation as an LP problem and try to enumerate some feasible solutions to it. The maximum profit is $18,310. How close did you get to this solution? (*Hint:* Add a row to account for unsatisfied demand. In the optimal solution, 20 units of D's demand is unmet and 30 units of G's demand is not satisfied.)

17-7. Four manufacturing plants must be built in four different geographic regions, with only one plant in each location. The costs of building each plant in the various locations are as follows, in millions of dollars:

	Location			
Plant	1	2	3	4
1	60	51	32	32
2	48	×	37	43
3	39	26	×	33
4	40	×	51	30

In this matrix an × indicates that, because of the need for highly specialized labor, it is impractical to assign a plant to a particular location. Formulate this as an assignment problem and generate, by trial and error, several feasible solutions to it.

17-8. Refer to Problem 7-23 of Chapter 7. Formulate this capital budgeting problem as a linear integer programming problem.

17-9. Refer to Problem 7-24 of Chapter 7. Set this problem up as a linear integer programming problem.

chapter 18

critical
path economy

In carrying out many types of projects in industry and government, a large number of interdependent tasks must be planned, scheduled, and controlled. These component tasks may be interdependent with respect to their timing, manpower requirements, equipment needs, and so on. Examples of large and complex one-time projects include development of a prototype space vehicle and construction of a high-rise apartment building.

Typically, these jobs require many months or years for completion, and many of the component tasks can be completed in varying amounts of time and cost, depending on how different amounts of resources are assigned to them. Consequently, the total job will involve different amounts of time for completion, and will involve different costs, depending on how the component tasks are done. However, in most instances only a few of the tasks are critical in that they will affect the actual outcome in time and cost. Several *critical path procedures* have been developed that enable such projects to be managed in an effective and economical manner. In most cases the ultimate goal is to determine when each task must begin to assure that the overall project will be finished within a specified amount of time.

These procedures have been widely adopted by industry, but they have also been found useful in many nonindustrial activities, such as planning a national meeting of a technical society. In most cases they have an economic objective, and present economy studies and decisions are involved in their application. The purpose of this chapter is to describe several project management techniques in which trade-offs between task timing and costs are possible.

the Gantt chart

Critical path methods are expansions of the old, well-known *Gantt charts*, which have been used since the early 1900s. Figure 18-1 is a simple Gantt chart that depicts the time schedule of the major activities that are involved in building a

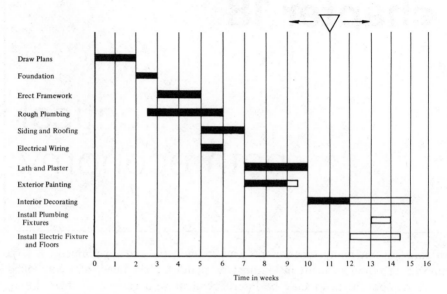

FIGURE 18-1. Gantt chart of work tasks involved in building a house.

house. Such a chart shows the required activities, the time period over which each activity will extend, and the time at which each activity should start in order that each prerequisite activity will be completed, or have progressed sufficiently, so that the following activities can start. In addition, a time indicator usually is provided, which can be moved to indicate the current time. As work progresses, the activity bars are filled in so that at any time we can determine whether the projected schedule is being met. In Figure 18-1, for example, some

FIGURE 18-2. Activity—elapsed-time chart.

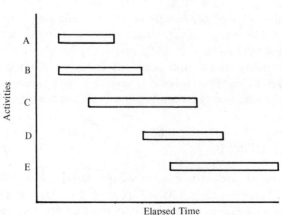

portion of the exterior painting is behind schedule and the interior decorating is ahead of schedule.

As the normal Gantt, or bar, chart is used, it has two rather serious limitations. First, it does not clearly indicate details regarding the progress of activities. This is particularly bothersome for activities that extend over relatively long periods of time. For example, in Figure 18-1 the interior decorating is scheduled to extend over 5 weeks. Although the decoration as a whole—perhaps in man-hours—is ahead of schedule, there is not any certainty that it is being done in the proper sequence so that the bathrooms will be decorated before the trim on the plumbing fixtures must be installed. The second, and more important, deficiency is that such charts do not give a clear indication as to what portions of any activity are specifically prerequisite to following activities or to dependent activities that may overlap.

These deficiencies of bar charts may be eliminated, to a large degree, by the procedures shown in Figures 18-2 and 18-3, wherein the activities are broken into subparts and the interdependence is shown by means of connecting arrows. These procedures lead directly to project network techniques.

activity–event networks

An activity–event network portrays the interrelationships between the activities and events that comprise a project or job, where an *event* is the start or end of a

FIGURE 18-3. Activity—elapsed-time chart, with interdependency shown by arrows.

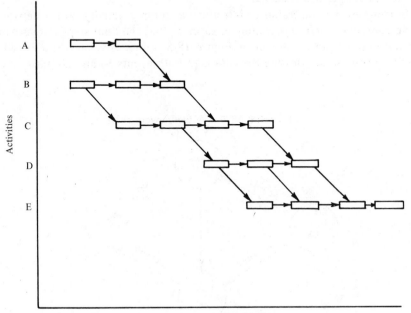

Elapsed Time

task and thus represents a point in time, and an *activity* is the work required to complete a specific event. Activities require time and resources for their accomplishment and represent recognizable parts of the project. In an activity–event network, as shown in Figure 18-4, events commonly are represented by circles or ovals and activities by arrows. Thus events are related to one another by activities. Figure 18-4 illustrates a simple activity-on-arrow diagram.

Events ordinarily are described by two- or three-word phases, with the first word of the phrase being *start* or *complete*. Thus "start drawings," "start electrical design," "complete foundation," and "complete drawings" are typical descriptions. Because, except for initial events and a few other cases, an event represents the completion of an activity, most events utilize *complete* as the first word of the description. Activities are described by brief phrases, such as "prepare mechanical drawings" or "build cabinets."

The events are numbered 10, 20, 30, and so on when the network is first drawn. Because the passage of time progresses from left to right on a network, early events are given low numbers and later events are assigned higher numbers. If it becomes necessary to alter the network by adding more events, these can be assigned numbers, 11, 21, 31, and so on. Activities usually are referred to by showing the events that they connect: activity 10–20, activity 20–30, and so forth.

In activity–event networks, no events, except an initial one, can be started until the completion of all the activities that are connected to it by *incoming* arrows. Thus, in Figure 18-4, event 30 cannot occur until both activity 10–30 and activity 20–30 are completed. Similarly, no activity can start until the event from which it emerges has occurred.

Sometimes it is desirable to introduce a dummy activity into a network to have consistent logic. For example, suppose that the final steps in overhauling an automobile are as shown in Figure 18-5. Obviously, activities 10–30 and 20–30 are the same—testing the car—and both events 10 and 20 must be com-

FIGURE 18-4.   Activity–event chart.

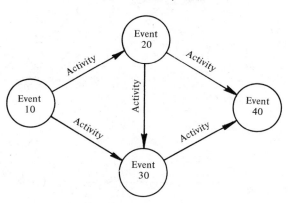

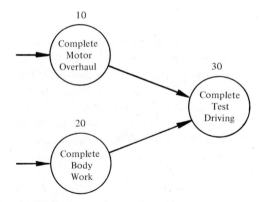

FIGURE 18-5. Activity–event sequence with the final event dependent on two duplicate activities (10–30 and 20–30).

pleted before the test driving can start. This type of situation can be resolved by use of a dummy activity, 10–20, as shown in Figure 18-6. A dummy activity requires no time and no resources.

Another important use of a dummy activity is to indicate that one of two independent events must precede the other. Thus, according to the logic depicted in Figure 18-5, the completions of the motor overhaul and the body work could occur simultaneously. But the use of the dummy activity in Figure 18-6 specifies that the completion of the motor overhaul (event 10) must precede the completion of the body work (event 20). Similarly, dummy activities are used in showing that one activity must precede another, as might be necessitated by the same piece of equipment being required for both activities. Although the dummy activity does not require any time, the resulting change in the sequence of the activities and/or events may extend the time required for a project and may alter the resource requirements.

FIGURE 18-6. Use of "dummy" activity (10–20) to avoid redundant activities portrayed in Figure 18-5.

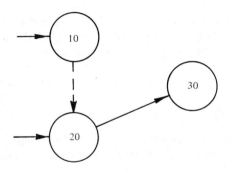

CPM and PERT

Several variations of activity–event networks have been developed and are in widespread use. Probably the two most common are CPM (critical path method) and PERT (program evaluation and review technique). The du Pont Company developed CPM in the late 1950s as a technique for planning and controlling complex engineering projects with respect to their time and costs. About the same time, PERT was developed by the United States Navy as a means of managing the Polaris missile program. The successful completion of the Polaris program some 2 years ahead of the original forecast was attributed to use of the PERT technique.

CPM and PERT are based on the same concepts, the fundamental one being identification of a critical sequence of activities to be closely monitored and controlled in meeting a specified completion date. The main difference is the manner in which activity–event networks are graphically constructed.

CPM is applied, primarily, to projects that are more deterministic in nature, such as construction work, in which cost and time estimates can be predicted with considerable certainty due to the existence of past experience. PERT, on the other hand, tends to be applied to one-time projects, such as research and development work, in which intellectual effort and first-time prototype manufacturing are involved, so that time and cost estimates tend to be quite uncertain. Consequently, probabilistic methods may be employed in connection with PERT.

When times that will be required for activities must be based solely on subjective estimates, it is common practice to obtain these estimates from several persons who are familiar with, or are to be involved in, each activity. In such cases some of the persons will give quite optimistic estimates, while others anticipate many possible difficulties and will give quite pessimistic estimates. In some cases the "pessimistic," "optimistic," and "probable" time estimates for activities may be included on a PERT chart. Common practice is to compute a single time estimate using the relationship

$$t_{\text{est}} = \frac{t_0 + 4t_m + t_p}{6} \tag{18-1}$$

where t_0 = optimistic time
t_m = most likely time
t_p = pessimistic time

Numerous variations of both CPM and PERT have been developed, so that in some cases the procedures are very similar. For further discussion in this chapter, CPM, with some modifications, is employed to illustrate the economy study aspects of these procedures.

critical paths

Assume that the network shown in Figure 18-7 represents activities and events that are required for carrying out a certain project. The numerals shown above

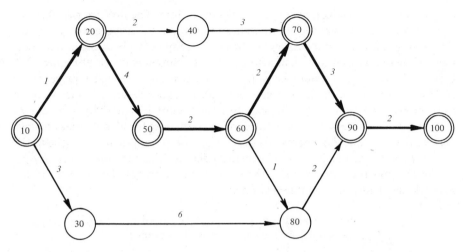

FIGURE 18-7. Typical activity–event network with critical path indicated by heavy arrows and critical events indicated by double circles.

each activity arrow are the number of working days estimated for the completion of each activity, when normal procedures are used.† From examination of the diagram it is clear that there are four paths between the first event (10) and the final event (100). Each of these paths must be undertaken in completing the overall project. If the total time along each path is determined, the results are as follows:

1. For path 10–20–40–70–90–100: 11 days.
2. For path 10–20–50–60–70–90–100: 14 days.
3. For path 10–20–50–60–80–90–100: 12 days.
4. For path 10–30–80–90–100: 13 days.

It thus is evident that path 10–20–50–60–70–90–100 determines the minimum time in which the project can be completed, and it is called the *critical path*. Hence, the critical path is a series of connected activities which, if delayed, will cause the entire project to be delayed.

For a simple project such as this one, the critical path can be determined quite quickly by enumerating all paths and evaluating the time required to complete each. However, networks for most real projects are more complex, and a considerable number of paths may exist. Therefore, a more systematic method of analysis is desirable. Such a procedure for making the critical path calculations is demonstrated below for the network of Figure 18-7. This procedure would also apply to more complex networks.

In view of the estimated activity times calculated with Equation 18-1, the critical path of the network and other information useful to managing a large

---

† Activity times are always shown in working periods—days, weeks, or months—not in calendar periods.

project can be determined by a two-phased procedure. The *forward pass* through a network commences at the initial node and proceeds toward the terminal node. Conversely, the *backward pass* moves from the terminal node to the initial node. Each of these phases is now demonstrated for the simple network of Figure 18-7.

In the forward pass a number is determined at each node that represents the earliest occurrence of the respective event. Activities are described in terms of an $(i-j)$ index, such that $i < j$. For example, in Figure 18-7 activity 4 would be designated 20–50. Let ES_i denote the earliest start times for all activities emanating from event i, and ES_j represents the completion time. When $i = 1$ (the initial event), E_1 is defined to be zero. Further, d_{ij} denotes the duration of activity $i-j$.

To carry out the forward pass, the earliest start time for the jth event in the network is calculated with this relationship:

$$ES_j = maximum \text{ value of } [ES_i + d_{ij}] \text{ for all set of activities}$$
$$emanating \text{ } from \text{ event } i \text{ that lead to event } j$$

This indicates that the earliest possible start time for a particular event is dependent on previous events plus the activity times of all preceding activities. When the relationship above is applied to the network of Figure 18-7, these early start times are obtained:

$$
\begin{aligned}
ES_{10} &= 0 \\
ES_{20} &= ES_{10} + d_{10-20} = 0 + 1 = 1 \\
ES_{30} &= ES_{10} + d_{10-30} = 0 + 3 = 3 \\
ES_{40} &= ES_{20} + d_{20-40} = 1 + 2 = 3 \\
ES_{50} &= ES_{20} + d_{20-50} = 1 + 4 = 5 \\
ES_{60} &= ES_{50} + d_{50-60} = 5 + 2 = 7 \\
ES_{70} &= max\,[ES_{40} + d_{40-70}, ES_{60} + d_{60-70}] \\
        &= max\,[3 + 3, 7 + 2] = 9 \\
ES_{80} &= max\,[ES_{30} + d_{30-80}, ES_{60} + d_{60-80}] \\
        &= max\,[3 + 6, 7 + 1] = 9 \\
ES_{90} &= max\,[ES_{70} + d_{70-90}, ES_{80} + d_{80-90}] \\
        &= max\,[9 + 3, 9 + 2] = 12 \\
ES_{100} &= ES_{90} + d_{90-100} = 12 + 2 = 14
\end{aligned}
$$

The earliest start time for event 100 (i.e., finished project) is, therefore, 14 days. This completes the forward pass through the network.

Now the backward pass is initiated at the terminal (end) event and the aim here is to compute the latest possible finish, or completion, times for all activities coming into each event. The latest finish time for event i is designated LF_i and for the terminal event $LF_i = ES_i$. As in the calculation of early start times, a simple relationship is utilized to determine the latest finish times for any given event i:

$$LF_i = minimum \text{ value of } [LF_j - d_{ij}] \text{ for all sets of activities emanating from event } i$$
$$\text{to event } j \text{ (recall that } i < j \text{ in numbering events from left to right on the network)}$$

To illustrate the use of this relationship, the latest finish times for events in Figure 18-7 are developed as follows:

$$LF_{100} = ES_{100} = 14$$
$$LF_{90} = LF_{100} - d_{90-100} = 14 - 2 = 12$$
$$LF_{80} = LF_{90} - d_{80-90} = 12 - 2 = 10$$
$$LF_{70} = LF_{90} - d_{70-90} = 12 - 3 = 9$$
$$LF_{60} = \min [LF_{70} - d_{60-70}, LF_{80} - d_{60-80}]$$
$$= \min [9 - 2, 10 - 1] = 7$$
$$LF_{50} = LF_{60} - d_{50-60} = 7 - 2 = 5$$
$$LF_{40} = LF_{70} - d_{40-70} = 9 - 3 = 6$$
$$LF_{30} = LF_{80} - d_{30-80} = 10 - 6 = 4$$
$$LF_{20} = \min [LF_{40} - d_{20-40}, LF_{50} - d_{20-50}]$$
$$= \min [6 - 2, 5 - 4] = 1$$
$$LF_{10} = \min [LF_{20} - d_{10-20}, LF_{30} - d_{10-30}]$$
$$= \min [1 - 1, 4 - 3] = 0$$

This completes the backward pass through the network.

Based on results of the forward and backward passes, the critical path activities can be easily identified. An i–j activity is part of the critical path if three conditions are satisfied:

1. $ES_i = LF_i$.
2. $ES_j = LF_j$.
3. $ES_j - ES_i = LF_j - LF_i = d_{ij}$.

Hence these conditions indicate that there is no *slack time* between the earliest time an activity can start and its latest finish time. When the slack for an event is zero, it means that any delay in the occurrence of that event will cause delay in starting subsequent activities. On the other hand, when there is positive slack, the activity could be delayed by the amount indicated without causing any delay in the total project. Table 18-1 summarizes the ES_i and LF_i times and shows the amount of slack associated with each event.

If the events having zero slack are connected, as shown by heavy arrows in Figure 18-7, the resulting path (or paths) is critical; such paths determine the

table 18-1

slack times for each event in figure 18-7

Event i	ES	LF	Slack (days)
10	0	0	0
20	1	1	0
30	3	4	1
40	3	6	3
50	5	5	0
60	7	7	0
70	9	9	0
80	9	10	1
90	12	12	0
100	14	14	0

minimum time required for completing the project, and delays in any of the events on the critical path will result in extending the completion time of the project. In Figure 18-7 the events having zero slack are shown as double circles. Readers should check path 10–20–50–60–70–90–100 to assure themselves that the three conditions listed above for a critical path have indeed been satisfied.

economy study aspects of CPM and PERT

Once a critical path is determined, it is apparent that in many cases economies can be obtained by shifting resources, or by changing procedures, so as to shorten the time required to complete a project. Getting the project completed and into operation at an earlier than predicted date can result in earlier and greater revenues. In construction projects, earlier completion may reduce or eliminate penalties for late completion or earn bonuses for early completion. Another possibility may be that money can be saved by reducing the allocation of resources to an activity on a noncritical path. However, reallocation of resources and changes in procedures usually involve some increases in costs. The added costs above normal costs by which shorter activity times can be achieved are frequently called *crash costs*. Obviously, there usually is a range of crash costs that must be considered. Thus the usual economy study problem of balancing costs versus revenues arises, and frequently alternative choices exist. It also is apparent that, because of the short time involved and the nature of the required expenditures, we are dealing with out-of-pocket costs in most instances, so that the economic analysis usually is one of present economy.

EXAMPLE 18-1

In the project depicted in Figure 18-7, it is found that a gross benefit of $150 per day can be obtained for each day saved, up to 2 days. Investigation reveals that the cost–time relationships for activities 20–50, 30–80, 50–60, 80–90, and 90–100 are as shown in Figure 18-8. What reallocation of resources should be made?

Solution

If 1 day's time were eliminated in the 20–50–60 path, the 10–20–50–60–70–90–100 path would be balanced with path 10–30–80–90–100; each would require 13 days. There then would be no advantage in further reducing the time required for the activities of the first path, unless reduction also was made for the time of the activities of the second path. It is apparent that the first reduction should be made in the 10–20, the 20–50, or the 50–60 activities. Since activity 10–20 presently requires only 1 day, it is not likely that 1 day can be saved there. What we wish to do, therefore, is to reduce the time for either activity 20–50 or activity 50–60 at minimum cost.

Considering activities 20–50 and 50–60, it will be noticed that the slopes of the two curves, in Figure 18-8, make it apparent that reduction in time can be obtained

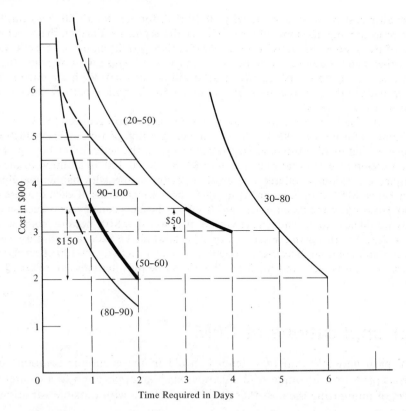

FIGURE 18-8. Cost–time relationships for several activities shown in Figure 18-7.

FIGURE 18-9. New activity–event network, resulting from eliminating 1 day for activity 20–50, as shown in Figure 18-7.

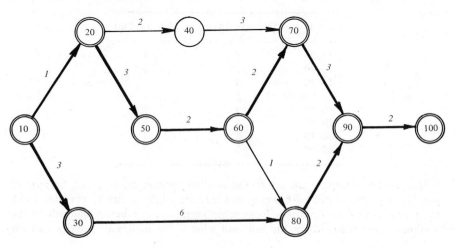

at smaller cost in the case of activity 20–50 than for activity 50–60. This can be seen by examining the slopes of the two heavy-line segments. Further, the marginal cost of 1 day's time for activity 50–60 is $150, thus permitting no net benefit. On the other hand, because 1 day can be saved in connection with activity 20–50 at a cost of only $50, a net saving of $100 will be achieved. With a reduction of 1 day made in the time required for activity 20–50, the critical path network is as shown in Figure 18-9.

With the total time for path 10–20–50–60–70–90–100 now reduced to 13 days— the same as for path 10–30–80–90–100—a consideration of any further reduction in time would involve the time–cost relationships of the activities of both paths. Investigation revealed that only activities 80–90 and 90–100, in addition to those considered previously, offered any possibilities for net benefits, having marginal costs below $150. As shown in Figure 18-8, the marginal costs of each of these is $100. However, if the time for activity 30–80 is reduced, a concurrent reduction of 1 day would have to be made in activity 20–50 at a cost of $100 because both paths are critical. For this plan the total marginal cost would be $200, and thus it is not an economic solution. Because a reduction of 1 day in activity 90–100 affects both paths, and can be achieved for $100, this would be the alternative to select. ▌

other applications of CPM

Critical path networks can be extended to aid in the economic solutions of numerous problems. Example 18-2 illustrates such an application to a common problem of minimizing the cost that may be associated with possible variations in crew size.

EXAMPLE 18-2

The activities required on a small construction job, and their durations and manpower requirements are

Activity	Duration (days)	Men per Day
10–20	1	1
20–30	2	3
10–40	4	3
30–50	3	2
40–50	3	1
10–60	4	3
50–70	2	3
60–70	3	2

It is desired to complete the job in the smallest number of days and, because of penalty costs associated with hiring and training workers, not to vary the work force any more than necessary from some minimum number. What will be the minimum time required for the job, and what is the minimum penalty cost for crew-size variation, assuming that the penalty costs are as shown below and that once an activity is started it must continue without interruption until completed?

Extra Men	Cost per Day
1	$10
2	15
3	25
4	30
5	35

Solution

Figure 18-10 is the activity–event network for this job, and Figure 18-11 is a modified slack chart, giving the solution to Example 18-2. It will be noted that the left-hand portion of this figure is a modified slack chart with activities listed, rather than events, and with T_{ES} representing the earliest time at which an activity can start and T_{LF} the latest time at which it can be completed according to the constraints of the network. It is evident that the earliest completion time for the project will be 9 days, determined by the critical path 10–40–50–70.

Figure 18-11 is a slack-manning chart useful for this type of problem. The horizontal line opposite each activity shows the days during which it could occur— extending from T_{ES} to T_{LF}. Thus activity 20–30 can start on the second day and may be finished as late as the fourth day. For any activity,

$$\text{slack} = T_{LF} - \text{duration} - T_{ES}$$

Consequently, for activity 20–30 there is 1 day of slack. Since the activity requires 2 days for completion, and assuming that it will go forward to completion without interruption when started, it could be accomplished either on the second and third days or on the third and fourth days. Where an activity has no slack, the space below the horizontal line on the chart has been filled in with crosshatching. The

FIGURE 18-10. Activity network for Example 18-2.

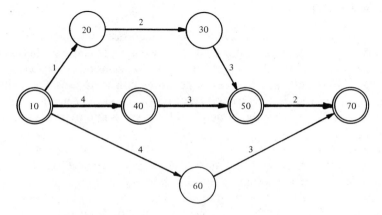

Activity	Duration	T_{ES}	T_{LF}	Slack	Men	1	2	3	4	5	6	7	8	9
10–20	1	0	2	1	1	1	1							
20–30	2	1	4	1	3		3	3	3					
10–40	4	0	4	0	3	3	3	3	3					
30–50	3	3	7	1	2				2	2	2	2		
40–50	3	4	7	0	1					1	1	1		
10–60	4	0	6	2	3	3	3	3	3	3	3			
50–70	2	7	9	0	3								3	3
60–70	3	4	9	2	2					2	2	2	2	2
Crew Size						4	6	9	6	6	6	5	5	5
Penalty Cost ($)						0	15	35	15	15	15	10	10	10

FIGURE 18-11. Slack chart and manning table for Example 18-2.

numerals above the horizontal lines show the number of men that will be required during each day for each activity that occurs.

Once such a chart is completed to this stage, the solution lies in trial-and-error adjustments to the manning table. This is done to cut off slack days from either the beginning or end of the horizontal lines so as to minimize crew-size variations. The days that are eliminated are shown crossed out. In this case, the minimum crew size is 4, and the total penalty cost is $125. ∎

This type of manning chart can be used advantageously to allocate limited work crews or craft types to projects so as to achieve minimum cost and to avoid unforseen delays.

conclusion

The applications of critical path networks that have been presented in this chapter show only a few of their uses. It is evident they can be of considerable assistance in solving several types of problems that involve economic decisions and that these decisions usually involve present economy considerations. Once the basic concepts and procedures of activity–event networks have been learned, there is little difficulty in applying the principles of present economy studies to specific problems.

The methods that have been discussed here for determining critical paths and slack times and for selecting the most economic alternatives are fairly routine and may readily be programmed for computers. A number of computer programs are available that permit quite complex networks to be solved with considerable ease. These greatly extend the usefulness of critical path methods.

problems

18-1. Construct a project network consisting of activities on arrows that has 12 activities, numbered 10 through 120. The following relationships must be satisfied in the network.

   (a) 10, 20, and 30 are the first activities of the project and are begun simultaneously.

   (b) 10 and 20 must precede 40.

   (c) 20 must precede 50 and 60.

   (d) 30 and 60 must precede 70.

   (e) 50 must precede 80 and 100.

   (f) 40 and 80 must precede 90.

   (g) 70 must precede 110 and 120.

   (h) 90 must precede 120.

   (i) 100, 110, and 120 are the terminal activities of the project.

Note that activities have been numbered here as opposed to events. Number or letter the nodes any way you desire as long as the relationships (network constraints) above are satisfied. You should find that two dummy activities are required.

18-2. For a minor construction project the activities and associated times are as follows:

Activity	Expected Time (days)
10–20	3
20–30	2
20–40	4
30–60	1
40–50	2
30–50	0
50–70	5
60–70	2
70–80	7

(a) Draw an activity–event diagram. (b) Determine the critical path and minimum time for the project. (c) Calculate and tabulate the earliest and latest times and the slack for each of the events.

18-3. The following activities and normal times are required for installing a large machine tool:

Activity	Normal Time (days)
00–10	2
10–20	4
10–30	3
10–40	2
20–60	2
30–50	4
40–50	3
50–70	3
60–80	5
70–80	2
80–90	1

(a) Draw the activity–event diagram. (b) Determine the critical path and minimum time that will be required for the installation. (c) Calculate the earliest and latest starting times and the slack for each event, and tabulate these values.

18-4. In Problem 18-3, assume that a bonus of $300 per day can be obtained for each day that the installation time can be reduced. Investigation shows that the times for activities 20–60, 30–50, and 70–80 can each be reduced, with the time–cost relationships being as portrayed by curves 50–60, 20–50, and 90–100, respectively, in Figure 18-8. Determine to what extent it would be economic to reduce the installation time and how it should be accomplished.

18-5. Refer to Figure 18-10.

(a) By using the forward pass and backward pass procedures described in the text, carry out the calculations required to determine the ES and LF times for each *event* and set up a table similar to that of Table 18-1.

(b) What is the critical path and minimum time required to complete the project?

(c) Check your answers with Figure 18-11 and indicate what differences exist between your table and the left side of Figure 18-11. Explain how you can easily convert from event-determined ES and LF times to T_{ES} and T_{LF} times for activities.

18-6. A company is planning a series of conferences for management personnel. These are to be held at a mountain-resort hotel in a sequence that will permit a series of interrelated decisions to be reached. Consequently, the seminar meetings can be considered to be activities leading to events (decisions). The following tabulation gives the duration of the seminars and the number of people who will be involved. It is desirable to know how to schedule the conferences so that a minimum total time will be required and so that the number of individual hotel rooms that will have to be reserved on any one night will be minimized. Assume that the personnel involved are nondependent.

Activity	Duration (days)	People
10–20	2	2
10–30	3	1
20–50	4	3
30–40	2	3
30–60	3	2
40–70	5	2
60–70	3	3
50–70	4	2
70–80	3	4

(a) Draw the activity–event diagram. (b) Determine the minimum number of days that will be required to complete the conference series. (c) Arrange the schedules so as to minimize the number of hotel rooms needed any night.

18-7. The activities, their durations, and manpower requirements on a small construction job are listed in the following table.

Activity	Duration (days)	People
10–20	1	1
10–40	4	3
10–60	4	3
20–30	2	3
30–50	3	2
40–50	3	1
50–70	2	3
60–70	3	2

It is desired to complete the job in the minimum time and, because of penalty costs associated with hiring and training workers, not to vary the size of the work force more than necessary from some minimum number. (a) Draw the activity–event diagram. (b) Determine the critical path and the minimum number of days for completing the job. (c) Assuming the penalty costs for crew-size variation to be

Extra Men	Cost per Day
1	$ 1
2	4
3	9
4	16
5	25

arrange the schedule so as to minimize the penalty cost.

appendix A
glossary of commonly used symbols and terminology

economic analysis methods and costs

A.C.	Annual cost
A.W.	Annual worth
A.W.-C.	Annual worth–cost (same as A.C.)
C.R.	Capital recovery cost (annual cost of depreciation plus interest on investment)
D_k	Net disbursements for kth year
E.R.R.	External rate of return
E.R.R.R.	Explicit reinvestment rate of return
I.R.R.	Internal rate of return
M.A.R.R.	Minimum attractive rate of return (same as $i*$)
M.R.P.	Minimum required profit (interest on investment)
P.W.	Present worth
P.W.-C.	Present worth–cost
R_k	Net receipts for kth year

compound interest symbols

e	Reinvestment rate (as used for E.R.R. and E.R.R.R. methods)
i	Effective interest rate per interest period
i'	Interest or rate of return (to be determined)
$i*$	Specified minimum attractive interest or rate of return
r	Nominal interest rate per year
$\underline{r}$	Nominal interest rate per year, compounded continuously
N	Number of compounding periods
P	Present sum of money (present worth); the equivalent worth of one or more cash flows at a relative point in time called the present
F	Future sum of money (future worth); the equivalent worth of one or more cash flows at a relative point in time called the future

A End-of-period cash flows (or equivalent end-of-period values) in a uniform series continuing for a specified number of periods

G Uniform period-by-period increase or decrease in cash flows or amounts (the arithmetic gradient)

$\bar{P}$ or $\bar{F}$ Amount of money (or equivalent value) flowing continuously and uniformly during a given period

$\bar{A}$ Amount of money (or equivalent value) flowing continuously and uniformly during each and every period continuing for a specific number of periods

functional forms (symbols) for compound interest factors[a]

Functional Format (symbol)	Name of Factor
ALL CASH FLOWS DISCRETE: END-OF-PERIOD COMPOUNDING	
$(F/P, i\%, N)$	Compound Amount Factor (Single Payment)
$(P/F, i\%, N)$	Present Worth Factor (Single Payment)
$(A/F, i\%, N)$	Sinking Fund Factor
$(A/P, i\%, N)$	Capital Recovery Factor
$(F/A, i\%, N)$	Compound Amount Factor (Uniform Series)
$(P/A, i\%, N)$	Present Worth Factor (Uniform Series)
$(A/G, i\%, N)$	Arithmetic Gradient Conversion Factor (to Uniform Series)
$(P/G, i\%, N)$	Arithmetic Gradient Conversion Factor (to Present Value)
ALL CASH FLOWS DISCRETE: CONTINUOUS COMPOUNDING	
$(F/P, r\%, N)$	Continuous Compounding: Compound Amount Factor (Single Payment)
$(P/F, r\%, N)$	Continuous Compounding: Present Worth Factor (Single Payment)
$(A/F, r\%, N)$	Continuous Compounding: Sinking Fund Factor
$(A/P, r\%, N)$	Continuous Compounding: Capital Recovery Factor
$(F/A, r\%, N)$	Continuous Compounding: Compound Amount Factor (Uniform Series)
$(P/A, r\%, N)$	Continuous Compounding: Present Worth Factor (Uniform Series)
CONTINUOUS, UNIFORM CASH FLOWS: CONTINUOUS COMPOUNDING (PAYMENTS DURING ONE PERIOD ONLY)	
$(P/\bar{F}, r$ or $i\%, N)$	Continuous Compounding: Present Worth Factor (Single, Continuous Payment)
$(F/\bar{P}, r$ or $i\%, N)$	Continuous Compounding: Compound Amount Factor (Single, Continuous Payment)
CONTINUOUS, UNIFORM CASH FLOWS: CONTINUOUS COMPOUNDING (PAYMENTS DURING A CONTINUOUS SERIES OF PERIODS)	
$(\bar{A}/F, r$ or $i\%, N)$	Continuous Compounding: Sinking Fund Factor (Continuous, Uniform Payments)
$(\bar{A}/P, r$ or $i\%, N)$	Continuous Compounding: Capital Recovery Factor (Continuous, Uniform Payments)
$(F/\bar{A}, r$ or $i\%, N)$	Continuous Compounding: Compound Amount Factor (Continuous, Uniform Payments)
$(P/\bar{A}, r$ or $i\%, N)$	Continuous Compounding: Present Worth Factor (Continuous, Uniform Payments)

[a] From *American National Standard Publication ANSI Z94.5–1972*, published by the American Society of Mechanical Engineers, New York, except that use of r is added for continuous compounding.

technical terms used in engineering economy†

Amortization—(a) (1) As applied to a capitalized asset, the distribution of the initial cost by periodic charges to operations as in depreciation. Most properly applies with indefinite life; (2) the reduction of a debt by either periodic or irregular payments; (b) a plan to pay off a financial obligation according to some prearranged program.

Annual Equivalent—(a) In *Time Value of Money* (*q.v.*), a uniform annual amount for a prescribed number of years that is equivalent in value to the present worth of any sequence of financial events for a given interest rate; (b) one of a sequence of equal end-of-year payments which would have the same financial effect when interest is considered as another payment or sequence of payments which are not necessarily equal in amount or equally spaced in time.

Annuity—(a) An amount of money payable to a beneficiary at regular intervals for a prescribed period of time out of a fund reserved for that purpose; (b) a series of equal payments occurring at equal periods of time.

Annuity Factor—The function of interest rate and time that determines the amount of periodic annuity that may be paid out of a given fund.

Annuity Fund—A fund that is reserved for payment of annuities. The present worth of funds required to support future annuity payments.

Annuity Fund Factor—The function of interest rate and time that determines the present worth of funds required to support a specified program of annuity payments.

Apportion—In accounting or budgeting, to assign a cost responsibility to a specific individual, organization unit, product, project, or order.

Average-Interest Method—A method of computing required return on investment based on the average book value of the asset during its life or during a specified study period.

Book Value—(a) The recorded current value of an asset. First cost less accumulated depreciation, amortization, or depletion; (b) original cost of an asset less the accumulated depreciation; (c) the worth of a property as shown on the accounting records of a company. It is ordinarily taken to mean the original cost of the property less the amounts that have been charged as depreciation expense.

Breakeven Chart—A graphic representation of the relation between total income and total costs for various levels of production and sales indicating areas of profit and loss.

Breakeven Point (S)—(a) (1) In business operations, the rate of operations, output, or sales at which income is sufficient to equal operating cost, or operating cost plus additional obligations that may be specified; (2) the operating condition, such as output, at which two alternatives are equal in economy. (b) The percentage of capacity operation of a manufacturing plant at which income will just cover expenses.

Capacity Factor—(a) The ratio of average load to maximum capacity; (b) the ratio between average load and the total capacity of the apparatus, which is the optimum load; (c) the ratio of the average actual use to the available capacity.

† From *American National Standard Publications ANSI Z94.5–1972*, published by the American Society of Mechanical Engineers, New York.

Capital—(a) The financial resources involved in establishing and sustaining an enterprise or project (see *Investment and Working Capital*); (b) a term describing wealth which may be utilized to economic advantage. The form that this wealth takes may be as cash, land, equipment, patents, raw materials, finished product, etc.

Capital Recovery—(a) Charging periodically to operations amounts that will ultimately equal the amount of capital expenditure (see *Amortization, Depletion, and Depreciation*); (b) the replacement of the original cost of an asset plus interest; (c) the process of regaining the net investment in a project by means of revenue in excess of the costs from the project. (Usually implies amortization of principal plus interest on the diminishing unrecovered balance.)

Capital Recovery Factor—A factor used to calculate the sum of money required at the end of each of a series of periods to regain the net investment of a project plus the compounded interest on the unrecovered balance.

Capitalized Cost—(a) The present worth of a uniform series of periodic costs that continue for an indefinitely long time (hypothetically infinite). Not to be confused with a capitalized expenditure; (b) the value at the purchase date of the first life of the asset of all expenditures to be made in reference to this asset over an infinite period of time. This cost can also be regarded as the sum of capital which, if invested in a fund earning a stipulated interest rate, will be sufficient to provide for all payments required to maintain the asset in perpetual service.

Cash Flow—(a) The flowback of profit plus depreciation from a given project; (b) the real dollars passing into and out of the treasury of a financial venture.

Common Costs—Costs which cannot be identified with a given output of products, operations, or services.

Compound Amount—The future worth of a sum invested (or loaned) at compound interest.

Compound Amount Factor—(a) The function of interest rate and time that determines the compound amount from a stated initial sum; (b) a factor which when multiplied by the single sum or uniform series of payments will give the future worth at compound interest of such single sum or series.

Compound Interest—(a) The type of interest that is periodically added to the amount of investment (or loan) so that subsequent interest is based on the cumulative amount; (b) the interest charges under the condition that interest is charged on any previous interest earned in any time period, as well as on the principal.

Compounding, Continuous—(a) A compound interest situation in which the compounding period is zero and the number of periods infinitely great. A mathematical concept that is practical for dealing with frequent compounding and small interest rates; (b) a mathematical procedure for evaluating compound interest factors based on a continuous interest function rather than discrete interest periods.

Compounding Period—The time interval between dates at which interest is paid and added to the amount of an investment or loan. Designates frequency of compounding.

Decisions Under Certainty—Simple decisions that assume complete information and no uncertainty connected with the analysis of the decisions.

Decisions Under Risk—A decision problem in which the analyst elects to consider several possible futures, the probabilities of which can be estimated.

Decisions Under Uncertainty—A decision for which the analyst elects to consider several possible futures, the probabilities of which *cannot* be estimated.

Declining Balance Depreciation—Also known as *percent on diminishing value*. A method of computing depreciation in which the annual charge is a fixed percentage of the depreciated book value at the beginning of the year to which the depreciation applies.

Demand Factor—(a) The ratio of the maximum instantaneous production rate to the production rate for which the equipment was designed; (b) the ratio between the maximum power demand and the total connected load of the system.

Depletion—(a) A form of capital recovery applicable to extractive property (e.g., mines). Can be a unit-of-output basis the same as straight-line depreciation related to original or current appraisal of extent and value of deposit. (Known as *cost depletion*). Can also be a percentage of income received from extractions (known as *percentage depletion*). (b) A lessening of the value of an asset due to a decrease in the quantity available. It is similar to depreciation except that it refers to such natural resources as coal, oil, and timber in forests.

Depreciated Book Value—The first cost of the capitalized asset minus the accumulation of annual depreciation cost charges.

Depreciation—(a) (1) Decline in value of a capitalized asset; (2) a form of capital recovery applicable to a property with two or more years' life span, in which an appropriate portion of the asset's value is periodically charged to current operations. (b) The loss of value because of obsolescence or due to attrition. In accounting, depreciation is the allocation of this loss of value according to some plan.

Development Cost—The sum of all the costs incurred by an inventor or sponsor of a project up to the time that the project is accepted by those who will promote it.

Discounted Cash Flow—(a) The present worth of a sequence in time of sums of money when the sequence is considered as a flow of cash into and/or out of an economic unit; (b) an investment analysis which compares the present worth of projected receipts and disbursements occurring at designated future times in order to estimate the rate of return from the investment or project.

Earning Value—The present worth of an income producer's probable future net earnings, as prognosticated on the basis of recent and present expense and earnings and the business outlook.

Economic Return—The profit derived from a project or business enterprise without consideration of obligations to financial contributors and claims of others based on profit.

Economy—The cost or profit situation regarding a practical enterprise or project, as in *economy study, engineering economy, project economy*.

Effective Interest—The true value of interest rate computed by equations for compound interest rate for a 1-year period.

Endowment—A fund established for the support of some project or succession of donations or financial obligations.

Endowment Method—As applied to economy study, a comparison of alternatives based on the present worth of the anticipated financial events.

Engineering Economy—(a) The application of engineering or mathematical analysis and synthesis to economic decisions; (b) a body of knowledge and techniques concerned with the evaluation of the worth of commodities and services relative to their cost; (c) the economic analysis of engineering alternatives.

Estimate—The true magnitude as closely as it can be determined by the exercise of sound judgement based on approximate computations and is not to be confused with offhand approximations that are little better than outright guesses.

Exact Method—A method of computing required return combined with depreciation, based on the assumption that the present worth of these plus that of salvage value during the prescribed study period is equivalent to the first cost.

Expected Return—The profit anticipated from a venture.

Expected Yield—The ratio expected return/investment, usually expressed as a percentage on an annual basis.

First Cost—The initial cost of a capitalized property, including transportation, installation, preparation for service, and other related initial expenditures.

Future Worth—(a) The equivalent value at a designated future date based on *time value of money*; (b) the monetary sum, at a given future time, which is equivalent to one or more sums at given earlier times when interest is compounded at a given rate.

Going-Concern Value—The difference between the value of a property as it stands possessed of its going elements and the value of the property alone as it would stand at completion of construction as a bare or inert assembly of physical parts.

Good-Will Value—That element of value which inheres in the fixed and favorable consideration of customers arising from an established well-known and well-conducted business.

Increment Cost—The additional cost that will be incurred as the result of increasing the output one more unit. Conversely, it can be defined as the cost that will not be incurred if the output is reduced one unit. More technically, it is the variation in output resulting from a unit change in input. It is known as the marginal cost.

In-Place Value—A value of a physical property—market value plus costs of transportation to site and installation.

Intangibles—(a) In economy studies, conditions or economy factors that cannot be readily evaluated in *quantitative* terms as in money; (b) in accounting, the assets that cannot be reliably evaluated (e.g., *good will*).

Interest—(a) (1) Financial share in a project or enterprise; (2) periodic compensation for the lending of money; (3) in economy study, synonymous with *required return*, expected profit, or *charge* for the use of capital. (b) The cost for the use of capital. Sometimes referred to as the *Time Value of Money (q.v.)*.

Interest Rate—The ratio of the interest payment to the principal for a given unit of time and is usually expressed as a percentage of the principal.

Interest Rate, Effective—An interest rate for a stated period (per year unless otherwise specified) that is the equivalent of a smaller rate of interest that is more frequently compounded.

Interest Rate, Nominal—The customary type of interest rate designation on an annual basis without consideration of compounding periods. The usual basis for computing periodic interest payments.

Investment—(a) As applied to an enterprise as a whole, the cost (or present value) of all the properties and funds necessary to establish and maintain the enterprise as a going concern. The *capital* tied up in the enterprise or project. (b) Any expenditure which has substantial and enduring value (at least two years' anticipated life) and which is therefore capitalized.

Investor's Method—(see *Discounted Cash Flow*)

Irreducibles—(a) A term that may be used for the class of intangible conditions or economy factors that can only be *qualitatively* appraised (e.g., ethical considerations; (b) matters that cannot readily be reduced to estimated money receipts and disbursements.

Life—(a) Economic: that period of time after which a machine or facility should be discarded or replaced because of its excessive costs or reduced profitability. The economic impairment may be absolute or relative. (b) Physical: that period of time after which a machine or facility can no longer be repaired in order to perform its design function properly. (c) Service: the period of time that a machine or facility will satisfactorily perform its function without major overhaul.

Load Factor—(a) A ratio that applies to physical plant or equipment: average loan/maximum demand, usually expressed as a percentage. Equivalent to percent of capacity operation if facilities just accommodate the maximum demand. (b) Is defined as the ratio of average load to maximum load.

MAPI Method—(a) A procedure for replacement analysis sponsored by the Machinery and Allied Products Institute. (b) A method of capital investment analysis which has been formulated by the Machinery and Allied Products Institute. This method uses a fixed format and provides charts and graphs to facilitate calculations. A prominent feature of this method is that it explicitly includes obsolescence.

Marginal Analysis—An economic concept concerned with those elements of costs and revenue which are associated directly with a specific course of action, normally using available current costs and revenue as a base and usually independent of traditional accounting allocation procedures.

Marginal Cost—(a) The cost of one additional unit of production, activity, or service; (b) the rate of change of cost with production or output.

Matheson Formula—A title for the formula used for *Declining Balance Depreciation* (*q.v.*).

Multiple Straight Line Depreciation Method—A method of depreciation accounting in which two or more straight line rates are used. This method permits a predetermined portion of the asset to be written off in a fixed number of years. One common practice is to employ a straight line rate which will write off $\frac{3}{4}$ of the cost in the first half of the anticipated service life; with a second straight line rate to write off the remaining $\frac{1}{4}$ in the remaining half life.

Nominal Interest—The number employed loosely to describe the annual interest rate.

Obsolescence—(a) The condition of being out of date. A loss of value occasioned by new developments which place the older property at a competitive disadvantage. A factor in depreciation. (b) A decrease in the value of an asset brought about by the development of new and more economical methods, processes, and/or machinery. (c) The loss of usefulness or worth of a product or facility as a result of the appearance of better and/or more economical products, methods, or facilities.

Payoff Period—(a) Regarding an investment, the number of years (or months) required for the related profit or savings in operating cost to equal the amount of said investment; (b) the period of time at which a machine, facility, or other investment has produced sufficient net revenue to recover its investment costs.

Perpetual Endowment—An endowment with hypothetically infinite life. (See *Capitalized Cost and Endowment*.)

Present Worth—(a) The equivalent value at the present, based on *time value of money*. (b) (1) The monetary sum which is equivalent to a future sum(s) when interest is compounded at a given rate. (2) The discounted value of future sums.

Present Worth Factor—(a) A mathematical expression also known as the present value of an annuity of one; (b) one of a set of mathematical formulas used to facilitate calculation of present worth in economic analyses involving compound interest.

Profitability Index—The rate of return in an economy study or investment decision when calculated by the *Discounted Cash Flow Method* or the *Investor's Method* (*q.v.*).

Promotion Cost—The sum of all the expenses found to be necessary to arrange for the financing and organizing of the business unit which will build and operate the project.

Rate of Return—(a) The interest rate at which the present worth of the cash flows on a project is zero; (b) the interest rate earned by an investment.

Replacement Policy—A set of decision rules (usually optimal) for the replacement of facilities that wear out, deteriorate, or fail over a period of time. Replacement models are generally concerned with weighing the increasing operating costs (and possibly decreasing revenues) associated with aging equipment against the net proceeds from alternative equipment.

Replacement Study—An economic analysis involving the comparison of an existing facility and a facility proposed to supplant the existing facility.

Required Return—The *minimum* return or profit necessary to justify an investment. Often termed *interest*, *expected* return or profit, or *charge for the use of capital*. It is the minimum acceptable percentage, no more and no less.

Required Yield—The ratio of required return over amount of investment, usually expressed as a percentage on an annual basis.

Retirement of Debt—The termination of a debt obligation by appropriate settlement with lender—understood to be in full amount unless partial settlement is specified.

Salvage Value—(a) The cost recovered or which could be recovered from a used property when removed, sold, or scrapped. A factor in appraisal of property value and in computing depreciation. (b) The market value of a machine or facility at any point in time. Normally, an estimate of an asset's net market value at the end of its estimated life.

Sensitivity—The relative magnitude of the change in one or more elements of an engineering economy problem that will reverse a decision among alternatives.

Simple Interest—(a) Interest that is not compounded—is not added to the income-producing investment or loan; (b) the interest charges under the condition that interest in any time is only charged on the principal.

Sinking Fund—(a) A fund accumulated by period deposits and reserved exclusively for a specific purpose, such as retirement of a debt or replacement of a property; (b) a fund created by making periodic deposits (usually equal) at compound interest in order to accumulate a given sum at a given future time for some specific purpose.

Sinking Fund Deposit Factor—The function of interest rate and time that determines the periodic deposit required to accumulate a specific future amount. Reciprocal of *sinking fund factor* (*q.v.*).

Sinking Fund Depreciation—(a) A method of computing depreciation in which the periodic amount is presumed to be deposited in a *sinking fund* that earns interest at a specified rate. Sinking fund may be real but is usually hypothetical. (b) A method of depreciation where a fixed sum of money is regularly deposited at compound interest in a real or imaginary fund in order to accumulate an amount equal to the total depreciation of an asset at the end of the asset's estimated life. The sinking fund depreciation in any year equals the sinking fund deposit plus interest in the sinking fund balance.

Sinking Fund Factor—(a) The function of interest rate and time that determines the cumulative amount of a sinking fund resulting from specified periodic deposits. Future worth per unit of uniform periodic amounts. (b) The mathematical formulas used to facilitate sinking fund calculations.

Straight Line Depreciation—Method of depreciation whereby the amount to be recovered (written off) is spread uniformly over the estimated life of the asset in terms of time periods or units of output. May be designated *percent of initial value*.

Study Period—In economy study, the length of time that is presumed to be covered in the schedule of events and appraisal of results. Often the anticipated life of the project under consideration, but a shorter time may be more appropriate for decision making.

Sum-of-Digits Method—Also known as sum-of-the-years'-digits method. A method of computing depreciation in which the amount for any year is based on the ratio: (years of remaining life)/$(1 + 2 + 3 + \cdots + n)$, being the total anticipated life.

Sunk Cost—(a) The unrecovered balance of an investment. It is a cost, already paid, that is not relevant to the decision concerning the future that is being made. Capital already invested that for some reason cannot be retrieved. (b) A past cost which has no relevance with respect to future receipts and disbursements of a facility undergoing an engineering economy study. This concept implies that since a past outlay is the same regardless of the alternative selected, it should not influence the choice between alternatives.

Tangibles—Things that can be *quantitatively* measured or valued, such as items of cost and physical assets.

Time Value of Money—(a) The cumulative effect of elapsed time on the money value of an event, based on the earning power of equivalent invested funds (see *Future Worth* and *Present Worth*). (b) The expected interest rate that capital should or will earn.

Traceable Costs—Costs which can be identified with a given product, operation, or service.

Valuation or Appraisal—The art of estimating the fair-exchange value of specific properties.

Working Capital—(a) That portion of investment represented by *current assets* (assets that are not capitalized) less the *current liabilities*. The capital necessary to sustain operations. (b) Those funds that are required to make the enterprise or project a going concern.

Yield—The ratio of return or profit over the associated investment, expressed as a percentage or decimal usually on an annual basis.

appendix B
the MAPI method for replacement and general investment analysis[†]

This appendix is intended to provide a brief summary of the main provisions, strengths, and weaknesses of the current (third) version of the MAPI method and to illustrate its use for a typical project analysis. The MAPI method or system for investment analysis has been evolved through several major works over the past 20 years by George Terborgh, Research Director of the Machinery and Allied Products Institute. The first version, which applied only to replacement problems, was described in *Dynamic Equipment Policy* and the *MAPI Replacement Manual*, both published in 1950. The second, and more-general-purpose MAPI system was published in *Business Investment Policy* in 1958. The current (third) version was summarized in 1967 in *Business Investment Management*. All were published by the Machinery and Allied Products Institute, Washington, D.C.

Basically, the MAPI method provides a series of charts and forms to facilitate investment analysis computation. One of its main features is the inclusion of provision for consideration of obsolescence and deterioration, assumed to affect operating results as a linear function of time. The MAPI charts provide for ease in determining the percentage retention value and are computed for various service lives, salvage values, and tax write-off methods. They are available for straight line, double declining balance, sum-of-the-years'-digits, and current expense depreciation methods for either a 1-year or longer-than-1-year comparison period—a total of eight charts. Figures B-1 and B-2 illustrate two such charts as given for a wide range of service lives and salvage value ratios. Assumptions on which the charts are based include 50% income tax rate, a 25%–75% debt-to-equity capital ratio, an average debt capital interest rate of 3% (after-tax), and an after-tax equity return of 10%.

† This discussion of the MAPI method was prepared primarily by Jack Turvaville, Professor of Industrial Engineering, Tennessee Technological University, and is used with his permission. It is based largely on pp. 149–158 of *Business Investment Management*, by George Terborgh, Machinery and Allied Products Institute, 1967, and reproduced by permission of the publisher.

520

MAPI CHART No. 1A
(ONE-YEAR COMPARISON PERIOD AND SUM-OF-DIGITS TAX DEPRECIATION)

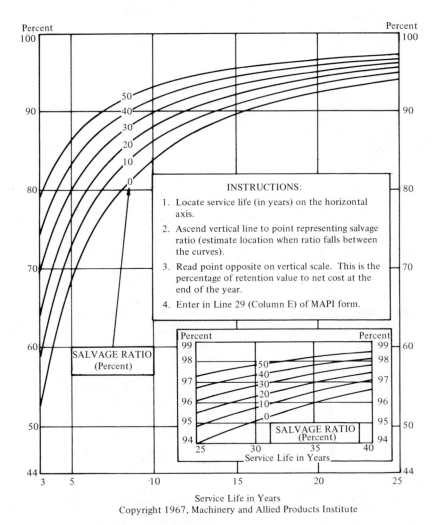

Service Life in Years

FIGURE B-1. Example of a Machinery and Allied Products Institute (MAPI) chart.

It should be recognized that allowance for deterioration and obsolescence is built into the chart retention values. According to Terborgh, the computation for the retention values projects a stream of pre-tax earnings over the estimated service life that conforms in shape to the projection pattern and in size to the following requirements: (1) It includes the income tax, payable at the specified rate, on the excess of the earnings over the deductions provided by tax depreciation and interest; (2) the remainder of the after-tax earnings suffices (with

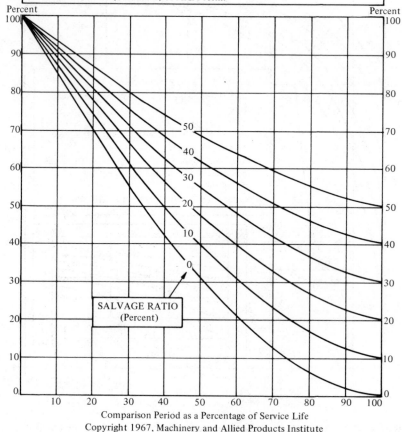

MAPI CHART No. 1B

(LONGER THAN ONE-YEAR COMPARISON PERIODS
AND SUM-OF-DIGITS TAX DEPRECIATION)

INSTRUCTIONS:
1. Locate on horizontal axis percentage which comparison period is of service life.
2. Ascend vertical line to point representing salvage ratio (estimate location when ratio falls between the curves).
3. Read point opposite on vertical scale. This is the percentage of retention value to net cost at end of comparison period.
4. Enter in line 29 (Column E) of MAPI form.

Comparison Period as a Percentage of Service Life
Copyright 1967, Machinery and Allied Products Institute

FIGURE B-2. Another example of a MAPI chart.

terminal salvage, if any) to permit a full recovery of the investment over the service life and to provide a return throughout on the unrecovered balance of equity at the prescribed after-tax return rate and on the unrecovered balance of debt at the prescribed interest rate.

The use of the MAPI method is consummated through a standard form as shown in Figures B-3 and B-4 with example amounts entered therein. The

PROJECT NO. _____   SHEET 1

MAPI SUMMARY FORM
(AVERAGING SHORTCUT)

PROJECT_____
ALTERNATIVE_____
COMPARISON PERIOD (YEARS)   (P)_____1_____
ASSUMED OPERATING RATE OF PROJECT (HOURS PER YEAR)   1,200

I. OPERATING ADVANTAGE

(NEXT-YEAR FOR A 1-YEAR COMPARISON PERIOD,* ANNUAL AVERAGES FOR LONGER PERIODS)

A. EFFECT OF PROJECT ON REVENUE

		INCREASE	DECREASE	
1	FROM CHANGE IN QUALITY OF PRODUCTS	$	$	1
2	FROM CHANGE IN VOLUME OF OUTPUT			2
3	TOTAL	$ X	$ Y	3

B. EFFECT ON OPERATING COSTS

		INCREASE	DECREASE	
4	DIRECT LABOR	$ 900	$	4
5	INDIRECT LABOR	150		5
6	FRINGE BENEFITS	190		6
7	MAINTENANCE	200		7
8	TOOLING	80		8
9	MATERIALS AND SUPPLIES		16,800	9
10	INSPECTION			10
11	ASSEMBLY			11
12	SCRAP AND REWORK			12
13	DOWN TIME			13
14	POWER	40		14
15	FLOOR SPACE		1,000	15
16	PROPERTY TAXES AND INSURANCE	320		16
17	SUBCONTRACTING			17
18	INVENTORY		1,100	18
19	SAFETY			19
20	FLEXIBILITY			20
21	OTHER			21
22	TOTAL	$ 1,880 Y	$ 18,900 X	22

C. COMBINED EFFECT

23	NET INCREASE IN REVENUE (3X−3Y)	$	23
24	NET DECREASE IN OPERATING COSTS (22X−22Y)	$ 17,020	24
25	ANNUAL OPERATING ADVANTAGE (23 + 24)	$ 17,020	25

*Next year means the first year of project operation. For projects with a significant break-in period, use performance after break-in.

FIGURE B-3. MAPI Summary Form, Sheet 1 (with entries for example project).

example amounts are based upon the following example project taken from p. 156 of Terborgh's *Business Investment Management.*†

An analyst desires to investigate whether it would be more economical for the company to make its own corrugated containers. He finds that to do this the company will have to purchase a large box machine and a box stitcher at a combined cost of $29,800.

It is estimated that direct labor cost will be increased by $900 a year, indirect labor (supervision) by $50, and fringe benefits by $190. Maintenance will be higher

† Published by Machinery and Allied Products Institute, 1967.

II. INVESTMENT AND RETURN

SHEET 2

A. INITIAL INVESTMENT

26	INSTALLED COST OF PROJECT	$ 29,000		
	MINUS INITIAL TAX BENEFIT OF	$ 2,100	(Net Cost)	$ 27,700	26
27	INVESTMENT IN ALTERNATIVE			
	CAPITAL ADDITIONS MINUS INITIAL TAX BENEFIT	$		
	PLUS: DISPOSAL VALUE OF ASSETS RETIRED			
	BY PROJECT*	$ 4,000	$ 4,000	27
28	INITIAL NET INVESTMENT (26–27)		23,700	28

B. TERMINAL INVESTMENT

29 RETENTION VALUE OF PROJECT AT END OF COMPARISON PERIOD
(ESTIMATE FOR ASSETS, IF ANY, THAT CANNOT BE DEPRECIATED OR EXPENSED, FOR OTHERS, ESTIMATE OR USE MAPI CHARTS.)

Item or Group	Installed Cost, Minus Initial Tax Benefit (Net Cost) A	Service Life (Years) B	Disposal Value, End of Life (Percent of Net Cost) C	MAPI Chart Number D	Chart Percentage E	Retention Value $\left(\dfrac{A \times E}{100}\right)$ F
Box Machine and Stitcher	$ 27,700	13	10	1A	89.4	$ 24,760

	ESTIMATED FROM CHARTS (TOTAL OF COL. F)	$ 24,760		
	PLUS: OTHERWISE ESTIMATED	$	$ 24,760	29
30	DISPOSAL VALUE OF ALTERNATIVE AT END OF PERIOD *		$ 4,000	30
31	TERMINAL NET INVESTMENT (29–30)		$ 20,760	31

C. RETURN

32	AVERAGE NET CAPITAL CONSUMPTION $\left(\dfrac{28-31}{P}\right)$	$	2,940	32
33	AVERAGE NET INVESTMENT $\left(\dfrac{28+31}{2}\right)$	$	22,230	33
34	BEFORE-TAX RETURN $\left(\dfrac{25-32}{33} \times 100\right)$	%	63.3	34
35	INCREASE IN DEPRECIATION AND INTEREST DEDUCTIONS	$	4,190	35
36	TAXABLE OPERATING ADVANTAGE (25–35)	$	12,830	36
37	INCREASE IN INCOME TAX (36 X TAX RATE)	$	6,415	37
38	AFTER-TAX OPERATING ADVANTAGE (25–37)	$	10,605	38
39	AVAILABLE FOR RETURN ON INVESTMENT (38–32)	$	7,665	39
40	AFTER-TAX RETURN $\left(\dfrac{39}{33} \times 100\right)$	%	34.5	40

*After terminal tax adjustments.

FIGURE B-4. MAPI Summary Form, Sheet 2 (with entries for example project).

by $200, tool costs by $80, power consumption by $40, and property taxes and insurance by $320. On the other hand, there will be a saving of $16,800 in the cost of purchased materials, $1,100 in inventory carrying costs (other than floor space), and $1,000 in floor space. The net cost reduction is therefore $17,020.

In addition to this operating advantage, the equipment will permit a reduction of

$4,000 in inventory investment. After consulting with operating officials and others. the analyst comes up with the following stipulations:

Comparison period	1 year
Project operating rate	1,200 hours
Service life	13 years
Terminal salvage ratio	10 percent of net cost
Tax depreciation method	sum-of-the-years'-digits
Tax rate	50 percent
Debt ratio	30 percent
Debt interest rate	5 percent
Investment credit	7 percent

The results shown on Line 40 of Figure B-4 indicate an after-tax return of 34.5% for the project above.

In reviewing the application of the MAPI method to the example project, it should be observed that all the considerations given to the determination of the operating advantage would be necessary regardless of the technique used in developing a measure of comparison. It must be further noted that line 29 of Figure B-4 states that an estimate of the retention value does not have to come from the MAPI charts. If the analyst has a better estimate, he can use it.

An after-tax cash flow diagram using the information from the MAPI forms in Figures B-3 and B-4 would be

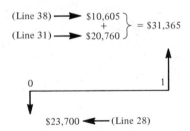

Upon solving, $-\$23,700 + \$31,365(P/F, i\%, 1) = 0$, and from interest tables, i can be interpolated to be 32.4%. This is close to the 34.5% return by the MAPI method. Generally, if the MAPI assumptions are not violated substantially, the return calculated by the MAPI method is a good approximation of the rate of return calculated by the discounted cash flow method. The MAPI method can be applied to problems involving a comparison period of more than 1 year in a manner very similar to the example above.

To most people trained in engineering economy principles, the use of a formula approach such as the MAPI method is too rigid. The assumptions built into the method can be significantly inappropriate and cause error in the analysis result. Further, the development of the MAPI charts is complicated to understand. However, the MAPI method does provide for obsolescence and deterioration, offers an excellent checklist, will provide consistent results, and is relatively easy to use.

appendix C

a computer program for calculating internal rate of return

general description

The computer program listed below will determine an internal rate of return (I.R.R.) on a net cash flow. It assumes the lender's point of view in that the initial cash flow (year 0) must be negative and that the total cash flow represents a positive return on investment. If the cash flow does not yield a positive rate of return, the program indicates that an interest rate cannot be found. The program is written in BASIC language and is designed to be used on a time-sharing system in a conversational mode.

program listing

```
5  DIM C(100)
10 LET N = 0
30 PRINT "ENTER CASH FLOW, FIRST PERIOD, LAST PERIOD"
40 PRINT "WHEN THROUGH, ENTER A CASH FLOW OF −999999
   AT TIME 0"
50 INPUT C1, J1, J2
55 IF C1 = −999999 THEN 210
60 FOR I = J1 TO J2
70 LET C(I + 1) = C(I + 1) + C1
80 NEXT I
85 IF J2 < = N THEN 50
87 LET N = J2
90 GO TO 50
210 LET E = .001
220 LET E = E/100
230 LET I1 = 0
240 LET I = 0
```

526

```
250 LET I2 = 1
260 GO SUB 400
270 LET D = 0.1
272 LET Z1 = ABS(P − C(1))
273 IF ABS(C(1)) < Z1 THEN 280
274 PRINT "NO INTEREST RATE CAN BE FOUND"
276 GO TO 510
280 LET P1 = P
290 FOR I = I1 TO I2 STEP D
300 GO SUB 400
310 LET V = SGN(P1) * SGN(P)
320 IF V < 0 THEN 350
330 LET P1 = P
340 NEXT I
350 IF D < = E THEN 480
360 LET I1 = I − D
370 LET I2 = I
380 D = D/10
390 GO TO 290
400 LET F1 = 1/(1 + I)
410 LET F = 1
420 LET P = 0
430 FOR X = 1 TO N + 1
440 LET P = P + C(X) * F
450 LET F = F * F1
460 NEXT X
470 RETURN
380 LET A = (I − D) * 100
490 LET B = I * 100
500 PRINT "INTEREST RATE IS BETWEEN"; A; "AND"; B; "PERCENT."
510 PRINT "DO YOU WISH TO ENTER NEW DATA".
520 INPUT Q$
530 IF Q$ = "NO" THEN 600
540 FOR I = 1 TO 100
550 LET C(I) = 0
560 NEXT I
565 LET N = 0
570 GO TO 50
600 END
```

data input

The data consist of the value of the net cash flow, the period in which the cash flow starts, and the period in which the cash flow ends. These entries are separated by commas, as illustrated below. The end of the cash flow and the signal to initiate the calculation of a rate of return is

$$− 999999,0,0$$

EXAMPLE PROBLEMS

1.

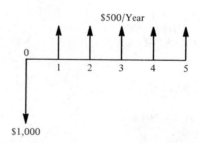

$500/Year

$1,000

Machine query: ENTER CASH FLOW FOR FIRST PERIOD THROUGH
                 THE LAST PERIOD.
User input: ? −1000,0,0 (must be −1000; not −1,000)
                 ? 500,1,5
                 ? −999999,0,0 (this signifies end of user data input)
Machine output: INTEREST RATE IS BETWEEN 41.041 AND 41.042%. DO
                 YOU WISH TO ENTER NEW DATA?
Write "YES" to continue. Write "NO" to end the run.

Illustrated are some further examples of correct data input. The computer-calculated
interest rate is shown in the lower right-hand corner for each.

2.

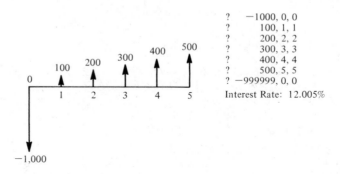

?	−1000, 0, 0
?	100, 1, 1
?	200, 2, 2
?	300, 3, 3
?	400, 4, 4
?	500, 5, 5
?	−999999, 0, 0

Interest Rate: 12.005%

3.

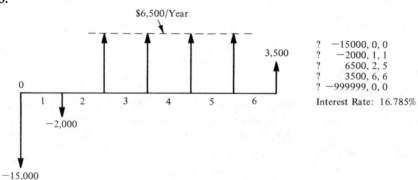

?	−15000, 0, 0
?	−2000, 1, 1
?	6500, 2, 5
?	3500, 6, 6
?	−999999, 0, 0

Interest Rate: 16.785%

appendix D

selected references

CANADA, J. R., *Intermediate Economic Analysis for Management and Engineering*. Englewood Cliffs, N.J.: Prentice-Hall, Inc., 1971.

Engineering Economy, 3rd ed. New York: Engineering Department, American Telephone and Telegraph Company. New York: McGraw-Hill Book Company, 1977.

FLEISCHER, G. A., *Capital Allocation Theory*. New York: Appleton-Century-Crofts, 1969.

GRANT, E. L., W. G. IRESON, and R. S. LEAVENWORTH, *Principles of Engineering Economy*, 6th ed. New York: The Ronald Press Company, 1976.

HAPPEL, J., *Chemical Process Economics*. New York: John Wiley & Sons, Inc., 1968.

JELEN, F. C., *Cost and Optimization Engineering*. New York: McGraw-Hill Book Company, 1970.

JEYNES, P. H., *Profitability and Economic Choice*. Ames, Iowa: The Iowa State University Press, 1968.

MILES, L. D., *Techniques of Value Analysis and Engineering*. New York: McGraw-Hill Book Company, 1961.

MODER, J. J., and C. R. PHILLIPS, *Project Management with CPM and PERT*. New York: Van Nostrand Reinhold Company, 1964.

NEWNAN, D. H., *Engineering Economic Analysis*. San José, Calif.: Engineering Press, 1976.

REISMAN, A., *Managerial and Engineering Economics*. Boston: Allyn and Bacon, Inc., 1971.

RIGGS, J. L., *Engineering Economics*. New York: McGraw-Hill Book Company, 1977.

SMITH, G. A., *Engineering Economy*, 2nd ed. Ames, Iowa: The Iowa State University Press, 1973.

SULLIVAN, W. G., and W. W. CLAYCOMBE, *Fundamentals of Forecasting*. Reston, Va.: Reston Publishing Company, Inc., 1977.

TARQUIN, A. J., and L. T. BLANK, *Engineering Economy: A Behavioral Approach*. New York: McGraw-Hill Book Company, 1976.

TAYLOR, G. A., *Managerial and Engineering Economy*, 2nd ed. New York: Van Nostrand Reinhold Company, 1975.

TERBORGH, G., *Business Investment Management*. Washington, D.C.: Machinery and
Applied Products Institute, 1967.

THUESEN, H. G., W. J. FABRYCKY, and G. J. THUESEN, *Engineering Economy*, 5th ed.
Englewood Cliffs, N.J.: Prentice-Hall, Inc., 1977.

WOODS, D. R., *Financial Decision Making in the Process Industry*. Englewood Cliffs,
N.J.: Prentice-Hall, Inc., 1975.

appendix E

interest and annuity tables for discrete compounding

(For various common values of i from $\frac{1}{4}\%$ to 50%)

$i =$ interest rate per period (usually 1 year)

$N =$ number of periods

$$(F/P, i\%, N) = (1 + i)^N$$

$$(P/F, i\%, N) = \frac{1}{(1 + i)^N}$$

$$(F/A, i\%, N) = \frac{(1 + i)^N - 1}{i}$$

$$(P/A, i\%, N) = \frac{(1 + i)^N - 1}{i(1 + i)^N}$$

$$(A/F, i\%, N) = \frac{i}{(1 + i)^N - 1}$$

$$(A/P, i\%, N) = \frac{i(1 + i)^N}{(1 + i)^N - 1}$$

$$(P/G, i\%, N) = \frac{1}{i}\left[\frac{(1 + i)^N - 1}{i(1 + i)^N} - \frac{N}{(1 + i)^N}\right]$$

$$(A/G, i\%, N) = \left[\frac{1}{i} - \frac{N}{(1 + i)^N - 1}\right]$$

table E-1

Discrete Compounding; $i = 1/4\%$

	SINGLE PAYMENT		UNIFORM SERIES				
	Compound Amount Factor	Present Worth Factor	Compound Amount Factor	Present Worth Factor	Sinking Fund Factor	Capital Recovery Factor	
	To find F Given P	To find P Given F	To find F Given A	To find P Given A	To find A Given F	To find A Given P	
N	F/P	P/F	F/A	P/A	A/F	A/P	N
1	1.0025	0.9975	1.0000	0.9975	1.0000	1.0025	1
2	1.0050	0.9950	2.0025	1.9925	0.4994	0.5019	2
3	1.0075	0.9925	3.0075	2.9851	0.3325	0.3350	3
4	1.0100	0.9901	4.0150	3.9751	0.2491	0.2516	4
5	1.0126	0.9876	5.0250	4.9627	0.1990	0.2015	5
6	1.0151	0.9851	6.0376	5.9478	0.1656	0.1681	6
7	1.0176	0.9827	7.0527	6.9305	0.1418	0.1443	7
8	1.0202	0.9802	8.0703	7.9107	0.1239	0.1264	8
9	1.0227	0.9778	9.0905	8.8885	0.1100	0.1125	9
10	1.0253	0.9753	10.1132	9.8638	0.0989	0.1014	10
11	1.0278	0.9729	11.1385	10.8367	0.0898	0.0923	11
12	1.0304	0.9705	12.1663	11.8072	0.0822	0.0847	12
13	1.0330	0.9681	13.1967	12.7753	0.0758	0.0783	13
14	1.0356	0.9656	14.2297	13.7409	0.0703	0.0728	14
15	1.0382	0.9632	15.2653	14.7041	0.0655	0.0680	15
16	1.0408	0.9608	16.3035	15.6650	0.0613	0.0638	16
17	1.0434	0.9584	17.3442	16.6234	0.0577	0.0602	17
18	1.0460	0.9561	18.3876	17.5795	0.0544	0.0569	18
19	1.0486	0.9537	19.4335	18.5331	0.0515	0.0540	19
20	1.0512	0.9513	20.4821	19.4844	0.0488	0.0513	20
21	1.0538	0.9489	21.5333	20.4333	0.0464	0.0489	21
22	1.0565	0.9466	22.5872	21.3799	0.0443	0.0468	22
23	1.0591	0.9442	23.6436	22.3241	0.0423	0.0448	23
24	1.0618	0.9418	24.7027	23.2659	0.0405	0.0430	24
25	1.0644	0.9395	25.7645	24.2054	0.0388	0.0413	25
26	1.0671	0.9371	26.8289	25.1425	0.0373	0.0398	26
27	1.0697	0.9348	27.8959	26.0773	0.0358	0.0383	27
28	1.0724	0.9325	28.9657	27.0098	0.0345	0.0370	28
29	1.0751	0.9302	30.0381	27.9399	0.0333	0.0358	29
30	1.0778	0.9278	31.1132	28.8678	0.0321	0.0346	30
35	1.0913	0.9163	36.5291	33.4723	0.0274	0.0299	35
40	1.1050	0.9050	42.0130	38.0197	0.0238	0.0263	40
45	1.1189	0.8937	47.5659	42.5107	0.0210	0.0235	45
50	1.1330	0.8826	53.1884	46.9460	0.0188	0.0213	50
55	1.1472	0.8717	58.8817	51.3262	0.0170	0.0195	55
60	1.1616	0.8609	64.6464	55.6521	0.0155	0.0180	60
65	1.1762	0.8502	70.4836	59.9244	0.0142	0.0167	65
70	1.1910	0.8396	76.3941	64.1436	0.0131	0.0156	70
75	1.2059	0.8292	82.3788	68.3105	0.0121	0.0146	75
80	1.2211	0.8189	88.4388	72.4257	0.0113	0.0138	80
85	1.2364	0.8088	94.5748	76.490	0.0106	0.0131	85
90	1.2520	0.7987	100.788	80.504	0.0099	0.0124	90
95	1.2677	0.7888	107.079	84.467	0.0093	0.0118	95
100	1.2836	0.7790	113.45	88.382	0.0088	0.0113	100
∞				400.0		0.0025	∞

table E-2

Discrete Compounding; $i = 1/2\%$

	SINGLE PAYMENT		UNIFORM SERIES				
	Compound Amount Factor	Present Worth Factor	Compound Amount Factor	Present Worth Factor	Sinking Fund Factor	Capital Recovery Factor	
N	To find F Given P F/P	To find P Given F P/F	To find F Given A F/A	To find P Given A P/A	To find A Given F A/F	To find A Given P A/P	N
1	1.0050	0.9950	1.0000	0.9950	1.0000	1.0050	1
2	1.0100	0.9901	2.0050	1.9851	0.4988	0.5038	2
3	1.0151	0.9851	3.0150	2.9702	0.3317	0.3367	3
4	1.0202	0.9802	4.0301	3.9505	0.2481	0.2531	4
5	1.0253	0.9754	5.0502	4.9259	0.1980	0.2030	5
6	1.0304	0.9705	6.0755	5.8964	0.1646	0.1696	6
7	1.0355	0.9657	7.1059	6.8621	0.1407	0.1457	7
8	1.0407	0.9609	8.1414	7.8229	0.1228	0.1278	8
9	1.0459	0.9561	9.1821	8.7790	0.1089	0.1139	9
10	1.0511	0.9513	10.2280	9.7304	0.0978	0.1028	10
11	1.0564	0.9466	11.2791	10.6770	0.0887	0.0937	11
12	1.0617	0.9419	12.3355	11.6189	0.0811	0.0861	12
13	1.0670	0.9372	13.3972	12.5561	0.0746	0.0796	13
14	1.0723	0.9326	14.4642	13.4887	0.0691	0.0741	14
15	1.0777	0.9279	15.5365	14.4166	0.0644	0.0694	15
16	1.0831	0.9233	16.6142	15.3399	0.0602	0.0652	16
17	1.0885	0.9187	17.6973	16.2586	0.0565	0.0615	17
18	1.0939	0.9141	18.7857	17.1727	0.0532	0.0582	18
19	1.0994	0.9096	19.8797	18.0823	0.0503	0.0553	19
20	1.1049	0.9051	20.9791	18.9874	0.0477	0.0527	20
21	1.1104	0.9006	22.0839	19.8879	0.0453	0.0503	21
22	1.1160	0.8961	23.1944	20.7840	0.0431	0.0481	22
23	1.1216	0.8916	24.3103	21.6756	0.0411	0.0461	23
24	1.1272	0.8872	25.4319	22.5628	0.0393	0.0443	24
25	1.1328	0.8828	26.5590	23.4456	0.0377	0.0427	25
26	1.1385	0.8784	27.6918	24.3240	0.0361	0.0411	26
27	1.1442	0.8740	28.8303	25.1980	0.0347	0.0397	27
28	1.1499	0.8697	29.9744	26.0676	0.0334	0.0384	28
29	1.1556	0.8653	31.1243	26.9330	0.0321	0.0371	29
30	1.1614	0.8610	32.2799	27.7940	0.0310	0.0360	30
35	1.1907	0.8398	38.1453	32.0353	0.0262	0.0312	35
40	1.2208	0.8191	44.1587	36.1721	0.0226	0.0276	40
45	1.2516	0.7990	50.3240	40.2071	0.0199	0.0249	45
50	1.2832	0.7793	56.6450	44.1427	0.0177	0.0227	50
55	1.3156	0.7601	63.1256	47.9813	0.0158	0.0208	55
60	1.3488	0.7414	69.7698	51.7254	0.0143	0.0193	60
65	1.3829	0.7231	76.5818	55.3773	0.0131	0.0181	65
70	1.4178	0.7053	83.5658	58.9393	0.0120	0.0170	70
75	1.4536	0.6879	90.7262	62.4135	0.0110	0.0160	75
80	1.4903	0.6710	98.0674	65.8022	0.0102	0.0152	80
85	1.5280	0.6545	105.594	69.107	0.0095	0.0145	85
90	1.5666	0.6383	113.311	72.331	0.0088	0.0138	90
95	1.6061	0.6226	121.22	75.475	0.0082	0.0132	95
100	1.6467	0.6073	129.33	78.542	0.0077	0.0127	100
∞				200.0		0.0050	∞

table E-3

Discrete Compounding; $i = 3/4\%$

	SINGLE PAYMENT		UNIFORM SERIES				
	Compound Amount Factor	Present Worth Factor	Compound Amount Factor	Present Worth Factor	Sinking Fund Factor	Capital Recovery Factor	
N	To find F Given P F/P	To find P Given F P/F	To find F Given A F/A	To find P Given A P/A	To find A Given F A/F	To find A Given P A/P	N
1	1.0075	0.9926	1.0000	0.9926	1.0000	1.0075	1
2	1.0151	0.9852	2.0075	1.9777	0.4981	0.5056	2
3	1.0227	0.9778	3.0226	2.9556	0.3308	0.3383	3
4	1.0303	0.9706	4.0452	3.9261	0.2472	0.2547	4
5	1.0381	0.9633	5.0756	4.8894	0.1970	0.2045	5
6	1.0459	0.9562	6.1136	5.8456	0.1636	0.1711	6
7	1.0537	0.9490	7.1595	6.7946	0.1397	0.1472	7
8	1.0616	0.9420	8.2132	7.7366	0.1218	0.1293	8
9	1.0696	0.9350	9.2748	8.6716	0.1078	0.1153	9
10	1.0776	0.9280	10.3443	9.5996	0.0967	0.1042	10
11	1.0857	0.9211	11.4219	10.5207	0.0876	0.0951	11
12	1.0938	0.9142	12.5076	11.4349	0.0800	0.0875	12
13	1.1020	0.9074	13.6014	12.3423	0.0735	0.0810	13
14	1.1103	0.9007	14.7034	13.2430	0.0680	0.0755	14
15	1.1186	0.8940	15.8136	14.1370	0.0632	0.0707	15
16	1.1270	0.8873	16.9322	15.0243	0.0591	0.0666	16
17	1.1354	0.8807	18.0592	15.9050	0.0554	0.0629	17
18	1.1440	0.8742	19.1947	16.7791	0.0521	0.0596	18
19	1.1525	0.8676	20.3386	17.6468	0.0492	0.0567	19
20	1.1612	0.8612	21.4912	18.5080	0.0465	0.0540	20
21	1.1699	0.8548	22.6523	19.3628	0.0441	0.0516	21
22	1.1787	0.8484	23.8222	20.2112	0.0420	0.0495	22
23	1.1875	0.8421	25.0009	21.0533	0.0400	0.0475	23
24	1.1964	0.8358	26.1884	21.8891	0.0382	0.0457	24
25	1.2054	0.8296	27.3848	22.7187	0.0365	0.0440	25
26	1.2144	0.8234	28.5902	23.5421	0.0350	0.0425	26
27	1.2235	0.8173	29.8046	24.3594	0.0336	0.0411	27
28	1.2327	0.8112	31.0282	25.1707	0.0322	0.0397	28
29	1.2420	0.8052	32.2609	25.9758	0.0310	0.0385	29
30	1.2513	0.7992	33.5028	26.7750	0.0298	0.0373	30
35	1.2989	0.7699	39.8537	30.6826	0.0251	0.0326	35
40	1.3483	0.7416	46.4464	34.4469	0.0215	0.0290	40
45	1.3997	0.7145	53.2900	38.0731	0.0188	0.0263	45
50	1.4530	0.6883	60.3941	41.5664	0.0166	0.0241	50
55	1.5083	0.6630	67.7686	44.9315	0.0148	0.0223	55
60	1.5657	0.6387	75.4239	48.1733	0.0133	0.0208	60
65	1.6253	0.6153	83.3706	51.2962	0.0120	0.0195	65
70	1.6871	0.5927	91.6198	54.3045	0.0109	0.0184	70
75	1.7514	0.5710	100.183	57.2026	0.0100	0.0175	75
80	1.8180	0.5500	109.072	59.9943	0.0092	0.0167	80
85	1.8872	0.5299	118.300	62.6837	0.0085	0.0160	85
90	1.9591	0.5104	127.879	65.275	0.0078	0.0153	90
95	2.0337	0.4917	137.822	67.770	0.0073	0.0148	95
100	2.1111	0.4737	148.14	70.175	0.0068	0.0143	100
∞				133.333		0.0075	∞

table E-4

Discrete Compounding; $i = 1\%$

| | SINGLE PAYMENT | | UNIFORM SERIES | | | | |
| | Compound Amount Factor | Present Worth Factor | Compound Amount Factor | Present Worth Factor | Sinking Fund Factor | Capital Recovery Factor | |
N	To find F Given P F/P	To find P Given F P/F	To find F Given A F/A	To find P Given A P/A	To find A Given F A/F	To find A Given P A/P	N
1	1.0100	0.9901	1.0000	0.9901	1.0000	1.0100	1
2	1.0201	0.9803	2.0100	1.9704	0.4975	0.5075	2
3	1.0303	0.9706	3.0301	2.9410	0.3300	0.3400	3
4	1.0406	0.9610	4.0604	3.9020	0.2463	0.2563	4
5	1.0510	0.9515	5.1010	4.8534	0.1960	0.2060	5
6	1.0615	0.9420	6.1520	5.7955	0.1625	0.1725	6
7	1.0721	0.9327	7.2135	6.7282	0.1386	0.1486	7
8	1.0829	0.9235	8.2857	7.6517	0.1207	0.1307	8
9	1.0937	0.9143	9.3685	8.5660	0.1067	0.1167	9
10	1.1046	0.9053	10.4622	9.4713	0.0956	0.1056	10
11	1.1157	0.8963	11.5668	10.3676	0.0865	0.0965	11
12	1.1268	0.8874	12.6825	11.2551	0.0788	0.0888	12
13	1.1381	0.8787	13.8093	12.1337	0.0724	0.0824	13
14	1.1495	0.8700	14.9474	13.0037	0.0669	0.0769	14
15	1.1610	0.8613	16.0969	13.8650	0.0621	0.0721	15
16	1.1726	0.8528	17.2578	14.7178	0.0579	0.0679	16
17	1.1843	0.8444	18.4304	15.5622	0.0543	0.0643	17
18	1.1961	0.8360	19.6147	16.3982	0.0510	0.0610	18
19	1.2081	0.8277	20.8109	17.2260	0.0481	0.0581	19
20	1.2202	0.8195	22.0190	18.0455	0.0454	0.0554	20
21	1.2324	0.8114	23.2391	18.8570	0.0430	0.0530	21
22	1.2447	0.8034	24.4715	19.6603	0.0409	0.0509	22
23	1.2572	0.7954	25.7162	20.4558	0.0389	0.0489	23
24	1.2697	0.7876	26.9734	21.2434	0.0371	0.0471	24
25	1.2824	0.7798	28.2431	22.0231	0.0354	0.0454	25
26	1.2953	0.7720	29.5256	22.7952	0.0339	0.0439	26
27	1.3082	0.7644	30.8208	23.5596	0.0324	0.0424	27
28	1.3213	0.7568	32.1290	24.3164	0.0311	0.0411	28
29	1.3345	0.7493	33.4503	25.0657	0.0299	0.0399	29
30	1.3478	0.7419	34.7848	25.8077	0.0287	0.0387	30
35	1.4166	0.7059	41.6602	29.4085	0.0240	0.0340	35
40	1.4889	0.6717	48.8863	32.8346	0.0205	0.0305	40
45	1.5648	0.6391	56.4809	36.0945	0.0177	0.0277	45
50	1.6446	0.6080	64.4630	39.1961	0.0155	0.0255	50
55	1.7285	0.5785	72.8523	42.1471	0.0137	0.0237	55
60	1.8167	0.5505	81.6695	44.9550	0.0122	0.0222	60
65	1.9094	0.5237	90.9364	47.6265	0.0110	0.0210	65
70	2.0068	0.4983	100.676	50.1684	0.0099	0.0199	70
75	2.1091	0.4741	110.912	52.5870	0.0090	0.0190	75
80	2.2167	0.4511	121.671	54.8881	0.0082	0.0182	80
85	2.3298	0.4292	132.979	57.0776	0.0075	0.0175	85
90	2.4486	0.4084	144.86	59.161	0.0069	0.0169	90
95	2.5735	0.3886	157.35	61.143	0.0064	0.0164	95
100	2.7048	0.3697	170.48	63.029	0.0059	0.0159	100
∞				100.000		0.0100	∞

table E-5

Discrete Compounding; $i = 1\ 1/2\%$

	SINGLE PAYMENT		UNIFORM SERIES				
	Compound Amount Factor	Present Worth Factor	Compound Amount Factor	Present Worth Factor	Sinking Fund Factor	Capital Recovery Factor	
N	To find F Given P F/P	To find P Given F P/F	To find F Given A F/A	To find P Given A P/A	To find A Given F A/F	To find A Given P A/P	N
1	1.0150	0.9852	1.0000	0.9852	1.0000	1.0150	1
2	1.0302	0.9707	2.0150	1.9559	0.4963	0.5113	2
3	1.0457	0.9563	3.0452	2.9122	0.3284	0.3434	3
4	1.0614	0.9422	4.0909	3.8544	0.2444	0.2594	4
5	1.0773	0.9283	5.1523	4.7826	0.1941	0.2091	5
6	1.0934	0.9145	6.2295	5.6972	0.1605	0.1755	6
7	1.1098	0.9010	7.3230	6.5982	0.1366	0.1516	7
8	1.1265	0.8877	8.4328	7.4859	0.1186	0.1336	8
9	1.1434	0.8746	9.5593	8.3605	0.1046	0.1196	9
10	1.1605	0.8617	10.7027	9.2222	0.0934	0.1084	10
11	1.1779	0.8489	11.8632	10.0711	0.0843	0.0993	11
12	1.1956	0.8364	13.0412	10.9075	0.0767	0.0917	12
13	1.2136	0.8240	14.2368	11.7315	0.0702	0.0852	13
14	1.2318	0.8118	15.4504	12.5434	0.0647	0.0797	14
15	1.2502	0.7999	16.6821	13.3432	0.0599	0.0749	15
16	1.2690	0.7880	17.9323	14.1312	0.0558	0.0708	16
17	1.2880	0.7764	19.2013	14.9076	0.0521	0.0671	17
18	1.3073	0.7649	20.4893	15.6725	0.0488	0.0638	18
19	1.3270	0.7536	21.7967	16.4261	0.0459	0.0609	19
20	1.3469	0.7425	23.1236	17.1686	0.0432	0.0582	20
21	1.3671	0.7315	24.4705	17.9001	0.0409	0.0559	21
22	1.3876	0.7207	25.8375	18.6208	0.0387	0.0537	22
23	1.4084	0.7100	27.2251	19.3308	0.0367	0.0517	23
24	1.4295	0.6995	28.6335	20.0304	0.0349	0.0499	24
25	1.4509	0.6892	30.0630	20.7196	0.0333	0.0483	25
26	1.4727	0.6790	31.5139	21.3986	0.0317	0.0467	26
27	1.4948	0.6690	32.9866	22.0676	0.0303	0.0453	27
28	1.5172	0.6591	34.4814	22.7267	0.0290	0.0440	28
29	1.5400	0.6494	35.9986	23.3761	0.0278	0.0428	29
30	1.5631	0.6398	37.5386	24.0158	0.0266	0.0416	30
35	1.6839	0.5939	45.5920	27.0756	0.0219	0.0369	35
40	1.8140	0.5513	54.2678	29.9158	0.0184	0.0334	40
45	1.9542	0.5117	63.6141	32.5523	0.0157	0.0307	45
50	2.1052	0.4750	73.6827	34.9997	0.0136	0.0286	50
55	2.2679	0.4409	84.5294	37.2714	0.0118	0.0268	55
60	2.4432	0.4093	96.2145	39.3802	0.0104	0.0254	60
65	2.6320	0.3799	108.803	41.3378	0.0092	0.0242	65
70	2.8355	0.3527	122.364	43.1548	0.0082	0.0232	70
75	3.0546	0.3274	136.97	44.8416	0.0073	0.0223	75
80	3.2907	0.3039	152.71	46.4073	0.0065	0.0215	80
85	3.5450	0.2821	169.66	47.8607	0.0059	0.0209	85
90	3.8189	0.2619	187.93	49.2098	0.0053	0.0203	90
95	4.1141	0.2431	207.61	50.4622	0.0048	0.0198	95
100	4.4320	0.2256	228.80	51.6247	0.0044	0.0194	100
∞				66.667		0.0150	∞

536

table E-6

Discrete Compounding; $i = 2\%$

	SINGLE PAYMENT		UNIFORM SERIES				
	Compound Amount Factor	Present Worth Factor	Compound Amount Factor	Present Worth Factor	Sinking Fund Factor	Capital Recovery Factor	
N	To find F Given P F/P	To find P Given F P/F	To find F Given A F/A	To find P Given A P/A	To find A Given F A/F	To find A, Given P A/P	N
1	1.0200	0.9804	1.0000	0.9804	1.0000	1.0200	1
2	1.0404	0.9612	2.0200	1.9416	0.4950	0.5150	2
3	1.0612	0.9423	3.0604	2.8839	0.3268	0.3468	3
4	1.0824	0.9238	4.1216	3.8077	0.2426	0.2626	4
5	1.1041	0.9057	5.2040	4.7135	0.1922	0.2122	5
6	1.1262	0.8880	6.3081	5.6014	0.1585	0.1785	6
7	1.1487	0.8706	7.4343	6.4720	0.1345	0.1545	7
8	1.1717	0.8535	8.5830	7.3255	0.1165	0.1365	8
9	1.1951	0.8368	9.7546	8.1622	0.1025	0.1225	9
10	1.2190	0.8203	10.9497	8.9826	0.0913	0.1113	10
11	1.2434	0.8043	12.1687	9.7868	0.0822	0.1022	11
12	1.2682	0.7885	13.4121	10.5753	0.0746	0.0946	12
13	1.2936	0.7730	14.6803	11.3484	0.0681	0.0881	13
14	1.3195	0.7579	15.9739	12.1062	0.0626	0.0826	14
15	1.3459	0.7430	17.2934	12.8493	0.0578	0.0778	15
16	1.3728	0.7284	18.6393	13.5777	0.0537	0.0737	16
17	1.4002	0.7142	20.0121	14.2919	0.0500	0.0700	17
18	1.4282	0.7002	21.4123	14.9920	0.0467	0.0667	18
19	1.4568	0.6864	22.8405	15.6785	0.0438	0.0638	19
20	1.4859	0.6730	24.2974	16.3514	0.0412	0.0612	20
21	1.5157	0.6598	25.7833	17.0112	0.0388	0.0588	21
22	1.5460	0.6468	27.2990	17.6580	0.0366	0.0566	22
23	1.5769	0.6342	28.8449	18.2922	0.0347	0.0547	23
24	1.6084	0.6217	30.4218	18.9139	0.0329	0.0529	24
25	1.6406	0.6095	32.0303	19.5234	0.0312	0.0512	25
26	1.6734	0.5976	33.6709	20.1210	0.0297	0.0497	26
27	1.7069	0.5859	35.3443	20.7069	0.0283	0.0483	27
28	1.7410	0.5744	37.0512	21.2813	0.0270	0.0470	28
29	1.7758	0.5631	38.7922	21.8444	0.0258	0.0458	29
30	1.8114	0.5521	40.5681	22.3964	0.0246	0.0446	30
35	1.9999	0.5000	49.9944	24.9986	0.0200	0.0400	35
40	2.2080	0.4529	60.4019	27.3555	0.0166	0.0366	40
45	2.4379	0.4102	71.8927	29.4902	0.0139	0.0339	45
50	2.6916	0.3715	84.5793	31.4236	0.0118	0.0318	50
55	2.9717	0.3365	98.5864	33.1748	0.0101	0.0301	55
60	3.2810	0.3048	114.051	34.7609	0.0088	0.0288	60
65	3.6225	0.2761	131.126	36.1975	0.0076	0.0276	65
70	3.9996	0.2500	149.978	37.4986	0.0067	0.0267	70
75	4.4158	0.2265	170.792	38.6771	0.0059	0.0259	75
80	4.8754	0.2051	193.772	39.7445	0.0052	0.0252	80
85	5.3829	0.1858	219.144	40.7113	0.0046	0.0246	85
90	5.9431	0.1683	247.16	41.5869	0.0040	0.0240	90
95	6.5617	0.1524	278.08	42.3800	0.0036	0.0236	95
100	7.2446	0.1380	312.23	43.0983	0.0032	0.0232	100
∞				50.0000		0.0200	∞

table E-7

Discrete Compounding; $i = 2\ 1/2\%$

	SINGLE PAYMENT		UNIFORM SERIES				
	Compound Amount Factor	Present Worth Factor	Compound Amount Factor	Present Worth Factor	Sinking Fund Factor	Capital Recovery Factor	
N	To find F Given P F/P	To find P Given F P/F	To find F Given A F/A	To find P Given A P/A	To find A Given F A/F	To find A Given P A/P	N
1	1.0250	0.9756	1.0000	0.9756	1.0000	1.0250	1
2	1.0506	0.9518	2.0250	1.9274	0.4938	0.5188	2
3	1.0769	0.9286	3.0756	2.8560	0.3251	0.3501	3
4	1.1038	0.9060	4.1525	3.7620	0.2408	0.2658	4
5	1.1314	0.8839	5.2563	4.6458	0.1902	0.2152	5
6	1.1597	0.8623	6.3877	5.5081	0.1566	0.1816	6
7	1.1887	0.8413	7.5474	6.3494	0.1325	0.1575	7
8	1.2184	0.8207	8.7361	7.1701	0.1145	0.1395	8
9	1.2489	0.8007	9.9545	7.9709	0.1005	0.1255	9
10	1.2801	0.7812	11.2034	8.7521	0.0893	0.1143	10
11	1.3121	0.7621	12.4835	9.5142	0.0801	0.1051	11
12	1.3449	0.7436	13.7955	10.2578	0.0725	0.0975	12
13	1.3785	0.7254	15.1404	10.9832	0.0660	0.0910	13
14	1.4130	0.7077	16.5189	11.6909	0.0605	0.0855	14
15	1.4483	0.6905	17.9319	12.3814	0.0558	0.0808	15
16	1.4845	0.6736	19.3802	13.0550	0.0516	0.0766	16
17	1.5216	0.6572	20.8647	13.7122	0.0479	0.0729	17
18	1.5597	0.6412	22.3863	14.3534	0.0447	0.0697	18
19	1.5986	0.6255	23.9460	14.9789	0.0418	0.0668	19
20	1.6386	0.6103	25.5446	15.5892	0.0391	0.0641	20
21	1.6796	0.5954	27.1833	16.1845	0.0368	0.0618	21
22	1.7216	0.5809	28.8628	16.7654	0.0346	0.0596	22
23	1.7646	0.5667	30.5844	17.3321	0.0327	0.0577	23
24	1.8087	0.5529	32.3490	17.8850	0.0309	0.0559	24
25	1.8539	0.5394	34.1577	18.4244	0.0293	0.0543	25
26	1.9003	0.5262	36.0117	18.9506	0.0278	0.0528	26
27	1.9478	0.5134	37.9120	19.4640	0.0264	0.0514	27
28	1.9965	0.5009	39.8598	19.9649	0.0251	0.0501	28
29	2.0464	0.4887	41.8563	20.4535	0.0239	0.0489	29
30	2.0976	0.4767	43.9027	20.9303	0.0228	0.0478	30
35	2.3732	0.4214	54.9282	23.1452	0.0182	0.0432	35
40	2.6851	0.3724	67.4025	25.1028	0.0148	0.0398	40
45	3.0379	0.3292	81.5161	26.8330	0.0123	0.0373	45
50	3.4371	0.2909	97.4843	28.3623	0.0103	0.0353	50
55	3.8888	0.2572	115.551	29.7140	0.0087	0.0337	55
60	4.3998	0.2273	135.992	30.9087	0.0074	0.0324	60
65	4.9780	0.2009	159.118	31.9646	0.0063	0.0313	65
70	5.6321	0.1776	185.284	32.8979	0.0054	0.0304	70
75	6.3722	0.1569	214.888	33.7227	0.0047	0.0297	75
80	7.2096	0.1387	248.383	34.4518	0.0040	0.0290	80
85	8.1570	0.1226	286.278	35.0962	0.0035	0.0285	85
90	9.2288	0.1084	329.15	35.6658	0.0030	0.0280	90
95	10.4416	0.0958	377.66	36.1692	0.0026	0.0276	95
100	11.8137	0.0846	432.55	36.6141	0.0023	0.0273	100
∞				40.0000		0.0250	∞

538

table E-8

Discrete Compounding; $i = 3\%$

	SINGLE PAYMENT		UNIFORM SERIES				
	Compound Amount Factor	Present Worth Factor	Compound Amount Factor	Present Worth Factor	Sinking Fund Factor	Capital Recovery Factor	
N	To find F Given P F/P	To find P Given F P/F	To find F Given A F/A	To find P Given A P/A	To find A Given F A/F	To find A Given P A/P	N
1	1.0300	0.9709	1.0000	0.9709	1.0000	1.0300	1
2	1.0609	0.9426	2.0300	1.9135	0.4926	0.5226	2
3	1.0927	0.9151	3.0909	2.8286	0.3235	0.3535	3
4	1.1255	0.8885	4.1836	3.7171	0.2390	0.2690	4
5	1.1593	0.8626	5.3091	4.5797	0.1884	0.2184	5
6	1.1941	0.8375	6.4684	5.4172	0.1546	0.1846	6
7	1.2299	0.8131	7.6625	6.2303	0.1305	0.1605	7
8	1.2668	0.7894	8.8923	7.0197	0.1125	0.1425	8
9	1.3048	0.7664	10.1591	7.7861	0.0984	0.1284	9
10	1.3439	0.7441	11.4639	8.5302	0.0872	0.1172	10
11	1.3842	0.7224	12.8078	9.2526	0.0781	0.1081	11
12	1.4258	0.7014	14.1920	9.9540	0.0705	0.1005	12
13	1.4685	0.6810	15.6178	10.6349	0.0640	0.0940	13
14	1.5126	0.6611	17.0863	11.2961	0.0585	0.0885	14
15	1.5580	0.6419	18.5989	11.9379	0.0538	0.0838	15
16	1.6047	0.6232	20.1569	12.5611	0.0496	0.0796	16
17	1.6528	0.6050	21.7616	13.1661	0.0460	0.0760	17
18	1.7024	0.5874	23.4144	13.7535	0.0427	0.0727	18
19	1.7535	0.5703	25.1168	14.3238	0.0398	0.0698	19
20	1.8061	0.5537	26.8703	14.8775	0.0372	0.0672	20
21	1.8603	0.5375	28.6765	15.4150	0.0349	0.0649	21
22	1.9161	0.5219	30.5367	15.9369	0.0327	0.0627	22
23	1.9736	0.5067	32.4528	16.4436	0.0308	0.0608	23
24	2.0328	0.4919	34.4264	16.9355	0.0290	0.0590	24
25	2.0938	0.4776	36.4592	17.4131	0.0274	0.0574	25
26	2.1566	0.4637	38.5530	17.8768	0.0259	0.0559	26
27	2.2213	0.4502	40.7096	18.3270	0.0246	0.0546	27
28	2.2879	0.4371	42.9309	18.7641	0.0233	0.0533	28
29	2.3566	0.4243	45.2188	19.1884	0.0221	0.0521	29
30	2.4273	0.4120	47.5754	19.6004	0.0210	0.0510	30
35	2.8139	0.3554	60.4620	21.4872	0.0165	0.0465	35
40	3.2620	0.3066	75.4012	23.1148	0.0133	0.0433	40
45	3.7816	0.2644	92.7197	24.5187	0.0108	0.0408	45
50	4.3839	0.2281	112.797	25.7298	0.0089	0.0389	50
55	5.0821	0.1968	136.071	26.7744	0.0073	0.0373	55
60	5.8916	0.1697	163.053	27.6756	0.0061	0.0361	60
65	6.8300	0.1464	194.332	28.4529	0.0051	0.0351	65
70	7.9178	0.1263	230.594	29.1234	0.0043	0.0343	70
75	9.1789	0.1089	272.630	29.7018	0.0037	0.0337	75
80	10.6409	0.0940	321.362	30.2008	0.0031	0.0331	80
85	12.3357	0.0811	377.856	30.6311	0.0026	0.0326	85
90	14.3004	0.0699	443.35	31.0024	0.0023	0.0323	90
95	16.5781	0.0603	519.27	31.3227	0.0019	0.0319	95
100	19.2186	0.0520	607.29	31.5989	0.0016	0.0316	100
∞				33.3333		0.0300	∞

table E-9

Discrete Compounding; $i = 4\%$

	SINGLE PAYMENT		UNIFORM SERIES				
	Compound Amount Factor	Present Worth Factor	Compound Amount Factor	Present Worth Factor	Sinking Fund Factor	Capital Recovery Factor	
N	To find F Given P F/P	To find P Given F P/F	To find F Given A F/A	To find P Given A P/A	To find A Given F A/F	To find A Given P A/P	N
1	1.0400	0.9615	1.0000	0.9615	1.0000	1.0400	1
2	1.0816	0.9246	2.0400	1.8861	0.4902	0.5302	2
3	1.1249	0.8890	3.1216	2.7751	0.3203	0.3603	3
4	1.1699	0.8548	4.2465	3.6299	0.2355	0.2755	4
5	1.2167	0.8219	5.4163	4.4518	0.1846	0.2246	5
6	1.2653	0.7903	6.6330	5.2421	0.1508	0.1908	6
7	1.3159	0.7599	7.8983	6.0021	0.1266	0.1666	7
8	1.3686	0.7307	9.2142	6.7327	0.1085	0.1485	8
9	1.4233	0.7026	10.5828	7.4353	0.0945	0.1345	9
10	1.4802	0.6756	12.0061	8.1109	0.0833	0.1233	10
11	1.5395	0.6496	13.4863	8.7605	0.0741	0.1141	11
12	1.6010	0.6246	15.0258	9.3851	0.0666	0.1066	12
13	1.6651	0.6006	16.6268	9.9856	0.0601	0.1001	13
14	1.7317	0.5775	18.2919	10.5631	0.0547	0.0947	14
15	1.8009	0.5553	20.0236	11.1184	0.0499	0.0899	15
16	1.8730	0.5339	21.8245	11.6523	0.0458	0.0858	16
17	1.9479	0.5134	23.6975	12.1657	0.0422	0.0822	17
18	2.0258	0.4936	25.6454	12.6593	0.0390	0.0790	18
19	2.1068	0.4746	27.6712	13.1339	0.0361	0.0761	19
20	2.1911	0.4564	29.7781	13.5903	0.0336	0.0736	20
21	2.2788	0.4388	31.9692	14.0292	0.0313	0.0713	21
22	2.3699	0.4220	34.2480	14.4511	0.0292	0.0692	22
23	2.4647	0.4057	36.6179	14.8568	0.0273	0.0673	23
24	2.5633	0.3901	39.0826	15.2470	0.0256	0.0656	24
25	2.6658	0.3751	41.6459	15.6221	0.0240	0.0640	25
26	2.7725	0.3607	44.3117	15.9828	0.0226	0.0626	26
27	2.8834	0.3468	47.0842	16.3296	0.0212	0.0612	27
28	2.9987	0.3335	49.9676	16.6631	0.0200	0.0600	28
29	3.1187	0.3207	52.9663	16.9837	0.0189	0.0589	29
30	3.2434	0.3083	56.0849	17.2920	0.0178	0.0578	30
35	3.9461	0.2534	73.6522	18.6646	0.0136	0.0536	35
40	4.8010	0.2083	95.0255	19.7928	0.0105	0.0505	40
45	5.8412	0.1712	121.029	20.7200	0.0083	0.0483	45
50	7.1067	0.1407	152.667	21.4822	0.0066	0.0466	50
55	8.6464	0.1157	191.159	22.1086	0.0052	0.0452	55
60	10.5196	0.0951	237.991	22.6235	0.0042	0.0442	60
65	12.7987	0.0781	294.968	23.0467	0.0034	0.0434	65
70	15.5716	0.0642	364.290	23.3945	0.0027	0.0427	70
75	18.9452	0.0528	448.631	23.6804	0.0022	0.0422	75
80	23.0498	0.0434	551.245	23.9154	0.0018	0.0418	80
85	28.0436	0.0357	676.090	24.1085	0.0015	0.0415	85
90	34.1193	0.0293	827.98	24.2673	0.0012	0.0412	90
95	41.5113	0.0241	1012.78	24.3978	0.0010	0.0410	95
100	50.5049	0.0198	1237.62	24.5050	0.0008	0.0408	100
∞				25.0000		0.0400	∞

table E-10

Discrete Compounding; $i = 5\%$

	SINGLE PAYMENT		UNIFORM SERIES				
	Compound Amount Factor	Present Worth Factor	Compound Amount Factor	Present Worth Factor	Sinking Fund Factor	Capital Recovery Factor	
N	To find F Given P F/P	To find P Given F P/F	To find F Given A F/A	To find P Given A P/A	To find A Given F A/F	To find A Given P A/P	N
1	1.0500	0.9524	1.0000	0.9524	1.0000	1.0500	1
2	1.1025	0.9070	2.0500	1.8594	0.4878	0.5378	2
3	1.1576	0.8638	3.1525	2.7232	0.3172	0.3672	3
4	1.2155	0.8227	4.3101	3.5460	0.2320	0.2820	4
5	1.2763	0.7835	5.5256	4.3295	0.1810	0.2310	5
6	1.3401	0.7462	6.8019	5.0757	0.1470	0.1970	6
7	1.4071	0.7107	8.1420	5.7864	0.1228	0.1728	7
8	1.4775	0.6768	9.5491	6.4632	0.1047	0.1547	8
9	1.5513	0.6446	11.0266	7.1078	0.0907	0.1407	9
10	1.6289	0.6139	12.5779	7.7217	0.0795	0.1295	10
11	1.7103	0.5847	14.2068	8.3064	0.0704	0.1204	11
12	1.7959	0.5568	15.9171	8.8633	0.0628	0.1128	12
13	1.8856	0.5303	17.7130	9.3936	0.0565	0.1065	13
14	1.9799	0.5051	19.5986	9.8986	0.0510	0.1010	14
15	2.0789	0.4810	21.5786	10.3797	0.0463	0.0963	15
16	2.1829	0.4581	23.6575	10.8378	0.0423	0.0923	16
17	2.2920	0.4363	25.8404	11.2741	0.0387	0.0887	17
18	2.4066	0.4155	28.1324	11.6896	0.0355	0.0855	18
19	2.5269	0.3957	30.5390	12.0853	0.0327	0.0827	19
20	2.6533	0.3769	33.0659	12.4622	0.0302	0.0802	20
21	2.7860	0.3589	35.7192	12.8212	0.0280	0.0780	21
22	2.9253	0.3418	38.5052	13.1630	0.0260	0.0760	22
23	3.0715	0.3256	41.4305	13.4886	0.0241	0.0741	23
24	3.2251	0.3101	44.5020	13.7986	0.0225	0.0725	24
25	3.3864	0.2953	47.7271	14.0939	0.0210	0.0710	25
26	3.5557	0.2812	51.1134	14.3752	0.0196	0.0696	26
27	3.7335	0.2678	54.6691	14.6430	0.0183	0.0683	27
28	3.9201	0.2551	58.4026	14.8981	0.0171	0.0671	28
29	4.1161	0.2429	62.3227	15.1411	0.0160	0.0660	29
30	4.3219	0.2314	66.4388	15.3725	0.0151	0.0651	30
35	5.5160	0.1813	90.3203	16.3742	0.0111	0.0611	35
40	7.0400	0.1420	120.800	17.1591	0.0083	0.0583	40
45	8.9850	0.1113	159.700	17.7741	0.0063	0.0563	45
50	11.4674	0.0872	209.348	18.2559	0.0048	0.0548	50
55	14.6356	0.0683	272.713	18.6335	0.0037	0.0537	55
60	18.6792	0.0535	353.584	18.9293	0.0028	0.0528	60
65	23.8399	0.0419	456.798	19.1611	0.0022	0.0522	65
70	30.4264	0.0329	588.528	19.3427	0.0017	0.0517	70
75	38.8327	0.0258	756.653	19.4850	0.0013	0.0513	75
80	49.5614	0.0202	971.228	19.5965	0.0010	0.0510	80
85	63.2543	0.0158	1245.09	19.6838	0.0008	0.0508	85
90	80.7303	0.0124	1594.61	19.7523	0.0006	0.0506	90
95	103.035	0.0097	2040.69	19.8059	0.0005	0.0505	95
100	131.501	0.0076	2610.02	19.8479	0.0004	0.0504	100
∞				20.0000		0.0500	∞

table E-11

Discrete Compounding; $i = 6\%$

	SINGLE PAYMENT		UNIFORM SERIES				
	Compound Amount Factor	Present Worth Factor	Compund Amount Factor	Present Worth Factor	Sinking Fund Factor	Capital Recovery Factor	
N	To find F Given P F/P	To find P Given F P/F	To find F Given A F/A	To find P Given A P/A	To find A Given F A/F	To find A Given P A/P	N
1	1.0600	0.9434	1.0000	0.9434	1.0000	1.0600	1
2	1.1236	0.8900	2.0600	1.8334	0.4854	0.5454	2
3	1.1910	0.8396	3.1836	2.6730	0.3141	0.3741	3
4	1.2625	0.7921	4.3746	3.4651	0.2286	0.2886	4
5	1.3382	0.7473	5.6371	4.2124	0.1774	0.2374	5
6	1.4185	0.7050	6.9753	4.9173	0.1434	0.2034	6
7	1.5036	0.6651	8.3938	5.5824	0.1191	0.1791	7
8	1.5938	0.6274	9.8975	6.2098	0.1010	0.1610	8
9	1.6895	0.5919	11.4913	6.8017	0.0870	0.1470	9
10	1.7908	0.5584	13.1808	7.3601	0.0759	0.1359	10
11	1.8983	0.5268	14.9716	7.8869	0.0668	0.1268	11
12	2.0122	0.4970	16.8699	8.3838	0.0593	0.1193	12
13	2.1329	0.4688	18.8821	8.8527	0.0530	0.1130	13
14	2.2609	0.4423	21.0151	9.2950	0.0476	0.1076	14
15	2.3966	0.4173	23.2760	9.7122	0.0430	0.1030	15
16	2.5404	0.3936	25.6725	10.1059	0.0390	0.0990	16
17	2.6928	0.3714	28.2129	10.4773	0.0354	0.0954	17
18	2.8543	0.3503	30.9056	10.8276	0.0324	0.0924	18
19	3.0256	0.3305	33.7600	11.1581	0.0296	0.0896	19
20	3.2071	0.3118	36.7856	11.4699	0.0272	0.0872	20
21	3.3996	0.2942	39.9927	11.7641	0.0250	0.0850	21
22	3.6035	0.2775	43.3923	12.0416	0.0230	0.0830	22
23	3.8197	0.2618	46.9958	12.3034	0.0213	0.0813	23
24	4.0489	0.2470	50.8155	12.5504	0.0197	0.0797	24
25	4.2919	0.2330	54.8645	12.7834	0.0182	0.0782	25
26	4.5494	0.2198	59.1563	13.0032	0.0169	0.0769	26
27	4.8223	0.2074	63.7057	13.2105	0.0157	0.0757	27
28	5.1117	0.1956	68.5281	13.4062	0.0146	0.0746	28
29	5.4184	0.1846	73.6397	13.5907	0.0136	0.0736	29
30	5.7435	0.1741	79.0581	13.7648	0.0126	0.0726	30
35	7.6861	0.1301	111.435	14.4982	0.0090	0.0690	35
40	10.2857	0.0972	154.762	15.0463	0.0065	0.0665	40
45	13.7646	0.0727	212.743	15.4558	0.0047	0.0647	45
50	18.4201	0.0543	290.336	15.7619	0.0034	0.0634	50
55	24.6503	0.0406	394.172	15.9905	0.0025	0.0625	55
60	32.9876	0.0303	533.128	16.1614	0.0019	0.0619	60
65	44.1449	0.0227	719.082	16.2891	0.0014	0.0614	65
70	59.0758	0.0169	967.931	16.3845	0.0010	0.0610	70
75	79.0568	0.0126	1300.95	16.4558	0.0008	0.0608	75
80	105.796	0.0095	1746.60	16.5091	0.0006	0.0606	80
85	141.579	0.0071	2342.98	16.5489	0.0004	0.0604	85
90	189.464	0.0053	3141.07	16.5787	0.0003	0.0603	90
95	253.546	0.0039	4209.10	16.6009	0.0002	0.0602	95
100	339.301	0.0029	5638.36	16.6175	0.0002	0.0602	100
∞				18.182		0.0600	∞

table E-12

Discrete Compounding; $i = 8\%$

	SINGLE PAYMENT		UNIFORM SERIES				
	Compound Amount Factor	Present Worth Factor	Compound Amount Factor	Present Worth Factor	Sinking Fund Factor	Capital Recovery Factor	
N	To find F Given P F/P	To find P Given F P/F	To find F Given A F/A	To find P Given A P/A	To find A Given F A/F	To find A Given P A/P	N
1	1.0800	0.9259	1.0000	0.9259	1.0000	1.0800	1
2	1.1664	0.8573	2.0800	1.7833	0.4808	0.5608	2
3	1.2597	0.7938	3.2464	2.5771	0.3080	0.3880	3
4	1.3605	0.7350	4.5061	3.3121	0.2219	0.3019	4
5	1.4693	0.6806	5.8666	3.9927	0.1705	0.2505	5
6	1.5869	0.6302	7.3359	4.6229	0.1363	0.2163	6
7	1.7138	0.5835	8.9228	5.2064	0.1121	0.1921	7
8	1.8509	0.5403	10.6366	5.7466	0.0940	0.1740	8
9	1.9990	0.5002	12.4876	6.2469	0.0801	0.1601	9
10	2.1589	0.4632	14.4866	6.7101	0.0690	0.1490	10
11	2.3316	0.4289	16.6455	7.1390	0.0601	0.1401	11
12	2.5182	0.3971	18.9771	7.5361	0.0527	0.1327	12
13	2.7196	0.3677	21.4953	7.9038	0.0465	0.1265	13
14	2.9372	0.3405	24.2149	8.2442	0.0413	0.1213	14
15	3.1722	0.3152	27.1521	8.5595	0.0368	0.1168	15
16	3.4259	0.2919	30.3243	8.8514	0.0330	0.1130	16
17	3.7000	0.2703	33.7502	9.1216	0.0296	0.1096	17
18	3.9960	0.2502	37.4502	9.3719	0.0267	0.1067	18
19	4.3157	0.2317	41.4463	9.6036	0.0241	0.1041	19
20	4.6610	0.2145	45.7620	9.8181	0.0219	0.1019	20
21	5.0338	0.1987	50.4229	10.0168	0.0198	0.0998	21
22	5.4365	0.1839	55.4567	10.2007	0.0180	0.0980	22
23	5.8715	0.1703	60.8933	10.3711	0.0164	0.0964	23
24	6.3412	0.1577	66.7647	10.5288	0.0150	0.0950	24
25	6.8485	0.1460	73.1059	10.6748	0.0137	0.0937	25
26	7.3964	0.1352	79.9544	10.8100	0.0125	0.0925	26
27	7.9881	0.1252	87.3507	10.9352	0.0114	0.0914	27
28	8.6271	0.1159	95.3388	11.0511	0.0105	0.0905	28
29	9.3173	0.1073	103.966	11.1584	0.0096	0.0896	29
30	10.0627	0.0994	113.283	11.2578	0.0088	0.0888	30
35	14.7853	0.0676	172.317	11.6546	0.0058	0.0858	35
40	21.7245	0.0460	259.056	11.9246	0.0039	0.0839	40
45	31.9204	0.0313	386.506	12.1084	0.0026	0.0826	45
50	46.9016	0.0213	573.770	12.2335	0.0017	0.0817	50
55	68.9138	0.0145	848.923	12.3186	0.0012	0.0812	55
60	101.257	0.0099	1253.21	12.3766	0.0008	0.0808	60
65	148.780	0.0067	1847.25	12.4160	0.0005	0.0805	65
70	218.606	0.0046	2720.08	12.4428	0.0004	0.0804	70
75	321.204	0.0031	4002.55	12.4611	0.0002	0.0802	75
80	471.955	0.0021	5886.93	12.4735	0.0002	0.0802	80
85	693.456	0.0014	8655.71	12.4820	0.0001	0.0801	85
90	1018.92	0.0010	12723.9	12.4877	a	0.0801	90
95	1497.12	0.0007	18071.5	12.4917	a	0.0801	95
100	2199.76	0.0005	27484.5	12.4943	a	0.0800	100
∞				12.5000		0.0800	∞

a Less than 0.0001.

table E-13

Discrete Compounding; $i = 10\%$

	SINGLE PAYMENT		UNIFORM SERIES				
	Compound Amount Factor	Present Worth Factor	Compound Amount Factor	Present Worth Factor	Sinking Fund Factor	Capital Recovery Factor	
N	To find F Given P F/P	To find P Given F P/F	To find F Given A F/A	To find P Given A P/A	To find A Given F A/F	To find A Given P A/P	N
1	1.1000	0.9091	1.0000	0.9091	1.0000	1.1000	1
2	1.2100	0.8264	2.1000	1.7355	0.4762	0.5762	2
3	1.3310	0.7513	3.3100	2.4869	0.3021	0.4021	3
4	1.4641	0.6830	4.6410	3.1699	0.2155	0.3155	4
5	1.6105	0.6209	6.1051	3.7908	0.1638	0.2638	5
6	1.7716	0.5645	7.7156	4.3553	0.1296	0.2296	6
7	1.9487	0.5132	9.4872	4.8684	0.1054	0.2054	7
8	2.1436	0.4665	11.4359	5.3349	0.0874	0.1874	8
9	2.3579	0.4241	13.5795	5.7590	0.0736	0.1736	9
10	2.5937	0.3855	15.9374	6.1446	0.0627	0.1627	10
11	2.8531	0.3505	18.5312	6.4951	0.0540	0.1540	11
12	3.1384	0.3186	21.3843	6.8137	0.0468	0.1468	12
13	3.4523	0.2897	24.5227	7.1034	0.0408	0.1408	13
14	3.7975	0.2633	27.9750	7.3667	0.0357	0.1357	14
15	4.1772	0.2394	31.7725	7.6061	0.0315	0.1315	15
16	4.5950	0.2176	35.9497	7.8237	0.0278	0.1278	16
17	5.0545	0.1978	40.5447	8.0216	0.0247	0.1247	17
18	5.5599	0.1799	45.5992	8.2014	0.0219	0.1219	18
19	6.1159	0.1635	51.1591	8.3649	0.0195	0.1195	19
20	6.7275	0.1486	57.2750	8.5136	0.0175	0.1175	20
21	7.4002	0.1351	64.0025	8.6487	0.0156	0.1156	21
22	8.1403	0.1228	71.4027	8.7715	0.0140	0.1140	22
23	8.9543	0.1117	79.5430	8.8832	0.0126	0.1126	23
24	9.8497	0.1015	88.4973	8.9847	0.0113	0.1113	24
25	10.8347	0.0923	98.3470	9.0770	0.0102	0.1102	25
26	11.9182	0.0839	109.182	9.1609	0.0092	0.1092	26
27	13.1100	0.0763	121.100	9.2372	0.0083	0.1083	27
28	14.4210	0.0693	134.210	9.3066	0.0075	0.1075	28
29	15.8631	0.0630	148.631	9.3696	0.0067	0.1067	29
30	17.4494	0.0573	164.494	9.4269	0.0061	0.1061	30
35	28.1024	0.0356	271.024	9.6442	0.0037	0.1037	35
40	45.2592	0.0221	442.592	9.7791	0.0023	0.1023	40
45	72.8904	0.0137	718.905	9.8628	0.0014	0.1014	45
50	117.391	0.0085	1163.91	9.9148	0.0009	0.1009	50
55	189.059	0.0053	1880.59	9.9471	0.0005	0.1005	55
60	304.481	0.0033	3034.81	9.9672	0.0003	0.1003	60
65	490.370	0.0020	4893.71	9.9796	0.0002	0.1002	65
70	789.746	0.0013	7887.47	9.9873	0.0001	0.1001	70
75	1271.89	0.0008	12708.9	9.9921	a	0.1001	75
80	2048.40	0.0005	20474.0	9.9951	a	0.1000	80
85	3298.97	0.0003	32979.7	9.9970	a	0.1000	85
90	5313.02	0.0002	53120.2	9.9981	a	0.1000	90
95	8556.67	0.0001	85556.7	9.9988	a	0.1000	95
100	13780.6	a	137796	9.9993	a	0.1000	100
∞				10.0000		0.1000	∞

a Less than 0.0001.

table E-14

Discrete Compounding; $i = 12\%$

	SINGLE PAYMENT		UNIFORM SERIES				
	Compound Amount Factor	Present Worth Factor	Compound Amount Factor	Present Worth Factor	Sinking Fund Factor	Capital Recovery Factor	
N	To find F Given P F/P	To find P Given F P/F	To find F Given A F/A	To find P Given A P/A	To find A Given F A/F	To find A Given P A/P	N
1	1.1200	0.8929	1.0000	0.8929	1.0000	1.1200	1
2	1.2544	0.7972	2.1200	1.6901	0.4717	0.5917	2
3	1.4049	0.7118	3.3744	2.4018	0.2963	0.4163	3
4	1.5735	0.6355	4.7793	3.0373	0.2092	0.3292	4
5	1.7623	0.5674	6.3528	3.6048	0.1574	0.2774	5
6	1.9738	0.5066	8.1152	4.1114	0.1232	0.2432	6
7	2.2107	0.4523	10.0890	4.5638	0.0991	0.2191	7
8	2.4760	0.4039	12.2997	4.9676	0.0813	0.2013	8
9	2.7731	0.3606	14.7757	5.3282	0.0677	0.1877	9
10	3.1058	0.3220	17.5487	5.6502	0.0570	0.1770	10
11	3.4785	0.2875	20.6546	5.9377	0.0484	0.1684	11
12	3.8960	0.2567	24.1331	6.1944	0.0414	0.1614	12
13	4.3635	0.2292	28.0291	6.4235	0.0357	0.1557	13
14	4.8871	0.2046	32.3926	6.6282	0.0309	0.1509	14
15	5.4736	0.1827	37.2797	6.8109	0.0268	0.1468	15
16	6.1304	0.1631	42.7533	6.9740	0.0234	0.1434	16
17	6.8660	0.1456	48.8837	7.1196	0.0205	0.1405	17
18	7.6900	0.1300	55.7497	7.2497	0.0179	0.1379	18
19	8.6128	0.1161	63.4397	7.3658	0.0158	0.1358	19
20	9.6463	0.1037	72.0524	7.4694	0.0139	0.1339	20
21	10.8038	0.0926	81.6987	7.5620	0.0122	0.1322	21
22	12.1003	0.0826	92.5026	7.6446	0.0108	0.1308	22
23	13.5523	0.0738	104.603	7.7184	0.0096	0.1296	23
24	15.1786	0.0659	118.155	7.7843	0.0085	0.1285	24
25	17.0001	0.0588	133.334	7.8431	0.0075	0.1275	25
26	19.0401	0.0525	150.334	7.8957	0.0067	0.1267	26
27	21.3249	0.0469	169.374	7.9426	0.0059	01.259	27
28	23.8839	0.0419	190.699	7.9844	0.0052	0.1252	28
29	26.7499	0.0374	214.583	8.0218	0.0047	0.1247	29
30	29.9599	0.0334	241.333	8.0552	0.0041	0.1241	30
35	52.7996	0.0189	431.663	8.1755	0.0023	0.1223	35
40	93.0509	0.0107	767.091	8.2438	0.0013	0.1213	40
45	163.988	0.0061	1358.23	8.2825	0.0007	0.1207	45
50	289.002	0.0035	2400.02	8.3045	0.0004	0.1204	50
55	509.320	0.0020	4236.00	8.3170	0.0002	0.1202	55
60	897.596	0.0011	7471.63	8.3240	0.0001	0.1201	60
65	1581.87	0.0006	13173.9	8.3281	a	0.1201	65
70	2787.80	0.0004	23223.3	8.3303	a	0.1200	70
75	4913.05	0.0002	40933.8	8.3316	a	0.1200	75
80	8658.47	0.0001	72145.6	8.3324	a	0.1200	80
∞				8.333		0.1200	∞

^{a} Less than 0.0001.

table E-15

Discrete Compounding; $i = 15\%$

	SINGLE PAYMENT		UNIFORM SERIES				
	Compound Amount Factor	Present Worth Factor	Compound Amount Factor	Present Worth Factor	Sinking Fund Factor	Capital Recovery Factor	
N	To find F Given P F/P	To find P Given F P/F	To find F Given A F/A	To find P Given A P/A	To find A Given F A/F	To find A Given P A/P	N
1	1.1500	0.8696	1.0000	0.8696	1.0000	1.1500	1
2	1.3225	0.7561	2.1500	1.6257	0.4651	0.6151	2
3	1.5209	0.6575	3.4725	2.2832	0.2880	0.4380	3
4	1.7490	0.5718	4.9934	2.8550	0.2003	0.3503	4
5	2.0114	0.4972	6.7424	3.3522	0.1483	0.2983	5
6	2.3131	0.4323	8.7537	3.7845	0.1142	0.2642	6
7	2.6600	0.3759	11.0668	4.1604	0.0904	0.2404	7
8	3.0590	0.3269	13.7268	4.4873	0.0729	0.2229	8
9	3.5179	0.2843	16.7858	4.7716	0.0596	0.2096	9
10	4.0456	0.2472	20.3037	5.0188	0.0493	0.1993	10
11	4.6524	0.2149	24.3493	5.2337	0.0411	0.1911	11
12	5.3502	0.1869	29.0017	5.4206	0.0345	0.1845	12
13	6.1528	0.1625	34.3519	5.5831	0.0291	0.1791	13
14	7.0757	0.1413	40.5047	5.7245	0.0247	0.1747	14
15	8.1371	0.1229	47.5804	5.8474	0.0210	0.1710	15
16	9.3576	0.1069	55.7175	5.9542	0.0179	0.1679	16
17	10.7613	0.0929	65.0751	6.0472	0.0154	0.1654	17
18	12.3755	0.0808	75.8363	6.1280	0.0132	0.1632	18
19	14.2318	0.0703	88.2118	6.1982	0.0113	0.1613	19
20	16.3665	0.0611	102.444	6.2593	0.0098	0.1598	20
21	18.8215	0.0531	118.810	6.3125	0.0084	0.1584	21
22	21.6447	0.0462	137.632	6.3587	0.0073	0.1573	22
23	24.8915	0.0402	159.276	6.3988	0.0063	0.1563	23
24	28.6252	0.0349	184.168	6.4338	0.0054	0.1554	24
25	32.9189	0.0304	212.793	6.4641	0.0047	0.1547	25
26	37.8568	0.0264	245.712	6.4906	0.0041	0.1541	26
27	43.5353	0.0230	283.569	6.5135	0.0035	0.1535	27
28	50.0656	0.0200	327.104	6.5335	0.0031	0.1531	28
29	57.5754	0.0174	377.170	6.5509	0.0027	0.1527	29
30	66.2118	0.0151	434.745	6.5660	0.0023	0.1523	30
35	133.176	0.0075	881.170	6.6166	0.0011	0.1511	35
40	267.863	0.0037	1779.09	6.6418	0.0006	0.1506	40
45	538.769	0.0019	3585.13	6.6543	0.0003	0.1503	45
50	1083.66	0.0009	7217.71	6.6605	0.0001	0.1501	50
55	2179.62	0.0005	14524.1	6.6636	a	0.1501	55
60	4384.00	0.0002	29220.0	6.6651	a	0.1500	60
65	8817.78	0.0001	58778.5	6.6659	a	0.1500	65
70	17735.7	a	118231	6.6663	a	0.1500	70
75	35672.8	a	237812	6.6665	a	0.1500	75
80	71750.8	a	478332	6.6666	a	0.1500	80
∞				6.667		0.1500	∞

a Less than 0.0001.

table E-16

Discrete Compounding; $i = 20\%$

	SINGLE PAYMENT		UNIFORM SERIES				
	Compound Amount Factor	Present Worth Factor	Compound Amount Factor	Present Worth Factor	Sinking Fund Factor	Capital Recovery Factor	
N	To find F Given P F/P	To find P Given F P/F	To find F Given A F/A	To find P Given A P/A	To find A Given F A/F	To find A Given P A/P	N
1	1.2000	0.8333	1.0000	0.8333	1.0000	1.2000	1
2	1.4400	0.6944	2.2000	1.5278	0.4545	0.6545	2
3	1.7280	0.5787	3.6400	2.1065	0.2747	0.4747	3
4	2.0736	0.4823	5.3680	2.5887	0.1863	0.3863	4
5	2.4883	0.4019	7.4416	2.9906	0.1344	0.3344	5
6	2.9860	0.3349	9.9299	3.3255	0.1007	0.3007	6
7	3.5832	0.2791	12.9159	3.6046	0.0774	0.2774	7
8	4.2998	0.2326	16.4991	3.8372	0.0606	0.2606	8
9	5.1598	0.1938	20.7989	4.0310	0.0481	0.2481	9
10	6.1917	0.1615	25.9587	4.1925	0.0385	0.2385	10
11	7.4301	0.1346	32.1504	4.3271	0.0311	0.2311	11
12	8.9161	0.1122	39.5805	4.4392	0.0253	0.2253	12
13	10.6993	0.0935	48.4966	4.5327	0.0206	0.2206	13
14	12.8392	0.0779	59.1959	4.6106	0.0169	0.2169	14
15	15.4070	0.0649	72.0351	4.6755	0.0139	0.2139	15
16	18.4884	0.0541	87.4421	4.7296	0.0114	0.2114	16
17	22.1861	0.0451	105.931	4.7746	0.0094	0.2094	17
18	26.6233	0.0376	128.117	4.8122	0.0078	0.2078	18
19	31.9480	0.0313	154.740	4.8435	0.0065	0.2065	19
20	38.3376	0.0261	186.688	4.8696	0.0054	0.2054	20
21	46.0051	0.0217	225.026	4.8913	0.0044	0.2044	21
22	55.2061	0.0181	271.031	4.9094	0.0037	0.2037	22
23	66.2474	0.0151	326.237	4.9245	0.0031	0.2031	23
24	79.4968	0.0126	392.484	4.9371	0.0025	0.2025	24
25	95.3962	0.0105	471.981	4.9476	0.0021	0.2021	25
26	114.475	0.0087	567.377	4.9563	0.0018	0.2018	26
27	137.371	0.0073	681.853	4.9636	0.0015	0.2015	27
28	164.845	0.0061	819.223	4.9697	0.0012	0.2012	28
29	197.814	0.0051	984.068	4.9747	0.0010	0.2010	29
30	237.376	0.0042	1181.88	4.9789	0.0008	0.2008	30
35	590.668	0.0017	2948.34	4.9915	0.0003	0.2003	35
40	1469.77	0.0007	7343.85	4.9966	0.0001	0.2001	40
45	3657.26	0.0003	18281.3	4.9986	a	0.2001	45
50	9100.43	0.0001	45497.2	4.9995	a	0.2000	50
55	22644.8	a	113219	4.9998	a	0.2000	55
60	56347.5	a	281732	4.9999	a	0.2000	60
∞				5.0000		0.2000	∞

a Less than 0.0001.

table E-17

Discrete Compounding; i = 25%

	SINGLE PAYMENT		UNIFORM SERIES				
	Compound Amount Factor	Present Worth Factor	Compound Amount Factor	Present Worth Factor	Sinking Fund Factor	Capital Recovery Factor	
N	To find F Given P F/P	To find P Given F P/F	To find F Given A F/A	To find P Given A P/A	To find A Given F A/F	To find A Given P A/P	N
1	1.2500	0.8000	1.0000	0.8000	1.0000	1.2500	1
2	1.5625	0.6400	2.2500	1.4400	0.4444	0.6944	2
3	1.9531	0.5120	3.8125	1.9520	0.2623	0.5123	3
4	2.4414	0.4096	5.7656	2.3616	0.1734	0.4234	4
5	3.0518	0.3277	8.2070	2.6893	0.1218	0.3718	5
6	3.8147	0.2621	11.2588	2.9514	0.0888	0.3388	6
7	4.7684	0.2097	15.0735	3.1611	0.0663	0.3163	7
8	5.9605	0.1678	19.8419	3.3289	0.0504	0.3004	8
9	7.4506	0.1342	25.8023	3.4631	0.0388	0.2888	9
10	9.3132	0.1074	33.2529	3.5705	0.0301	0.2801	10
11	11.6415	0.0859	42.5661	3.6564	0.0235	0.2735	11
12	14.5519	0.0687	54.2077	3.7251	0.0184	0.2684	12
13	18.1899	0.0550	68.7596	3.7801	0.0145	0.2645	13
14	22.7374	0.0440	86.9495	3.8241	0.0115	0.2615	14
15	28.4217	0.0352	109.687	3.8593	0.0091	0.2591	15
16	35.5271	0.0281	138.109	3.8874	0.0072	0.2572	16
17	44.4089	0.0225	173.636	3.9099	0.0058	0.2558	17
18	55.5112	0.0180	218.045	3.9279	0.0046	0.2546	18
19	69.3889	0.0144	273.556	3.9424	0.0037	0.2537	19
20	86.7362	0.0115	342.945	3.9539	0.0029	0.2529	20
21	108.420	0.0092	429.681	3.9631	0.0023	0.2523	21
22	135.525	0.0074	538.101	3.9705	0.0019	0.2519	22
23	169.407	0.0059	673.626	3.9764	0.0015	0.2515	23
24	211.758	0.0047	843.033	3.9811	0.0012	0.2512	24
25	264.698	0.0038	1054.79	3.9849	0.0009	0.2509	25
26	330.872	0.0030	1319.49	3.9879	0.0008	0.2508	26
27	413.590	0.0024	1650.36	3.9903	0.0006	0.2506	27
28	516.988	0.0019	2063.95	3.9923	0.0005	0.2505	28
29	646.235	0.0015	2580.94	3.9938	0.0004	0.2504	29
30	807.794	0.0012	3227.17	3.9950	0.0003	0.2503	30
35	2465.19	0.0004	9856.76	3.9984	0.0001	0.2501	35
40	7523.16	0.0001	30088.7	3.9995	a	0.2500	40
45	22958.9	a	91831.5	3.9998	a	0.2500	45
50	70064.9	a	280256	3.9999	a	0.2500	50
∞				4.0000		0.2500	∞

a Less than 0.0001.

table E-18

Discrete Compounding; $i = 30\%$

	SINGLE PAYMENT		UNIFORM SERIES				
	Compound Amount Factor	Present Worth Factor	Compound Amount Factor	Present Worth Factor	Sinking Fund Factor	Capital Recovery Factor	
N	To find F Given P F/P	To find P Given F P/F	To find F Given A F/A	To find P Given A P/A	To find A Given F A/F	To find A Given P A/P	N
1	1.3000	0.7692	1.000	0.769	1.0000	1.3000	1
2	1.6900	0.5917	2.300	1.361	0.4348	0.7348	2
3	2.1970	0.4552	3.990	1.816	0.2506	0.5506	3
4	2.8561	0.3501	6.187	2.166	0.1616	0.4616	4
5	3.7129	0.2693	9.043	2.436	0.1106	0.4106	5
6	4.8268	0.2072	12.756	2.643	0.0784	0.3784	6
7	6.2749	0.1594	17.583	2.802	0.0569	0.3569	7
8	8.1573	0.1226	23.858	2.925	0.0419	0.3419	8
9	10.604	0.0943	32.015	3.019	0.0312	0.3312	9
10	13.786	0.0725	42.619	3.092	0.0235	0.3235	10
11	17.922	0.0558	56.405	3.147	0.0177	0.3177	11
12	23.298	0.0429	74.327	3.190	0.0135	0.3135	12
13	30.287	0.0330	97.625	3.223	0.0102	0.3102	13
14	39.374	0.0254	127.91	3.249	0.0078	0.3078	14
15	51.186	0.0195	167.29	3.268	0.0060	0.3060	15
16	66.542	0.0150	218.47	3.283	0.0046	0.3046	16
17	86.504	0.0116	285.01	3.295	0.0035	0.3035	17
18	112.46	0.0089	371.52	3.304	0.0027	0.3027	18
19	146.19	0.0068	483.97	3.311	0.0021	0.3021	19
20	190.05	0.0053	630.16	3.316	0.0016	0.3016	20
21	247.06	0.0040	820.21	3.320	0.0012	0.3012	21
22	321.18	0.0031	1067.3	3.323	0.0009	0.3009	22
23	417.54	0.0024	1388.5	3.325	0.0007	0.3007	23
24	542.80	0.0018	1806.0	3.327	0.0005	0.3005	24
25	705.64	0.0014	2348.8	3.329	0.0004	0.3004	25
26	917.33	0.0011	3054.4	3.330	0.0003	0.3003	26
27	1192.5	0.0008	3971.8	3.331	0.0003	0.3003	27
28	1550.3	0.0006	5164.3	3.331	0.0002	0.3002	28
29	2015.4	0.0005	6714.6	3.332	0.0002	0.3002	29
30	2620.0	0.0004	8730.0	3.332	0.0001	0.3001	30
31	3406.0	0.0003	11350.	3.332	a	0.3001	31
32	4427.8	0.0002	14756.	3.333	a	0.3001	32
33	5756.1	0.0002	19184.	3.333	a	0.3001	33
34	7483.0	0.0001	24940.	3.333	a	0.3000	34
35	9727.8	0.0001	32423.	3.333	a	0.3000	35
∞				3.333		0.3000	∞

a Less than 0.0001

549

table E-19

Discrete Compounding; i = 40%

	SINGLE PAYMENT		UNIFORM SERIES				
	Compound Amount Factor	Present Worth Factor	Compound Amount Factor	Present Worth Factor	Sinking Fund Factor	Capital Recovery Factor	
N	To find F Given P F/P	To find P Given F P/F	To find F Given A F/A	To find P Given A P/A	To find A Given F A/F	To find A Given P A/P	N
1	1.4000	0.7143	1.000	0.714	1.000	1.4000	1
2	1.9600	0.5102	2.400	1.224	0.4167	0.8167	2
3	2.7440	0.3644	4.360	1.589	0.2294	0.6294	3
4	3.8416	0.2603	7.104	1.849	0.1408	0.5408	4
5	5.3782	0.1859	10.946	2.035	0.0934	0.4914	5
6	7.5295	0.1328	16.324	2.168	0.0613	0.4613	6
7	10.541	0.0949	23.853	2.263	0.0419	0.4419	7
8	14.758	0.0678	34.395	2.331	0.0291	0.4291	8
9.	20.661	0.0484	49.153	2.379	0.0203	0.4203	9
10	28.925	0.0346	69.814	2.414	0.0143	0.4143	10
11	40.496	0.0247	98.739	2.438	0.0101	0.4101	11
12	56.694	0.0176	139.23	2.456	0.0072	0.4072	12
13	79.371	0.0126	195.93	2.469	0.0051	0.4051	13
14	111.12	0.0090	275.30	2.478	0.0036	0.4036	14
15	155.57	0.0064	386.42	2.484	0.0026	0.4026	15
16	217.80	0.0046	541.99	2.489	0.0018	0.4019	16
17	304.91	0.0033	759.78	2.492	0.0013	0.4013	17
18	426.88	0.0023	1064.7	2.494	0.0009	0.4009	18
19	597.63	0.0017	1491.6	2.496	0.0007	0.4007	19
20	836.68	0.0012	2089.2	2.497	0.0005	0.4005	20
21	1171.4	0.0009	2925.9	2.498	0.0003	0.4003	21
22	1639.9	0.0006	4097.2	2.498	0.0002	0.4002	22
23	2295.9	0.0004	5737.1	2.499	0.0002	0.4002	23
24	3214.2	0.0003	8033.0	2.499	0.0001	0.4001	24
25	4499.9	0.0002	11247.	2.499	a	0.4001	25
26	6299.8	0.0002	15747.	2.500	a	0.4001	26
27	8819.8	0.0001	22047.	2.500	a	0.4000	27
28	12348.	0.0001	30867.	2.500	a	0.4000	28
29	17287.	0.0001	43214.	2.500	a	0.4000	29
30	24201.	a	60501.	2.500	a	0.4000	30
∞				2.500		0.4000	∞

a Less than 0.0001

table E-20

Discrete Compounding; $i = 50\%$

	SINGLE PAYMENT		UNIFORM SERIES				
	Compound Amount Factor	Present Worth Factor	Compound Amount Factor	Present Worth Factor	Sinking Fund Factor	Capital Recovery Factor	
N	To find F Given P F/P	To find P Given F P/F	To find F Given A F/A	To find P Given A P/A	To find A Given F A/F	To find A Given P A/P	N
1	1.5000	0.6667	1.000	0.667	1.0000	1.5000	1
2	2.2500	0.4444	2.500	1.111	0.4000	0.9000	2
3	3.3750	0.2963	4.750	1.407	0.2101	0.7105	3
4	5.0625	0.1975	8.125	1.605	0.1231	0.6231	4
5	7.5938	0.1317	13.188	1.737	0.0758	0.5758	5
6	11.391	0.0878	20.781	1.824	0.0481	0.5481	6
7	17.086	0.0585	32.172	1.883	0.0311	0.5311	7
8	25.629	0.0390	49.258	1.922	0.0203	0.5203	8
9	38.443	0.0260	74.887	1.948	0.0134	0.5134	9
10	57.665	0.0173	113.33	1.965	0.0088	0.5088	10
11	86.498	0.0116	171.00	1.977	0.0059	0.5059	11
12	129.75	0.0077	257.49	1.985	0.0039	0.5039	12
13	194.62	0.0051	387.24	1.990	0.0026	0.5026	13
14	291.93	0.0034	581.86	1.993	0.0017	0.5017	14
15	437.89	0.0023	873.79	1.995	0.0011	0.5011	15
16	656.84	0.0015	1311.7	1.997	0.0008	0.5008	16
17	985.26	0.0010	1968.5	1.998	0.0005	0.5005	17
18	1477.9	0.0007	2953.8	1.999	0.0003	0.5003	18
19	2216.8	0.0005	4431.7	1.999	0.0002	0.5002	19
20	3325.3	0.0003	6648.5	1.999	0.0002	0.5002	20
21	4987.9	0.0002	9973.8	2.000	0.0001	0.5001	21
22	7481.8	0.0001	14962.	2.000	a	0.5001	22
23	11223.	0.0001	22443.	2.000	a	0.5000	23
24	16834.	0.0001	33666.	2.000	a	0.5000	24
25	25251.	a	50500.	2.000	a	0.5000	25
∞				2.000		0.5000	∞

a Less than 0.0001

table E-21 gradient to present worth conversion factor for discrete compounding (to find P, given G)

$$(P/G, i\%, N) = \frac{1}{i}\left[\frac{(1+i)^N - 1}{i(1+i)^N} - \frac{N}{(1+i)^N}\right]$$

n	1%	2%	4%	6%	8%	10%	12%	15%	20%	25%	n
1	0.00	0.00	0.00	0.00	0.00	0.00	0.00	0.00	0.00	0.00	1
2	0.98	0.96	0.92	0.89	0.86	0.83	0.80	0.76	0.69	0.64	2
3	2.92	2.85	2.70	2.57	2.45	2.33	2.22	2.07	1.85	1.66	3
4	5.80	5.62	5.27	4.95	4.65	4.38	4.13	3.79	3.30	2.89	4
5	9.61	9.24	8.55	7.93	7.37	6.86	6.40	5.78	4.91	4.20	5
6	14.32	13.68	12.50	11.46	10.52	9.68	8.93	7.94	6.58	5.51	6
7	19.92	18.90	17.07	15.45	14.02	12.76	11.64	10.19	8.26	6.77	7
8	26.38	24.88	22.18	19.84	17.81	16.03	14.47	12.48	9.88	7.95	8
9	33.69	31.57	27.80	24.58	21.81	19.42	17.36	14.75	11.43	9.02	9
10	41.84	38.95	33.88	29.60	25.98	22.89	20.25	16.98	12.89	9.99	10
11	50.80	47.00	40.38	34.87	30.27	26.40	23.13	19.13	14.23	10.85	11
12	60.57	55.67	47.25	40.34	34.63	29.90	25.95	21.18	15.47	11.60	12
15	94.48	85.20	69.74	57.55	47.89	40.15	33.92	26.69	18.51	13.33	15
20	165.46	144.60	111.56	87.23	69.09	55.41	44.97	33.58	21.74	14.89	20
25	252.89	214.26	156.10	115.97	87.80	67.70	53.10	38.03	23.43	15.56	25
30	355.00	291.72	201.06	142.36	103.46	77.08	58.78	40.75	24.26	15.83	30
35	470.15	374.88	244.88	165.74	116.09	83.99	62.61	42.36	24.66	15.94	35
40	596.85	461.99	286.53	185.96	126.04	88.95	65.12	43.28	24.85	15.98	40
45	733.70	551.56	325.40	203.11	133.73	92.45	66.73	43.81	24.93	15.99	45
50	879.41	642.36	361.16	217.46	139.59	94.89	67.76	44.10	24.97	15.56	50
60	1192.80	823.70	423.00	239.04	147.30	97.70	68.81	44.34	24.99	—	60
70	1528.64	999.83	472.48	253.33	151.53	98.99	69.21	44.42	—	—	70
80	1879.87	1166.79	511.12	262.55	153.80	99.56	69.36	44.47	—	—	80
90	2240.55	1322.17	540.77	268.39	154.99	99.81	—	—	—	—	90
100	2605.76	1464.75	563.12	272.05	155.61	99.92	—	—	—	—	100

table E-22 gradient to uniform series conversion factor for discrete compounding (to find A, given G)

$$(A/G, i\%, N) = \left[\frac{1}{i} - \frac{N}{(1+i)^N - 1}\right]$$

n	1%	2%	4%	6%	8%	10%	12%	15%	20%	25%	n
1	0.0001	0.0000	0.0000	0.0000	0.0000	0.0000	0.0000	0.0000	0.0000	0.0000	1
2	0.4974	0.4950	0.4902	0.4854	0.4808	0.4762	0.4717	0.4651	0.4545	0.4444	2
3	0.9932	0.9868	0.9739	0.9612	0.9487	0.9366	0.9246	0.9071	0.8791	0.8525	3
4	1.4874	1.4752	1.4510	1.4272	1.4040	1.3812	1.3589	1.3263	1.2742	1.2249	4
5	1.9799	1.9604	1.9216	1.8836	1.8465	1.8101	1.7746	1.7228	1.6405	1.5631	5
6	2.4708	2.4422	2.3857	2.3304	2.2763	2.2236	2.1720	2.0972	1.9788	1.8683	6
7	2.9600	2.9208	2.8433	2.7676	2.6937	2.6216	2.5515	2.4498	2.2902	2.1424	7
8	3.4476	3.3961	3.2944	3.1952	3.0985	3.0045	2.9131	2.7813	2.5756	2.3872	8
9	3.9335	3.8680	3.7391	3.6133	3.4910	3.3724	3.2574	3.0922	2.8364	2.6048	9
10	4.4177	4.3367	4.1773	4.0220	3.8713	3.7255	3.5847	3.3832	3.0739	2.7971	10
11	4.9003	4.8021	4.6090	4.4213	4.2395	4.0641	3.8953	3.6549	3.2893	2.9663	11
12	5.3813	5.2642	5.0343	4.8113	4.5957	4.3884	4.1897	3.9082	3.4841	3.1145	12
15	6.8141	6.6309	6.2721	5.9260	5.5945	5.2789	4.9803	4.5650	3.9588	3.4530	15
20	9.1692	8.8433	8.2091	7.6051	7.0369	6.5081	6.0202	5.3651	4.4643	3.7667	20
25	11.4829	10.9744	9.9925	9.0722	8.2254	7.4580	6.6708	5.8834	4.7352	3.9052	25
30	13.7555	13.0251	11.6274	10.3422	9.1897	8.1762	7.2974	6.2066	4.8731	3.9628	30
35	15.9869	14.9961	13.1198	11.4319	9.9611	8.7086	7.6577	6.4019	4.9406	3.9858	35
40	18.1774	16.8885	14.4765	12.3590	10.5699	9.0962	7.8988	6.5168	4.9728	3.9947	40
45	20.3271	18.7033	15.7047	13.1413	11.0447	9.3740	8.0572	6.5830	4.9877	3.9980	45
50	22.4362	20.4420	16.8122	13.7964	11.4107	9.5704	8.1597	6.6205	4.9945	3.9993	50
60	26.5331	23.6961	18.6972	14.7909	11.9015	9.8023	8.2664	6.6530	4.9989	—	60
70	30.4701	26.6632	20.1961	15.4613	12.1783	9.9113	8.3082	6.6627	—	—	70
80	34.2490	29.3572	21.3718	15.9033	12.3301	9.9609	8.3241	6.6656	—	—	80
90	37.8723	31.7929	22.2826	16.1891	12.4116	9.9831	—	—	—	—	90
100	41.3424	33.9863	22.9800	16.3711	12.4545	9.9927	—	—	—	—	100

appendix F

interest and annuity tables for continuous compounding

(For various common values of r from 1% to 25%)

r = nominal interest rate per year, compounded continuously

y = number of compounding periods

$$(F/P, r\%, N) = e^{rN}$$

$$(P/F, r\%, N) = e^{-rN} = \frac{1}{e^{rN}}$$

$$(F/A, r\%, N) = \frac{e^{rN} - 1}{e^r - 1}$$

$$(P/A, r\%, N) = \frac{e^{rN} - 1}{e^{rN}(e^r - 1)}$$

$$(F/\bar{A}, i\%, N) = \frac{e^{rN} - 1}{r}$$

$$(P/\bar{A}, i\%, N) = \frac{e^{rN} - 1}{re^{rN}}$$

table F-1

Continuous Compounding; $r = 1\%$

	Discrete Flows				Continuous Flows		
	SINGLE PAYMENT		UNIFORM SERIES		UNIFORM SERIES		
	Compound Amount Factor	Present Worth Factor	Compound Amount Factor	Present Worth Factor	Compound Amount Factor	Present Worth Factor	
N	To find F Given P F/P	To find P Given F P/F	To find F Given A F/A	To find P Given A P/A	To find F Given $\overline{A}$ $F/\overline{A}$	To find P Given $\overline{A}$ $P/\overline{A}$	N
1	1.0101	0.9900	1.0000	0.9900	1.0050	0.9950	1
2	1.0202	0.9802	2.0101	1.9703	2.0201	1.9801	2
3	1.0305	0.9704	3.0303	2.9407	3.0455	2.9554	3
4	1.0408	0.9608	4.0607	3.9015	4.0811	3.9211	4
5	1.0513	0.9512	5.1015	4.8527	5.1271	4.8771	5
6	1.0618	0.9418	6.1528	5.7945	6.1837	5.8235	6
7	1.0725	0.9324	7.2146	6.7269	7.2508	6.7606	7
8	1.0833	0.9231	8.2871	7.6500	8.3287	7.6884	8
9	1.0942	0.9139	9.3704	8.5639	9.4174	8.6069	9
10	1.1052	0.9048	10.4646	9.4688	10.5171	9.5163	10
11	1.1163	0.8958	11.5698	10.3646	11.6278	10.4166	11
12	1.1275	0.8869	12.6860	11.2515	12.7497	11.3080	12
13	1.1388	0.8781	13.8135	12.1296	13.8828	12.1905	13
14	1.1503	0.8694	14.9524	12.9990	15.0274	13.0642	14
15	1.1618	0.8607	16.1026	13.8597	16.1834	13.9292	15
16	1.1735	0.8521	17.2645	14.7118	17.3511	14.7856	16
17	1.1853	0.8437	18.4380	15.5555	18.5305	15.6335	17
18	1.1972	0.8353	19.6233	16.3908	19.7217	16.4730	18
19	1.2092	0.8270	20.8205	17.2177	20.9250	17.3041	19
20	1.2214	0.8187	22.0298	18.0365	22.1403	18.1269	20
21	1.2337	0.8106	23.2512	18.8470	23.3678	18.9416	21
22	1.2461	0.8025	24.4849	19.6496	24.6077	19.7481	22
23	1.2586	0.7945	25.7309	20.4441	25.8600	20.5466	23
24	1.2712	0.7866	26.9895	21.2307	27.1249	21.3372	24
25	1.2840	0.7788	28.2608	22.0095	28.4025	22.1199	25
26	1.2969	0.7711	29.5448	22.7806	29.6930	22.8948	26
27	1.3100	0.7634	30.8417	23.5439	30.9964	23.6621	27
28	1.3231	0.7558	32.1517	24.2997	32.3130	24.4216	28
29	1.3364	0.7483	33.4748	25.0480	33.6427	25.1736	29
30	1.3499	0.7408	34.8113	25.7888	34.9859	25.9182	30
35	1.4191	0.7047	41.6976	29.3838	41.9068	29.5312	35
40	1.4918	0.6703	48.9370	32.8034	49.1825	32.9680	40
45	1.5683	0.6376	56.5476	36.0563	56.8312	36.2372	45
50	1.6487	0.6065	64.5483	39.1505	64.8721	39.3469	50
55	1.7333	0.5769	72.9593	42.0939	73.3253	42.3050	55
60	1.8221	0.5488	81.8015	44.8936	82.2119	45.1188	60
65	1.9155	0.5220	91.0971	47.5569	91.5541	47.7954	65
70	2.0138	0.4966	100.869	50.0902	101.375	50.3415	70
75	2.1170	0.4724	111.143	52.5000	111.700	52.7633	75
80	2.2255	0.4493	121.942	54.7923	122.554	55.0671	80
85	2.3396	0.4274	133.296	56.9727	133.965	57.2585	85
90	2.4596	0.4066	145.232	59.0468	145.960	59.3430	90
95	2.5857	0.3867	157.780	61.0198	158.571	61.3259	95
100	2.7183	0.3679	170.971	62.8965	171.828	63.2121	100

table F-2

Continuous Compounding; $r = 2\%$

	Discrete Flows				Continuous Flows		
	SINGLE PAYMENT		UNIFORM SERIES		UNIFORM SERIES		
	Compound Amount Factor	Present Worth Factor	Compound Amount Factor	Present Worth Factor	Compound Amount Factor	Present Worth Factor	
N	To find F Given P F/P	To find P Given F P/F	To find F Given A F/A	To find P Given A P/A	To find F Given $\overline{A}$ $F/\overline{A}$	To find P Given $\overline{A}$ $P/\overline{A}$	N
1	1.0202	0.9802	1.0000	0.9802	1.0101	0.9901	1
2	1.0408	0.9608	2.0202	1.9410	2.0405	1.9605	2
3	1.0618	0.9418	3.0610	2.8828	3.0918	2.9118	3
4	1.0833	0.9231	4.1228	3.8059	4.1644	3.8442	4
5	1.1052	0.9048	5.2061	4.7107	5.2585	4.7581	5
6	1.1275	0.8869	6.3113	5.5976	6.3748	5.6540	6
7	1.1503	0.8694	7.4388	6.4670	7.5137	6.5321	7
8	1.1735	0.8521	8.5891	7.3191	8.6755	7.3928	8
9	1.1972	0.8353	9.7626	8.1544	9.8609	8.2365	9
10	1.2214	0.8187	10.9598	8.9731	11.0701	9.0635	10
11	1.2461	0.8025	12.1812	9.7756	12.3038	9.8741	11
12	1.2712	0.7866	13.4273	10.5623	13.5625	10.6686	12
13	1.2969	0.7711	14.6985	11.3333	14.8465	11.4474	13
14	1.3231	0.7558	15.9955	12.0891	16.1565	12.2108	14
15	1.3499	0.7408	17.3186	12.8299	17.4929	12.9591	15
16	1.3771	0.7261	18.6685	13.5561	18.8564	13.6925	16
17	1.4049	0.7118	20.0456	14.2678	20.2474	14.4115	17
18	1.4333	0.6977	21.4505	14.9655	21.6665	15.1162	18
19	1.4623	0.6839	22.8839	15.6494	23.1142	15.8069	19
20	1.4918	0.6703	24.3461	16.3197	24.5912	16.4840	20
21	1.5220	0.6570	25.8380	16.9768	26.0981	17.1477	21
22	1.5527	0.6440	27.3599	17.6208	27.6354	17.7982	22
23	1.5841	0.6313	28.9126	18.2521	29.2037	18.4358	23
24	1.6161	0.6188	30.4967	18.8709	30.8037	19.0608	24
25	1.6487	0.6065	32.1128	19.4774	32.4361	19.6735	25
26	1.6820	0.5945	33.7615	20.0719	34.1014	20.2740	26
27	1.7160	0.5827	35.4435	20.6547	35.8003	20.8626	27
28	1.7507	0.5712	37.1595	21.2259	37.5336	21.4395	28
29	1.7860	0.5599	38.9102	21.7858	39.3019	22.0051	29
30	1.8221	0.5488	40.6962	22.3346	41.1059	22.5594	30
35	2.0138	0.4966	50.1824	24.9199	50.6876	25.1707	35
40	2.2255	0.4493	60.6663	27.2591	61.2770	27.5336	40
45	2.4596	0.4066	72.2528	29.3758	72.9802	29.6715	45
50	2.7183	0.3679	85.0578	31.2910	85.9141	31.6060	50
55	3.0042	0.3329	99.2096	33.0240	100.208	33.3564	55
60	3.3201	0.3012	114.850	34.5921	116.006	34.9403	60
65	3.6693	0.2725	132.135	36.0109	133.465	36.3734	65
70	4.0552	0.2466	151.238	37.2947	152.760	37.6702	70
75	4.4817	0.2231	172.349	38.4564	174.084	38.8435	75
80	4.9530	0.2019	195.682	39.5075	197.652	39.9052	80
85	5.4739	0.1827	221.468	40.4585	223.697	40.8658	85
90	6.0496	0.1653	249.966	41.3191	252.482	41.7351	90
95	6.6859	0.1496	281.461	42.0978	284.295	42.5216	95
100	7.3891	0.1353	316.269	42.8023	319.453	43.2332	100

table F-3

Continuous Compounding; $r = 4\%$

	Discrete Flows				Continuous Flows	
	SINGLE PAYMENT		UNIFORM SERIES		UNIFORM SERIES	
	Compound Amount Factor	Present Worth Factor	Compound Amount Factor	Present Worth Factor	Compound Amount Factor	Present Worth Factor
N	To find F Given P F/P	To find P Given F P/F	To find F Given A F/A	To find P Given A P/A	To find F Given $\overline{A}$ $F/\overline{A}$	To find P Given $\overline{A}$ $P/\overline{A}$	N
1	1.0408	0.9608	1.0000	0.9608	1.0203	0.9803	1
2	1.0833	0.9231	2.0408	1.8839	2.0822	1.9221	2
3	1.1275	0.8869	3.1241	2.7708	3.1874	2.8270	3
4	1.1735	0.8521	4.2516	3.6230	4.3378	3.6964	4
5	1.2214	0.8187	5.4251	4.4417	5.5351	4.5317	5
6	1.2712	0.7866	6.6465	5.2283	6.7812	5.3343	6
7	1.3231	0.7558	7.9178	5.9841	8.0782	6.1054	7
8	1.3771	0.7261	9.2409	6.7103	9.4282	6.8463	8
9	1.4333	0.6977	10.6180	7.4079	10.8332	7.5581	9
10	1.4918	0.6703	12.0513	8.0783	12.2956	8.2420	10
11	1.5527	0.6440	13.5432	8.7223	13.8177	8.8991	11
12	1.6161	0.6188	15.0959	9.3411	15.4019	9.5304	12
13	1.6820	0.5945	16.7119	9.9356	17.0507	10.1370	13
14	1.7507	0.5712	18.3940	10.5068	18.7668	10.7198	14
15	1.8221	0.5488	20.1446	11.0556	20.5530	11.2797	15
16	1.8965	0.5273	21.9668	11.5829	22.4120	11.8177	16
17	1.9739	0.5066	23.8632	12.0895	24.3469	12.3346	17
18	2.0544	0.4868	25.8371	12.5763	26.3608	12.8312	18
19	2.1383	0.4677	27.8916	13.0439	28.4569	13.3083	19
20	2.2255	0.4493	30.0298	13.4933	30.6385	13.7668	20
21	2.3164	0.4317	32.2554	13.9250	32.9092	14.2072	21
22	2.4109	0.4148	34.5717	14.3398	35.2725	14.6304	22
23	2.5093	0.3985	36.9826	14.7383	37.7323	15.0370	23
24	2.6117	0.3829	39.4919	15.1212	40.2924	15.4277	24
25	2.7183	0.3679	42.1036	15.4891	42.9570	15.8030	25
26	2.8292	0.3535	44.8219	15.8425	45.7304	16.1636	26
27	2.9447	0.3396	47.6511	16.1821	48.6170	16.5101	27
28	3.0649	0.3263	50.5958	16.5084	51.6214	16.8430	28
29	3.1899	0.3135	53.6607	16.8219	54.7483	17.1628	29
30	3.3201	0.3012	56.8506	17.1231	58.0029	17.4701	30
35	4.0552	0.2466	74.8626	18.4609	76.3800	18.8351	35
50	4.9530	0.2019	96.8625	19.5562	98.8258	19.9526	40
45	6.0496	0.1653	123.733	20.4530	126.241	20.8675	45
50	7.3891	0.1353	156.553	21.1872	159.726	21.6166	50
55	9.0250	0.1108	196.640	21.7883	200.625	22.2299	55
60	11.0232	0.0907	245.601	22.2804	250.579	22.7321	60
65	13.4637	0.0743	305.403	22.6834	311.593	23.1432	65
70	16.4446	0.0608	378.445	23.0133	386.116	23.4797	70
75	20.0855	0.0498	467.659	23.2834	477.138	23.7553	75
80	24.5325	0.0408	576.625	23.5045	588.313	23.9809	80
85	29.9641	0.0334	709.717	23.6856	724.102	24.1657	85
90	36.5982	0.0273	872.275	23.8338	889.956	24.3169	90
95	44.7012	0.0224	1070.82	23.9552	1092.530	24.4407	95
100	54.5982	0.0183	1313.33	24.0545	1339.954	24.5421	100

table F-4

Continuous Compounding; $r = 5\%$

	Discrete Flows				Continuous Flows		
	SINGLE PAYMENT		UNIFORM SERIES		UNIFORM SERIES		
	Compound Amount Factor	Present Worth Factor	Compound Amount Factor	Present Worth Factor	Compound Amount Factor	Present Worth Factor	
N	To find F Given P F/P	To find P Given F P/F	To find F Given A F/A	To find P Given A P/A	To find F Given $\overline{A}$ $F/\overline{A}$	To find P Given $\overline{A}$ $P/\overline{A}$	N
1	1.0513	0.9512	1.0000	0.9512	1.0254	0.9754	1
2	1.1052	0.9048	2.0513	1.8561	2.1034	1.9033	2
3	1.1618	0.8607	3.1564	2.7168	3.2367	2.7858	3
4	1.2214	0.8187	4.3183	3.5355	4.4281	3.6254	4
5	1.2840	0.7788	5.5397	4.3143	5.6805	4.4240	5
6	1.3499	0.7408	6.8237	5.0551	6.9972	5.1836	6
7	1.4191	0.7047	8.1736	5.7598	8.3814	5.9062	7
8	1.4918	0.6703	9.5926	6.4301	9.8365	6.5936	8
9	1.5683	0.6376	11.0845	7.0678	11.3662	7.2474	9
10	1.6487	0.6065	12.6528	7.6743	12.9744	7.8694	10
11	1.7333	0.5769	14.3015	8.2512	14.6651	8.4610	11
12	1.8221	0.5488	16.0347	8.8001	16.4424	9.0238	12
13	1.9155	0.5220	17.8569	9.3221	18.3108	9.5591	13
14	2.0138	0.4966	19.7724	9.8187	20.2751	10.0683	14
15	2.1170	0.4724	21.7862	10.2911	22.3400	10.5527	15
16	2.2255	0.4493	23.9032	10.7404	24.5108	11.0134	16
17	2.3396	0.4274	26.1287	11.1678	26.7929	11.4517	17
18	2.4596	0.4066	28.4683	11.5744	29.1921	11.8686	18
19	2.5857	0.3867	30.9279	11.9611	31.7142	12.2652	19
20	2.7183	0.3679	33.5137	12.3290	34.3656	12.6424	20
21	2.8577	0.3499	36.2319	12.6789	37.1530	13.0012	21
22	3.0042	0.3329	39.0896	13.0118	40.0833	13.3426	22
23	3.1582	0.3166	42.0938	13.3284	43.1639	13.6673	23
24	3.3201	0.3012	45.2519	13.6296	46.4023	13.9761	24
25	3.4903	0.2865	48.5721	13.9161	49.8069	14.2699	25
26	3.6693	0.2725	52.0624	14.1887	53.3859	14.5494	26
27	3.8574	0.2592	55.7317	14.4479	57.1485	14.8152	27
28	4.0552	0.2466	59.5891	14.6945	61.1040	15.0681	28
29	4.2631	0.2346	63.6443	14.9291	65.2623	15.3086	29
30	4.4817	0.2231	67.9074	15.1522	69.6338	15.5374	30
35	5.7546	0.1738	92.7346	16.1149	95.0921	16.5245	35
40	7.3891	0.1353	124.613	16.8646	127.781	17.2933	40
45	9.4877	0.1054	165.546	17.4484	169.755	17.8920	45
50	12.1825	0.0821	218.105	17.9032	223.650	18.3583	50
55	15.6426	0.0639	285.592	18.2573	292.853	18.7214	55
60	20.0855	0.0498	372.247	18.5331	381.711	19.0043	60
65	25.7903	0.0388	483.515	18.7479	495.807	19.2245	65
70	33.1155	0.0302	626.385	18.9152	642.309	19.3961	70
75	42.5211	0.0235	809.834	19.0455	830.422	19.5296	75
80	54.5981	0.0183	1045.39	19.1469	1071.963	19.6337	80
85	70.1054	0.0143	1347.84	19.2260	1382.108	19.7147	85
90	90.0171	0.0111	1736.20	19.2875	1780.342	19.7778	90
95	115.584	0.0087	2234.87	19.3354	2291.686	19.8270	95
100	148.413	0.0067	2875.17	19.3727	2948.263	19.8652	100

table F-5

Continuous Compounding; $r = 8\%$

	Discrete Flows				Continuous Flows		
	SINGLE PAYMENT		UNIFORM SERIES		UNIFORM SERIES		
	Compound Amount Factor	Present Worth Factor	Compound Amount Factor	Present Worth Factor	Compound Amount Factor	Present Worth Factor	
N	To find F Given F F/P	To find P Given P P/F	To find F Given A F/A	To find P Given A P/A	To find F Given $\overline{A}$ $F/\overline{A}$	To find P Given $\overline{A}$ $P/\overline{A}$	N
1	1.0833	0.9231	1.0000	0.9231	1.0411	0.9610	1
2	1.1735	0.8521	2.0833	1.7753	2.1689	1.8482	2
3	1.2712	0.7866	3.2568	2.5619	3.3906	2.6672	3
4	1.3771	0.7261	4.5280	3.2880	4.7141	3.4231	4
5	1.4918	0.6703	5.9052	3.9584	6.1478	4.1210	5
6	1.6161	0.6188	7.3970	4.5771	7.7009	4.7652	6
7	1.7507	0.5712	9.0131	5.1483	9.3834	5.3599	7
8	1.8965	0.5273	10.7637	5.6756	11.2060	5.9088	8
9	2.0544	0.4868	12.6602	6.1624	13.1804	6.4156	9
10	2.2255	0.4493	14.7147	6.6117	15.3193	6.8834	10
11	2.4109	0.4148	16.9402	7.0265	17.6362	7.3152	11
12	2.6117	0.3829	19.3511	7.4094	20.1462	7.7138	12
13	2.8292	0.3535	21.9628	7.7629	22.8652	8.0818	13
14	3.0649	0.3263	24.7920	8.0891	25.8107	8.4215	14
15	3.3201	0.3012	27.8569	8.3903	29.0015	8.7351	15
16	3.5966	0.2780	31.1770	8.6684	32.4580	9.0245	16
17	3.8962	0.2567	34.7736	8.9250	36.2024	9.2917	17
18	4.2207	0.2369	38.6698	9.1620	40.2587	9.5384	18
19	4.5722	0.2187	42.8905	9.3807	44.6528	9.7661	19
20	4.9530	0.2019	47.4627	9.5826	49.4129	9.9763	20
21	5.3656	0.1864	52.4158	9.7689	54.5694	10.1703	21
22	5.8124	0.1720	57.7813	9.9410	60.1555	10.3494	22
23	6.2965	0.1588	63.5938	10.0998	66.2067	10.5148	23
24	6.8120	0.1466	69.8903	10.2464	72.7620	10.6674	24
25	7.3891	0.1353	76.7113	10.3817	79.8632	10.8083	25
26	8.0045	0.1249	84.1003	10.5067	87.5559	10.9384	26
27	8.6711	0.1153	92.1048	10.6220	95.8892	11.0584	27
28	9.3933	0.1065	100.776	10.7285	104.917	11.1693	28
29	10.1757	0.0983	110.169	10.8267	114.696	11.2716	29
30	11.0232	0.0907	120.345	10.9174	125.290	11.3660	30
35	16.4446	0.0608	185.439	11.2765	193.058	11.7399	35
40	24.5325	0.0408	282.547	11.5172	294.157	11.9905	40
45	36.5982	0.0273	427.416	11.6786	444.978	12.1585	45
50	54.5982	0.0183	643.535	11.7868	669.977	12.2711	50
55	81.4509	0.0123	965.947	11.8593	1005.64	12.3465	55
60	121.510	0.0082	1446.93	11.9079	1506.38	12.3971	60
65	181.272	0.0055	2164.47	11.9404	2253.40	12.4310	65
70	270.426	0.0037	3234.91	11.9623	3367.83	12.4538	70
75	403.429	0.0025	4831.83	11.9769	5030.36	12.4690	75
80	601.845	0.0017	7214.15	11.9867	7510.56	12.4792	80
85	897.847	0.0011	10768.1	11.9933	11210.6	12.4861	85
90	1339.43	0.0007	16070.1	11.9977	16730.4	12.4907	90
95	1998.20	0.0005	23979.7	12.0007	24964.9	12.4937	95
100	2980.96	0.0003	35779.3	12.0026	37249.5	12.4958	100

table F-6

Continuous Compounding; $r = 10\%$

	Discrete Flows				Continuous Flows		
	SINGLE PAYMENT		UNIFORM SERIES		UNIFORM SERIES		
	Compound Amount Factor	Present Worth Factor	Compound Amount Factor	Present Worth Factor	Compound Amount Factor	Present Worth Factor	
N	To find F Given P F/P	To find P Given F P/F	To find F Given A F/A	To find P Given A P/A	To find F Given $\overline{A}$ $F/\overline{A}$	To find P Given $\overline{A}$ $P/\overline{A}$	N
1	1.1052	0.9048	1.0000	0.9048	1.0517	0.9516	1
2	1.2214	0.8187	2.1052	1.7236	2.2140	1.8127	2
3	1.3499	0.7408	3.3266	2.4644	3.4986	2.5918	3
4	1.4918	0.6703	4.6764	3.1347	4.9182	3.2968	4
5	1.6487	0.6065	6.1683	3.7412	6.4872	3.9347	5
6	1.8221	0.5488	7.8170	4.2900	8.2212	4.5119	6
7	2.0138	0.4966	9.6391	4.7866	10.1375	5.0341	7
8	2.2255	0.4493	11.6528	5.2360	12.2554	5.5067	8
9	2.4596	0.4066	13.8784	5.6425	14.5960	5.9343	9
10	2.7133	0.3679	16.3380	6.0104	17.1828	6.3212	10
11	3.0042	0.3329	19.0563	6.3433	20.0417	6.6713	11
12	3.3201	0.3012	22.0604	6.6445	23.2012	6.9881	12
13	3.6693	0.2725	25.3806	6.9170	26.6930	7.2747	13
14	4.0552	0.2466	29.0499	7.1636	30.5520	7.5340	14
15	4.4817	0.2231	33.1051	7.3867	34.8169	7.7687	15
16	4.9530	0.2019	37.5867	7.5886	39.5303	7.9810	16
17	5.4739	0.1827	42.5398	7.7713	44.7395	8.1732	17
18	6.0496	0.1653	48.0137	7.9366	50.4965	8.3470	18
19	6.6859	0.1496	54.0634	8.0862	56.8589	8.5043	19
20	7.3891	0.1353	60.7493	8.2215	63.8906	8.6466	20
21	8.1662	0.1225	68.1383	8.3440	71.6617	8.7754	21
22	9.0250	0.1108	76.3045	8.4548	80.2501	8.8920	22
23	9.9742	0.1003	85.3295	8.5550	89.7418	8.9974	23
24	11.0232	0.0907	95.3037	8.6458	100.232	9.0928	24
25	12.1825	0.0821	106.327	8.7278	111.825	9.1791	25
26	13.4637	0.0743	118.509	8.8021	124.637	9.2573	26
27	14.8797	0.0672	131.973	8.8693	138.797	9.3279	27
28	16.4446	0.0608	146.853	8.9301	154.446	9.3919	28
29	18.1741	0.0550	163.298	8.9852	171.741	9.4498	29
30	20.0855	0.0498	181.472	9.0349	190.855	9.5021	30
35	33.1155	0.0302	305.364	9.2212	321.154	9.6980	35
40	54.5981	0.0183	509.629	9.3342	535.982	9.8168	40
45	90.0171	0.0111	846.404	9.4027	890.171	9.8889	45
50	148.413	0.0067	1401.65	9.4443	1474.13	9.9326	50
55	244.692	0.0041	2317.10	9.4695	2436.92	9.9591	55
60	403.429	0.0025	3826.43	9.4848	4024.29	9.9752	60
65	665.142	0.0015	6314.88	9.4940	6641.42	9.9850	65
70	1096.63	0.0009	10417.6	9.4997	10956.3	9.9909	70
75	1808.04	0.0006	17182.0	9.5031	18070.7	9.9945	75
80	2980.96	0.0003	28334.4	9.5051	29799.6	9.9966	80
85	4914.77	0.0002	46721.7	9.5064	49137.7	9.9980	85
90	8103.08	0.0001	77037.3	9.5072	81020.8	9.9988	90
95	13359.7	a	127019	9.5076	133587	9.9993	95
100	22026.5	a	209425	9.5079	220255	9.9995	100

a Less than 0.0001.

table F-7

Continuous Compounding; $r = 15\%$

	Discrete Flows				Continuous Flows		
	SINGLE PAYMENT		UNIFORM SERIES		UNIFORM SERIES		
	Compound Amount Factor	Present Worth Factor	Compound Amount Factor	Present Worth Factor	Compound Amount Factor	Present Worth Factor	
N	To find F Given P F/P	To find P Given F P/F	To find F Given A F/A	To find P Given A P/A	To find F Given $\overline{A}$ $F/\overline{A}$	To find P Given $\overline{A}$ $P/\overline{A}$	N
1	1.1618	0.8607	1.0000	0.8607	1.0789	0.9286	1
2	1.3499	0.7408	2.1618	1.6015	2.3324	1.7279	2
3	1.5683	0.6376	3.5117	2.2392	3.7887	2.4158	3
4	1.8221	0.5488	5.0800	2.7880	5.4808	3.0079	4
5	2.1170	0.4724	6.9021	3.2603	7.4467	3.5176	5
6	2.4596	0.4066	9.0191	3.6669	9.7307	3.9562	6
7	2.8577	0.3499	11.4787	4.0168	12.3843	4.3337	7
8	3.3201	0.3012	14.3364	4.3180	15.4674	4.6587	8
9	3.8574	0.2592	17.6565	4.5773	19.0495	4.9384	9
10	4.4817	0.2231	21.5139	4.8004	23.2113	5.1791	10
11	5.2070	0.1920	25.9956	4.9925	28.0465	5.3863	11
12	6.0496	0.1653	31.2026	5.1578	33.6643	5.5647	12
13	7.0287	0.1423	37.2522	5.3000	40.1913	5.7182	13
14	8.1662	0.1225	44.2809	5.4225	47.7745	5.8503	14
15	9.4877	0.1054	52.4471	5.5279	56.5849	5.9640	15
16	11.0232	0.0907	61.9348	5.6186	66.8212	6.0619	16
17	12.8071	0.0781	72.9580	5.6967	78.7140	6.1461	17
18	14.8797	0.0672	85.7651	5.7639	92.5315	6.2186	18
19	17.2878	0.0578	100.645	5.8217	108.585	6.2810	19
20	20.0855	0.0498	117.933	5.8715	127.237	6.3348	20
21	23.3361	0.0429	138.018	5.9144	148.907	6.3810	21
22	27.1126	0.0369	161.354	5.9513	174.084	6.4208	22
23	31.5004	0.0317	188.467	5.9830	203.336	6.4550	23
24	36.5982	0.0273	219.967	6.0103	237.322	6.4845	24
25	42.5211	0.0235	256.565	6.0338	276.807	6.5099	25
26	49.4024	0.0202	299.087	6.0541	322.683	6.5317	26
27	57.3975	0.0174	348.489	6.0715	375.983	6.5505	27
28	66.6863	0.0150	405.886	6.0865	437.909	6.5667	28
29	77.4785	0.0129	472.573	6.0994	509.856	6.5806	29
30	90.0171	0.0111	550.051	6.1105	593.448	6.5926	30
35	190.566	0.0052	1171.36	6.1467	1263.78	6.6317	35
40	403.429	0.0025	2486.67	6.1638	2682.86	6.6501	40
45	854.059	0.0012	5271.19	6.1719	5687.06	6.6589	45
50	1808.04	0.0006	11166.0	6.1757	12046.9	6.6630	50
55	3827.63	0.0003	23645.3	6.1775	25510.8	6.6649	55
60	8103.08	0.0001	50064.1	6.1784	54013.9	6.6658	60
65	17154.2	a	105993	6.1788	114355	6.6663	65
70	36315.5	a	224393	6.1790	242097	6.6665	70
75	76879.9	a	475047	6.1791	512526	6.6666	75
80	162755	a	1005680	6.1791	1085030	6.6666	80

a Less than 0.0001.

Continuous Compounding; $r = 20\%$

	Discrete Flows				Continuous Flows		
	SINGLE PAYMENT		UNIFORM SERIES		UNIFORM SERIES		
	Compound Amount Factor	Present Worth Factor	Compound Amount Factor	Present Worth Factor	Compound Amount Factor	Present Worth Factor	
	To find F Given P	To find P Given F	To find F Given A	To find P Given A	To find F Given $\overline{A}$	To find P Given $\overline{A}$	
N	F/P	P/F	F/A	P/A	$F/\overline{A}$	$P/\overline{A}$	N
1	1.2214	0.8187	1.0000	0.8187	1.1070	0.9063	1
2	1.4918	0.6703	2.2214	1.4891	2.4591	1.6484	2
3	1.8221	0.5488	3.7132	2.0379	4.1106	2.2559	3
4	2.2255	0.4493	5.5353	2.4872	6.1277	2.7534	4
5	2.7183	0.3679	7.7609	2.8551	8.5914	3.1606	5
6	3.3201	0.3012	10.4792	3.1563	11.6006	3.4940	6
7	4.0552	0.2466	13.7993	3.4029	15.2760	3.7670	7
8	4.9530	0.2019	17.8545	3.6048	19.7652	3.9905	8
9	6.0496	0.1653	22.8075	3.7701	25.2482	4.1735	9
10	7.3891	0.1353	28.8572	3.9054	31.9453	4.3233	10
11	9.0250	0.1108	36.2462	4.0162	40.1251	4.4460	11
12	11.0232	0.0907	45.2712	4.1069	50.1159	4.5464	12
13	13.4637	0.0743	56.2944	4.1812	62.3187	4.6286	13
14	16.4446	0.0608	69.7581	4.2420	77.2232	4.6959	14
15	20.0855	0.0498	86.2028	4.2918	95.4277	4.7511	15
16	24.5325	0.0408	106.288	4.3325	117.663	4.7962	16
17	29.9641	0.0334	130.821	4.3659	144.820	4.8331	17
18	36.5982	0.0273	160.785	4.3932	177.991	4.8634	18
19	44.7012	0.0224	197.383	4.4156	218.506	4.8881	19
20	54.5981	0.0183	242.084	4.4339	267.991	4.9084	20
21	66.6863	0.0150	296.682	4.4489	328.432	4.9250	21
22	81.4509	0.0123	363.369	4.4612	402.254	4.9386	22
23	99.4843	0.0101	444.820	4.4713	492.422	4.9497	23
24	121.510	0.0082	544.304	4.4795	602.552	4.9589	24
25	148.413	0.0067	665.814	4.4862	737.066	4.9663	25
26	181.272	0.0055	814.227	4.4917	901.361	4.9724	26
27	221.406	0.0045	995.500	4.4963	1102.03	4.9774	27
28	270.426	0.0037	1216.91	4.5000	1347.13	4.9815	28
29	330.299	0.0030	1487.33	4.5030	1646.50	4.9849	29
30	403.429	0.0025	1817.63	4.5055	2012.14	4.9876	30
35	1096.63	0.0009	4948.60	4.5125	5478.17	4.9954	35
40	2980.96	0.0003	13459.4	4.5151	14899.8	4.9983	40
45	8103.08	0.0001	36594.3	4.5161	40510.4	4.9994	45
50	22026.5	a	99481.4	4.5165	110127	4.9998	50
55	59874.1	a	270426	4.5166	299366	4.9999	55
60	162755	a	735103	4.5166	813769	5.0000	60

a Less than 0.0001.

table F-9

Continuous Compounding; $r = 25\%$

	Discrete Flows				Continuous Flows		
	SINGLE PAYMENT		UNIFORM SERIES		UNIFORM SERIES		
	Compound Amount Factor	Present Worth Factor	Compound Amount Factor	Present Worth Factor	Compound Amount Factor	Present Worth Factor	
N	To find F Given P F/P	To find P Given F P/F	To find F Given A F/A	To find P Given A P/A	To find F Given $\overline{A}$ $F/\overline{A}$	To find P Given $\overline{A}$ $P/\overline{A}$	N
1	1.2840	0.7788	1.0000	0.7788	1.1361	0.8848	1
2	1.6487	0.6065	2.2840	1.3853	2.5949	1.5739	2
3	2.1170	0.4724	3.9327	1.8577	4.4680	2.1105	3
4	2.7183	0.3679	6.0497	2.2256	6.8731	2.5285	4
5	3.4903	0.2865	8.7680	2.5121	9.9614	2.8540	5
6	4.4817	0.2231	12.2584	2.7352	13.9268	3.1075	6
7	5.7546	0.1738	16.7401	2.9090	19.0184	3.3049	7
8	7.3891	0.1353	22.4947	3.0443	25.5562	3.4587	8
9	9.4877	0.1054	29.8837	3.1497	33.9509	3.5784	9
10	12.1825	0.0821	39.3715	3.2318	44.7300	3.6717	10
11	15.6426	0.0639	51.5539	3.2957	58.5705	3.7443	11
12	20.0855	0.0498	67.1966	3.3455	76.3421	3.8009	12
13	25.7903	0.0388	87.2821	3.3843	99.1614	3.8449	13
14	33.1155	0.0302	113.073	3.4145	128.462	3.8792	14
15	42.5211	0.0235	146.188	3.4380	166.084	3.9059	15
16	54.5982	0.0183	188.709	3.4563	214.393	3.9267	16
17	70.1054	0.0143	243.307	3.4706	276.422	3.9429	17
18	90.0171	0.0111	313.413	3.4817	356.068	3.9556	18
19	115.584	0.0087	403.430	3.4904	458.337	3.9654	19
20	148.413	0.0067	519.014	3.4971	589.653	3.9730	20
21	190.566	0.0052	667.427	3.5023	758.265	3.9790	21
22	244.692	0.0041	857.993	3.5064	974.768	3.9837	22
23	314.191	0.0032	1102.69	3.5096	1252.76	3.9873	23
24	403.429	0.0025	1416.88	3.5121	1609.72	3.9901	24
25	518.013	0.0019	1820.30	3.5140	2068.05	3.9923	25
26	665.142	0.0015	2338.31	3.5155	2656.57	3.9940	26
27	854.059	0.0012	3003.46	3.5167	3412.23	3.9953	27
28	1096.63	0.0009	3857.52	3.5176	4382.53	3.9964	28
29	1408.10	0.0007	4954.15	3.5183	5628.42	3.9972	29
30	1808.04	0.0006	6362.26	3.5189	7228.17	3.9978	30
35	6310.69	0.0002	22215.2	3.5203	25238.8	3.9994	35
40	22026.5	a	77547.5	3.5207	88101.9	3.9998	40
45	76879.9	a	270676	3.5208	307516	3.9999	45
50	268337	a	944762	3.5208	1073350	4.0000	50

a Less than 0.0001.

answers to even-numbered problems

Chapter 1
1-18. Industrial ratios: (a) 1.61   (b) 0.69   (c) 9.05
1-20. 141%

Chapter 2
2-14. $D^* = 200$
2-16. (a) 0.7 ($700,000)   (b) 0.617   (c) 0.7

Chapter 3
3-2. Choose charter service.
3-4. (a) Conventional equipment   (b) 13.2 miles
3-6. Select aluminum.
3-8. Use the first method.
3-10. A three-man crew is less expensive.
3-12. The net cost per aluminum can = $0.04425, but glass bottles are cheaper.
3-14. (a) Wood containers are less expensive only for 25-lb containers.
   (b) Wood is less expensive for the 25- and 50-lb containers.

Chapter 4
4-2. $500
4-4. $3,702, difference = $422
4-6. $2,410.80
4-8. $17,946
4-10. (a) 0.1025   (b) 0.1038   (c) 0.1052
4-12. (a) 8.24%   (b) $6340.64
4-14. $N = 89$ months
4-16. $N = 30$ months
4-18. $471
4-20. 5.17%
4-22. 13.76%
4-24. $887.10
4-26. $3,500
4-28. (a) $N = 5$ years   (b) $656

4-30. $8,441

4-32. Do not buy the ranch.

4-34. $F = -$314.70$

4-36. $5,205.57

4-38. $-$17,409$

4-40. $-$217,182.50$

4-42. $28,652

4-44. $26,442

4-46. Select plan B.

4-48. Pay year by year.

4-50. (a) $13,094   (b) $18,914   (c) 9.14%

4-52. (a) $3,296,800   (b) $40,260.60   (c) $9,257

Chapter 5

5-2. Add the space heaters.

5-4. CR cost = $1,713

5-6. Choose motor A.

5-8. Recommend alternative 3.

5-10. Select copper piping.

5-12. (a) Alternative A   (b) $i' = 19.7\%$

5-14. (a) and (b) Select motor R.

5-16. Adopt the proposed system.

5-18. $i'/\text{year} = 14\%$, $i'/\text{year} = 6.6\%$

5-20. (a) Select B.   (b) Select B.

5-22. $i' = 21.4\%$

5-28. $P \leq $31,138$

5-30. $r \simeq 11\% < 12\%$

5-32. (a) 100%, 200%   (b) $i = 53.8\%$

Chapter 6

6-2. F_2 could be increased by as much as $8,024.70 without reversing the initial decision.

6-4. 1.26¢/kWh

6-6. Build the four-lane bridge now.

6-8. Choose the four-lane bridge.

6-10. 3,159 units/year

6-12. Recommend alternative T.

6-14. Do not purchase the equipment.

6-16. (a) Alternative C   (b) Alternative W   (c) Alternative U

6-18. Offer B is best.

Chapter 7

7-2. (a), (b), and (c) Select the apartment house.

7-4. (a) and (b) Select B.

7-6. (a) and (b) Recommend alternative C.

7-8. Choose the second alternative.

7-10. They could spend up to $482,800 for the jet.

7-12. (a) Recommend wet tower natural draft systems.

7-14. Install the cable now.

7-16. Select the diesel unit.
7-18. Select plan A.
7-20. Choose motor R.
7-22. Adopt alternative C.
7-24. (b) Proposals A1, B1, and C1 should be implemented.

Chapter 8
8-6. $407,050
8-8. $6,080.70
8-10. One would probably select the certificate of deposit.
8-12. $4,000
8-14. Develop export business and produce 30,000 pairs/year.
8-16. (a) Total cost = $775,300 (b) Total cost = $625,300
8-18. Existing profit = $85,000/month; new profit = $72,940
8-20. For a 25% return, unit price = $14.76
8-22. New selling price = $186.53, price adjustment = $27.78
8-24. $5,000

Chapter 9
9-8. (a) BV = $45,000 (b) BV = $35,294 (c) BV = $30,775 (d) BV = $48,093
9-10. (a) BV = $960 (b) BV = $1,158 (c) BV = $508 (d) BV = $640
9-12. CR cost = $1,052
9-14. (a) $1,900 long-term loss (b) $809 long-term loss
9-16. (a) $16,700 (b) $5,600 loss

Chapter 10
10-2. (a) Income taxes = $5,314 (b) $1,120 (c) Income taxes = $4,985 and effective rate = 36%
10-4. (a) $i' = 0$ (b) $i' = 8.7\%$ (c) $i' = 6.8\%$
10-6. Before taxes, $i' = 21.5\%$; after taxes, $i' = 14.4\%$
10-8. (a) $1,965 (b) 5.46% (c) $500 gain (d) greater than 50%
10-10. (a) $2,426 (b) $3,046 (c) $2,580 (d) D.D.B.
10-12. (a) $-$768 (b) $-$760
10-14. $i' = 5\%$
10-16. $1,251
10-18. PW = $2,610.85; purchase the equipment.
10-22. $n = 8+$ years
10-24. (a) $i' = 63\%$ (b) $i' = 67\%$

Chapter 11
11-2. Select plan B.
11-4. (a) Keep present machine (b) Doublecounting occurs and the time value of money is ignored.
11-6. Select plan III.
11-8. Keep the car 4 years.
11-10. The economic life is 1 year.
11-12. Buy power from outside source.
11-14. The present machine should be kept.
11-16. The incremental rate of return is 7.8% (after taxes).

11-18. The incremental rate of return is 8.2% (after taxes).
11-20. Install the terminals.
11-22. The old warehouse should be kept.

Chapter 12
12-6. (a) The new bridge costs less.   (b) B/C ratio = 1.41.
12-8. Private vehicles should pay $0.21.
12-10. (a) The time value of money is ignored.   (b) Retain the steel pier.
12-12. (a) Alternative C   (b) Alternative B
12-14. Improvements should be made; the B/C ratio is 5.53.

Chapter 13
13-10. It is more economical to wait 3 years.
13-12. Recommend the hydroelectric plant.
13-14. The charge per kWh: residential, $0.0371; commercial, $0.0397; industrial, $0.0258

Chapter 14
14-6. (a) $8,744.56   (b) $154,400
14-8. i/year = 8.16%
14-10. Bonds retired at end of first year = $6,800,000; bonds retired at end of fifteenth year = $15,368,172
14-12. PW purchase option = $-6,163$; PW lease option = $-6,255$. Purchase the equipment.
14-14. Cost of capital before taxes = 8.64%; cost of capital after taxes = 16.62%
14-16. P.W. = $-5,876$
14-18. The best combination is M, S, and T.

Chapter 15
15-4. (a) The lot size drops by $1/\sqrt{2}$.   (b) The lot size is reduced by $1/\sqrt{3}$.   (c) Q should be reduced.
15-6. 1,789 cases
15-8. 2,357 fixtures
15-10. Have your company manufacture the parts.
15-12. Group replacement should occur after the third time period.

Chapter 17
17-2. The solution is $X_1 = 251$, $X_2 = 100$ with a maximum profit of $600.
17-6. The maximum profit is $18,310.
17-8. Maximize $10.75B1 + 6.45B2 - 1.85C1 + 0.19C2 + 8.22D$; subject to $C1 + C2 \leq 1$, $C1 + C2 \leq B2$, $D \leq 1$.

Chapter 18
18-2. (b) The critical path is 10-20-40-50-70-80. The minimum time is 21 days.
18-4. One day can be saved on 30-50 and another on either 70-80 or 30-50. The total dollar savings would be $300.
18-6. (b) The minimum time is 13 days.

index